D1158377

determination of pH

THEORY AND PRACTICE

determination of pH

THEORY AND PRACTICE

Roger G. Bates, PROFESSOR OF CHEMISTRY
UNIVERSITY OF FLORIDA, GAINESVILLE

SECOND EDITION, 1973

A WILEY-INTERSCIENCE PUBLICATION

JOHN WILEY & SONS
New York · London · Sydney · Toronto

Library of Congress Cataloging in Publication Data

Bates, Roger Gordon, 1912-
  Determination of pH.

" A Wiley-Interscience production"
  Originally published in 1954 under title:
Electrometric pH determinations.
  Includes bibliographical references.
  1. Hydrogen-ion concentration—Measurement.
2. Electrochemistry. I. Title.

QD561.B32 1973   541'.3728   72-8779
ISBN 0-471-05647-2

Printed in the United States of America

10 9 8 7 6 5

preface

When Sørensen defined the pH unit and outlined the method for determining pH values in his classic papers of 1909, the wide utilization of pH measurements in the research laboratory, the control laboratory, the clinic, and the factory could hardly have been envisioned. The far-reaching application of pH measurements in modern commerce and industry was made possible by the discovery of the hydrogen ion function of glass membranes which led to the development of convenient, practical glass electrodes, pH meters, and controllers that permit the pH of process solutions to be regulated automatically. As so often happens, the technology of pH instrumentation has outdistanced the slow progress toward an adequate understanding and interpretation of the numbers that pH meters furnish so easily, and often with reproducibility and precision of a high order.

With the perfection of chemical thermodynamics, it became evident that Sørensen's experimental method did not, in fact, yield hydrogen ion concentrations. Although the numbers obtained depended in a complex manner on the *activity* of the electrolytes in the "test" solution, they were not an exact measure of the hydrogen ion activity and, indeed, could never be made so. Thus the door was opened to a plurality of "theoretical" pH units, none of which could be matched exactly with the experimental number. The realization that the pH scale must have a defined or *conventional* basis later added to the confusion, for no single convention was generally adopted.

Happily, the situation is somewhat improved today, and substantial agreement on a practical approach to the standardization of pH measurements has been achieved. Regulatory groups in several countries, as well as the International Union of Pure and Applied Chemistry, have endorsed the operational definition of the practical pH value and a standard scale fixed by one or more standard buffer solutions whose assigned pH values are formally consistent with the thermodynamic properties of the solutions and with a single conventional definition of the individual ionic activity. The pH determination is essentially a determination of a difference, and the meaning of the experimental pH value is largely that of the number assigned to the

v

standard.   Any interpretation of the pH, if justifiable at all, must begin with a consideration of the significance of the standard pH.

The determination of pH values is a branch of applied electrochemistry, and the latter is inextricably tied to chemical thermodynamics.   Yet the pH value is a thermodynamic quantity only in respect to its form; its substance is derived from extrathermodynamic considerations.   If there is justification for a theoretical section in a book on pH determinations, it is because the pH unit lacks precise fundamental definition and because it is essential for users of pH numbers to understand the meaning and limitations of the unit in order to employ pH measurements most advantageously.   For this reason, an attempt is made here to indicate how far the exact thermodynamic approach can be pursued before empirical standardization begins, and to shed light on the practical consequences of arbitrary conventions in the area where theory can no longer serve as a guide.

A part of this book is a second revision of material that appeared in the volume *Electrometric pH Determinations*, first published in 1954.   The work was expanded considerably in 1964 and the title changed to *Determination of pH*.   At that time, close attention was given to acid-base behavior in non-aqueous and mixed solvents, and a chapter on indicator pH measurements was added.

The theory and practice of electrometric pH determinations continue to be the chief concerns of this book.   Wherever possible the treatment of buffer solutions, dissociation equilibria, salt effects, and medium effects is thermodynamically rigorous instead of "classical".   In general, this means simply a strict adherence to the thermodynamic formulation of chemical equilibrium.   Most of the earlier material has been brought up to date.   Chapter 9 is new and Chapters 7 and 8 largely so.   Chapters 4 and 11 have also been extensively revised.

The first four chapters are devoted to an examination of the definitions of pH scales, to conventions which permit an acceptable compromise between theory and experiment to be achieved, and to pH standards.   In Chapters 5 and 6, buffer solutions and indicator methods are discussed from the viewpoint of both theory and practice.   Chapters 7 and 8 are concerned with our rapidly growing knowledge of acid-base phenomena and measurement scales in nonaqueous and mixed solvents as well as with recent attempts to evaluate medium effects for individual ionic species.   The new Chapter 9 examines electrometric determination of hydrogen ion concentrations in constant ionic media, a subject of special importance in coordination chemistry.   Chapter 10 describes properties of hydrogen ion electrodes, liquid junctions, and reference electrodes, while a separate chapter (Chapter 11) reviews the properties and behavior of glass electrodes.

Developments in pH instrumentation and in the application of pH measurements to industrial control now clearly exceed the scope of this book. Nevertheless, a general summary of the experimental and theoretical basis for measurement of the electromotive force of pH cells, the general characteristics of pH meters, and the principles of pH control is given in Chapters 12 and 13. No longer is an attempt made to compare features of the many commercial instruments available. In this connection, it should be noted that this book is addressed to chemists rather than to engineers.

Inevitably, the passage of two decades has brought a change in the demands placed on pH measurements and in the areas where such measurements are of critical concern. In this edition, for example, special needs for accurate pH measurements in clinical chemistry and biomedical research, as well as in nonaqueous solutions, are acknowledged. An attempt has nonetheless been made to introduce new contributions and new emphasis without slighting the orderly development of ideas from which present-day concepts have emerged.

ROGER G. BATES

Gainesville, Florida
July 1972

acknowledgments

I acknowledge gratefully the cooperation of the many colleagues who have assisted me in various aspects of the preparation of this book. I am particularly indebted to my associate, Professor R. A. Robinson, who read and criticized a large portion of the manuscript. In addition, Dr. A. K. Covington, Mr. C. G. Malmberg, and Professor W. C. Vosburgh have rendered generous assistance through their comments on specific chapters in this or earlier editions.

A number of the tables and figures have appeared earlier in other publications, and I am grateful for permission to reproduce them. Figures 4–4, 4–5, and 4–8 appeared in *The Analyst* and are reproduced with acknowledgment to the Society of Public Analysts and Other Analytical Chemists. Table 8 of the Appendix and Figs. 5–2, 5–3, and 11–2 were taken from *Analytical Chemistry*. Tables 2–3 and 6–10 and Figs. 2–2 and 4–1 are from *Chemical Reviews*. Figure 13–3 is from *The Industrial Chemist*. Table 3–3 and Figs. 7–3 and 10–1 appeared first in the *Journal of the American Chemical Society*. Figure 11–6 is from the *Journal of the Electrochemical Society*. Figure 3–1 was published in the *Journal of Physical Chemistry*, and Fig. 10–10 is from *Science*.

Finally, I am pleased to acknowledge the cooperation of the manufacturers of pH equipment and electrical instruments who have permitted the use of illustrations and tabular material from their bulletins. I am particularly indebted to Beckman Instruments, Inc.; Coleman Instruments, Inc.; Electronic Instruments Ltd., Leeds and Northrup Co.; Minneapolis-Honeywell Regulator Co.; North American Philips Co., Inc.; Radiometer A/S; the Rubicon Company; and W. A. Taylor and Co.

R.G.B.

contents

fundamental principles

and conventions

The electrometric determination of pH is essentially the evaluation of a quasi-thermodynamic constant from a measurement of the electromotive force of a suitable galvanic cell. It is accordingly necessary to examine the cell process in the light of thermodynamic concepts in order to comprehend fully the meaning of the experimental pH value and its limitations. It should be noted also that pH measurements made with indicator dyes relate indirectly to the pH scale defined in terms of galvanic cells. Some of the basic principles of electromotive force measurements and of the thermodynamics of electrolytic solutions needed for an understanding of the theory and practice of pH measurements are considered briefly in this introductory chapter.[1]

Reversible Cells and Potential

When current is drawn from a galvanic cell, an oxidation-reduction process called the *cell reaction* always takes place, and chemical energy is transformed into electrical energy. Oxidation-reduction reactions can conveniently be regarded as the sum of two half-reactions, one an oxidation and the other

---

[1] Symbols and terminology relating to pH and its measurement are in accord with recommendations accepted by the International Union of Pure and Applied Chemistry; see R. G. Bates and E. A. Guggenheim, *Pure Appl. Chem.*, **1**, 163 (1960), and *Manual of Symbols and Terminology for Physicochemical Quantities and Units*, IUPAC, Butterworths, London, 1970. In general, the International System of Units (SI) is used. The few exceptions include the atmosphere (atm) as a unit of pressure and the symbol m to represent both the quantity molality and its unit (mol kg^{-1}). Thus a solution containing 0.1 mol kg^{-1} of HCl will be designated either by HCl, 0.1 m or by HCl ($m = 0.1$).

a reduction, the former providing the electrons utilized by the latter. The sites of the half-reactions are the *electrodes*, and, if these are suitably placed, the passage of electrons, necessary for the reaction to occur, can take place in significant amounts only through a circuit external to the cell. When this circuit is closed, oxidation will occur spontaneously at one electrode and reduction at the other; electrons will continue to pass from one electrode to the other until a state of equilibrium has been attained. The electrode at which spontaneous oxidation takes place and electrons appear is the negative electrode; that at which reduction occurs spontaneously, consuming electrons, is the positive electrode.

Any reversible electrode can suffer an oxidation or a reduction, that is, can either supply or remove electrons. The *potential* or tendency of a given electrode to supply electrons depends on the Gibbs energies of the oxidized and reduced states of the chemical substances, the temperature, and the pressure. Indeed, it is often convenient to regard the electron as a chemical element, a reactant or product in the oxidation or reduction process.[2]

The potentials of the two electrodes with respect to each other determine the direction of the flow of current. The potential difference, or *electromotive force* (e.m.f., E) is the driving force of the reaction. If the e.m.f. of a reversible cell is balanced externally by an equal and oppositely directed e.m.f., there will be no chemical change within the cell. However, if the external e.m.f. is then increased by an infinitesimal amount, a small current will flow and a small amount of reaction will occur. When, on the other hand, the applied e.m.f. is decreased infinitesimally, current begins to flow in the opposite direction and the chemical reaction is reversed. The measurement of the e.m.f. of a galvanic cell by the compensation method, where the e.m.f. supplied by the potentiometer balances almost exactly that of the cell, approaches very closely the condition of thermodynamic reversibility; hence the principles of equilibrium thermodynamics can be applied to these systems.

ENERGY OF THE CELL REACTION

Measurement of the potential difference between two reversible electrodes provides one of the most accurate means of studying the driving force of the reaction taking place within the cell and the change in Gibbs energy

---

[2] M. Randall, *Amer. Phys. Teacher*, **6**, 291 (1938); **7**, 292 (1939). The concepts of "electrode potential" and "electromotive force" will be used in this book instead of "tension" as recommended by the International Committee for Electrochemical Thermodynamics and Kinetics (now International Society of Electrochemistry, ISE), because the author believes these terms and concepts to be more familiar to the general reader. The ISE recommendations have been summarized by P. Van Rysselberghe, *Electrochim. Acta*, **5**, 28 (1961); *J. Electroanal. Chem.*, **2**, 265 (1961). The unit of e.m.f. and electrode potential used consistently in this book is that formerly termed the "absolute volt."

as the reaction proceeds. Let us consider, for example, the following important reaction,

$$\tfrac{1}{2}H_2(g) + AgCl(s) = Ag(s) + HCl(m) \tag{1}$$

The letters g and s in this equation represent gas phase and solid phase, respectively, and the letter m indicates, first, that the hydrochloric acid is present in solution and, second, that its molality (moles per kilogram of solvent) is m. This is an oxidation-reduction process which proceeds spontaneously from left to right at room temperature when the hydrogen pressure is 1 atm and the solution of hydrochloric acid is less than $9m$.[3]

When the reaction proceeds as written, hydrogen atoms are oxidized and silver chloride is reduced. In the cell[4]

$$Pt; H_2(g), HCl(m), AgCl(s); Ag(s) \tag{2}$$

the "oxidizing agent" is isolated from the "reducing agent." If diffusion through the cell is negligible, no appreciable reduction of silver chloride by hydrogen can ensue. To sustain the process, a transfer of electric charge must take place. If this transfer is not possible, chemical change at the electrodes ceases. Nevertheless, the difference of electrical potential between the platinum and silver electrodes reveals the *tendency* for reaction 1 to take place.

The reversible electromotive force E multiplied by the quantity of electric charge that must be transferred to bring about the unit amount of reaction gives the maximum electrical work[5] that this amount of chemical reaction is capable of producing at constant temperature, pressure, and concentration. This is the decrease in Gibbs energy $-\Delta G$ which accompanies the reaction, namely

$$-\Delta G = nFE \tag{3}$$

where F is the Faraday constant and n is an integer.[6]

---

[3] G. Akerlof and J. W. Teare, *J. Amer. Chem. Soc.*, **59**, 1855 (1937).

[4] Finely divided platinum, which catalyzes the dissociation of hydrogen molecules into atoms and the oxidation of the latter, is the common site for the hydrogen electrode.

[5] In accordance with the recommendations of the IUPAC Manual (reference 1), the symbol G is used for the Gibbs energy, $H - TS$, where H and S are the heat content (enthalpy) and entropy, respectively, and T is the thermodynamic temperature in kelvins.

[6] For E in volts, F in coulombs per mole, and n in faradays per reaction unit as written, $-\Delta G$ is in joules. The value of n is readily determined by separating the cell reaction into its oxidation and reduction parts; n is then seen to be the number of moles of electrons necessary to balance each half-reaction electrically. Energy in thermochemical calories is obtained by the defined relationship 1 cal = 4.184 J.

ABSOLUTE POTENTIALS

The only possible way that atoms can furnish electrons to metallic conductors (or vice versa) is through a change in charge, that is, a change of valence. Hence the passage of current between metallic and electrolytic conductors is accompanied by a chemical reaction. Thermodynamics can furnish information only about the complete reaction and the *differences* of potential between the electrodes. The peculiar demands of pH measurements, however, focus attention on the half-reaction that occurs at the electrode reversible to hydrogen ion and on the half-cell potential. The "absolute" scale of single electrode potentials has long been sought by both experimental and theoretical means, but these efforts have thus far not met with success.

It should be noted that an absolute scale of potential is not essential for the establishment of a single scale of acidity for all solvent media. One needs rather a scale of potential for each solvent relative to a conventional scale in one solvent (for example, in water). Proposals for interrelating hydrogen electrode potentials[7] in a variety of solvent media are discussed in Chapters 7 and 8.

THE HYDROGEN SCALE

In spite of the progress that has been made, the validity and accuracy of a basis for an absolute scale of potential have not been established. Fortunately, a numerical scale of potential that meets nearly all practical requirements in the aqueous medium can be based on an arbitrary conventional reference point. The arbitrary zero of potential is defined as follows. The potential of the reversible hydrogen electrode in contact with hydrogen gas at 1 atm partial pressure and immersed in a solution containing hydrogen ions at unit activity,

$$\text{Pt}; \text{H}_2(g, 1 \text{ atm}), \text{H}^+(a = 1) \tag{4}$$

is taken as zero at all temperatures. The scale of potentials based on this zero point is called the *hydrogen scale*.

The potential of a single electrode on the hydrogen scale is evidently the e.m.f. of a complete cell consisting of the electrode in question combined with this standard hydrogen electrode. The e.m.f. should not include the diffusion potentials that arise at the liquid-liquid boundary between the two half-cells. Inasmuch as the standard hydrogen electrode is not a convenient reference for practical measurements, potentials on the hydrogen scale are

---

[7] V. A. Pleskov, *Usp. Khim.*, **16**, 254 (1947); H. M. Koepp, H. Wendt, and H. Strehlow, *Z. Elektrochem.*, **64**, 483 (1960); N. A. Izmailov, *Dokl. Akad. Nauk SSSR*, **127**, 104 (1959). The subject of absolute and relative electrode potentials has been reviewed by G. Milazzo and G. Bombara, *J. Electroanal. Chem.*, **1**, 265 (1959–60).

usually calculated from the measured e.m.f. of other electrode combinations —for example, from the potential of the given electrode with respect to the calomel reference element. The sign of the single electrode potential must be decided by convention. This matter will be considered in a later section.

Chemical Potential and Activity

E.M.F. AND THE CELL REACTION

Equation 3 expresses the relationship between the electromotive force of a galvanic cell and the decrease of Gibbs energy when the cell reaction proceeds reversibly. Both the (Gibbs) free energy change and the e.m.f. measure the tendency for the cell reaction to take place. It is appropriate to consider next the effect on the e.m.f. of changes in the concentration or activity of the individual reactants or products.[8] For this purpose we shall write a cell reaction in general terms, as follows:

$$iI + jJ + \cdots = uU + vV + \cdots \tag{5}$$

The Gibbs energy of the system represented by equation 5 is a function of the temperature T, the pressure P, and the composition, expressed in terms of the numbers of moles n_I, n_J, n_U, n_V, etc., of each of the components of the system. When the system is subjected to small changes in these variables, the change dG in free energy is (at constant charge)

$$dG = -S\,dT + V\,dP + \sum \mu_X\,dn_X \tag{6}$$

where S and V are the entropy and volume of the system, respectively, and μ_X is a quantity called the *partial molar free energy* or *chemical potential* of the species (X) indicated by the subscript. The summation is made for all species of which the system is composed. Mathematically, the chemical potential of the species X is the partial derivative of the free energy with respect to moles of species X, when the temperature, pressure, and amounts of the other substances remain constant:

$$\mu_X = \left(\frac{\partial G}{\partial n_X} \right)_{T, P, n} \tag{7}$$

---

[8] Excellent authoritative discussions of the thermodynamics of cells and of electrolytic solutions have been given by D. A. MacInnes, *The Principles of Electrochemistry*, Reinhold Publishing Corp., New York, 1939; Dover Publications, New York, 1961; S. Glasstone, *Introduction to Electrochemistry*, D. Van Nostrand Co., New York, 1942; H. S. Harned and B. B. Owen, *The Physical Chemistry of Electrolytic Solutions*, 3rd ed., Reinhold Publishing Corp., New York, 1958; and R. A. Robinson and R. H. Stokes, *Electrolyte Solutions*, 2nd ed. revised, Butterworths, London, 1970.

Physically, μ_X represents the change in free energy, at constant temperature and pressure, caused by the reversible addition of 1 mole of species X to such a large amount of the system that no appreciable change in composition results.

The measured changes of free energy for chemical reactions correspond to a variety of states for the substances taking part in the reaction. These states are chosen by the investigator to meet the demands of the particular experimental conditions; for example, one substance may be in a gaseous state, another a constituent of a concentrated solution, and so forth. To be of greatest use in subsequent calculations, the changes of free energy for a given reaction measured under all these varied conditions must be expressed in terms of the change associated with the reaction when each of the substances is in a standard reference state. This state is identified by affixing a small superscript circle to the appropriate symbol. Because the system is usually not ideal, it is often convenient to express the chemical potential of each species in any particular state in terms of a quantity a, the *activity* of that species, instead of its concentration:

$$\mu_X - \mu_X{}^\circ = RT \ln a_X \tag{8}$$

where R is the gas constant. Evidently $\mu_X - \mu_X{}^\circ$ is zero in the standard state, and a_X is unity.

Let us suppose that reaction 5 occurs at constant temperature and pressure and consider the change of free energy that accompanies the conversion of i moles of I and j moles of J, etc., into u moles of U and v moles of V, etc. If the reaction interval is so small that the chemical potentials of the reactants and products remain constant, we have, from equations 6 and 8,

$$\Delta G = u\mu_U + v\mu_V + \cdots - i\mu_I - j\mu_J - \cdots$$

$$= \Delta G^\circ + RT \ln \frac{a_U{}^u a_V{}^v \cdots}{a_I{}^i a_J{}^j \cdots} \tag{9}$$

where ΔG°, the change of free energy when all the substances are in their standard states (and each activity is unity by definition), is given by

$$\Delta G^\circ = u\mu_U{}^\circ + v\mu_V{}^\circ + \cdots - i\mu_I{}^\circ - j\mu_J{}^\circ - \cdots \tag{10}$$

Two important conclusions emerge directly from equation 9. The first is the relationship between the reversible e.m.f. of a galvanic cell, in which reaction 5 takes place, and the activities of the reactants and products. This relationship is obtained by combining equations 3 and 9:

$$E = E^\circ - \frac{RT}{nF} \ln \frac{a_U{}^u a_V{}^v \cdots}{a_I{}^i a_J{}^j \cdots} \tag{11}$$

The second conclusion concerns the condition of chemical equilibrium. One of the properties of the equilibrium state at constant temperature and pressure is that $\Delta G = 0$. Hence, from equation 9,

$$\Delta G^\circ = -RT \ln\left(\frac{a_U{}^u a_V{}^v \cdots}{a_I{}^i a_J{}^j \cdots}\right)_{\text{eq.}} = -RT \ln K \tag{12}$$

where K is the equilibrium constant for the reaction, and the activities are those in a mixture at equilibrium (eq.). It is clear from these equations that E°, the standard potential, relates to the equilibrium state,

$$E^\circ = \frac{RT}{nF} \ln K \tag{13}$$

and that the e.m.f. E becomes zero as equilibrium is attained. Values of $(RT \ln 10)/F$ are given in Table 1 of the Appendix.

THE STANDARD STATE

The choice of the standard state of unit activity is to a large extent arbitrary; hence the activity function is more restricted than the chemical potential. As equation 8 demonstrates, the numerical value of the activity of a given substance varies with the reference state selected. The practical advantages of a free choice of the standard state best adapted to a particular problem nonetheless outweigh the confusion that necessarily arises from the lack of an inflexible rule.

For cell measurements at a pressure no greater than 1 atm it is usually convenient to employ the following standard states. For gases, the activity at any temperature is taken equal to the partial pressure of the gaseous component. For pure solids and pure liquids at atmospheric pressure, the activity is taken to be unity at all temperatures. For liquid solutions, the standard state is usually chosen so that $a/m = 1$ when $m = 0$, that is, at zero molality or infinite dilution of the solute, for every temperature and pressure.

The selection of this standard state for solutes reflects a desire to express the activity and the molality on a common basis, so that the activity can be taken equal to the molality for many purposes. The chosen state is a hypothetical one that possesses not only unit activity but some of the thermodynamic properties of an infinitely dilute solution. It should be emphasized that actual solutions in which a substance may possess unit activity cannot be regarded as solutions of the substance in the standard state. The activity would not remain at unity with changes of temperature and pressure, for the thermodynamic properties of the actual state are not those of an infinitely dilute solution.[9]

---

[9] The nature of the standard state is discussed by R. A. Robinson and R. H. Stokes, *op. cit.*, Chapter 2.

CHANGES OF TEMPERATURE AND PRESSURE

The change of the Gibbs free energy ΔG for the cell reaction at constant temperature can be expressed in terms of the changes of enthalpy (ΔH) and of entropy (ΔS) by

$$\Delta G = \Delta H - T \Delta S \tag{14}$$

It can be shown from equation 6 that

$$\left(\frac{\partial \Delta G}{\partial T} \right)_P = -\Delta S \tag{15}$$

Hence, by combination of equations 14 and 15 with equation 3, we obtain the Gibbs-Helmholtz equation

$$\Delta H = -nF \left[E - T \left(\frac{\partial E}{\partial T} \right)_P \right] \tag{16}$$

from which the change of enthalpy for the cell reaction can be obtained.

The change of the activity a_i of a component i with changes of temperature (pressure constant) and with changes of pressure (temperature constant) is given by the equations

$$\left(\frac{\partial \ln a_i}{\partial T} \right)_P = - \frac{\bar{H}_i - \bar{H}_i^\circ}{RT^2} \tag{17}$$

and

$$\left(\frac{\partial \ln a_i}{\partial P} \right)_T = \frac{\bar{V}_i - \bar{V}_i^\circ}{RT} \tag{18}$$

where $\bar{H}_i - \bar{H}_i^\circ$ and $\bar{V}_i - \bar{V}_i^\circ$ are, respectively, the relative partial molar heat content and the relative partial molar volume of the component i.

ACTIVITY COEFFICIENTS

Even though the usual standard state for aqueous solutions is adopted, the activities of substances in actual solutions are not usually equal to the molalities, particularly if the substances are electrolytes. The ratio of the activity of species i to its molality is the *activity coefficient* γ_i,

$$\gamma_i = \frac{a_i}{m_i} \tag{19}$$

It is convenient, when other scales of concentration are used, to define the standard state in such a way that a_i/c_i or a_i/x_i becomes unity when c_i or x_i, the molar concentration or mole fraction, respectively, becomes zero. There are therefore two other activity coefficients, y_i and f_i, corresponding to these reference states. We shall have no need for the latter. Activities and activity coefficients on the molality (m) and concentration (c) scales are related by the equations

$$a_c = a_m \, d° \tag{20}$$

and

$$y = \gamma \, d° \frac{m}{c} \tag{21}$$

where $d°$ is the density of the solvent.[10]

Equation 19 applies formally to ions, and one may write $\gamma_+ = a_+/m_+$ for positive ions and $\gamma_- = a_-/m_-$ for negative ions. However, the activity coefficients of individual ions cannot be defined uniquely or measured; only certain quotients or products of single ionic activity coefficients can be determined exactly. Nevertheless, it is advantageous to express certain thermodynamic properties in terms of hypothetical or "conventional" ionic activity coefficients. The arbitrary means of evaluating these coefficients will be considered in a later chapter.

From thermodynamic studies of electrolytic solutions one always obtains mean activity coefficients $\gamma_\pm$ for combinations of ions. If the electrolyte (molality m, activity a) dissociates into a total of v ions, of which v_+ are positive ions and v_- are negative ions, the mean ionic activity $a_\pm$ is given by

$$a_\pm = (a_+{}^{v_+} a_-{}^{v_-})^{1/v} = a^{1/v} \tag{22}$$

The mean ionic molality $m_\pm$ is

$$m_\pm = m(v_+{}^{v_+} v_-{}^{v_-})^{1/v} \tag{23}$$

and the mean ionic activity coefficient is

$$\gamma_\pm = \frac{a_\pm}{m_\pm} = (\gamma_+{}^{v_+} \gamma_-{}^{v_-})^{1/v} \tag{24}$$

As defined in the IUPAC Manual,[1] $a_\pm$ is a relative activity and therefore dimensionless. Similarly, the mean molality is defined relative to a hypothetical solution of unit molality. Through this device, both the activity and activity coefficient become dimensionless.

Although ionic activity coefficients cannot be measured individually, they can be calculated by theoretical means at concentrations so low that the interionic forces depend primarily on the charge, radius, and distribution of the ions and the dielectric constant of the medium, rather than on the

---

[10] Equation 20 permits the relationship between K_c and K_m, the equilibrium constants on the two scales of concentration, to be determined readily.

chemical properties of the ions. At a given temperature the activity co-efficients of ions in aqueous solutions of low concentration vary chiefly with the distances between the ions and with the number of charges borne by the ions. These factors are combined in the *ionic strength I*, defined by

$$I = \tfrac{1}{2} \sum m_i z_i^2 \tag{25}$$

where m_i and z_i are the molality and charge of each ionic species and the summation is made for all the ionic species present in the solution.

An equation for the activity coefficient γ_i of the ionic species i was developed from electrostatic and statistical theory by Debye and Hückel.[11] It may be written in the form

$$-\log \gamma_i = \frac{A z_i^2 \sqrt{I}}{1 + \rho \sqrt{I}} \tag{26}$$

In this equation z_i is the valence of the ion i; ρ is $B\mathring{a}$, where $\mathring{a}$ is the "ion-size parameter" or "mean distance of closest approach" of the ions. The latter is of the same order as the ionic diameter. The Debye-Hückel limiting law is obtained by setting $\rho = 0$ in equation 26. The values of A and B vary with the temperature and dielectric constant of the solvent; values for the water medium ($\mathring{a}$ in angstrom units) are listed in Table 4 of the Appendix. In dilute solutions, I may be expressed in terms of either molar concentration or molality, provided the corresponding values of A and B are used.

The Debye-Hückel equation actually relates the activity coefficient f_i on the mole fraction scale with the square root of the concentration (c_i). Equation 26 preserves the original form, and differences between the concentration scales are allowed for with considerable success in the values of A and ρ. This equation is most useful as an interpolation formula to which experimental values of mean activity coefficients are fitted by a suitable adjustment of $\mathring{a}$. For this purpose, equation 26 is combined with equation 24.

At moderate ionic strengths a considerable improvement in the fit is effected by subtracting a term bI from the right side of equation 26; b is an adjustable parameter. This two-parameter equation is of the Hückel form.[12] When I is greater than 0.2, a still better fit is obtained when a third term allowing for the difference between $f_{\pm}$ and $\gamma_{\pm}$ is also added.

---

[11] P. Debye and E. Hückel, *Physik. Z.*, **24**, 185 (1923). For details of the derivation of this important equation, the reader is referred to D. A. MacInnes, *The Principles of Electro-chemistry*, Chapter 7, Reinhold Publishing Corp., New York, 1939; Dover Publications, New York, 1961; and R. A. Robinson and R. H. Stokes, *Electrolyte Solutions*, 2nd ed. revised, Chapter 4, Butterworths, London, 1970.
[12] E. Hückel, *Physik, Z.*, **26**, 93 (1925).

Hydrogen Ion Equilibria

The state of the dissolved hydrogen ion is not known with certainty, yet it is safe to say that free protons do not exist in any significant amounts in aqueous solutions.[13] Nevertheless, the stability of the hydrogen electrode shows that the proton activity in these media has a definite value. Although water does not take part in the electrode process, it is not unlikely that direct exchange of protons between the electrode and the various solvated species present can occur. Hence it is usually convenient and justifiable to use the term "hydrogen ion" and the symbol H^+ as "a shorthand for the whole group of hydrates."[14]

In a strict sense, the reactions of acids and bases with water may be a dissociation or an ionization.[15] Ionization is evidenced by the enhanced conductance of the solution, and dissociation (of salts, for example) may be proved analytically by the detection of foreign species. The reaction of NH_3 with water to produce NH_4^+ is demonstrably an ionization process but, as usually formulated, is not a dissociation.

The determination of thermodynamic ionization (dissociation) constants cannot distinguish among the various possible explanations of the properties of solutions of acids and bases or the mechanisms of the processes.[16] These constants, indeed, should be regarded as overall equilibrium constants which may include a variety of effects, such as solvation and tautomerization of the solute species as well as the actual ionization.

In this book acid-base reactions are treated from the point of view of the Brønsted-Lowry definition.[17] The term "acid" is restricted to proton donors and the term "base" to substances capable of accepting a proton. Acid-base reactions in general, including the "dissociation" of acids and bases in water, are regarded as proton-transfer processes. The thermodynamic equilibrium constant for the process

$$HA + H_2O \;\rightleftharpoons\; H_3O^+ + A \tag{27}$$

will be called the *acidic dissociation constant*, K_a, of the acid HA, and that for

$$B + H_2O \;\rightleftharpoons\; BH^+ + OH^- \tag{28}$$

will be called the *basic dissociation constant*, K_b, of the base B.

---

[13] According to I. M. Kolthoff, *Rec. Trav. Chim.*, **49**, 401 (1930), the equilibrium constant for the dissociation of $H_2O \cdot H^+$ into protons and water molecules is about 10^{-130}.

[14] G. N. Lewis, *J. Franklin Inst.*, **226**, 293 (1938).

[15] J. E. Ricci, *Hydrogen Ion Concentration*, Chapters 1 and 6, Princeton University Press, Princeton, N.J., 1952.

[16] J. E. Ricci, *J. Amer. Chem. Soc.*, **70**, 109 (1948).

[17] J. N. Brønsted, *Rec. Trav. Chim.*, **42**, 718 (1923); *Chem. Rev.*, **5**, 231 (1928); T. M. Lowry, *Chem. Ind.* (*London*), **42**, 43 (1923); *Trans. Faraday Soc.*, **20**, 13 (1924).

Conventions of E.M.F. Measurements

REPRESENTATION OF CELLS

The following conventions for the representation of galvanic cells are used in this book.[18]

1. A metal-electrolyte boundary is marked by a semicolon (;).
2. Two or more solutes in the same solution are separated by commas (,).
3. The location of a liquid-liquid boundary is marked by a vertical line ($|$).
4. The pH-responsive membrane of the glass electrode is located at a double vertical line ($\|$).
5. The physical state of a component is indicated by (s), (l), or (g), representing solid, liquid, and gas, respectively.
6. The metal pieces (for example, copper wires) attached to the electrodes during the measurement of the e.m.f. of the cell are omitted from the scheme.

These rules may be illustrated by writing the scheme for the cell with glass and saturated calomel electrodes commonly employed for pH determinations:

$$\text{Glass} \parallel \text{Soln. X} \mid \text{KCl (satd.)},\text{Hg}_2\text{Cl}_2(s);\text{Hg}(l) \qquad (29)$$

This cell contains both a glass membrane and a liquid junction. The comma indicates that potassium chloride and calomel are both present in the liquid phase. Calomel is also present as a solid phase in contact with liquid mercury. The composition of the reference cell within the glass electrode is not shown in this scheme.

SIGN OF THE E.M.F.

A completely unambiguous way to indicate the polarity of a galvanic cell would be to affix plus and minus signs to the two electrodes in the scheme of the cell, and this is often done.[19] Then the e.m.f. assigned to the cell need only indicate the magnitude of the potential difference and need convey no information regarding the direction in which the cell reaction takes place on spontaneous discharge. It is clearly evident that oxidation occurs at the negative electrode and reduction at the positive electrode.

The following convention has nonetheless been adopted rather generally, in order to relate the representation of the cell, the form in which the cell

---

[18] This system is essentially that adopted by MacInnes in his monograph, with minor modifications.

[19] Likewise, a sign convention would be unnecessary if cell reactions were always written in the form indicating the direction of the spontaneous reaction when current is drawn from the cell; see J. B. Ramsey, *J. Electrochem. Soc.*, **104**, 691 (1957). The reader is also referred to the excellent discussion of conventions contained in W. M. Clark's monograph, *Oxidation-Reduction Potentials of Organic Systems*, Chapter 3, The Williams and Wilkins Co., Baltimore, Md., 1960.

reaction is written, and the Gibbs energy change for that reaction (equation 3) in a completely consistent manner.

1. The e.m.f. is given a positive sign if the electrode on the left of the scheme is negative and that on the right is positive. If the negative electrode is on the right of the cell scheme, the e.m.f. is given a negative sign.
2. The cell reaction is always written *as if* oxidation occurred spontaneously at the left electrode and reduction at the right; hence, a positive e.m.f. signifies that the cell reaction proceeds from left to right, and a negative e.m.f. means that the reaction occurs spontaneously in the reverse direction.
3. The change of free energy for the cell reaction, written as prescribed in paragraph 2, is given by $\Delta G = -nFE$, where the sign of the e.m.f. is determined as prescribed in specification 1.

In 1953 the International Union of Pure and Applied Chemistry, meeting in Stockholm, recommended this convention in the following words.[20]

The sign and magnitude of the e.m.f. of a cell are identical with the sign and magnitude of the electrical potential of the metallic lead to the electrode written on the right of the diagram with respect to a lead of similar metal connected to the electrode written on the left, the potential of the latter being taken as zero.

These rules may be illustrated by applying them to cell 2. When the partial pressure of hydrogen is 1 atm and the molality of hydrochloric acid is 0.1, platinum is the negative electrode and the potential difference between the electrodes is 0.3523 V at 25 °C; hence E is $+0.3523$ V. The cell reaction is written as if hydrogen atoms were being oxidized spontaneously:

$$\tfrac{1}{2}H_2(g, \text{ 1 atm}) + AgCl(s) = Ag(s) + HCl(0.1m) \tag{30}$$

Inasmuch as the e.m.f. is positive, the reaction proceeds spontaneously from left to right and the free energy change for this reaction is negative:

$$\Delta G = -nFE = -1 \times 96{,}487 \times 0.3523 = -34{,}000 \text{ joules} \tag{31}$$

SIGN OF THE ELECTRODE POTENTIAL

Although the hydrogen scale of potentials has found practically universal acceptance, there is, unfortunately, no general agreement on the sign to be given the single electrode potential. As we have seen, the potential of a single half-cell is considered equal to the potential of a complete cell in which the electrode or half-cell in question is combined with the standard hydrogen half-cell outlined in equation 4. The rules for determining the sign of the

[20] J. A. Christiansen and M. Pourbaix, *Compt. Rend. 17th Conf., Union Intern. Chim. Pure Appl.*, Stockholm, 83 (1953); J. A. Christiansen, *J. Amer. Chem. Soc.*, **82**, 5517 (1960). For a restatement of this international convention, see the IUPAC *Manual* (reference 1).

e.m.f. of complete cells were set forth in the preceding section. It is only necessary, therefore, to agree on the position of the standard hydrogen half-cell in the cell scheme to fix the sign of the single potential.

The Stockholm convention for the sign of the electrode potential (E_e) provides that the standard hydrogen electrode be written on the left. The half-cell reaction corresponding to the electrode potential is thus always formulated as a reduction, namely Oxid. $+ ne =$ Red. For the silver-silver chloride electrode,

$$AgCl + e \; \overrightarrow{\longleftarrow} \; Ag + Cl^- \tag{32}$$

and

$$E_e = E_e^{\circ} - \frac{RT}{nF} \ln \frac{a_{red.}}{a_{oxid.}} = E_e^{\circ} - \frac{RT}{F} \ln a_{Cl^-} \tag{33}$$

where E_e° is the *standard electrode potential*. It follows that the e.m.f. of the cell is obtained by subtracting the electrode potential of the left electrode from that of the right:

$$E = E_e(r) - E_e(l) \tag{34}$$

A positive electrode potential signifies that the electrode system is a better oxidizing agent than hydrogen ion ($a = 1$) in contact with hydrogen ($p_{H_2} = 1$ atm) on platinum; a negative electrode potential means that the system is a better reducing agent than hydrogen on platinum under the same standard conditions.[21]

The Stockholm convention for the sign of the electrode potential was formerly known as the " European convention," as opposed to that associated with the G. N. Lewis school, which was sometimes called the " American convention."[22] The former, which is now being adopted generally,[23] makes the sign of the electrode potential the same as the polarity of the electrode in a cell the other electrode of which is the standard hydrogen electrode.

---

[21] The concept of the electrode potential and the rules for the combination of electrode potentials and half-reactions have been considered by R. G. Bates, *Treatise on Analytical Chemistry*, I. M. Kolthoff and P. J. Elving, eds., Part I, Vol. 1, Chapter 9, Interscience Publishers, New York, 1959.

[22] These two opposing conventions have been compared in detail by A. J. de Béthune, *J. Electrochem. Soc.*, **102**, 288C (1955); T. S. Licht and A. J. de Béthune, *J. Chem. Educ.*, **34**, 433 (1957); and T. F. Young, *American Institute of Physics Handbook*, p. 5–268, McGraw-Hill Book Co., New York, 1957. J. B. Ramsey, *J. Electrochem. Soc.*, **104**, 255 (1957), has pointed out the usefulness of the concept of the chemical potential or escaping tendency of electrons in choosing the sign of the electrode potential.

[23] This new convention has, in fact, found use in the revised edition of Lewis and Randall's classic work on thermodynamics: G. N. Lewis, M. Randall, K. S. Pitzer, and L. Brewer, *Thermodynamics*, 2nd ed., McGraw-Hill Book Co., New York, 1961.

CHAPTER 2

pH scales

The relationship between the concentrations of hydrogen and hydroxyl ions in an aqueous medium is fixed by the equilibrium between these ions and water molecules:[1]

$$H_2O \rightleftharpoons H^+ + OH^- \tag{1}$$

If solutes are present in such small concentrations that the activity of water is practically unity,[2] the equilibrium constant can be formulated as follows:

$$K_w = m_H m_{OH} \gamma_H \gamma_{OH}, \tag{2}$$

where m is molality and K_w is the dissociation constant or ion product constant of water. The value of K_w at temperatures from 0 to 60 °C is listed in the Appendix, Table 2. The process represented by reaction 1 has a large heat effect; accordingly, the temperature coefficient of K_w is large.

In the interest of a simplified notation, the charges of ions appearing in subscripts will usually be omitted, as in equation 2, when no confusion is likely to result. Thus we shall write m_{OH}, a_H, γ_{Cl}, and so forth.

The dissociation of pure water is extremely small, and the activity coefficients in equation 2 are practically unity unless solutes are present. Hence, at 25 °C where the value[3] of K_w is 1.008×10^{-14},

$$m_H = m_{OH} = \sqrt{K_w} = 1.004 \times 10^{-7} \tag{3}$$

---

[1] A systematic, rigorous treatment of the many complex reactions by which solutes disturb the hydrogen and hydroxyl ion concentrations of water has been presented by J. E. Ricci, *Hydrogen Ion Concentration*, Princeton University Press, Princeton, N.J., 1952.

[2] The activity of water in an aqueous solution is $p/p°$, where p and $p°$ are the pressures of water vapor in equilibrium with the solution and with pure water, respectively, at the temperature in question, provided that p and $p°$ are so low that water vapor can be considered an ideal gas.

[3] H. S. Harned and R. A. Robinson, *Trans. Faraday Soc.*, **36**, 973 (1940).

Accepting for the moment the definition pH $\equiv -\log m_H$, we see that the *neutral point*, or point at which the hydrogen and hydroxyl ions are present in equivalent concentrations, is located at pH 7. Inasmuch as K_w is a function of temperature, however, the neutral point will vary as the temperature is changed. The neutral pH is 7.5 at 0 °C and 6.5 at 60 °C and is altered by salt effects. It is evident that the range of the pH scale also depends on the magnitude of K_w. In view of the fact that concentrations of hydrogen and hydroxyl ions rarely exceed 10 moles per liter, the approximate practical range of the pH scale is from -1 to 15 at ordinary temperatures.

Rational Scales of Acidity

Although the acidity, or reaction, of aqueous solutions is customarily expressed in the pH notation, it is of interest to examine briefly some other systems that have been proposed but have failed to receive wide adoption. In general, these are "rational" scales; that is, in contrast to the pH, the index is a larger number the more highly acidic the solution.

The pR unit of Giribaldo[4] is defined by

$$pR \equiv \log \frac{c_H}{c_{OH}} \tag{4}$$

where c represents the concentration in moles per liter. Thus pR is $+14$ for a 1 M solution of hydrochloric acid and -14 for a solution of sodium hydroxide of the same concentration. Gerstle[5] has suggested a unit of acidity called the "hydron." According to his proposal, the degree of acidity or alkalinity would be expressed as a number of hydrons N_h or of decihydrons N_{dh}, where $N_{dh} = 10N_h$. The N_h of solution 1 is defined relative to a standard of reference (solution 2) by

$$N_h \equiv \log \frac{(a_H)_1}{(a_H)_2} \tag{5}$$

where a_H is the activity of hydrogen ion. It was suggested by Gerstle that pure water be chosen as a standard. Catani's rA (*reaction actual*)[6] is the number of millimoles of ionic hydrogen per liter below pH 7 (with a positive sign) or millimoles of hydroxyl ion (with a negative sign) above pH 7.

These scales are compared with the pH scale in Table 2–1. In the sections

---

[4] A detailed comparison of pR and pH is found in D. Giribaldo's booklet, *Ventajas del pR sobre el pH en la Expresión de la Reacción,* a collection of papers from *Anales Fac. Quim. y Farm.,* **2**, Montevideo, 1938.

[5] J. Gerstle, *Trans. Electrochem. Soc.,* **73**, 183 (1938).

[6] R. A. Catani, *Rev. Brasil. Quim. (São Paulo),* **15**, 264 (1943).

TABLE 2-1. COMPARISON OF pH AND RATIONAL SCALES OF ACIDITY

c_H or a_H	pH	pR	N_h	rA
10	−1	+16	+8	+10,000
1	0	+14	+7	+ 1,000
1×10^{-3}	3	+ 8	+4	+ 1
1×10^{-6}	6	+ 2	+1	+ 0.001
1×10^{-7}	7	0	0	± 0.0001
1×10^{-8}	8	− 2	−1	− 0.001
1×10^{-11}	11	− 8	−4	− 1
1×10^{-14}	14	−14	−7	− 1,000
1×10^{-15}	15	−16	−8	−10,000

to follow, the early history of the pH scale and various approaches to a precise definition of the pH unit will be considered.[7] Approximate pH values of some common reagent solutions are summarized in the Appendix, Table 6.

Hydrogen Ion Concentration

The unique influence of the hydrogen ion concentration on biochemical reactions was described by the Danish chemist Sørensen in 1909. In two important papers, totaling some 170 pages and published simultaneously in German and French,[8] Sørensen compared the usefulness of the *degree of acidity* with that of the *total acidity*, proposed the hydrogen ion exponent, set up standard methods for the determination of hydrogen ion concentrations by both electrometric and colorimetric means, with a description of suitable buffers and indicators, and discussed in detail the application of pH measurements to enzymatic studies.

In the study of the pepsin hydrolysis, or digestion, of proteins in acid-salt mixtures, for example, the hydrogen ion concentration is found to be an important variable. Yet the effective concentration of hydrogen ion cannot readily be calculated from the compositions of the mixtures, for the basic groups of the protein molecule bind a part of the free acid, and the salt

---

[7] The reader is referred to the following review articles: G. Kortüm, *Z. Elektrochem.*, **48**, 145 (1942; F. Müller and H. Reuther, *ibid.*, **48**, 288 (1942); R. G. Bates, *Chem, Rev.*, **42**, 1 (1948); A. C. Schuffelen, *Chem. Weekbl.*, **46**, 898 (1950); K. Schwabe, *Chimia*, **13**, 385 (1959); L. Laloi, *Bull. Soc. Chim. France*, 1663 (1961).

[8] S. P. L. Sørensen, "Enzymstudien. II. Über die Messung und die Bedeutung der Wasserstoffionenkonzentration bei enzymatischen Prozessen," *Biochem. Z.*, **21**, 131 (1909); continuation, **21**, 201 (1909); *Compt. Rend. Trav. Lab. Carlsberg*, **8**, 1 (1909).

present influences the degree of ionization. The effective concentration c_H of hydrogen ion is indeed often very small and is commonly expressed, for convenience, by exponential arithmetic. Thus a molar concentration of 0.01 becomes 10^{-2}, and 0.00002 becomes 2×10^{-5} or $10^{-4.7}$. In general,

$$c_H = 10^{-p} = \frac{1}{10^p} \tag{6}$$

where p, the initial letter of the words *Potenz*, *puissance*, and *power*, was called the "hydrogen ion exponent" by Sørensen and was written p_H. .[9]

Doubtless for convenience in typesetting, the original Sørensen symbol for the hydrogen ion exponent has gradually been superseded by "pH." In the discussion of the present chapter, in order to distinguish among the several definitions of pH that have come into being, we shall modify this symbol further. Thus the pH value based on hydrogen ion concentration (or molality) will be designated pc_H (or pm_H):[10]

$$pc_H \equiv -\log c_H \tag{7}$$

and

$$pm_H \equiv -\log m_H \tag{7a}$$

The Sørensen pH Scale

As a means of measuring pH values Sørensen chose the cell

$$\text{Pt}; \text{H}_2, \text{Soln. X} \,|\, \text{Salt bridge} \,|\, \text{0.1 M Calomel electrode} \tag{8}$$

where the vertical lines represent liquid-liquid boundaries, and followed essentially the techniques that had been outlined earlier by Bjerrum.[11] An attempt was made to eliminate the diffusion potentials at the liquid junctions by the "Bjerrum extrapolation."[12] For this purpose, two measurements of the e.m.f. E of cell 8, with bridge solutions of 3.5 M and 1.75 M potassium chloride interposed between the two half-cells, were made for each solution X. The observed difference in potential was added or subtracted (see Fig. 2–1)

---

[9] "... die Zahl der Grammatome Wasserstoffionen pro Liter ... kann gleich 10^{-p} gesetzt werden. Für die Zahl p schlage ich den Namen 'Wasserstoffionenexponent' und die Schreibweise p_H. vor. Unter dem Wasserstoffexponenten (p_H.) einer Lösung wird dann der Briggsche Logarithmus der reziproken Wertes des Wasserstoffionen bezogenen Normalitätsfaktors der Lösung verstanden."

[10] In accordance with the IUPAC recommendations: R. G. Bates and E. A. Guggenheim, *Pure Appl. Chem.*, **1**, 163 (1960).

[11] N. Bjerrum, *Kgl. Danske Videnskab. Selskab. Skr.*, [7] **4**, 13 (1906).

[12] N. Bjerrum, *Z. Physik. Chem.*, **53**, 428 (1905); *Z. Elektrochem.*, **17**, 389 (1911).

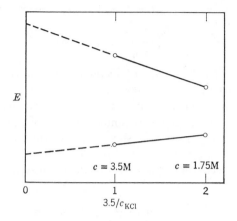

FIG. 2-1. The Bjerrum extrapolation.

from the e.m.f. of the cell with the more concentrated bridge solution as necessary to obtain the hypothetical potential corresponding to a bridge solution of infinite concentration ($1/c = 0$) which, presumably, would nullify the boundary potentials completely. There is evidence, however, that this procedure reduces the liquid-junction potential to a negligible value only when the observed difference of e.m.f. is small.[13] Michaelis[14] regarded the Bjerrum extrapolation as advantageous when the concentration of hydrogen ion or hydroxyl ion is the test solution exceeds 0.001 mole per liter.

When Sørensen first defined the pH scale, the properties of electrolytes and the relationship between the e.m.f. and the chemical reaction of a galvanic cell were customarily formulated in a manner consistent with the classical concepts of Arrhenius and Nernst. The difference in e.m.f., $E_1 - E_2$, between two cells of type 8 was accordingly written in terms of the hydrogen ion concentrations $(c_H)_1$ and $(c_H)_2$ of the two solutions, as follows:

$$E_1 - E_2 = \frac{RT}{F} \ln \frac{(c_H)_2}{(c_H)_1} \tag{9}$$

Evidently if the hydrogen ion concentration $(c_H)_2$ in the second solution

---

[13] A. H. W. Aten and J. van Dalfsen, *Rec. Trav. Chim.*, **45**, 177 (1926); N. Bjerrum and A. Unmack, *Kgl. Danske Videnskab. Selskab. Mat.-Fys. Medd.*, **9**, No. 1 (1929).
[14] L. Michaelis, *Hydrogen Ion Concentration* (translated by W. Perlzweig), Chapter 7, The Williams and Wilkins Co., Baltimore, Md., 1926.

were fixed, at unity for example, E_2 would have a definite value at each particular temperature. Designating this standard potential $E^{\circ\prime}$, we have

$$E_1 = E^{\circ\prime} + \frac{RT}{F} \ln \frac{1}{(c_H)_1} = E^{\circ\prime} + \frac{RT \ln 10}{F} pc_H \qquad (10)$$

With a knowledge of $E^{\circ\prime}$ this equation would appear to furnish a means of determining hydrogen ion concentrations in "unknown" solutions.

It may be seen from equation 10 that $E^{\circ\prime}$ can be derived readily from a measurement of E_1 for cells of type 8 containing solutions of known hydrogen ion concentration $(c_H)_1$. For these standards, Sørensen chose solutions of hydrochloric acid and mixtures of hydrochloric acid with sodium chloride. He assumed that the concentration of hydrogen ion in each of these solutions is given by

$$(c_H)_1 = \alpha_1 c_1 \qquad (11)$$

where c_1 is the molar concentration of hydrochloric acid and α_1 is the degree of dissociation of the acid computed in the classical manner from measurements of electrolytic conductance. By this procedure, $E^{\circ\prime}$ is found to be 0.3377 V at 18 °C, or 0.3380 V when the cell potentials are corrected to a standard pressure of 1 atm of dry hydrogen. Clark[15] has considered the temperature coefficient of $E^{\circ\prime}$ for the Sørensen scale. His values for the standard potential at temperatures from 18 to 40 °C are given in Table 2–2.

Two reasons for the failure of this experimental method to furnish the hydrogen ion concentrations of unknown mixtures are now readily apparent.

TABLE 2–2. STANDARD POTENTIAL OF CELL 8 (SØRENSEN SCALE)

Temperature °C	$E^{\circ\prime}$ V
	(After Clark)
18	0.3380
20	0.3379
25	0.3376
30	0.3371
35	0.3365
38	0.3361
40	0.3358

[15] W. M. Clark, The Determination of Hydrogen Ions, 3rd ed., Chapter 22, The Williams and Wilkins Co., Baltimore, Md., 1928.

They may be regarded as defects in equations 9 and 11. The first of these equations holds only for ideal solutions and only for strict equality of the two liquid-junction potentials, that is, only when the residual junction potential is zero. In the modern form of this equation, activities replace concentrations. Furthermore, it is now realized that the hydrogen ion concentration in a solution of a strong monobasic acid (concentration c_1) is equal to c_1, rather than $\alpha_1 c_1$ as equation 11 indicates.

For these reasons, Sørensen's pH unit, which we designate psH, is a measure of neither the concentration nor the activity of hydrogen ion. Rather, this scale is a conventional one, defined in terms of operations and a specified formula. Thus, at 25 °C,

$$\mathrm{p}s\mathrm{H} \equiv \frac{E - 0.3376}{0.05916} \tag{12}$$

In spite of the fact that psH bears no simple direct relationship to chemical equilibria, this scale has been widely used, and extensive tables of psH values for buffer mixtures are available.[16] The original suggestion of Sørensen that corrections be applied for the liquid-junction potential has been abandoned in later outlines of standard procedure, and bridge solutions of 3.5 M or saturated potassium chloride are utilized almost universally.

Hydrogen Ion Activity

The modern formulation of the concepts of thermodynamics and the newer theories of electrolytic solutions have clarified the meaning of psH in some degree but have not provided an exact measurement of acidity. It was recognized that the e.m.f. of galvanic cells reveals changes of activity rather than of concentration, and Sørensen and Linderstrøm-Lang[17] proposed a new pH unit, termed pa_H:

$$\mathrm{p}a_\mathrm{H} \equiv -\log a_\mathrm{H} = -\log m_\mathrm{H} \gamma_\mathrm{H} \tag{13}$$

It appears that these authors did not intend that pa_H should supplant psH but that both scales would find use.

The quantity a_H in equation 13 is the activity of hydrogen ion, defined as $y_\mathrm{H} c_\mathrm{H}$ or $\gamma_\mathrm{H} m_\mathrm{H}$, where y and γ are activity coefficients corresponding to

---

[16] W. M. Clark, *The Determination of Hydrogen Ions*, 3rd ed., Chapter 9, The Williams and Wilkins Co., Baltimore, Md., 1928; H. T. S. Britton, *Hydrogen Ions*, 4th ed., Vol. I, Chapter 17, D. Van Nostrand Co., New York, 1956.

[17] S. P. L. Sørensen and K. Linderstrøm-Lang, *Compt. Rend. Trav. Lab. Carlsberg*, **15**, No. 6 (1924).

the two scales of concentration.[18] The fact that the activity of a single ionic species is a concept lacking unique physical definition does not preclude the establishment of a reasonable scale of pa_H, but this scale must be a conventional one. In other words, the values of a_H, although not unique hydrogen ion activities, will nonetheless be numbers which, inserted in equations involving a_H, will furnish equilibrium data consistent with those obtained by rigorous thermodynamic methods.

Let us turn our attention to the formal relationship between the hydrogen ion activity and the e.m.f. of cell 8, in order to examine the conditions necessary for an accurate measurement of pa_H. The cell reaction is written

$$\tfrac{1}{2}H_2(g) + \tfrac{1}{2}Hg_2Cl_2 = Hg(l) + H^+ \text{ (in soln. X)}$$
$$+ Cl^- \text{ (in 0.1 M KCl)} \pm \text{ion transfer} \quad (14)$$

The observed e.m.f. E, corrected to a partial pressure of 760 mm of hydrogen, is given by

$$E = E^\circ - \frac{RT}{F} \ln a_H a_{Cl} + E_j \qquad (15)$$

or

$$pa_H = \frac{F(E - E^\circ - E_j)}{RT \ln 10} + \log a_{Cl} \qquad (15a)$$

where E_j is the algebraic sum of the potentials across the liquid-liquid boundaries. It should be noted that E° is the e.m.f. of a hypothetical cell of type 8, corrected for the liquid-junction potential, in which each of the reactants and the products of the cell reaction is present at an activity of 1. In other words, this standard potential refers to the standard state of unit activity instead of unit concentration, as in the Sørensen definition.

The quantity a_H in equation 15 is the activity of hydrogen ion in solution X, whereas a_{Cl} is the activity of chloride ion in the solution of potassium chloride in contact with the calomel electrode. Hence, $a_H a_{Cl}$ is not accessible to direct measurement, and E_j cannot be obtained from equation 15 or 15a. An exact determination of pa_H would require a knowledge not only of the liquid-junction potentials but also of the activity coefficient of chloride ion in a 0.1 M solution of potassium chloride. There is no completely rigorous

---

[18] The standard state requires that ionic activities on the scales of concentration (c) and molality (m) shall be related by $a_c = a_m d^\circ$, where d° is the density of water. Thus

$$(pa_H)_m - (pa_H)_c = \log d^\circ$$

At temperatures below 25 °C, pa_H values on the scales differ by 0.001 unit or less. At 60 °C, the difference is 0.007, and at 100 °C, 0.019. The value of the standard potential also reflects a change in standard state and scale of concentration.

means of obtaining either of these quantities. However, if E_j is hopefully regarded as varying but slightly from one measurement to another and is joined with the other constant elements of equation 15a, we have at 25 °C

$$\mathrm{p}a_\mathrm{H} = \frac{E - (E^{\circ\prime} + E_j)}{0.05916} \tag{16}$$

The form of equation 16 is identical with that of equation 12, and it is clear that psH and pa_H differ by a constant amount. The value of $E^{\circ\prime} + E_j$ is about 0.3356 V. Hence the psH value is nearly equal to $-\log 1.1a_\mathrm{H}$,[19] and

$$\mathrm{p}a_\mathrm{H} = \mathrm{p}s\mathrm{H} + 0.04 \tag{17}$$

This convenient relationship may be used to convert to approximate pa_H values the extensive volume of psH numbers appearing in the literature. It is noteworthy that there can be no constant difference between pa_H and pm_H. These two quantities differ by $\log \gamma_\mathrm{H}$, which is a function of the ionic strength.

pH Scales Based on Mean Ionic Activities

THE ptH VALUE

It is now well recognized that the activity of a single ionic species plays no real part in the development of the e.m.f. of a galvanic cell, whether or not that cell is of the type with a liquid junction. It is probable that the same is true of other phenomena influenced by hydrogen ion. Hence, a quantity that may be termed ptH, defined by[20]

$$\mathrm{p}t\mathrm{H} \equiv -\log m_\mathrm{H}\gamma_\pm \tag{18}$$

is formally less objectionable than pa_H. The symbol $\gamma_\pm$ represents the mean activity coefficient of a uni-univalent electrolyte in the solution.

In a strict sense, this definition is either inadequate or ambiguous when the solution contains no electrolyte of the 1:1 valence type, or more than one. Accordingly, it is customary to regard $\gamma_\pm$ as the mean activity coefficient of an "average" uni-univalent electrolyte in a mixture of similar composition. It thus becomes apparent that there is no real difference between ptH and pa_H, for some of the common conventions identify the single ionic activity coefficient with the mean activity coefficient of an average uni-univalent electrolyte.

---

[19] S. P. L. Sørensen and K. Linderstrøm-Lang, *Compt. Rend. Trav. Lab. Carlsberg*, **15**, No. 6 (1924); I. M. Kolthoff and W. Bosch, *Rec. Trav. Chim.*, **46**, 430 (1927); M. Kilpatrick and M. L. Kilpatrick, *J. Chem. Educ.*, **9**, 1010 (1932).
[20] See D. A. MacInnes, D. Belcher, and T. Shedlovsky, *J. Amer. Chem. Soc.*, **60**, 1094 (1938); I. Prigogine and R. Defay, *Thermodynamique Chimique*, Vol. II, p. 197, Dunod, Paris, 1946.

p($a_H \gamma_{Cl}$) AND p($a_H \gamma_A$)

There have been other attempts to define scales of acidity that would be free from the theoretical objections of psH and pa_H and yet be well adapted to experimental measurements. Guggenheim[21] and Hitchcock[22] have called attention to the advantages of a unit of acidity we shall call p($a_H \gamma_{Cl}$):

$$p(a_H \gamma_{Cl}) \equiv -\log(m_H \gamma_H \gamma_{Cl}) \tag{19}$$

Unlike pa_H, this quantity is physically defined at all ionic strengths. It can be determined exactly from measurements of cells without liquid junction comprising electrodes reversible to hydrogen and chloride ions. Probably the most reproducible cell of this type is

$$\text{Pt}; H_2, \text{Soln. } X + \text{MCl}, \text{AgCl}; \text{Ag} \tag{20}$$

The p($a_H \gamma_{Cl}$) is computed from the e.m.f. E of cell 20 by the expression

$$p(a_H \gamma_{Cl}) = \frac{F(E - E^\circ)}{RT \ln 10} + \log m_{Cl} \tag{21}$$

Another unit of acidity, similar to p($a_H \gamma_{Cl}$) but formulated in a slightly more general manner, is the quantity p($a_H \gamma_A$). Acidities are expressed in terms of this unit through measurements of the e.m.f. of cells of a type analogous to 20, in which the silver-silver chloride electrode is replaced by an electrode reversible to the ion A and a salt of HA is substituted for MCl.

Hitchcock envisioned practical measurements of acidity, expressed in p($a_H \gamma_{Cl}$) units, by cells of type 20. The unknown solution must contain chloride ion in an amount known within 1 per cent, if an accuracy within a few thousandths of a p($a_H \gamma_{Cl}$) unit is to be obtained. It must also be free of the many substances that disturb the potentials of silver-silver chloride electrodes. However, the experimental difficulties of the hydrogen electrode can be avoided by replacement of this element with the glass electrode.

Entirely apart from these problems of a purely experimental nature, the p($a_H \gamma_{Cl}$) value does not appear to be ideal for the practical expression of the degree of acidity. For example, a combination of equation 19 with the mass law equation for the dissociation of a weak acid HA gives

$$p(a_H \gamma_{Cl}) = -\log K - \log \frac{m_{HA}}{m_A} - \log \frac{\gamma_{HA} \gamma_{Cl}}{\gamma_A} \tag{22}$$

in which K is the thermodynamic dissociation constant of the acid HA. When HA is a monobasic acid, the activity coefficient term is small. If,

---

[21] E. A. Guggenheim, *J. Phys. Chem.*, **34**, 1758 (1930).
[22] D. I. Hitchcock, *J. Amer. Chem. Soc.*, **58**, 855 (1936).

further, the buffer ratio is unity, $p(a_H \gamma_{Cl})$ is approximately equal to $-\log K$ at all concentrations. Hence, significant changes in c_H and a_H may scarcely be reflected in the $p(a_H \gamma_{Cl})$ values for buffer solutions of this class.

This constancy is evident in Fig. 2–2, where the relationships among pm_H, psH, pa_H, and $p(a_H \gamma_{Cl})$ for buffer mixtures composed of equal molal amounts of acetic acid and sodium acetate are plotted as a function of

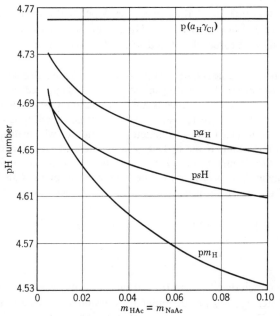

FIG. 2-2. Acidity of acetate buffer solutions at 25°C on four different scales, as a function of molality.

molality.[23] It is clear that a change of 30 per cent in the hydrogen ion molality (0.16 in pm_H) has no perceptible effect on $p(a_H \gamma_{Cl})$. When HA is the primary or secondary anion of a dibasic or tribasic acid, $p(a_H \gamma_{Cl})$ changes with ionic strength, but less sharply than do pm_H, psH, and pa_H. At infinite dilution, pm_H and pa_H are identical. Before that limit is reached, however, the solute is so attenuated that the acid may be highly dissociated and the

[23] R. G. Bates, *Chem. Rev.*, **42**, 1 (1948); R. G. Bates and E. R. Smith, *J. Wash. Acad. Sci.*, **38**, 61 (1948).

buffering action may largely disappear. The pH of the solution then approaches that of pure water.

Nevertheless, the $p(a_H \gamma_{Cl})$ unit is very useful in the determination of equilibrium data. When it is desired, for example, to measure pK values of indicators, combination of $p(a_H \gamma_{Cl})$ for buffer solutions with absorbance data obtained in these solutions permits one to avoid the uncertainties of pH measurements.[24] Extensive tables of $p(a_H \gamma_{Cl})$ for buffer solutions have been compiled by Bates and Gary,[24] and the uses of this acidity function have been described. Some values of this function are summarized in Table 7 of the Appendix.

Guggenheim points out that all processes dependent on acid-base reactions are controlled by either c_H or $c_H f_{\pm}^2$. His preference for these quantities over pa_H was clearly and forcefully stated in the following words:[25]

In particular, those engaged in "pH measurements" are liable sometimes to attach great importance to the "hydrogen ion activity" as opposed to the "hydrogen ion concentration." Such an attitude is however due to a misconception even more fundamental than the failure to realise that the former quantity is physically undefined. It is due in fact to an incomplete realisation of why hydrogen ion determinations are important. If the hydrogen ion content of an indicator solution or any other solution is important, it is presumably because the colour of the former or some equally important property of the latter depends on the state of some acid-base equilibrium, which in turn is controlled by hydrogen ion.

If the property of the solution depends entirely on c_A/c_B in the acid-base system,

$$A \rightleftharpoons B + H^+ \tag{23}$$

Guggenheim concludes, ". . . in the former case the determining factor is c_H and in the latter $c_H f_{\pm}^2$ . . . but under no possible circumstances can it ever become $c_H f_H$. Thus without assuming that the 'hydrogen ion activity' $c_H f_H$ is physically undefined, we have proved that, even if it were defined, it could never have the importance of either c_H or $c_H f_{\pm}^2$."

Choice of a Standard pH Unit

The choice of a pH scale must take into account both the theoretical and the experimental aspects. Unfortunately, no convenient experimental

---

[24] R. G. Bates and G. Schwarzenbach, *Helv. Chim. Acta*, **37**, 1069 (1954); R. G. Bates and R. Gary, *J. Res. Nat. Bur. Stand.*, **65A**, 495 (1961).

[25] E. A. Guggenheim, *J. Phys. Chem.*, **34**, 1758 (1930); the symbol f here represents the activity coefficient on the molar scale.

method exists for the routine measurement of pH values on the scales that are the most satisfactory in theory. Furthermore, the pH obtained by the convenient experimental techniques has no simple exact meaning.

As yet there has been devised no practical experimental method sufficiently convenient and versatile to supplant the electrometric determination of acidity by the Sørensen cell (type 8) and its counterpart with a glass electrode and a saturated calomel reference electrode. As Guggenheim[25] has pointed out, however, cells of these types actually respond to a quantity that, to indicate its indefinite nature, might be designated $m_H \gamma_?$, where $\gamma_?$ is much more complicated than $\gamma_\pm$, for it depends to some extent on the transference numbers of the ions in the cell. Neither γ_H nor $\gamma_\pm$ can be derived from $\gamma_?$, and any practical standardization of pa_H or ptH will be theoretically unsatisfactory. The experimental method is so firmly established, however, that the selection of a pH scale is in actuality a search for the hydrogen ion function most nearly consistent with the e.m.f. response of this particular cell that is used so widely. This approach excludes pm_H and $p(a_H \gamma_{Cl})$ from further consideration.

The simple expressions for computing psH and pa_H (equations 12 and 16) are very much alike. It is evident from these expressions and from Fig. 2–2 that a change of $E^{o\prime} + E_j$ is necessary and sufficient to shift the pH number from one of these scales to the other. That is, by choice of a suitable value for this standard potential, the pH number computed from the same measured value of the e.m.f. can be made to conform to the scale of either psH or pa_H. Table 2–3 lists these values of $E^{o\prime} + E_j$ for cell 8 and for the corresponding cell with a saturated calomel reference electrode.[26] Inasmuch

TABLE 2–3. VALUES OF $E^{o\prime} + E_j$ AT 25 °C FOR THE CELL
H_2; SOLN. X | SATD. KCl | CALOMEL REFERENCE ELECTRODE

Scale Unit	0.1M Calomel Reference V	Satd. Calomel Reference V
pm_H	About 0.3400[a]	About 0.2488[a]
psH	0.3376	0.2464
pa_H	0.3353–0.3358	0.2441–0.2446

[a] For 0.1 M HCl and 0.01 M HCl, 0.09 M KCl.

[26] R. G. Bates, *Chem. Rev.*, **42**, 1 (1948).

as the cell does not respond in a simple manner to changes of hydrogen ion concentration, $E^{\circ\prime} + E_j$ does not have a constant value for the pm_H scale.

The fact that the psH (Sørensen) scale has virtually no fundamental meaning does not prevent it from being useful for reproducible comparisons of acidity. The pa_H value has in itself no meaning in terms of physical reality, yet its role in chemical equilibria can be simply and unequivocally, although conventionally, defined. This unit possesses, therefore, all the virtues of psH and, in addition, the possibility, under ideal experimental conditions, of a limited amount of theoretical interpretation.

The shift from psH to pa_H involves merely a change in the standard potential by an amount (about 2.3 mV) that corresponds approximately to an increase of 0.04 in pH. There can be little question of the value of making such a shift in reference. In deciding to do so, we locate a new *area* of reference in the vicinity of $E^{\circ\prime} + E_j = 0.3353$ V (0.1 M calomel electrode). To choose a *point* of reference within this area, that is, to define precisely a particular activity scale, it is necessary to adopt a conventional definition of ionic activity coefficients and to assign a pH value to a standard reference solution. A value of $E^{\circ\prime} + E_j$ can then be calculated from the e.m.f. measured with this standard solution in the cell. It is immaterial whether the activity unit be defined formally according to equation 13 (pa_H) or equation 18 (ptH). The pH standards established by Hitchcock and Taylor, by MacInnes, Belcher, and Shedlovsky, and by the National Bureau of Standards[27] are consistent with these definitions.

Hereafter in this book the scale unit fixed by the reference solutions (pH standards) will be regarded as $-\log a_H$ or $-\log m_H \gamma_H$, where a_H is a conventional hydrogen ion activity, in accordance with the procedure followed at the National Bureau of Standards. The standard scale is properly termed a *conventional activity scale*. The nature of this scale will be considered further in Chapter 4.

With the precise definition of a conventional pa_H, it becomes possible to establish standards of pH and hence to compute the standard potential $E^{\circ\prime} + E_j$ of the cell used for practical pH measurements. As will be shown in succeeding chapters, this standard potential remains satisfactorily constant for dilute aqueous solutions of intermediate acidity, pH 3 to 11. At high acidities and high alkalinities, however, the highly mobile hydrogen and hydroxyl ions alter the value of the liquid-junction potential to such an extent that the standard potential is changed appreciably.[28]

---

[27] D. I. Hitchcock and A. C. Taylor, *J. Amer. Chem. Soc.*, **59**, 1812 (1937); **60**, 2710 (1938); D. A. MacInnes, D. Belcher, and T. Shedlovsky, *ibid.*, **60**, 1094 (1938); R. G. Bates, *Analyst*, **77**, 653 (1952); *J. Res. Nat. Bur. Stand.* **66A**, 179 (1962).

[28] R. G. Bates, G. D. Pinching, and E. R. Smith, *J. Res. Nat. Bur. Stand.* **45**, 418 (1950).

Operational Definition of the Measured pH

In the previous section we were concerned principally with the standard pH of reference solutions. For some time there has been a growing realization that a fundamental definition of the "practical" measured pH is of little value. The great diversity of pH measurements in modern commerce and industry demands an experimental definition, framed in terms of operations and calculations.[29] In the words of MacInnes, "In possibly all but one case in a thousand it is not necessary to consider the meaning of pH in terms of solution theory at all, but only to accept the numbers as a practical scale of acidity and alkalinity."

Considerations such as these have led to the wide acceptance of an operational definition of the measured pH, formulated in the following manner:

$$pH(X) = pH(S) + \frac{(E_X - E_S)F}{RT \ln 10} \qquad (24)$$

This definition has been endorsed by standardizing groups in many countries including the United States, Great Britain, and Japan and has been recommended by the International Union of Pure and Applied Chemistry.[30]

In equation 24 pH(S) is the assigned pH of the standard, whereas E_X and E_S are the values of the e.m.f. of a pH cell with the electrodes immersed in the unknown fluid X and the standard S, respectively. The cell may be of type 8, or for convenience other reference electrodes and other electrodes reversible to hydrogen ion may be chosen. The assembly consisting of glass and saturated calomel electrodes and a bridge of 3.5 M or saturated potassium chloride is the arrangement most commonly employed. When the glass electrode is used, corrections to $E_X - E_S$ at low and high pH values may be necessary to assure that electrodes of different types will furnish the same pH.

---

[29] See, for example, W. M. Clark, *Ind. Eng. Chem.*, **28**, 620 (1936); D. S. McKinney, P. Fugassi, and J. C. Warner, *Symposium on pH Measurement*, p. 19, ASTM Tech. Publ. 73, Philadelphia, Pa., 1947; D. A. MacInnes, *Science*, **108**, 693 (1948).

[30] *Standard Method of Test for pH of Aqueous Solutions with the Glass Electrode*, ASTM Method E70-68, Philadelphia, Pa., 1968; *Specification for pH Scale*, British Standard 1647, British Standards Institution, London, 1961; *Method for Determination of pH*, Japanese Industrial Standard Z8802, Japanese Standards Assn., Tokyo, 1958; R. G. Bates and E. A. Guggenheim, *Pure Appl. Chem.*, **1**, 163 (1960); J. A. Christiansen, *J. Amer. Chem. Soc.*, **82**, 5517 (1960); *Manual of Symbols and Terminology for Physicochemical Quantities and Units*, International Union of Pure and Applied Chemistry, Butterworths, London, 1970.

Limitations of the pH

By comparison of the operational definition with equation 16, it will be seen that the necessary condition for the measured pH to fall on the scale defined by the activity standards is that $E^{\circ\prime} + E_j$, the standard potential, remain unchanged when the standard solution is replaced by the unknown. These conditions are approximately fulfilled only when the unknown is a fairly dilute aqueous solution of simple solutes and its acidity matches closely that of the standard solution selected. Unfortunately, the great majority of test solutions will not meet these stringent requirements, and the measured pH cannot then justifiably be regarded as an approximate measure of the (conventional) hydrogen ion activity of the test medium.

The establishment of a conventional scale of hydrogen ion activity at a given temperature is equivalent to fixing the single potential E_H of the hydrogen electrode in those solutions to which pH(S) values were assigned. It will be remembered that the arbitrary zero of potential that forms the basis for the hydrogen scale of electrode potentials is the standard potential E_H° of the reversible hydrogen electrode. Thus

$$E_H = E_H^{\circ} + \frac{RT}{F} \ln a_H = 0 + \frac{RT}{F} \ln a_H \tag{25}$$

This standard potential is, by definition, zero at all temperatures. As a consequence, measurements of hydrogen electrode potentials can give no exact comparison between the "true" hydrogen ion activity at two different temperatures. In order to set up numerical scales of activity, therefore, one must choose a standard state at each temperature. Thus we have, inescapably, several different pH scales, and the pH at one temperature has no quantitative meaning relative to that at another.

The concentration of hydrogen ion is, of course, a well-defined concept, and the conventional activity of hydrogen ion is usually made to approach the concentration of this ion in solutions of low ionic strength at all temperatures. Yet it may well be that the increased rate of an acid-catalyzed reaction at elevated temperatures and the rapid dissolution of a metal, for example, indicate an increase in "absolute" hydrogen ion activity that the conventional pH scale cannot reflect.

CHAPTER 3

liquid-junction potentials
and ionic activities

The subject of pH standardization in its broadest sense comprises all the techniques and procedures that permit accurate and reproducible pH determinations to be made. The subject has two aspects. The first is concerned with such purely experimental and manipulative details as specifications for reference electrodes, the maintenance of temperature equilibrium, and adequate washing of the electrodes. The more fundamental aspect, and one which largely determines the interpretation, if any, that may justifiably be placed on the measured pH values, is the assignment of pH values to the buffer solutions with which the pH equipment is standardized. This important phase of pH standardization will be considered in the present chapter and the one to follow.

The impossibility of determining *experimentally* a pH value with exact fundamental definition and of defining simply the quantity measured with the pH meter necessitates a compromise. Inasmuch as the pH cannot in every instance, or even in a majority of instances, be an exact numerical expression of a single quantity, the fundamental meaning of the practical scale of pH is limited. The selection of a single standard suffices to establish a reproducible pH scale with qualitative, ordinal significance.[1] The adoption of a more elaborate set of pH standards can be justified only if these standards make possible the useful interpretation of measured pH values, or at least an important segment of them.

As we have seen, the formal definition

$$\text{pH(S)} \equiv -\log a_\text{H} = -\log m_\text{H}\, \gamma_\text{H} \tag{1}$$

[1] D. A. MacInnes, *Science*, **108**, 693 (1948).

where a_H is the activity of the hydrogen ion and γ_H its activity coefficient, provides a simple unified basis for a series of selected pH standards which can, in turn, establish a useful operational pH scale (equation 24, Chapter 2). Although the individual ionic activity and activity coefficient are not uniquely and rigorously defined by thermodynamics, it is possible to set up a *conventional scale* of activity coefficients for single ions, that is to select arbitrarily a set of activity coefficients that will meet all practical requirements. This is what must be done in order to utilize the simple formal definition of equation 1, and thereby the nature and dimensions of the standard pH unit become completely prescribed. The practical pH, the number furnished by the assembly adjusted to the conventional activity scale, approaches the dimensions of the standard unit in dilute aqueous solutions. Under these conditions, which may be regarded as ideal, the measured pH possesses a definite meaning in terms of chemical equilibria. However, as the residual liquid-junction potential will rarely be zero in the practical measurement, the measured pH will almost never fall *exactly* on the conventional activity scale defined by the standards.

Individual Ionic Activity Coefficients

Is there any possible means, still awaiting discovery, of determining $-\log a_H$ exactly and thus endowing the pH with the full significance of a true physical constant? Thermodynamics leads us to believe that there can be none. A consideration of any cell with a hydrogen ion electrode and a liquid junction underscores the fundamental difficulties of such a determination.

Let us examine further the cell so important in the determination of pH values:

$$\text{Pt}; H_2(g), \text{Soln. } X \mid \text{KCl (satd.)}, Hg_2Cl_2(s); Hg(l) \tag{2}$$

The cell reaction is

$$\tfrac{1}{2}H_2(g) + \tfrac{1}{2}Hg_2Cl_2(s) = Hg(l) + Cl^- \text{ (in satd. KCl)} + H^+$$
$$\text{(in soln. X)} \pm \text{ion transfer} \tag{3}$$

The activity of hydrogen ion in solution X can be related to the total e.m.f. E of the cell, the standard potential $E^{\circ\prime}$, and the potential E_j across the liquid junction (vertical line) as follows:

$$-\ln a_H = \frac{(E - E^{\circ\prime} - E_j)F}{RT} \tag{4}$$

The symbol for the standard potential is given a prime mark here to indicate that the potential in question is not referred to the state in which $a_H a_{Cl} = 1$, as customary, but is the potential of the cell for unit activity of hydrogen ion and the activity of chloride ion that exists in a saturated solution of potassium chloride.

Of the quantities appearing in equation 4, E, R, T, and F are all physically defined. It is evident that $-\log a_H$ lacks physical definition just because $E^{\circ\prime} + E_j$ is also not experimentally obtainable. The liquid-junction potential E_j is a complicated function of the activities $m_i \gamma_i$ and transference numbers t_i of the several ionic species i in the transition layers:[2]

$$E_j = -\frac{RT}{F} \int_1^2 \sum \frac{t_i}{z_i} \, d \ln m_i \gamma_i \tag{5}$$

where z_i is the ionic charge and 1 and 2 represent the two end solutions between which the boundary is formed. It is clear that an exact evaluation of E_j is not possible without a knowledge of a_H which, in turn, is the object of the calculation. As many have emphasized, this is a dilemma from which there is apparently no escape.[3]

Values for E_j can be obtained by integration of equation 5, provided that (1) the distribution of the ionic concentrations in the transition layers between 1 and 2 is known, and (2) values of γ_i are estimated with the aid of a nonthermodynamic assumption or, alternatively, the solutions are assumed to be ideal ($\gamma_i = 1$). It is convenient to separate equation 5 into its ideal and nonideal parts, as follows:

$$E_j = -\frac{RT}{F} \int_1^2 \sum \frac{t_i}{z_i} \, d \ln m_i - \frac{RT}{F} \int_1^2 \sum \frac{t_i}{z_i} \, d \ln \gamma_i \tag{5a}$$

or

$$E_j = E_{jm} + E_{jf} \tag{5b}$$

For certain boundaries of relatively simple structure, a steady state of diffusion, characterized by a well-defined distribution of ions in the transition

---

[2] These are the Hittorf transference numbers, corrected for the conductance of the solvent; see F. O. Koenig, *J. Phys. Chem.*, **44**, 101 (1940).

[3] In the words of H. S. Harned, Vol. II, p. 782, of *A Treatise on Physical Chemistry*, edited by H. S. Taylor, D. Van Nostrand Co., New York, 1924: "We are thus confronted with the interesting perplexity that it is not possible to compute liquid junction potentials without a knowledge of individual ion activities, and it is not possible to determine individual ion activities without an exact knowledge of liquid junction potentials. For the solution of this difficult problem, it is necessary to go outside the domain of exact thermodynamics." The reader is also referred to H. S. Harned, *J. Phys. Chem.*, **30**, 433 (1926); P. B. Taylor, *ibid.*, **31**, 1478 (1927); E. A. Guggenheim, *ibid.*, **33**, 842 (1929); *J. Amer. Chem. Soc.*, **52**, 1315 (1930); D. A. MacInnes and L. G. Longsworth, *Cold Spring Harbor Symposia Quant. Biol.*, **4**, 18 (1936); and D. A. MacInnes, D. Belcher, and T. Shedlovsky, *J. Amer. Chem. Soc.*, **60**, 1094 (1938).

layers, is achieved, and E_{jm} can be evaluated. We shall have more to say about the physical structures of liquid-liquid boundaries in Chapter 10. If the junction is of the continuous-mixture type, for example, the equation of Henderson[4] furnishes the ideal term, E_{jm}. Similarly, the Planck equation[4] yields E_{jm} for junctions of the constrained-diffusion type. Neither equation gives the complete liquid-junction potential as defined in equation 5b, for E_{jf} is not specifically recognized.

It should be noted that these two boundary structures were not chosen by Henderson and Planck for analysis because of their practical importance but because integrations of the first term on the right of equation 5a were possible for these types. A third type, the free-diffusion junction, is not amenable to simple mathematical treatment although it is encountered very commonly in practice.

Guggenheim has done much to dispel the confusion that arises from a lack of understanding of the exact relationships between liquid-junction potentials and the activity coefficients of individual ions.[5] These principles, although completely general, may be simply illustrated in terms of a cell consisting of two hydrogen electrodes immersed in two different solutions separated by an intermediate bridge solution:

$$\text{Pt}; \text{H}_2, \text{Soln. 1} \,|\text{Bridge}|\, \text{Soln. 2}, \text{H}_2; \text{Pt} \qquad (6)$$

The e.m.f. E of cell 6 is the sum of the potential differences E_e at the electrodes and E_j at the liquid junctions:

$$E = E_e + E_j \qquad (7)$$

The term E_j is the algebraic sum of the potentials at the two junctions: Soln. 1 |Bridge, and Bridge| Soln. 2. The electrode potential term can be subdivided into an ideal and a nonideal part, as has already been done in equation 5b for E_j:

$$E_e = \frac{RT}{F} \ln \frac{m_{\text{H}(2)}}{m_{\text{H}(1)}} + \frac{RT}{F} \ln \frac{\gamma_{\text{H}(2)}}{\gamma_{\text{H}(1)}} \qquad (8)$$

where the subscripts (1) and (2) refer to solutions 1 and 2, or

$$E_e = E_{em} + E_{ef} \qquad (9)$$

---

[4] P. Henderson, Z. Physik. Chem., **59**, 118 (1907); **63**, 325 (1908); M. Planck, Ann. Physik, [3]**39**, 161 (1890); **40**, 561 (1890). The Henderson equation has been remodeled by A. K. Airola, Svensk Kem. Tidskr., **50**, 128, 235, 278 (1938), and Ann. Acad. Sci. Fennicae, A**55**, No. 5 (1940). L. G. Sillén, Physik. Z., **40**, 466 (1939), has given a modern derivation of the Planck equation.

[5] E. A. Guggenheim, Phil. Mag., [7]**22**, 983 (1936). See also the earlier analysis of the problem by P. B. Taylor, J. Phys. Chem., **31**, 1478 (1927), and the summary given by H. S. Harned and B. B. Owen, The Physical Chemistry of Electrolytic Solutions, 3rd ed., Chapter 10, Reinhold Publishing Corp., New York, 1958.

By combining equations 5b, 7, and 9, we have

$$E = E_{em} + E_{jm} + (E_{ef} + E_{jf}) \qquad (10)$$

The first term on the right can be computed from the compositions of solutions 1 and 2, if only strong electrolytes are present. If the boundaries are of the continuous-mixture or constrained-diffusion types and the transference numbers in the intermediate transition layers are known, the second term can be calculated by the Henderson or Planck equation. But to calculate a ratio of hydrogen ion activity ($m_H \gamma_H$) by equation 8, one needs to know $E_{em} + E_{ef}$. Harned[6] was one of the first to show that it is not sufficient to subtract E_{jm} from the measured E, ignoring E_{jf}, as has so often been done.

It is evident, then, that the *sum* of the two nonideal terms, E_{ef} and E_{jf}, can be determined, although each member is itself indeterminate.[7] Furthermore, this sum can always be transformed into an expression containing only mean activity coefficients. The manner in which this transformation is accomplished in a typical case can be illustrated by deriving a complete expression for the e.m.f. E of the cell

$$\text{Ag};\text{AgCl},\text{HCl}(m_1) \,|\, \text{KCl}(m_2),\text{AgCl};\text{Ag} \qquad (11)$$

in terms of molalities m, transference numbers t, mean activity coefficients γ_{MCl}, and no physically undefined quantities.[8]

From equations 5 and 7 and the expression for the electrode potentials E_e of cell 11 in terms of the chloride ion activities, a_1 and a_2, in the two solutions, namely

$$E_e = \frac{RT}{F} (\ln a_1 - \ln a_2) \qquad (12)$$

one can write

$$E = \frac{RT}{F} \left(\ln a_1 - \ln a_2 - \int_1^2 t_H \, d \ln a_H - \int_1^2 t_K \, d \ln a_K + \int_1^2 t_{Cl} \, d \ln a_{Cl} \right) \qquad (13)$$

Inasmuch as $t_{Cl} = 1 - t_H - t_K$, and, by definition, $a_H a_{Cl} = m_1{}^2 \gamma_{HCl}^2$ and $a_K a_{Cl} = m_2{}^2 \gamma_{KCl}^2$, equation 13 becomes

$$-E = \frac{RT}{F} \int_1^2 \frac{t_H}{m_1{}^2 \gamma_{HCl}^2} \, d(m_1{}^2 \gamma_{HCl}^2) + \frac{RT}{F} \int_1^2 \frac{t_K}{m_2{}^2 \gamma_{KCl}^2} \, d(m_2{}^2 \gamma_{KCl}^2) \qquad (14)$$

[6] H. S. Harned, *J. Phys. Chem.*, **30**, 433 (1926).

[7] E. A. Guggenheim, *Phil. Mag.*, [7] **22**, 983 (1936); A. Unmack and E. A. Guggenheim, *Kgl. Danske Videnskab. Selskab, Mat.-Fys. Medd.*, **10**, Nos. 8 and 14 (1930, 1931).

[8] D. A. MacInnes and L. G. Longsworth, *Cold Spring Harbor Symposia Quant. Biol.*, **4**, 18 (1936).

No individual ionic activity coefficients appear in equation 14. MacInnes and Longsworth integrated this equation by a graphical method and obtained values of E in excellent agreement with experiment.

The activity coefficient of a single ionic species is thus not an operational concept and is completely unnecessary for an adequate treatment of the thermodynamics of the cell with liquid junction. Conversely, this type of cell can furnish no exact information regarding the activity of a single ionic species.

Nevertheless, the view expressed by Guggenheim[9] that the ionic activity coefficient can never be more than a mathematical device has been regarded as extreme. For example, Brønsted, Delbanco, and Volqvartz[10] expressed the opinion that the activity coefficient of an ion has a meaning as well defined as the activity coefficient of an uncharged molecule. MacInnes[11] has pointed out the usefulness of individual activity coefficients in visualizing cell mechanisms, and Kortüm[12] has made a strong case for the validity, at ionic strengths below 0.01, of individual activity coefficients which, for uni-univalent electrolytes, are taken to be equal to the mean activity coefficients.

Having disposed of the ionic activity coefficient and the liquid-junction potential as lacking independent definition, we find it appropriate to consider possible conventions which will permit a useful scale of ionic activity coefficients to be set up. There are two general approaches. The first involves elimination of the liquid-junction potential or its estimation, and the second is concerned with defining relationships between measurable combinations or averages of activity coefficients and the activity coefficient of each individual ionic species.

Calculation of the Liquid-Junction Potential

The integration of equation 5 by the method of Henderson[13] is based on the simple assumption that the junction consists of a continuous series of mixtures of the two boundary solutions 1 and 2. At any given point in the boundary, therefore, the composition is given by α(Soln. 2) $+ (1 - \alpha)$ (Soln. 1), where α is the mixing fraction, varying linearly with distance through the boundary. The concentration c_i of each ion can then be expressed

---

[9] E. A. Guggenheim, *J. Phys. Chem.*, **34**, 1758 (1930).

[10] J. N. Brønsted, A. Delbanco, and K. Volqvartz, *Z. Physik. Chem.*, **A162**, 128 (1932). See also H. S. Frank, *J. Phys. Chem.*, **67**, 1554 (1963), and R. N. Goldberg and H. S. Frank, *ibid.*, **76**, 1758 (1972).

[11] D. A. MacInnes, *The Principles of Electrochemistry*, Chapter 13, Reinhold Publishing Corp., New York, 1939; Dover Publications, New York, 1961.

[12] G. Kortüm, *Elektrolytlösungen*, Chapter 12, Becker und Erler Kom.-Ges., Leipzig, 1941.

[13] P. Henderson, *Z. Physik. Chem.*, **59**, 118 (1907); **63**, 325 (1908).

in terms of α and the ionic concentrations in the end solutions. It is also assumed that the activity of each ionic species is equal to the concentration of that species and that the mobility of each ion is constant in the range c_i' (Soln. 1) to c_i'' (Soln. 2). The transference number t_i is expressed in terms of concentrations and mobilities u_i as

$$t_i = \frac{c_i u_i}{\alpha \sum c_i'' u_i + (1 - \alpha) \sum c_i' u_i} \tag{15}$$

where the summation is for all the ions of which the boundary is composed.

Substitution in equation 5 and integration give the complete equation of Henderson:

$$E_{jm} = \frac{RT}{F} \frac{\sum (u_i/z_i)(c_i'' - c_i')}{\sum u_i (c_i'' - c_i')} \ln \frac{\sum c_i' u_i}{\sum c_i'' u_i} \tag{16}$$

in which z_i is the charge borne by the ionic species i. If the ionic mobilities in the two end solutions are taken equal to the mobilities at infinite dilution (or λ^0/F, where λ^0 is the limiting ionic conductivity), one can write[14]

$$E_{jm} = \frac{RT}{F} \frac{(U_1 - V_1) - (U_2 - V_2)}{(U_1' + V_1') - (U_2' + V_2')} \ln \frac{U_1' + V_1'}{U_2' + V_2'} \tag{17}$$

In this equation U is $\sum c_+ \lambda_+^0$ and V is $\sum c_- \lambda_-^0$ for the cations and anions in the end solutions indicated by the subscript, namely 1 or 2; U' is $\sum c_+ \lambda_+^0 |z_+|$ and V' is $\sum c_- \lambda_-^0 |z_-|$, where $|z|$ is the valence of the ion without regard to sign and c_+ and c_- are in mol dm^{-3}. When solution 2 is a saturated solution of potassium chloride (4.16 M) and the temperature is 25 °C, equation 17 becomes

$$E_{jm} = 0.059157 \frac{U_1 - V_1 + 11.6}{U_1' + V_1' - 623} \log \frac{U_1' + V_1'}{623} \tag{18}$$

The Henderson and Planck equations often yield useful semiquantitative estimates of liquid-junction potentials. The Henderson equation is of special interest for two reasons. First, it is possible that many freshly formed mechanically stable junctions should be regarded as continuous-mixture boundaries. Second, there is a rather wide belief that the junction potential computed by the complete Henderson equation does not reflect simply the magnitude of the ideal term E_{jm}, but that the introduction of the mobilities of the ions in the two end solutions may provide at least a partial compensation for the changes in activity coefficients which account for the term E_{jf}.

---

[14] Other special forms of the Henderson equation are discussed by D. A. MacInnes, *The Principles of Electrochemistry*, Reinhold Publishing Corp., New York, 1939.

The approximate potentials across some liquid-liquid boundaries of interest in pH measurements are given in Table 3–1. These were computed from limiting mobilities by equation 18.[15] Although not of high accuracy, the values in the table provide an index of sign and relative magnitude.

By combining equations 7 and 8, the e.m.f. of cell 6 is expressed as follows:

$$E = \frac{RT}{F} \ln \frac{(m_H \gamma_H)_2}{(m_H \gamma_H)_1} + E_j \tag{19}$$

TABLE 3–1. LIQUID-JUNCTION POTENTIALS, IN MILLIVOLTS AT 25 °C, COMPUTED FROM LIMITING IONIC MOBILITIES BY THE HENDERSON EQUATION[a]

	Junction: Soln. X | KCl (satd.)		
Solution X	E_j	Solution X	E_j
HCl, 1 M	14.1	CH₃COOH, 0.01 M;	
HCl, 0.1 M	4.6	CH₃COONa, 0.01 M	3.1
HCl, 0.01 M	3.0	KH₂PO₄, 0.025 M;	
HCl, 0.01 M; NaCl, 0.09 M	1.9	Na₂HPO₄, 0.025 M	1.9
HCl, 0.01 M; KCl, 0.09 M	2.1	NaHCO₃, 0.025 M;	
KCl, 0.1 M	1.8	Na₂CO₃, 0.025 M	1.8
KH₃(C₂O₄)₂, 0.1 M	3.8	Na₂CO₃, 0.025 M	2.0
KH₃(C₂O₄)₂, 0.05 M	3.3	Na₂CO₃, 0.01 M	2.4
KH₃(C₂O₄)₂, 0.01 M	3.0	Na₃PO₄, 0.01 M	1.8
KHC₂O₄, 0.1 M	2.5	NaOH, 0.01 M	2.3
KH phthalate, 0.05 M	2.6	NaOH, 0.05 M	0.7
KH₂ citrate, 0.1 M	2.7	NaOH, 0.1 M	−0.4
KH₂ citrate, 0.02 M	2.9	NaOH, 1 M	−8.6
CH₃COOH, 0.05 M;		KOH, 0.1 M	−0.1
CH₃COONa, 0.05 M	2.4	KOH, 1 M	−6.9

Junction : Sohn. X | KCl (0.1 M)	
Solution X	E_j
HCl, 0.1 M	26.9
HCl, 0.01 M	9.1
NaOH, 0.1 M	−19.2
NaOH, 0.01 M	−4.5

[a] Positive E_j signifies a boundary of polarity− | +.

[15] The limiting ionic conductivities were taken from R. A. Robinson and R. H. Stokes, *Electrolyte Solutions*, 2nd ed. revised, p. 465, Butterworths, London, 1970, and from H. S. Harned and B. B. Owen, *The Physical Chemistry of Electrolytic Solutions*, 3rd ed., p. 231, Reinhold Publishing Corp., New York, 1958. Values for the phthalate, citrate, and phosphate anions were estimated from data for ions of similar structures.

Evidently E_j can be obtained from the e.m.f. of cell 6 if an assumption from outside the realm of thermodynamics is introduced to permit $m_H \gamma_H$ in the two solutions to be evaluated. If these solutions contain only salts and strong acids, m_H is known from the compositions, but the activity coefficients must be estimated.

One of the two simple conventions that have been widely used for this purpose was suggested by MacInnes.[16] It asserts the equality of the activity coefficients of the ions of potassium chloride in a pure aqueous solution of the salt of any given concentration. The activity coefficients of potassium and chloride ions are then assumed to have these values in other mixtures where the same ionic strength and the same concentration of potassium or chloride ion are maintained. Thus the activity coefficient of chloride ion is regarded as having the same value in the following solutions: 0.1 M HCl; a mixture of HCl (0.01 M) and KCl (0.09 M); and 0.1 M NaCl; and to be equal in each case to the mean activity coefficient of potassium chloride in its 0.1 M solution. If γ_H, the activity coefficient of hydrogen ion in the mixture 0.01 M HCl, 0.09 M KCl, is needed, for example, it can be obtained with the aid of the definition

$$\gamma_{HCl} = (\gamma_H \gamma_{Cl})^{1/2} \tag{20}$$

in which γ_{HCl} is the measured mean ionic activity coefficient of hydrochloric acid in the mixture in question. The value of γ_{Cl} obtained with the aid of the MacInnes convention is substituted in equation 20, and γ_H is calculated.

Valensi and Maronny,[17] in search of an experimental realization of the concept of the "compensated activity coefficient,"[18] have in effect extended the MacInnes convention by applying the relationship

$$\gamma_{Cl} = \gamma_{KCl} \tag{21}$$

(where γ_{KCl} is the measured mean activity coefficient) to buffer solutions containing potassium chloride and no cations other than potassium and hydrogen. They have measured the mean activity coefficient γ_{KCl} of potassium chloride in these mixtures by means of cells with dilute potassium amalgam electrodes and silver-silver chloride electrodes.

According to the second simple and widely used convention, the activity coefficients of the ions of any binary electrolyte MX are considered to be equal. Thus

$$\gamma_M = \gamma_X = \gamma_{MX} \tag{22}$$

---

[16] D. A. MacInnes, *J. Amer. Chem. Soc.*, **41**, 1086 (1919).

[17] G. Maronny and G. Valensi, *J. Chim. Phys.*, **49**, C91 (1952); G. Maronny, *Electrochim. Acta*, **2**, 326 (1960). Another useful modification of the MacInnes convention has been proposed by I. Feldman, *Anal. Chem.*, **28**, 1859 (1956).

[18] G. Valensi, *Compt. Rend. 3me Réunion, Com. Int. Therm. et Cinétique Electrochim.* (Berne, 1951), p. 3.

At ionic strengths less than 0.01, this convention is a natural consequence of the fact that the mean activity coefficients of all uni-univalent electrolytes are substantially equal in this low range of concentrations. Inasmuch as $-\ln \gamma_i$ at a given ionic strength is nearly proportional to z_i^2, where z_i is the charge of the ion i, the convention of equation 22 can be expressed in the following form, applicable to the ions of strong electrolytes of all charge types:

$$\ln \gamma_i = \frac{vz_i^2}{v_+z_+^2 + v_-z_-^2} \ln \gamma_{M_{v_+}X_{y_-}} \tag{23}$$

In equation 23, v_+ is the number of cations and v_- the number of anions formed by the dissociation of one molecule, and $v = v_+ + v_-$. For the special case of binary electrolytes, equation 23 reduces to equation 22.

In 1930, Guggenheim proposed[19] a general convention for the activity coefficient of each of the ions in a mixture of strong electrolytes. This convention expresses the ionic activity coefficients in a mixture of several uni-univalent electrolytes (each represented by MX) in terms of the thermodynamically defined combinations $\gamma_{M_1}/\gamma_{M_2}$, $\gamma_{X_1}/\gamma_{X_2}$, and $\gamma_M\gamma_X$. If hydrochloric acid is one of the electrolytes MX, for example, one can write

$$\ln \gamma_H = \frac{\sum_M m_M \ln(\gamma_H/\gamma_M) + \sum_X m_X \ln(\gamma_H\gamma_X)}{\sum_M m_M + \sum_X m_X} \tag{24}$$

An analogous expression can be written for any other cation in the mixture. The corresponding convention for any one of the anions (for example Cl^-) is

$$\ln \gamma_{Cl} = \frac{\sum_M m_M \ln(\gamma_M\gamma_{Cl}) + \sum_X m_X \ln(\gamma_{Cl}/\gamma_X)}{\sum_M m_M + \sum_X m_X} \tag{25}$$

Unfortunately, the Guggenheim convention has little practical utility for mixed electrolytes because the values of the mean activity coefficients in the mixture are usually not known. For a single binary electrolyte, however, equations 24 and 25 yield the simple relationship of equation 22. It should be noted that it is not proper to refer to equation 22 as the "Guggenheim convention" if the former is applied to a single cation-anion pair in a mixture (for example, to hydrochloric acid in a buffer-chloride mixture), as has sometimes been done.

When the solutions are composed entirely of strong electrolytes, a convention can be employed directly to evaluate γ_H (equation 19), and E_j can then be computed, if desired, in a simple manner from the measured e.m.f. of cells of type 6. Harned[20] measured the e.m.f. of the cells

---

[19] E. A. Guggenheim, *J. Phys. Chem.*, **34**, 1758 (1930); see also R. G. Bates and E. A. Guggenheim, *Pure Appl. Chem.*, **1**, 163 (1960).
[20] H. S. Harned, *J. Phys. Chem.*, **30**, 433 (1926).

$$Hg; Hg_2Cl_2, HCl(m_0), MCl(m), H_2; Pt\text{--}Pt; H_2, HCl(m_0), Hg_2Cl_2; Hg \qquad (26)$$

and

$$Pt; H_2, HCl(m_0) \,|\, KCl \text{ (satd.)} \,|\, HCl(m_0), MCl(m), H_2; Pt \qquad (27)$$

in which M is potassium, sodium, or lithium, and of the corresponding cells in which an alkali hydroxide, MOH, replaced the hydrochloric acid. It will be noted that cell 27 is of the type already referred to as cell 6. Cell 26 is a concentration cell without transference. Its e.m.f. is given by

$$E_{\text{HCl}} = \frac{RT}{F} \ln \frac{(a_H a_{Cl})_s}{(a_H a_{Cl})_a} \qquad (28)$$

where the subscripts s and a refer to the salt-acid mixture and to the pure acid, respectively. By the use of the MacInnes convention, Harned evaluated the activities of chloride ion in the two solutions and, hence, E_e:

$$E_e = \frac{RT}{F} \ln \frac{(a_H)_s}{(a_H)_a} \qquad (29)$$

which is the "electrode potential" term for cell 6 when solution 2 is the salt mixture and solution 1 the pure acid. Indeed, the value of E_e determined by Harned may be regarded as the e.m.f. of cell 6 (or cell 27) with the liquid-junction potential eliminated (compare equation 19). It must be remembered, however, that this value was obtained by the use of a nonthermodynamic convention.

We are now in a position to compute E_j by equation 7 from the measured e.m.f. of cell 27 and to judge the effectiveness of the bridge solution, for which Harned chose saturated potassium chloride, in reducing the magnitude of E_j. In Fig. 3–1, Harned's results for mixtures of hydrochloric acid and alkali chlorides and for hydroxide-chloride mixtures obtained in an analogous manner,[21] are plotted as a function of the molality m of the added salt. The molality of acid or hydroxide was 0.1 in all cases. The electrode potential term E_e is represented by the solid lines, and the e.m.f. of cell 27, E_{27}, by dashed lines. The difference $E_{27} - E_e$, represented by the separation of the two curves, is the residual liquid-junction potential, $E_j(a) - E_j(s)$. The sign and approximate magnitude can be obtained from the Henderson equation (equation 18).

Some interesting qualitative conclusions can be drawn from Fig. 3–1. The anion in solutions of lithium, sodium, and potassium chlorides is known

---

[21] The reader is referred to Harned's paper for details of the calculation for alkali-salt mixtures.

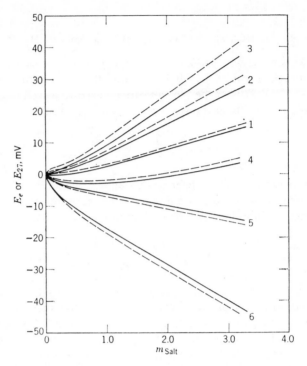

FIG. 3-1. Comparison of E_e (solid lines) and E_{27} (dashed lines) for acid-salt mixtures and hydroxide-salt mixtures as a function of the molality of salt (after Harned).

1. HCl (0.1m), KCl (m)	4. KOH (0.1m), KCl (m)
2. HCl (0.1m), NaCl (m)	5. NaOH (0.1m), NaCl (m)
3. HCl (0.1m), LiCl (m)	6. LiOH (0.1m), LiCl (m)

to have a greater mobility than the cation, although the difference is small for potassium chloride. Hence the two junction potentials between the potassium chloride solution and the two half-cell solutions are unequal. The residual effect of the presence of salt in the right half-cell and not in the left makes the e.m.f. of cell 27 greater than E_e. The differences between E_e and E_{27} are found to increase in the order potassium, sodium, lithium, as would be expected from the differences in the mobilities of the anion and cation of the chlorides of these alkali metals (compare equation 18).

The results for hydroxide-chloride mixtures illustrate the same principles, except for the mixture of potassium hydroxide and potassium chloride, for which E_e is smaller than E_{27}. Harned outlined a graphical method of integrating the last term on the right of equation 5a, obtaining E_{jf}, the nonideal part of the liquid-junction potential. The ionic mobilities were assumed to remain constant. In this way he was able to show that the apparently anomalous result for the mixture of potassium hydroxide and potassium chloride can be explained by the form of the plots of the transference number t_i as a function of log γ_i for the ions concerned.

It is clear that the sign of the liquid-junction potential is determined, as a first approximation, by the relative concentrations and mobilities of the positive and negative ions of which the transition layers are composed. This effect is embodied in equations such as those of Planck and Henderson, which give the ideal part of the liquid-junction potential, E_{jm}. However, in certain instances the nonideal term E_{jf}, determined by the single activity coefficients, may be large enough to change considerably the magnitude of the calculated potential, or even to change its sign.

The dangers of partial corrections for liquid-junction potentials, that is, corrections for only the ideal term, have been pointed out by Hamer and Acree,[22] who found that partial corrections sometimes introduce larger errors in pH values and dissociation constants than complete neglect of the liquid-junction potential. The calculations of Hamer[23] by Harned's graphical method showed that E_{jf} for junctions between hydrochloric acid (molality m) and saturated potassium chloride can amount to as much as one-third of E_j, when m is less than 1. However, for junctions formed between acetate buffer solutions and saturated potassium chloride, E_{jf} may be only 0.1 to 0.2 mV, or about 5 per cent of E_j.

Picknett has estimated by two different approaches the liquid-junction potentials between dilute buffer solutions and a saturated solution of potassium chloride.[24] First, the ionic compositions of the solutions were calculated from the dissociation constants, with the aid of activity coefficients derived from the Hückel equation (see equation 33). Theoretical values of E_j were then obtained by an integration of equation 5. When the concentration of the buffer was 0.01 M or less, the "theoretical" E_j was found to agree within 0.5 mV with "experimental" values obtained by equation 19 from the e.m.f. of cells of type 2 (with a glass electrode). In these dilute solutions, Picknett found E_j to vary linearly with the specific conductivity κ of the solution according to the equation $E_j = 0.05 - 1.02 \log \kappa$. Some of his values for two dilute buffer solutions and for dilute solutions of HCl, KOH, and KCl are summarized in Table 3-2.

[22] W. J. Hamer and S. F. Acree, *J. Res. Nat. Bur. Stand.* **17**, 605 (1936); see also J. O. Burton, W. J. Hamer, and S. F. Acree, *ibid.*, **16**, 575 (1936).

[23] W. J. Hamer, *Trans. Electrochem. Soc.*, **72**, 45 (1937).

[24] R. G. Picknett, *Trans. Faraday Soc.*, **64**, 1059 (1968).

TABLE 3-2. LIQUID-JUNCTION POTENTIALS BETWEEN DILUTE SOLUTIONS AND A
SATURATED SOLUTION OF POTASSIUM CHLORIDE AT 25 °C, IN MILLIVOLTS

	(After Picknett)				
Concentration (M)	Acetic Acid (M), NaAc (M)	KH Phthalate	HCl	KOH	KCl
10^{-2}	3.2	3.5	2.9	1.9	2.8
10^{-3}	4.1	4.1	4.0	3.2	3.9
10^{-4}	5.0	4.9	4.8	4.5	5.0
10^{-5}	5.8	5.8	5.7	5.7	6.1
10^{-6}	6.7	6.7	6.7	6.9	7.1

The value of E_j for the junction HCl(0.1 M) | KCl (satd.) has been found
to be 2.0 mV when the computation is based on the Guggenheim assumption,
and 2.8 to 2.9 mV when based on the MacInnes assumption.[25, 26] Thus it
appears to make a decided difference which convention is chosen. Nevertheless, a calculation of the potential of the complete cell by the separate evaluation of E_e and E_j (equation 7) leads to the same result, *provided that the same
convention is used for the computation of both quantities.* The result is also the
same as that obtained by integration of equation 14, which contains only
mean activity coefficients. Indeed, only mean activity coefficients (or other
combinations of ionic activity coefficients) can be derived from the e.m.f. of
galvanic cells by thermodynamic methods, whether or not a liquid junction
is present.

Elimination of the Liquid-Junction Potential

It is doubtful that the electric potential between two solutions of different
sorts of ions can ever be accurately computed or completely eliminated. If a
bridge solution that would reduce E_j to zero were found, the apparent
individual ionic activities obtained would still be complicated functions of
the activities and transference numbers of all the ions present, including those
of the salt bridge.[9] The experimental aspects of salt bridges are considered
in Chapter 10.

Ingenious extrapolation procedures for eliminating the *effect* of hetero-

---

[25] D. A. MacInnes and L. G. Longsworth, *Cold Spring Harbor Symposia Quant. Biol.*, **4**,
18 (1936).
[26] As compared with 4.6 mV given by the Henderson equation when mobilities at infinite
dilution are used (see Table 3-1).

ionic liquid-junction potentials in the determination of certain electro-chemical constants have been devised. In general, the desired property can only be evaluated in the limit of zero concentration of the particular substance under investigation in a solution of a neutral salt, as the results acquire thermodynamic significance only when the bounding solutions become identical. A second extrapolation to zero concentration of neutral salt, however, yields the value of the property at an ionic strength of zero.

This method has been applied successfully to determinations of the thermodynamic dissociation constants of weak acids and bases,[27] as well as of solubility product constants and electrode potentials.[28] The direct comparison of the potentials of the silver-silver chloride and silver-silver bromide electrodes by Owen and King illustrates the general procedure by which the liquid-junction potential is effectively eliminated.

The e.m.f. of the cell

$$\text{Ag;AgBr,} \quad \begin{array}{c} \text{KBr, } xm \\ \text{KNO}_3, (1-x)m \end{array} \left| \begin{array}{c} \text{KCl, } xm \\ \text{KNO}_3, (1-x)m \end{array} \right. \text{,AgCl;Ag} \qquad (30)$$

was measured. Six values of x, from 0.1 to 1.0 at $m = 0.03$ and $m = 0.05$, were chosen. For a given value of m and x,

$$E = E^\circ - \frac{RT}{F} \ln \frac{\gamma_{\text{Cl}^-}}{\gamma_{\text{Br}^-}} + E_j \qquad (31)$$

Two extrapolations were made. The first gave $E^{\circ\prime}$, the apparent standard potential of the cell, at $x = 0$, where E_j must also be zero. The second was a short linear extension of $E^{\circ\prime}$ to $m = 0$, where $\gamma_{\text{Cl}^-}/\gamma_{\text{Br}^-}$ becomes unity.[29] The final value of the standard potential of the cell was in excellent accord with that to be expected from the known values of the standard potentials of the two electrodes.

A similar device avoids to a large extent the effects of liquid junctions in the determination of hydrogen ion concentrations from the e.m.f. of certain cells. Consider once more cell 6, with a bridge consisting of a saturated

[27] H. S. Harned and B. B. Owen, *J. Amer. Chem. Soc.*, **52**, 5079 (1930); E. Larsson and B. Adell, *Z. Physik. Chem.*, **A156**, 352, 381 (1931); **A157**, 342 (1931); K. J. Pedersen, *Kgl. Danske Videnskab. Selskab, Mat.-Fys. Medd.*, **14**, No. 9 (1937); **15**, No. 3 (1937); D. H. Everett and W. F. K. Wynne-Jones, *Proc. Roy. Soc. (London)*, **169A**, 190 (1938); **177A**, 499 (1941); *Trans. Faraday Soc.*, **35**, 1380 (1939); **48**, 531 (1952).

[28] B. B. Owen, *J. Amer. Chem. Soc.*, **60**, 2229 (1938); B. B. Owen and S. R. Brinkley, Jr., *ibid.*, **60**, 2233 (1938); B. B. Owen and E. J. King, *ibid.*, **63**, 1711 (1941).

[29] The conditions of the extrapolation for three different limiting classes of junctions have been discussed by B. B. Owen and S. R. Brinkley, Jr., *J. Amer. Chem. Soc.*, **64**, 2071 (1942).

solution of potassium chloride. The e.m.f. of this cell can be obtained by combining the potentials of the two hydrogen electrode half-cells as measured with respect to a saturated calomel reference electrode. The analogy with the practical pH measurement is evident. The expression (equation 19) for the e.m.f. of cell 6 contains two unknowns, namely $(m_H \gamma_H)_2/(m_H \gamma_H)_1$ and E_j. If the two solutions are so constituted that the preponderant electrolytes are identical, the dissimilar molecular and ionic species are of low concentration, and the total ionic strengths are the same, there is reason to believe that E_j will be nearly zero and $\gamma_{H(2)}/\gamma_{H(1)}$ nearly unity. Under these conditions, therefore, the e.m.f. of cell 6 is an approximate measure of $m_{H(2)}/m_{H(1)}$. In this manner hydrogen ion concentrations in buffer solutions can be obtained.[30] In particular, constant-ionic media containing a high concentration of a neutral supporting electrolyte have been used extensively in studies of the stabilities of coordination complexes in order to control activity coefficients.[31]

These approaches to the elimination of the liquid-junction potential depend to a degree on the validity of a kinetic principle enunciated by Brønsted[32] and applied to acid-base equilibria by Güntelberg and Schiödt, Kilpatrick, Brodersen, Hitchcock, and others.[33] According to the Brønsted principle, the activity coefficients of substances present in small amounts in a mixture of electrolytes are virtually independent of the quantities of these substances and are determined solely by the salt present in large amount. For this reason, the use of buffers of low concentration in an aqueous salt medium of relatively high ionic strength may be advantageous when the main variable is to be the hydrogen ion concentration. Inasmuch as the activity coefficient of hydrogen ion has nearly the same value in different buffer solutions prepared with this same artificial solvent, it is possible also to measure relative hydrogen ion activities. However, the estimation of a reference hydrogen ion activity in any one of these solutions would be very difficult, in view of the relatively high ionic strength.

---

[30] Values of pm_H determined in this way have been used, for example, to evaluate stability constants of complexes; see G. Schwarzenbach, A. Willi, and R. O. Bach, *Helv. Chim. Acta*, **30**, 1303 (1947), and later articles in the series "Komplexone." A detailed discussion of this subject is found in Chapter 9.

[31] These methods have been surveyed by G. Biedermann and L. G. Sillén, *Arkiv Kemi*, **5**, 425 (1953), and by F. J. C. Rossotti and H. Rossotti, *The Determination of Stability Constants*, Chapter 2, McGraw-Hill Book Co., New York, 1961. See V. K. La Mer, *J. Phys. Chem.*, **66**, 973 (1962), for a critique of these procedures.

[32] J. N. Brønsted, *Trans. Faraday Soc.*, **23**, 430 (1927).

[33] E. Güntelberg and E. Schiödt, *Z. Physik. Chem.*, **A135**, 393 (1928); M. Kilpatrick and E. F. Chase, *J. Amer. Chem. Soc.*, **53**, 1732 (1931); M. Kilpatrick, E. F. Chase, and L. C. Riesch, *ibid.*, **56**, 2051 (1934); R. Brodersen, *Acta Physiol. Scand.*, **7**, 162 (1944); D. I. Hitchcock and R. Peters, *J. Amer. Chem. Soc.*, **68**, 1753 (1946).

Estimation of Ionic Activity Coefficients

CHOICE OF A CONVENTION

From a consideration of the studies mentioned and many others, one may conclude that no rigorous test of the validity of the MacInnes and Guggenheim conventions, or of other nonthermodynamic assumptions, can be expected. If we are forced to choose a formula or convention in order to define conventional scales of liquid-junction potentials, ionic activity coefficients, or pH values, the choice should be based for the most part on convenience and reasonableness in the light of what is known from the theory of ionic solutions.

The equality of the activity coefficients of the cation and anion in solutions of potassium chloride is reasonable, for example, because of the similar electronic structures and mobilities of the potassium and chloride ions. Furthermore, specific differences in the interaction of anions with different cations are of secondary importance in dilute solutions. Hence it is not unlikely that the chloride ion will have nearly the same activity coefficient in many different solutions in which its concentration and the ionic strength remain unaltered, and we may therefore regard the MacInnes convention as reasonable.[34]

Similarly, the Debye–Hückel theory lends support to the convention set forth in equation 22, so far as very dilute solutions are concerned. The studies of diffusion potentials by Hermans and by Szabó[35] are of particular interest for their bearing on the validity of this convention in moderately dilute solutions as well.[36] These authors set out to modify the classical equations by taking account of interionic effects. In somewhat the same approach as that used later by Picknett,[24] they calculated "experimental" and "theoretical" liquid-junction potentials and found them to agree well, in some instances at concentrations as high as 0.2 M. As the individual ionic activity coefficients were derived by equation 22 from mean activity coefficients, Szabó[37] concluded that this convention is valid considerably above the Debye–Hückel region.

---

[34] The MacInnes convention is favored by G. Valensi, *J. Chim. Phys.*, **49**, C84 (1952), in part because of the general use of the bridge of potassium chloride in the measurement of pH values.

[35] J. J. Hermans, *Rec. Trav. Chim.*, **56**, 635, 658 (1937); *Chem. Weekbl.*, **34**, 25 (1937); a review: *Naturwissenschaften*, **31**, 257 (1943); Z. Szabó, *Z. Physik. Chem.*, **A174**, 22, 33 (1935); **A176**, 125 (1936).

[36] The reader is referred to the excellent summary given by G. Kortüm, *Elektrolytlösungen*, Chapter 12, Becker und Erler Kom.-Ges., Leipzig, 1941.

[37] Z. Szabó, *Naturwissenschaften*, **24**, 539 (1936); *Z. Physik. Chem.*, **A181**, 169 (1938).

It must nonetheless be realized that the activity coefficients of individual ionic species, the potentials of single electrodes, liquid-junction potentials, and pH values derived with the use of these conventions are to be regarded also as conventional. No detailed verification of the validity of the convention is necessary or possible.

IONIC ACTIVITY COEFFICIENTS AND THE DEBYE–HÜCKEL EQUATION

Inasmuch as ionic activity coefficients are commonly estimated by relating them in some way to mean ionic activity coefficients, it is often convenient to consider them in terms of the Debye–Hückel equation, written in the form which allows for the restriction of attraction between ions due to their finite size,

$$-\log \gamma_i = \frac{A z_i^2 \sqrt{I}}{1 + \rho \sqrt{I}} \tag{32}$$

or the Hückel equation

$$-\log \gamma_i = \frac{A z_i^2 \sqrt{I}}{1 + \rho \sqrt{I}} - b_i I \tag{33}$$

where ρ is $B\mathring{a}$ (see Chapter 1). The experimental values of the mean activity coefficients of many uni-univalent electrolytes can be fitted reasonably well to equations of these forms up to ionic strengths I of 0.1 (equation 32) and 1.0 (equation 33).[38] The parameters $\mathring{a}$ and b_i are usually adjusted to fit a given set of data. The parameter $\mathring{a}$ has the dimension of length, is of the same order of magnitude as the ionic diameter, and is often referred to as the "ion-size parameter." The ion-size parameter for strong uni-univalent electrolytes in aqueous solution is greater than 3.5 Å, and values over 6 Å for electrolytes of this valence type are uncommon. This is in agreement with Bjerrum's calculation.[39] The parameter b_i may be regarded as an ionic inter-action coefficient.[40] The coefficients A and B are functions of the temperature and the dielectric constant of the solvent; their values from 0 to 100 °C for the water medium are given in the Appendix, Table 4.

---

[38] H. S. Harned and B. B. Owen, *The Physical Chemistry of Electrolytic Solutions*. 3rd ed., Chapter 12, Reinhold Publishing Corp., New York, 1958. To express log γ_{MX} for a 1–1 electrolyte, z_i^2 is set equal to unity. The theoretical development of the Hückel equation has been improved by G. Scatchard, *Physik. Z.*, **33**, 22 (1932). The activity coefficients of strong electrolytes have been interpreted successfully from the standpoint of ionic hydration and spatial requirements by R. H. Stokes and R. A. Robinson, *J. Amer. Chem. Soc.*, **70**, 1870 (1948), and E. Wicke and M. Eigen, *Z. Elektrochem.*, **56**, 551 (1952); *Z. Naturforsch.*, **8a**, 161 (1953).

[39] N. Bjerrum, *Kgl. Danske Videnskab. Selskab, Mat.-Fys. Medd.*, **7**, No. 9 (1926).

[40] E. A. Guggenheim and J. C. Turgeon, *Trans. Faraday Soc.*, **51**, 747 (1955).

Kielland[41] has estimated the value of $\mathring{a}$ (that is, ρ/B) in equation 32 for 130 inorganic and organic ions from data for the ionic mobilities, radii in the crystalline solid, deformability, and hydration numbers. The values of $\mathring{a}$ (in angstroms) and γ_i for some of the ions commonly present in buffer solutions are given in Table 3-3. The parameters for hydrogen and hydroxyl ions were

TABLE 3-3. INDIVIDUAL COEFFICIENTS OF IONS IN WATER

| | | (After Kielland) | | | |
| | | γ_i at Ionic Strength of | | | |
Ion	$\mathring{a}$ (in Å)	0.005	0.01	0.05	0.1
	Univalent Ions				
H^+	9	0.933	0.914	0.86	0.83
Li^+, $C_6H_5COO^-$	6	0.929	0.907	0.835	0.80
Na^+, HCO_3^-, $H_2PO_4^-$, CH_3COO^-	4.5	0.928	0.902	0.82	0.775
OH^-, F^-, ClO_4^-, $HCOO^-$, H_2 citrate$^-$	3.5	0.926	0.900	0.81	0.76
K^+, Cl^-, Br^-, I^-, NO_3^-	3	0.925	0.899	0.805	0.755
Rb^+, Cs^+, NH_4^+	2.5	0.924	0.898	0.80	0.75
	Bivalent Ions				
Mg^{2+}	8	0.755	0.69	0.52	0.45
Ca^{2+}, phthalate^{2-}	6	0.749	0.675	0.485	0.405
Sr^{2+}, Ba^{2+}, malonate^{2-}, succinate^{2-}, tartrate^{2-}	5	0.744	0.67	0.465	0.38
CO_3^{2-}, oxalate^{2-}, H citrate^{2-}	4.5	0.741	0.663	0.45	0.36
SO_4^{2-} HPO_4^{3-}	4	0.740	0.660	0.445	0.355
	Trivalent Ions				
Citrate^{3-}	5	0.51	0.405	0.18	0.115
PO_4^{3-}	4	0.505	0.395	0.16	0.095

not calculated independently, as were those for the other ions, but were taken to be 9 and 3.5 Å, respectively. Kielland's activity coefficients are consistent with values derived from cells with liquid junctions,[42] where a comparison can be made.

[41] J. Kielland, *J. Amer. Chem. Soc.*, **59**, 1675 (1937).

[42] N. Bjerrum and A. Unmack, *Kgl. Danske Videnskab. Selskab, Mat.-Fys. Medd.*, **9**, No. 1 (1929); K. Hass and K. Jellinek, *Z. Physik. Chem.*, **A162**, 153 (1932).

Ionic Activities and Cells without Liquid Junction

We have been concerned principally in the foregoing sections with useful ways of evaluating the liquid-junction potential or the single ionic activity coefficient with a view to deriving hydrogen ion activities from the e.m.f. of cells with a liquid junction. The theoretical difficulties are fundamentally the same if we choose a cell without liquid junction, although there are certain experimental advantages to be gained from this approach to the assignment of standard pH values, pH(S).

Consider, for example, the cell

$$\text{Pt};\text{H}_2,\text{Buffer soln.},\text{Cl}^-,\text{AgCl};\text{Ag} \tag{34}$$

The standard potential E° of this cell is known over a wide range of temperatures (see Chapter 10), and the chloride ion molality is known from the composition of the cell solution. Hence unambiguous values of the quantity $p(a_H \gamma_{Cl})$ are obtainable:

$$p(a_H \gamma_{Cl}) \equiv -\log (m_H \gamma_H \gamma_{Cl}) = \frac{(E - E^\circ)F}{RT \ln 10} + \log m_{Cl} \tag{35}$$

If desired, the value $p(a_H \gamma_{Cl})^0$ of this quantity for the buffer solution in the limit $m_{Cl} = 0$ can also be obtained. The hydrogen ion activity can be computed from the physically defined $p(a_H \gamma_{Cl})$ value only by estimating an individual ionic activity coefficient, that of chloride ion:[43]

$$pa_H \equiv -\log (m_H \gamma_H) = p(a_H \gamma_{Cl}) + \log \gamma_{Cl} \tag{36}$$

ACTIVITY COEFFICIENTS IN MIXTURES OF ELECTROLYTES

The manner in which the MacInnes convention should be applied is not always clear, and the application of the Guggenheim convention is often complex. This is illustrated by a consideration of a mixture of two or more electrolytes which have no common ion. It is difficult, for example, to relate the activity coefficient of hydrogen ion in an aqueous mixture of potassium hydrogen phthalate and sodium chloride to the mean activity coefficients of the four binary salts, none of which furnishes an appreciable amount of hydrogen ion to the solution. Furthermore, the mean activity coefficients in such mixtures are usually not known.

Several other reasonable procedures, all fundamentally alike, have been

---

[43] Practical pH determinations by means of cell 36 were envisioned by W. J. Hamer and S. F. Acree, *J. Res. Nat. Bur. Stand.*, **23**, 647 (1939). It was proposed that a soluble chloride be added in known amount to the test solution and that γ_{Cl} be estimated by equation 33. Unfortunately, the ionic strength is usually unknown.

employed, therefore, to define a conventional γ_{Cl} for the computation of pa_H values from the e.m.f. of hydrogen-silver chloride cells of type 34. Hamer and Acree,[44] for example, computed γ_{Cl} in the 0.05 M solution of potassium hydrogen phthalate ($I = 0.053$, $p(a_H \gamma_{Cl})^0 = 4.096$) by assuming that log γ_{Cl} varies linearly with $p(a_H \gamma_{Cl})$. By use of the convention of equation 22, the values of $-\log \gamma_{Cl}$ in 0.053 m solutions of hydrochloric acid and potassium hydroxide (extrapolated to zero concentration of potassium chloride) were derived from earlier data. These values were plotted as a function of $p(a_H \gamma_{Cl})$, connected by a straight line, and $-\log \gamma_{Cl}$ at $p(a_H \gamma_{Cl}) = 4.096$ read off. The activity coefficients of chloride ion found for pure solutions of hydrochloric acid and potassium hydroxide correspond to. $\mathring{a}$ values of 5.5 and 3.4 Å (equation 32), and hence the method postulates a dependence of $\mathring{a}$ on the acidity of the solution. Actually, these values of log γ_{Cl} in the acid and hydroxide solutions differ by only 0.01, and the interpolation is hardly necessary.

At ionic strengths below 0.05, the differences among the mean activity coefficients of strong electrolytes of the same valence type at a given concentration become rapidly smaller. It is therefore not unreasonable to take the activity coefficient of chloride ion in a given mixture equal to the mean activity coefficient of hydrochloric acid[45] or of sodium chloride[46] in their pure aqueous solutions of the same ionic strength as the mixture in question. These conventions have been considered elsewhere in more detail.[47]

Brønsted's principle of specific ionic interaction[48] offers some guidance in selecting the mean activity coefficients that can most reasonably be expected to represent γ_{Cl} in the particular mixture in question. This theory has made it possible to interpret successfully the properties of many mixtures of strong electrolytes. It may be summarized by stating its two postulates: (a) Ions of like charge repel each other so strongly that their short-range interactions are identical and negligible, and (b) each ion exerts a "salting-out effect," that is, an influence tending to increase the activity coefficient, on every other ion in the solution.

[44] W. J. Hamer and S. F. Acree, *J. Res. Nat. Bur. Stand.*, **32**, 215 (1944).

[45] R. G. Bates, G. L. Siegel, and S. F. Acree, *J. Res. Nat. Bur. Stand.*, **31**, 205 (1943); W. J. Hamer, G. D. Pinching, and S. F. Acree, *ibid.*, **31**, 291 (1943).

[46] G. G. Manov, N. J. DeLollis, and S. F. Acree, *J. Res. Nat. Bur. Stand.*, **33**, 287 (1944); G. G. Manov, N. J. DeLollis, P. W. Lindvall, and S. F. Acree, *ibid.*, **36**, 543 (1946).

[47] R. G. Bates, *Chem. Rev.*, **42**, 1 (1948).

[48] J. N. Brønsted, *J. Amer. Chem. Soc.*, **44**, 877 (1922); **45**, 2898 (1923). See also E. Güntelberg, *Z. Physik, Chem.*, **123**, 199 (1926); G. Scatchard and S. S. Prentiss, *J. Amer. Chem. Soc.*, **56**, 2320 (1934); H. S. Harned and B. B. Owen, *The Physical Chemistry of Electrolytic Solutions*, 3rd ed., Chapter 14, Reinhold Publishing Corp., New York, 1958; G. Scatchard and R. G. Breckenridge, *J. Phys. Chem.*, **58**, 596 (1954); and H. S. Harned and R. A. Robinson, *Multicomponent Electrolyte Solutions*, Pergamon Press, New York, 1968.

G. N. Lewis recognized that the quantity he termed the ionic strength is the primary factor determining the magnitude of activity coefficients in aqueous solutions. The Brønsted theory indicates what the secondary specific salt effects are likely to be at a fixed ionic strength. The first of Brønsted's postulates suggests, for example, that γ_{Cl} will be affected only slightly by the substitution of acetate or metaborate ion for chloride but will, on the contrary, be altered by changes of composition in which cations are involved. The theory also predicts that log $\gamma_{\pm}$ will vary in linear fashion with changes of composition, as long as the ionic strength remains constant and the kinds of ions unchanged. It is reasonable, then, to identify γ_{Cl} in a buffer mixture of acetic acid and sodium acetate with γ_{NaCl} in a sodium chloride solution of the same ionic strength as the acetate buffer, or with γ_{HCl}^{0} at the limit of zero concentration of hydrochloric acid in the same chloride solution.[49] In a solution of potassium hydrogen phthalate, either γ_{HCl}^{0} or γ_{KCl} in potassium chloride solutions might be regarded as a reasonable choice.

These mean activity coefficients, and thus the assigned numerical values of the activity coefficient of chloride ion, can be expressed up to an ionic strength of 0.2 by equation 32 with the $\mathring{a}$ values given in Table 3–4. It is

TABLE 3–4. VALUES OF $\mathring{a}$ (EQUATION 32) FOR CHLORIDE ION, IN Å, DERIVED FROM MEAN ACTIVITY COEFFICIENTS

	$\mathring{a}$ for Ionic Strength of		
Source	0.05	0.1	0.2
γ_{HCl} in HCl	5.5	6.0	6.4
γ_{HCl}^{0} in NaCl	5.0	5.2	5.7
γ_{HCl}^{0} in KCl	4.5	5.0	5.4
γ_{NaCl} in NaCl	4.6	4.6	4.6
γ_{KCl} in KCl	4.0	4.0	4.0

evident that a constant $\mathring{a}$ is not adequate in all cases to cover this range. In view of the assumptions necessarily involved in the estimation of γ_{Cl}, use of the two-parameter form (equation 33) is unjustified.[50]

---

[49] The medium effect of the acetic acid molecules on the activity coefficient is negligible at concentrations of acetic acid less than 0.1 M; see B. B. Owen, *J. Amer. Chem. Soc.*, **54**, 1758 (1932), and Chapter 9.

[50] A one-parameter form of the Debye-Hückel equation has been developed by R. H. Stokes and R. A. Robinson, *J. Amer. Chem. Soc.*, **70**, 1870 (1948), through introduction of hydration numbers for the ions.

There is, of course, no rigorous way to select the proper value of $\mathring{a}$ for the computation of an individual ionic activity coefficient. However, it may be concluded from Table 3-4 that γ_{Cl} in dilute solutions could reasonably be defined conventionally in terms of the Debye–Hückel equation in the form of equation 32, with $\mathring{a}$ values of 4 to 6 Å. If the ionic strengths of the buffer solutions are less than 0.1, a variation of $\mathring{a}$ from 4 to 6 Å corresponds to a change of less than 0.013 unit in the pa_H computed by equation 36.

In order to derive $p(m_H \gamma_H)$ or pa_H from $p(a_H \gamma_{Cl})$ by equation 36, Bates and Guggenheim[51] suggested that γ_{Cl} at ionic strengths not exceeding 0.1 be calculated by equation 32 with $\rho = 1.5\ kg^{1/2}\ mol^{-1/2}$. This value of ρ corresponds to $\mathring{a} = 4.56$ Å at 25 °C and therefore makes γ_{Cl} very nearly the same as the mean activity coefficient of sodium chloride in its pure aqueous solutions (see Table 3-4). The values of the pa_H of hydrochloric acid solutions and acid–chloride mixtures furnished by four conventions are compared in Table 3-5.

TABLE 3-5. pa_H OF HYDROCHLORIC ACID SOLUTIONS AND ACID-CHLORIDE MIXTURES AT 25 °C

Solution	MacInnes Convention	Guggenheim Convention	Equation 22	Bates- Guggenheim Convention
HCl, 0.01 M	2.042	2.043	2.043	2.044
HCl, 0.1 M	1.085	1.099	1.099	1.088
HCl, 0.01 M; NaCl, 0.09 M	2.092	2.097	2.103	2.096
HCl, 0.01 M; KCl, 0.09 M	2.095	2.095	2.104	2.098

SEPARATION OF γ_{Cl} FROM THE ACTIVITY-COEFFICIENT TERM

Another approach has been used for the estimation of γ_{Cl} in buffer solutions composed of the primary and secondary anions (HA^- and A^{2-}) of a dibasic or polybasic acid.[52] When the buffer solution is of this type and

[51] R. G. Bates and E. A. Guggenheim, *Pure Appl. Chem.*, **1**, 163 (1960).

[52] W. J. Hamer, J. O. Burton, and S. F. Acree, *J. Res. Nat. Bur. Stand.*, **24**, 269 (1940); W. J. Hamer and S. F. Acree, *ibid.*, **35**, 381, 539 (1945); R. G. Bates and S. F. Acree, *ibid.*, **30**, 129 (1943); **32**, 131 (1944); **34**, 373 (1945); R. G. Bates, G. L. Siegel, and S. F. Acree, *ibid.*, **31**, 205 (1943); G. D. Pinching and R. G. Bates, *ibid.*, **40**, 405 (1948).

contains chloride, the e.m.f. of cells of type 34 furnishes values of $\gamma_{HA} \cdot \gamma_{Cl^-}/\gamma_{A^{2-}}$, which will be designated γ_r:

$$\log \gamma_r = -\log K_2 - p(a_H \gamma_{Cl}) - \log \frac{m_{HA^-}}{m_{A^{2-}}} \tag{37}$$

This quantity is an unequivocally defined combination of activity coefficients. It can be obtained, provided the dissociation constant K_2 of the acid is known, by equation 37, which was derived from the mass law expression for the acidic dissociation. The values of γ_r at different ionic strengths were found to fit an equation of the Hückel form (compare equation 33):

$$\log \gamma_r \equiv \log \frac{\gamma_{HA} \cdot \gamma_{Cl^-}}{\gamma_{A^{2-}}} = \frac{2A\sqrt{I}}{1 + B\mathring{a}^*\sqrt{I}} + b^*I \tag{38}$$

The parameters $\mathring{a}^*$ and b^* are readily obtained by graphical methods.

Perhaps the simplest way to obtain the desired γ_{Cl} from the known γ_r is to assume the validity of the valence relationships of the Debye–Hückel equation, namely $\gamma_{Cl^-}^4 = \gamma_{HA^-}^4 = \gamma_{A^{2-}}$. Thus $\log \gamma_{Cl}$ would be set equal to $-\frac{1}{2} \log \gamma_r$. Slightly different results are obtained if the parameters $\mathring{a}^*$ and b^* and identified with $\mathring{a}$ and b_i, respectively, for the calculation of γ_{Cl} by equation 33.

The practical effect of these various arbitrary definitions of the individual activity coefficient of chloride ion on the value of pa_H computed by equation 36 will be considered in the following chapter. The differences among many of these formulations of the ionic activity coefficient are unimportant at low and moderate ionic strengths.[53] The elaborate treatment has no more validity in theory than the simple one and usually demands an unwarranted expenditure of effort.

The Residual Liquid-Junction Potential

The difference of pa_H between two solutions designated 1 and 2 is, according to equation 4,

$$(pa_H)_1 - (pa_H)_2 = \frac{(E_1 - E_2)F}{RT \ln 10} - \frac{(E_{j1} - E_{j2})F}{RT \ln 10} \tag{39}$$

where E_{j1} and E_{j2} represent the potentials at the junctions between solutions 1 and 2, respectively, and the salt bridge. The accurate measurement of pa_H depends on the constancy of the liquid-junction potential when the standard

---

[53] R. G. Bates, *Chem. Rev.*, **42**, 1 (1948); R. G. Bates and E. R. Smith, *J. Wash. Acad. Sci.*, **38**, 61 (1948).

solution is replaced by the unknown or test solution. If the two junction potentials cancel to a large extent, the difference of e.m.f., $E_1 - E_2$, is a close measure of the difference between the common logarithms of the (conventional) hydrogen ion activities of the two solutions.

The hydrogen and hydroxyl ions are highly mobile, and consequently the liquid-junction potential between the bridge of saturated potassium chloride and solutions of pH less than 3 or greater than 11 may be appreciably different from that between the bridge and a standard buffer of intermediate pH.[54] The difference, or residual liquid-junction potential, manifests itself as an error in the measured pH. The assignment of conventional pa_H values determined by methods independent of a liquid-junction error prepares the way for an estimation of the magnitude of this residual potential and the error in pH it causes.

Bates, Pinching, and Smith[55] determined the extent of the aberration of the practical pH scale near its ends by comparing pa_H for a number of buffer solutions and solutions of strong and weak acids and bases with the pH furnished by the cell with liquid junction. The values of pa_H were calculated by equation 36 from the e.m.f. of cells of type 34. The activity coefficient of chloride ion was defined by equation 32; the parameter $\mathring{a}$ was taken to be 4 to 6 Å. The cell with liquid junction, shown in Fig. 3–2, was of type 6. One of the two hydrogen electrode compartments c and c' was filled with the 0.025 M standard phosphate buffer [pH(S) = pa_H = 6.865 at 25 °C] and the other with the "test" solution. The liquid junctions were formed in the small bulbs directly below the two electrode compartments. As in the usual determination, the practical pH was calculated by

$$pH = pH(S) + \frac{(E_X - E_S)F}{RT \ln 10} \tag{40}$$

The differences pH(l.j.) $-$ pa_H between the pH (derived from the cell with liquid junction) and pa_H (from the cell without liquid junction) for 24 solutions are plotted in Fig. 3–3 as a function of pH. The ends of each horizontal line correspond to the pa_H calculated with $\mathring{a}$ values of 4 and 6 Å; hence the length of each line is a rough indication of the uncertainty in the calculation. Test solutions labeled S were strong electrolytes, those labeled W were weak electrolytes, and the remainder were buffer solutions.

---

[54] According to G. G. Manov, N. J. DeLollis, and S. F. Acree, *J. Res. Nat. Bur. Stand.*, **34**, 115 (1945), the apparent differences (referred to a neutral buffer solution) are 1.2 mV for 0.1 M hydrochloric acid and 1.0 mV for a solution of calcium hydroxide, pH 12.4, and are of opposite sign.

[55] R. G. Bates, G. D. Pinching, and E. R. Smith, *J. Res. Nat. Bur. Stand.*, **45**, 418 (1950).

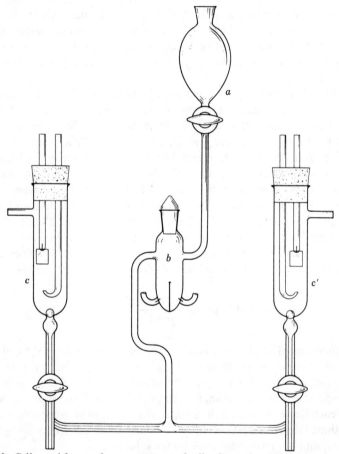

FIG. 3-2. Cell vessel for e.m.f. measurements of cells of type 6. *a*, KCl reservoir; *b*, calomel electrode; *c, c'*, hydrogen electrode compartments.

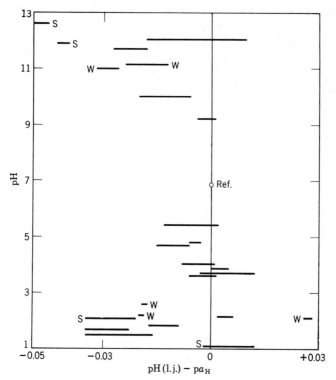

FIG. 3-3. Liquid-junction error of the practical pH scale.

pH at Surfaces

It is well known that aliphatic acids RCOOH and RNH_3^+ composed of a long-chain hydrophobic group R and an ionizable head group —COOH or —NH_3^+ form an electrical double layer at the interface between water and a nonpolar solvent such as benzene. A difference of potential ψ therefore exists between surface and bulk, altering the distribution of hydrogen ions in the aqueous phase. The resulting difference between the pH at the surface and that (pH_b) in the bulk solution is thus given by[56]

$$pH = pH_b + \frac{Ne\psi}{RT \ln 10} \qquad (41)$$

where N is Avogadro's number and e is the electronic charge.

[56] J. T. Davies and E. K. Rideal, *Interfacial Phenomena*, 2nd ed., Chapter 2, Academic Press, New York, 1963.

The potential ψ is positive for carboxylic acids and may lead to surface pH values that are three units higher than in the bulk. Similarly, negative surface potentials of about the same magnitude for amines cause surface pH values three to four units lower than those in the bulk aqueous phase. These differences are revealed, for example, by dyes adsorbed at the surface.[57] There is evidence that this phenomenon is also present at air–water interfaces, although the effects are much smaller.

---

[57] R. A. Peters, *Proc. Roy. Soc. (London)*, **A133**, 140 (1931).

CHAPTER 4

pH standards

The great majority of practical pH determinations depend on the measurement of the e.m.f. of cells of the type

$$\text{Pt}; \text{H}_2(g), \text{Soln. X} \,|\, \text{KCl (satd.)} \,|\, \text{Reference electrode} \qquad (1)$$

The glass, quinhydrone, or antimony electrodes, with which the hydrogen gas electrode is commonly replaced, are to be regarded as hydrogen electrodes of more or less imperfect response. The most useful reference electrodes are the calomel and silver-silver chloride electrodes in a 0.1 M, 1 M, 3.5 M, or saturated solution of potassium chloride.[1]

As was seen in the preceding chapter, the hydrogen ion activity a_H of solution X is formally related to the e.m.f. E of cell 1 by the expression

$$pa_H \equiv -\log a_H = \frac{[E - (E^{\circ\prime} + E_j)]F}{RT \ln 10} \qquad (2)$$

The selection of values for $E^{\circ\prime} + E_j$, which will be termed the standard potential of cell 1, is all that is necessary to permit conventional pH values or hydrogen ion activities to be derived from the e.m.f. of this cell. However, this procedure has been found to be impractical for modern pH measurements, particularly when glass electrodes are used. The standard potential of one cell assembly may differ appreciably from that of another of apparently identical design; it may also vary considerably with time. Hence a redetermination of $E^{\circ\prime} + E_j$ at frequent intervals is preferred. This standardization is accomplished by the measurement of E_S, the e.m.f. of cell 1 with a standard

---

[1] The 3.5 M solution of potassium chloride has been employed extensively as a salt bridge in Europe, whereas the saturated solution (4.16 M at 25 °C) is more common in America. When one of these bridge solutions is substituted for the other, operational pH values appear to be unaffected. The *structure* of the liquid-liquid boundary may be a matter of considerably greater importance (see Chapter 10).

solution [whose pa_H is denoted pH(S)] replacing solution X in the cell. The desired reference potential is then obtained from equation 2.

If the reference potential and the liquid-junction potential remain unchanged on replacement of the standard solution by the unknown, one finds with the aid of equation 2

$$pH = pH(S) + \frac{(E_X - E_S)F}{RT \ln 10} \tag{3}$$

which is the operational definition of the pH value. In the standardization of commercial pH assemblies, day-to-day changes in the cell potential are compensated by a manual adjustment of resistances in the meter circuit. In this way, the instrument is made to indicate the correct pH(S) of the buffer standard, and the value of $E^{\circ\prime} + E_j$ need not be ascertained.

Evidently the assignment of a value of pH(S) to a particular buffer solution fixes the value of $E^{\circ\prime} + E_j$; conversely, the selection of a standard potential defines the scale of pH. Both quantities can be evaluated only with the aid of nonthermodynamic conventions, some of which have been discussed in the foregoing chapter. We shall find it convenient here to consider both of these approaches to the standardization of pH measurements.

The Standard Potential

When cell 1 is allowed to discharge spontaneously, the chemical reaction given in equation 3 of Chapter 3 takes place. Hydrogen and chloride ions are formed in equal numbers, but not in the same medium. Hence it is advantageous to define a standard potential $E^{\circ\prime} + E_j$ which represents the e.m.f. that cell 1 would have if solution X were replaced by the hypothetical solution of unit hydrogen ion activity without altering the liquid-junction potential or the reference electrode. It will be remembered that the usual standard potential, designated E°, is conventionally referred to the state of unit activity of all the reactants and products of the cell reaction, with boundary potentials eliminated. Inasmuch as the activity of chloride ion at the calomel reference electrode is constant (although unknown and indeed undefined), it is convenient to introduce a quantity $E^{\circ\prime}$, related to E° in the following manner:

$$E^{\circ\prime} \equiv E^{\circ} - \frac{RT}{F} \ln a_{Cl} \tag{4}$$

The liquid-junction potential and means by which it can be evaluated or eliminated have already been considered. It is appropriate now to explore the possibility of determining $E^{\circ\prime}$, the other member of the standard potential term.

The value of $E°$ for the hydrogen–calomel cell is found to be 0.2679 V at 25 °C by adding 0.2223 V, the standard e.m.f. of the cell (see Table 10–6)

$$Pt;H_2, HCl(m),AgCl;Ag \tag{5}$$

to 0.0456 V, the e.m.f. of the cell[2]

$$Ag;AgCl,MCl,Hg_2Cl_2;Hg \tag{6}$$

Both the MacInnes convention and the Guggenheim convention permit the substitution of a_{KCl}, the activity of potassium chloride at the calomel reference electrode, for a_{Cl} in equation 4. The molality of potassium chloride is 0.1005 mol kg^{-1} in the 0.1 M solution and 4.804 in the solution saturated at 25 °C, and the activity coefficient is 0.769 and 0.588, respectively, in these solutions.[3] Hence $E°'$ is found to be 0.3340 V for the 0.1 M calomel electrode and 0.2415 V for the saturated electrode.

Sørensen and Linderstrøm–Lang,[4] who first defined the activity scale of pH, made extensive e.m.f. measurements of cells that contained mixtures of hydrochloric acid and sodium or potassium chloride. Each measured value of the e.m.f. was corrected for the liquid-junction potential by means of the Bjerrum extrapolation (see Chapter 2). The activity of hydrogen ion in these solutions was based on semitheoretical equations for activity coefficients, and a value of 0.3357 V for $E°'$ (0.1 M calomel electrode) at 18 °C was selected. At 25 °C, $E°'$ becomes 0.3353 V. In their later paper, these authors apparently abandoned their recommendation that the Bjerrum extrapolation be used in an attempt to eliminate the liquid-junction potential in all practical measurements of pH. Hence we may regard their value of $E°' + E_j$ also to be 0.3353 V. By subtraction of 0.0912 V, the e.m.f. of the cell

$$Hg;Hg_2Cl_2,KCl\ (satd.)\ |\ KCl(0.1\ M),Hg_2Cl_2;Hg \tag{7}$$

at 25 °C,[5] a value of 0.2441 V for the standard potential $E°' + E_j$ of the saturated calomel electrode is found.

[2] R. H. Gerke, *J. Amer. Chem. Soc.*, **44**, 1684 (1922); M. Randall and L. E. Young, *ibid.*, **50**, 989 (1928). By direct measurement of the e.m.f. of hydrogen-calomel cells, $E°$ has been found to be 0.26816 V at 25 °C: S. R. Gupta, G. J. Hills, and D. J. G. Ives, *Trans. Faraday Soc.*, **59**, 1874 (1963). See also Chapter 10.

[3] T. Shedlovsky and D. A. MacInnes, *J. Amer. Chem. Soc.*, **59**, 503 (1937); H. S. Harned and M. A. Cook, *ibid.*, **59**, 1290 (1937); G. Scatchard, W. J. Hamer, and S. E. Wood, *ibid.*, **60**, 3061 (1938).

[4] S. P. L. Sørensen and K. Linderstrøm-Lang, *Compt. Rend. Trav. Lab. Carlsberg*, **15**, No. 6 (1924); S. P. L. Sørensen, K. Linderstrøm-Lang, and L. Lund, *J. Gen. Physiol.*, **8**, 543 (1927).

[5] N. Bjerrum and A. Unmack, *Kgl. Danske Videnskab. Selskab, Mat.-Fys. Medd.*, **9**, No. 1 (1929); H. Riehm, *Z. Physik. Chem.*, **A160**, 1 (1932); D. A. MacInnes, D. Belcher, and T. Shedlovsky, *J. Amer. Chem. Soc.*, **60**, 1094 (1938).

Scatchard[6] assumed that the potential across the flowing junction between saturated potassium chloride and solutions of hydrochloric acid is unchanged as the concentration of hydrochloric acid is reduced from a molality of 0.1 mol kg^{-1} to zero, where γ_H becomes unity by definition. By a suitable extrapolation of e.m.f. data for the cell

$$Ag;AgCl,HCl(m) \,|\, KCl \,(satd.),Hg_2Cl_2;Hg \qquad (8)$$

to $m = 0$, and combination of the limiting potential with the known standard potential for the silver-silver chloride electrode. Scatchard obtained 0.2454 V at 25 °C for $E^{\circ\prime} + E_j$ of cell 1 with a saturated calomel reference electrode. This treatment did not require the adoption of a convention regarding individual ionic activity coefficients, although the constancy of the liquid-junction potential constituted an assumption not subject to proof. Thus he was able to compute experimental values of γ_H and γ_{Cl} in the acid solutions and found them to be in excellent accord with those computed by the MacInnes convention up to a concentration of 0.2 M.

Bjerrum and Unmack[7] based their selection of $E^{\circ\prime} = 0.3360$ V (0.1 M calomel electrode) on an extensive series of e.m.f. measurements of solutions of hydrochloric acid and acid-chloride mixtures and on a critical examination of the data of other investigators. Each value of the e.m.f. was corrected for the ideal part of the liquid-junction potential. Their data support the later observation of Hitchcock and Taylor[8] that the values of $E^{\circ\prime} + E_j$ furnished by solutions of hydrochloric acid and by mixtures of the acid with sodium chloride and with potassium chloride are somewhat divergent. The same conclusion is reached by a consideration of the liquid-junction potentials calculated by the Henderson equation (see Table 3–1).

A value of 0.3337 V for $E^{\circ\prime}$ (0.1 M calomel reference electrode) constitutes the basis of the scale of activity pH proposed by Guggenheim and Schindler.[9] This value was chosen because it was found to yield, in equation 2, values of a_H for 0.1 M hydrochloric acid and mixtures of 0.01 M hydrochloric acid with 0.09 M sodium and potassium chlorides that were nearly the same as those calculated with the aid of the MacInnes convention. The Henderson equation was used to evaluate E_j. From the e.m.f. and pH values given by Guggenheim and Schindler for three acetate and three phosphate buffer solutions, one finds, by equation 2, an average $E^{\circ\prime} + E_j$ of

$$0.3345 \pm 0.0005 \text{ V } (0.1 \text{ M calomel electrode}).$$

[6] G. Scatchard, *J. Amer. Chem. Soc.*, **47**, 696 (1925).

[7] N. Bjerrum and A. Unmack, *Kgl. Danske Videnskab. Selskab, Mat.-Fys. Medd.*, **9**, No. 1 (1929).

[8] D. I. Hitchcock and A. C. Taylor, *J. Amer. Chem. Soc.*, **59**, 1812 (1937); **60**, 2710 (1938).

[9] E. A. Guggenheim and T. D. Schindler, *J. Phys. Chem.*, **38**, 533 (1934).

Hamer[10] took into account the nonideal part of the liquid-junction potential, which he evaluated with the aid of the MacInnes convention and the graphical integration proposed by Harned.[11] He was able to show that a value of 0.3337 V for $E^{\circ\prime}$ (0.1 M calomel electrode) is consistent with e.m.f. data for acetate buffer solutions as well as for solutions of hydrochloric acid. Hamer's calculations confirmed, furthermore, that the liquid-junction potential for strong acids is sufficiently different from that for buffer solutions to warrant the use of at least two different values of $E^{\circ\prime} + E_j$. It is this effect that gives rise to the residual liquid-junction error we have already discussed. From the measurements of Manov, DeLollis, and Acree,[12] it may be concluded that still another value is necessary for strongly alkaline solutions.

The Cohn Method of pH Standardization

The values of the standard potential selected by Scatchard and by Guggenheim were regarded as reasonable because they furnished individual activity coefficients that were consistent with the MacInnes convention. In an extension of this same idea, Cohn, Heyroth, and Menkin[13] set out to choose a value of $E^{\circ\prime} + E_j$ that would be consistent with the known dissociation constants of weak acids. By analysis of e.m.f. data for buffer solutions of acetic acid and sodium acetate, these authors demonstrated the dependence of the value of $-\log K$ for acetic acid, computed from the measured pH by the mass law equation, on the value of $E^{\circ\prime} + E_j$ chosen. Several values of the standard potential were tried until the one yielding the accepted value of $-\log K$ was found.

The Cohn method has been elaborated by Kauko, Hitchcock, MacInnes, and their coworkers. Kauko and his associates Carlberg and Airola[14] studied the effect of changes in $E^{\circ\prime} + E_j$ on the value of $\log K$ for the first dissociation step of carbonic acid derived from e.m.f. measurements of cells of type 1. Inasmuch as the dissociation constant was known from independent measurements, they were able to select a value of $E^{\circ\prime} + E_j$ that would, by the customary mode of calculation, yield the correct result.

Hitchcock and Taylor[8, 15] and MacInnes, Belcher, and Shedlovsky[16, 17]

[10] W. J. Hamer, *Trans. Electrochem. Soc.*, **72**, 45 (1937).

[11] H. S. Harned, *J. Phys. Chem.*, **30**, 433 (1926).

[12] G. G. Manov, N. J. DeLollis, and S. F. Acree, *J. Res. Nat. Bur. Stand.*, **34**, 115 (1945).

[13] E. J. Cohn, F. F. Heyroth, and M. F. Menkin, *J. Amer. Chem. Soc.*, **50**. 696 (1928); see also the earlier paper of A. B. Hastings and J. Sendroy, Jr., *J. Biol. Chem.*, **65**, 445 (1925).

[14] Y. Kauko and J. Carlberg, *Soc. Sci. Fennica, Commentationes Phys.-Math.*, **8**, No. 23 (1936); Y. Kauko and A. Airola, *Z. Physik. Chem.*, **A178**, 437 (1937).

[15] D. I. Hitchcock, *Trans. Electrochem. Soc.*, **72**, 89 (1937).

[16] D. A. MacInnes, *Cold Spring Harbor Symposia Quant. Biol.*, **1**, 190 (1933).

[17] D. A. MacInnes, D. Belcher, and T. Shedlovsky, *J. Amer. Chem. Soc.*, **60**, 1094 (1938).

established the first scale of activity pH of demonstrated utility and set in motion a gradual shift from the Sørensen scale (psH) to the semithermo-dynamic basis. Hence their work deserves detailed consideration.

Hitchcock and Taylor measured the e.m.f. at 25 °C of cells of type 1 (saturated calomel reference) containing acetate, phosphate, borate, and glycolate buffer solutions at several dilutions. In addition, the e.m.f. of cells of the type

$$Pt;H_2,Soln. \; X \,|\,KCl \; (satd.) \,|\,HCl(0.1 \; M),H_2;Pt \tag{9}$$

was measured at 38 °C. Several concentrations of acetates, phosphates, and borax constituted the solutions used in the half-cell on the left. MacInnes, Belcher, and Shedlovsky studied acetate buffer solutions at 12, 25, and 38 °C by means of measurements of the e.m.f. of cell 1 in which the 0.1 M calomel electrode was employed as a reference. These data were supplemented by a series of measurements with chloracetate buffer solutions at 25 °C.

For buffer solutions composed of a weak monobasic acid (HA) and its salt (MA), the thermodynamic dissociation constant K and the "incomplete" dissociation constant K' are given by

$$-\log K' \equiv -\log K + \log \frac{\gamma_{A^-}}{\gamma_{HA}} = pa_H - \log \frac{m_{A^-}}{m_{HA}} \tag{10}$$

The last term of this equation can be obtained from the composition of the buffer solution, with a small hydrolysis correction, where necessary. Hence values of $-\log K'$ are readily obtained from a measurement of the pH or approximate pa_H. The form of the activity coefficient term suggests that $\log K'$ will probably be a linear function of the square root of the ionic strength in dilute solutions. An extrapolation to zero ionic strength, with or without the aid of approximate activity coefficients computed by the Debye-Hückel equation, furnishes the thermodynamic dissociation constant K. The best value of the standard potential is presumably that which yields pa_H values consistent with the known value of the dissociation constant. Alternatively, the limit at zero ionic strength of the apparent standard potential $(E^{\circ \prime} + E_j)'$ defined by a combination of equations 2 and 10,

$$(E^{\circ \prime} + E_j)' \equiv (E^{\circ \prime} + E_j) + k \log \frac{\gamma_{A^-}}{\gamma_{HA}} = E + k \log K - k \log \frac{m_{A^-}}{m_{HA}} \tag{11}$$

where k is written for $(RT \ln 10)/F$, can be determined by a suitable extrapolation.

By these procedures a practical pH scale with a certain amount of thermodynamic significance was established. This scale meets the demands of modern pH instrumentation, as it avoids awkward corrections for liquid-junction potentials. In so far as the experimental pH values fall on the pa_H

scale defined by the standards, their meaning in terms of chemical equilibria is clearly and explicitly set forth by relationships such as equation 10.

Nevertheless, the pH cannot be proved to be consistent with log K at any finite concentration. This proof becomes possible only when a conventional scale of activity coefficients is defined, in terms of the Debye-Hückel equation for example, so that the value of K' becomes fixed at a finite ionic strength. Furthermore, the potential across the liquid junction undoubtedly changes as the ionic strength of the buffer solution decreases. Hence $E^{o'} + E_j$ is not constant. The slopes of the plots of log K' as a function of the square root of the ionic strength are found to be somewhat larger than the Debye-Hückel slope at low concentrations.[17]

Contrary to a rather common belief, the potential at the liquid boundary between buffer solutions and a saturated solution of potassium chloride does not approach zero as the buffer is diluted, but actually increases in the very dilute range, as the data in Table 3–2 demonstrate.

Figure 4–1 illustrates the variation with ionic strength of an "apparent" standard potential for the cell H_2;Buffer soln. | Saturated calomel reference.[18] The plots are so constructed as to be straight lines (but not necessarily of zero slope) for a constant value of $E^{o'} + E_j$, and the absolute values of the plotted potentials are of no significance except at zero ionic strength. However, a departure from linearity or a change in slope denotes a changing standard potential. The dotted lines are drawn through points for acetate solutions (upper line), glycolate solutions (middle line), and chloroacetate solutions (lower line). The data were taken from the papers of Hitchcock (HT), MacInnes (MBS), and their associates, and from Guggenheim and Schindler (GS)[19] and Larsson and Adell (LA).[20]

When the ionic strength of the buffer solution is less than 0.01, an alteration in the slope of the line becomes apparent. This change probably signifies a variation of the liquid-junction potential as the solute on one side of the boundary becomes attenuated.

Inasmuch as the ionic strengths of the buffers studied by Hitchcock and Taylor exceeded 0.01, extrapolation of the data gave $E^{o'} + E_j = 0.2441$ V at 25 °C. On the other hand, the acetate and chloroacetate buffer solutions on which the standard potential chosen by MacInnes, Belcher, and Shedlovsky was based fell in the range below $I = 0.01$, and hence these authors obtained

[17] D. A. MacInnes, D. Belcher, and T. Shedlovsky, *J. Amer. Chem. Soc.* **60**, 1094 (1938).

[18] R. G. Bates, *Chem. Rev.*, **42**, 1 (1948). I. T. Oiwa, *Sci. Rpt. Tohoku Univ., First Ser.*, **41**, 129 (1957), has obtained similar results in measurements of cells containing solutions of hydrochloric acid in methanol-water solvents.

[19] E. A. Guggenheim and T. D. Schindler, *J. Phys. Chem.*, **38**, 533 (1934).

[20] E. Larsson and B. Adell, *Z. Physik. Chem.*, **A156**, 352, 381 (1931); **A157**, 342 (1931).

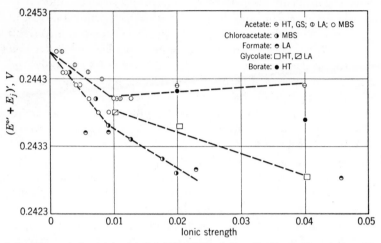

FIG. 4–1. Apparent standard potential $(E^{\circ\prime} + E_j)'$ of the cell: H_2; Buffer soln.|KCl (satd.), calomel as a function of ionic strength.

a value of 0.2446 V for $E^{\circ\prime} + E_j$. This difference in standard potential causes a difference of 0.008 unit in the pa_H values based on the two scales.

Table 4–1 summarizes the values of $E^{\circ\prime}$ (saturated calomel electrode) and of $E^{\circ\prime} + E_j$ for cell 1 (saturated and 0.1 M calomel reference electrodes) at 25 °C that have been proposed in order to define conventional scales of activity pH (pa_H). The values of $E^{\circ\prime} + E_j$ for the cell with 0.1 M calomel electrode (fourth column) are higher by 0.0912 V (the e.m.f. of cell 7 at 25 °C) than those for the saturated electrode.

pa_H of Reference Solutions

There are three primary reasons why reference solutions of known pH are preferred to the use of a fixed standard potential. First, saturated calomel reference electrodes are not highly reproducible, and this is particularly true of the small immersion-type electrodes; second, the potentials of commercial glass electrodes vary rather widely, and the asymmetry potentials may fluctuate from day to day; and third, the pH meter is usually calibrated to read directly in pH units. The selection of a value for the standard potential $E^{\circ\prime} + E_j$ permits the pa_H of useful standard reference solutions to be calculated by equation 2 from a measurement of e.m.f. With these solutions of known pa_H, the standard potential of the practical cell is, in effect redetermined each time a pH measurement is made. The value of the standard

TABLE 4–1. STANDARD POTENTIALS OF CELL 1 AT 25 °C. pa$_H$ OF THE MIXTURE HCl(0.01 M), KCl(0.09 M)

	Year	$E^{\circ\prime}$ (0.1 M) V	$E^{\circ\prime} + E_J$ (0.1 M)	$E^{\circ\prime} + E_J$ (satd.)	pa$_H$
Sørensen and Linderstrøm-Lang	1924	0.3353			
Scatchard	1925		0.3364	0.2452	2.061
Sørensen, Linderstrøm-Lang, and Lund	1927		0.3353	0.2441	2.079
Kolthoff and Bosch	1927				2.075
Cohn, Heyroth, and Menkin	1928		0.3361	0.2449	2.066
Bjerrum and Unmack	1929	0.3360	0.3364	0.2452	2.061
Guggenheim and Schindler	1934	0.3337	0.3345	0.2433	2.093
Hamer	1937	0.3337			
Strong acids			0.3369	0.2457	2.052
Buffer solutions			0.3346	0.2434	
Kauko and Airola	1937		0.3350	0.2438	2.085
Hitchcock and Taylor	1937				
Hydrochloric acid			0.3362	0.2450	2.064
Buffer solutions			0.3353	0.2441	
MacInnes, Belcher, and Shedlovsky	1938		0.3358	0.2446	2.071
Manov, DeLollis, and Acree	1945				
Strong acids			0.3358	0.2446	2.071
Buffer solutions			0.3346	0.2434	
Calcium hydroxide			0.3336	0.2424	

potential is then unimportant, and it need only remain constant for the time interval of the pH measurement.

The mixture of hydrochloric acid (0.01 M) and potassium chloride (0.09 M), often referred to as " Veibel's solution," [21] has found wide use in the standardization of pH assemblies, although it cannot be recommended for measurements above pH 3. When this mixture is placed in cell 1 (0.1 M calomel reference), an e.m.f. of 0.4583 V is measured. From this value of E, the pa$_H$ of the acid–salt mixture on the several scales discussed in the foregoing section has been computed and is given in the last column of Table 4–1.

---

[21] S. Veibel, *J. Chem. Soc.*, **123**, 2203 (1923). In a private communication, Professor Veibel states that credit for suggesting this reference solution should be given to Niels Bjerrum.

In 1927, Kolthoff and Bosch[22] assigned a pa_H value of 2.075 to this standard acid mixture, providing an early practical standardization of the activity scale of pH proposed by Sørensen and Linderstrøm-Lang[23] three years previously. This value was based on a comparison of the acidity of the acid-salt mixture with that of a 0.001 M solution of hydrochloric acid in a cell with liquid junction. The pa_H of the dilute acid solution was calculated by means of the reasonable assumption that the activity coefficient of hydrogen ion is equal to the mean activity coefficient of the acid at this low concentration.

Kauko and Airola[24] proposed a 0.0001 M solution of sodium bicarbonate saturated with pure carbon dioxide as a pH standard. The ionic strength of this buffer solution is so low that the activity coefficients can be computed accurately by the Debye–Hückel formula. The pa_H of this reference solution was found to be 4.038 at 0 °C, 4.181 at 25 °C, and 4.237 at 38 °C.

As a useful standard for pH near neutrality, Grove-Rasmussen[25] has recommended piperazine phosphate ($C_4H_{12}N_2HPO_4 \cdot H_2O$). The buffer capacity of the solution saturated at room temperature (0.065 M) is twice that of the 0.05 M solution of potassium hydrogen phthalate, and the change of pH on dilution with an equal volume of water is almost imperceptible. Hetzer, Robinson, and Bates[26] confirmed the favorable properties of piperazine phosphate solutions but did not recommend this material as a pH standard. Their values of pa_H (NBS scale) at molalities (m) of 0.02 and 0.05 mol kg^{-1} are as follows:

t, °C	$m = 0.02$	$m = 0.05$
0	6.582	6.591
25	6.285	6.295
50	6.056	6.065

Salts of a weak acid and a weak base are attractive as pH standards because of the small change in pH that occurs when their solutions are diluted. The acid salts of dicarboxylic acids with "overlapping" dissociation

[22] I. M. Kolthoff and W. Bosch, Rec. Trav. Chim., 46, 430 (1927); I. M. Kolthoff, ibid., 49, 401 (1930).

[23] S. P. L. Sørensen and K. Linderstrøm-Lang, Compt. Rend. Trav. Lab. Carlsberg, 15, No. 6 (1924).

[24] Y. Kauko and A. Airola, Z. Physik. Chem., A178, 437 (1937); A. Airola, Suomen Kemistilehti, 10A, 19, 29 (1937).

[25] K. V. Grove-Rasmussen, Acta Chem. Scand., 7, 231 (1953).

[26] H. B. Hetzer, R. A. Robinson, and R. G. Bates, Anal. Chem., 40, 634 (1968).

steps (for example, potassium dihydrogen citrate) are also logical choices. The labor of preparing solutions of known concentrations can be avoided by the selection of saturated solutions as secondary standards. For this purpose, Schaal and Souchay[27] have suggested that saturated solutions of the following three materials: potassium hydrogen tartrate, borax, and lithium carbonate be employed as pH standards.

The pa_H values at 25 and 38 °C of the standard reference solutions proposed by Hitchcock and Taylor[8] are summarized in Table 4–2. The pa_H standards of MacInnes, Belcher, and Shedlovsky[17] at 12, 25, and 38 °C are listed in Table 4–3.

The standard pa_H scale of Valensi and Maronny,[28] based on the concept of "compensated activity coefficients," identifies γ_{Cl} in the quantity $p(a_H \gamma_{Cl})$ with the mean activity coefficient of a standard reference electrolyte, potassium

TABLE 4–2. pa_H OF THE HITCHCOCK AND TAYLOR STANDARD SOLUTIONS

Composition	25 °C	28 °C
0.1 M HCl	1.085	1.082
0.01 M HCl, 0.09 M KCl	2.075	2.075
0.1 M K tetroxalate	1.480	1.495
0.05 M KH phthalate	4.010	4.025
0.1 M acetic acid, 0.1 M Na acetate	4.645	4.655
0.025 M KH_2PO_4, 0.025 M Na_2HPO_4	6.855	6.835
0.05 M $Na_2B_4O_7.10H_2O$	9.180	9.070

TABLE 4–3. pa_H OF THE MACINNES, BELCHER, AND SHEDLOVSKY STANDARD SOLUTIONS

Composition	12 °C	25 °C	38 °C
0.05 M KH phthalate	4.000	4.000	4.015
0.1 M acetic acid, 0.1 M Na acetate	4.650	4.640	4.635
0.01 M acetic acid, 0.01 M Na acetate	4.710	4.700	4.710

[27] R. Schaal and P. Souchay, *Mikrochim. Acta*, **3–4**, 371 (1954).
[28] G. Valensi, *J. Chim. Phys.*, **49**, C84 (1952); G. Maronny and G. Valensi, *ibid.*, **49**, C91 (1952); G. Valensi, *Corrosion Anti-corrosion*, **7**, 185, 253 (1959).

chloride. The latter is determined in each standard solution by measurement of cells with potassium amalgam and silver-silver chloride electrodes; consequently, sodium ions must be absent.

The standards contain an excess of potassium chloride, in general, and Valensi claims that the scale retains its validity up to ionic strengths of 1.0 or more. To derive the full benefit in practical pH measurements, the authors recommend that a conductivity measurement of each "unknown" be made and that a standard of matching conductivity be selected. This procedure, they believe, is effective in assuring that the residual liquid-junction potential is small. The pa_H values at 25 °C of one series of phosphate standards established by Maronny[29] are given in Table 4-4. It is interesting to note that the

TABLE 4-4.   PHOSPHATE-KCl pa_H STANDARDS OF MARONNY

KH$_2$PO$_4$ (0.025m), K$_2$HPO$_4$ (0.025m), KCl (x molal)	
x	pa_H at 25°
0.0500	6.811
0.1006	6.771
0.2017	6.730
0.3034	6.704
0.4057	6.684
0.5086	6.670
0.7166	6.647
1.0331	6.627

pa_H calculated by the Bates–Guggenheim convention (above $I = 0.1$, its intended limit) differs from that given in Table 4-4 by only 0.005 unit for the most dilute mixture and by only 0.013 unit for the most concentrated. Although this agreement can only be fortuitous, it demonstrates the essential congruence of two different approaches, one simple and the other elaborate, to the definition of the single ion activity.

Another means of basing standard pa_H values entirely on measured activities has been proposed by Van Rysselberghe.[30] According to this suggestion, known molalities of the cation j and the anion i would be added to the buffer solution and the quantities $a_H \gamma_i$ and a_H/γ_j determined by e.m.f. measurements of suitable cells without liquid junction. The geometric mean of these quantities, namely $a_H(\gamma_i/\gamma_j)^{1/2}$, should be approximately equal to a_H

[29] G. Maronny, *Electrochim. Acta*, **2**, 326 (1960).
[30] P. Van Rysselberghe, *Electrochim. Acta*, **2**, 337 (1960); *J. Electroanal. Chem.*, **4**, 314 (1962).

in dilute solutions. Van Rysselberghe suggested that measurements be made with a number of ji pairs, if possible, and that the geometric mean of these be taken as the true a_H in the buffer solution.

The Bates–Guggenheim convention (see page 75) has been adopted as the basis for the National Bureau of Standards method of assigning pa_H values to solutions of ionic strength not in excess of 0.1. The pa_H of certain reference solutions on the NBS scale is compared in Table 4–5 with the corresponding values assigned by Hitchcock and Taylor[8] and by MacInnes, Belcher, and Shedlovsky.[17] Inasmuch as a liquid-junction potential is involved in the data given in the second and third columns but not in the measurements on which the NBS scale is based, the differences are expected to be greatest at the ends of the pH scale. In general, the agreement is quite acceptable.

TABLE 4–5. COMPARISON OF THREE SCALES OF pa_H VALUES FOR REFERENCE SOLUTIONS AT 25 °C

Solution	Hitchcock and Taylor	MacInnes, Belcher, and Shedlovsky	NBS Scale
0.1 M HCl	1.085	—	1.088
0.01 M HCl, 0.09 M KCl	2.075	—	2.098
0.1 M K tetroxalate	1.480	—	1.518
0.03 M KH tartrate	3.567	—	3.569
0.1 M KH₂ citrate	3.719	—	3.713
0.05 M KH phthalate	4.010	4.000	4.008
0.1 M acetic acid, 0.1 M Na acetate	4.645	4.640	4.652
0.01 M acetic acid, 0.01 M Na acetate	4.714	4.700	4.718
0.025 M KH₂PO₄, 0.025 M Na₂HPO₄	6.855	—	6.865
0.01 M KH₂PO₄, 0.01 M Na₂HPO₄	6.963	—	6.961
0.05 M Na₂B₄O₇ · 10H₂O	9.180	—	9.196

Assignment of pH(S) Values to NBS Standards

In practice, the pH of unknown solutions is usually determined with an electrometric pH meter or with the aid of indicators whose color changes have been correlated with electrometric pH values. The practical importance of the electrometric procedure, together with the fact that the pH value lacks unique fundamental definition, have led to the adoption of an operational pH scale (equation 3). Accordingly, pH is defined in terms of the pH(S) of one or more selected standard solutions together with the observed *change* of e.m.f. when the standard is replaced by the unknown in a suitable cell.

In order to endow the measured pH with as much precise significance as possible in those relatively few instances where the residual liquid-junction potential is small, three measures have been adopted, the reasons for which should be apparent from the discussions in Chapters 1–3. The first is the selection of a conventional pa_H scale for the numerical values of pH(S), the second is the selection of reference solutions with ionic strengths less than 0.1, and the third is the restriction of primary standards to the pH range lying between the approximate limits 2.5 and 11.5, within which considerable uniformity of the liquid-junction potential and e.m.f./pH slope is to be expected.

Certified samples of buffer substances from which standard reference solutions of reproducible pH can be prepared are issued by the National Bureau of Standards. The assigned values of pH(S) for the seven primary standard solutions at temperatures from 0 to 95 °C are given in Table 4–6.[31] These are smoothed values calculated by the equation

$$pH(S) = \frac{A}{T} + B + CT + DT^2 \tag{12}$$

where T is the temperature in kelvins. A four-constant equation of this form was found to express satisfactorily the "observed" pa_H for these five solutions over the entire temperature range studied. The constants of equation 12 are summarized in Table 4–7, the last column of which gives the estimates of the standard deviation of a single value of pa_H.

The methods by which the NBS values of pH(S) were assigned will now be considered in detail. There are four steps in the assignment of the standard values. They are:

1. Determination of $p(a_H\gamma_{Cl})$ for three or more portions of the buffer solution with different small concentrations of added soluble chloride by measurement of the e.m.f. of hydrogen–silver chloride cells without liquid junction (cell 34 of Chapter 3). The $p(a_H\gamma_{Cl})$ is obtained from the measured e.m.f. E and known values of the standard e.m.f. $E°$ of the cell[32] by

$$p(a_H\gamma_{Cl}) = \frac{(E - E°)F}{RT \ln 10} + \log m_{Cl} \tag{13}$$

---

[31] R. G. Bates, *J. Res. Nat. Bur. Stand.*, **66A**, 179 (1962); B. R. Staples and R. G. Bates, *ibid.*, **73A**, 37 (1969).

[32] R. G. Bates and V. E. Bower, *J. Res. Nat. Bur. Stand.*, **53**, 283 (1954).

TABLE 4-6. pH(S) OF NBS PRIMARY STANDARDS FROM 0 TO 95 °C. [m = molality (mol kg^{-1})]

Temperature °C	KH tartrate (satd. at 25 °C)	KH$_2$ citrate ($m = 0.05$)	KH phthalate ($m = 0.05$)	KH$_2$PO$_4$ ($m = 0.025$), Na$_2$HPO$_4$ ($m = 0.025$)	KH$_2$PO$_4$ ($m = 0.008695$), Na$_2$HPO$_4$ ($m = 0.03043$)	Borax ($m = 0.01$)	NaHCO$_3$ ($m = 0.025$), Na$_2$CO$_3$ ($m = 0.025$)
0	—	3.863	4.003	6.984	7.534	9.464	10.317
5	—	3.840	3.999	6.951	7.500	9.395	10.245
10	—	3.820	3.998	6.923	7.472	9.332	10.179
15	—	3.802	3.999	6.900	7.448	9.276	10.118
20	—	3.788	4.002	6.881	7.429	9.225	10.062
25	3.557	3.776	4.008	6.865	7.413	9.180	10.012
30	3.552	3.766	4.015	6.853	7.400	9.139	9.966
35	3.549	3.759	4.024	6.844	7.389	9.102	9.925
38	3.548	3.755	4.030	6.840	7.384	9.081	9.903
40	3.547	3.753	4.035	6.838	7.380	9.068	9.889
45	3.547	3.750	4.047	6.834	7.373	9.038	9.856
50	3.549	3.749	4.060	6.833	7.367	9.011	9.828
55	3.554	—	4.075	6.834	—	8.985	—
60	3.560	—	4.091	6.836	—	8.962	—
70	3.580	—	4.126	6.845	—	8.921	—
80	3.609	—	4.164	6.859	—	8.885	—
90	3.650	—	4.205	6.877	—	8.850	—
95	3.674	—	4.227	6.886	—	8.833	—

TABLE 4-7. CONSTANTS OF EQUATION 12 FOR SEVEN pH STANDARD SOLUTIONS
m = molality (mol kg^{-1})

Solution	Temperature Range, °C	A	B	C	$10^5 D$	Standard Deviation
KH tartrate (satd. at 25 °C)	25—95	−1727.96	23.7406	−0.075947	9.2873	0.0016
KH$_2$ citrate (m = 0.05)	0—50	1280.4	−4.1650	0.012230	0	0.0010
KH phthalate (m = 0.05)	0—95	1678.30	−9.8357	0.034946	−2.4804	0.0027
KH$_2$PO$_4$ (m = 0.025), Na$_2$HPO$_4$ (m = 0.025)	0—95	3459.39	−21.0574	0.073301	−6.2266	0.0017
KH$_2$PO$_4$ (m = 0.008695), Na$_2$HPO$_4$ (m = 0.03043)	0—50	5706.61	−43.9428	0.154785	−15.6745	0.0011
Na$_2$B$_4$O$_7$ (m = 0.01)	0—95	5259.02	−33.1064	0.114826	−10.7860	0.0025
NaHCO$_3$ (m = 0.025), Na$_2$CO$_3$ (m = 0.025)	0—50	2557.1	−4.2846	0.019185	0	0.0026

2. Evaluation of $p(a_H\gamma_{Cl})^\circ$, the limit approached by $p(a_H\gamma_{Cl})$ as the concentration of added chloride in the buffer solution approaches zero. (Values of $p(a_H\gamma_{Cl})$ and $p(a_H\gamma_{Cl})^\circ$ from 0 to 60 °C for a number of solutions are collected in Table 7 of the Appendix.)
3. Computation of pa_H from $p(a_H\gamma_{Cl})^\circ$ by introduction of a conventional individual ionic activity coefficient:

$$pa_H = p(a_H\gamma_{Cl})^\circ + \log \gamma_{Cl} \tag{14}$$

The Bates–Guggenheim convention[33] is used to evaluate $\log \gamma_{Cl}$:

$$\log \gamma_{Cl} = - \frac{AI^{1/2}}{1 + 1.5I^{1/2}} \tag{15}$$

where A is the Debye–Hückel slope (see Appendix, Table 4).
4. Identification of the pa_H of certain selected reference solutions with pH(S). The primary standards have pH values in the range 2.5 to 11.5 and are chosen for their reproducibility, stability, buffer capacity, and ease of preparation. For these few solutions,

$$pH(S) \equiv pa_H \tag{16}$$

It is convenient for the moment to pass over the experimental aspects of e.m.f. measurements and to examine the second and third steps as they apply to buffer solutions containing equal molal amounts of potassium dihydrogen phosphate and disodium hydrogen phosphate.[34]

Equimolal Phosphate Buffer Solutions

DETERMINATION OF $p(a_H\gamma_{Cl})^\circ$

The compositions of the buffer-chloride mixtures are so chosen as to provide data for an easy and unambiguous extrapolation to zero concentration of the chloride. In the light of the Brønsted theory of specific ionic interaction, plots of $p(a_H\gamma_{Cl})$ as a function of composition of the mixture *at a constant total ionic strength* might be expected to be approximately linear, and this was indeed found to be the case for phosphate–chloride mixtures. However, the change of $p(a_H\gamma_{Cl})$ on addition of chloride *at a constant buffer concentration* is usually found to be small, particularly if the buffer concentration exceeds that of the added chloride. Hence $p(a_H\gamma_{Cl})^\circ$ can usually be determined in a simple manner by plotting $p(a_H\gamma_{Cl})$ for each buffer solution

---

[33] R. G. Bates and E. A. Guggenheim, *Pure Appl. Chem.*, **1**, 163 (1960). This convention is intended to apply at ionic strengths (I) not exceeding 0.1.
[34] R. G. Bates and S. F. Acree, *J. Res. Nat. Bur. Stand.*, **34**, 373 (1945); R. G. Bates, *ibid.*, **39**, 411 (1947); 60 to 95 °C: V. E. Bower and R. G. Bates, *ibid.*, **59**, 261 (1957).

as a function of the molality of added chloride and extending the straight lines to $m_{Cl} = 0.$[35]

The two types of extrapolation plot for the 1 : 1 phosphate buffer solution (KH_2PO_4 and Na_2HPO_4, each at a molality of 0.025 mol kg^{-1}) are illustrated in Fig. 4–2. The two lines furnish substantially the same value of $p(a_H\gamma_{Cl})^\circ$. In most instances, the extrapolation at constant buffer concentration appears to be the simpler and the more precise.

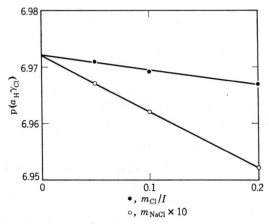

FIG. 4–2 Determination of $p(a_H\gamma_{Cl})^\circ$ for the 0.025m phosphate buffer solution. Upper line: extrapolation at constant ionic strength. Lower line: extrapolation at constant buffer concentration.

CALCULATION OF pa_H

Before the adoption of a single convention, namely equation 15, as a basis for the assignment of the pa_H values that were to serve as the pH(S) in the operational definition of the practical pH, considerable attention was devoted to procedures for estimating "reasonable" activity coefficients for chloride ion in the respective buffer mixtures. These studies are of continuing interest in so far as they demonstrate the practical effect of the choice of a convention on the pa_H obtained.

The hydrogen-silver chloride cell is highly reproducible. Furthermore, it has been applied extensively in the study of the dissociation of weak acids and bases, and its standard e.m.f. is known over a wider range of temperatures than that of any other comparable cell. It was logical therefore, that this

[35] W. J. Hamer and S. F. Acree, *J. Res. Nat. Bur. Stand.*, **32**, 215 (1944).

cell should be used to establish standard reference values of pH. One naturally wonders whether the same values of pa_H would be obtained if another cell were used, necessitating the estimation of the activity coefficient of bromide or iodide ion, for example, instead of that of chloride ion. The author undertook to provide an answer to this question by means of studies of the e.m.f. of hydrogen-silver halide cells containing equimolal phosphate buffer solutions with added sodium halide, NaX, where X was chloride, bromide, or iodide.[34]

The $p(a_H\gamma_X)$ value, defined in a general manner by

$$p(a_H\gamma_X) \equiv -\log(m_H\gamma_H\gamma_X) \tag{17}$$

was calculated from the e.m.f. of the three types of cell by equation 13, where E° is the standard potential of the particular cell used and m_{Cl} is replaced by m_X. The extrapolation of $p(a_H\gamma_X)$ to zero halide at five values of the ionic strength is shown in Fig. 4–3. The $p(a_H\gamma_X)$ at constant ionic strength is seen to be a linear function of the fractional contribution m_X/I of sodium

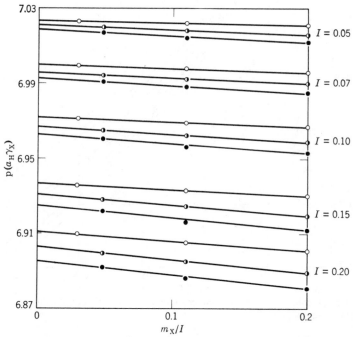

FIG. 4–3 Determination of $p(a_{HX}\gamma)^\circ$ for phosphate buffer solutions. Type of cell: ○ chloride ◑ bromide, ● iodide.

halide to the total ionic strength. The e.m.f. data for the chloride cells furnished the upper line of each group, the bromide data the middle line, and the iodide data the lower line.

When pa_H for the halide-free phosphate solutions at 25 °C was computed from the three sets of $p(a_H \gamma_X)$ values with five different definitions of the individual activity coefficients, the twelve curves of Fig. 4–4 were obtained.[36] The dotted lines mark the course of the pa_H curve derived from the e.m.f. of the chloride cells with chloride ion activity coefficients defined by the Debye-Hückel equation in the form

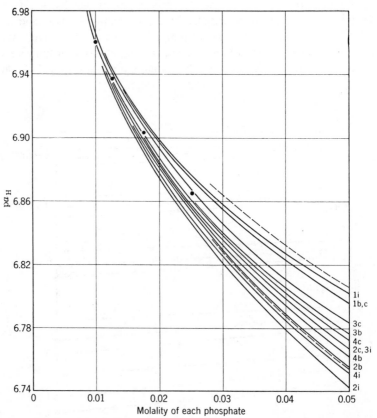

FIG. 4–4 pa_H if equimolal phosphate buffer solutions as a function of molality of each phosphate. Dots represent NBS values.

36 R. G. Bates, *Analyst*, **77**, 653 (1952).

$$-\log \gamma_i = \frac{A z_i^2 \sqrt{I}}{1 + B \mathring{a} \sqrt{I}} \qquad (18)$$

in which $\mathring{a}$ was assigned values of 8 Å (upper line) and 3 Å (lower line). The notation beside each of the other curves identifies both the convention on which the computation of the ionic activity coefficient was based and the type of cell (chloride, c; bromide, b; or iodide, i) with which $p(a_H \gamma_X)^0$ was obtained. When the ionic activity coefficient γ_X was taken equal to the mean activity coefficient γ_{HX} of the halogen acid at the same ionic strength, the curves labeled 1 resulted. If γ_X was computed from the measured activity coefficient term, $\gamma_X \gamma_{H_2PO_4^-}/\gamma_{HPO_4^-}$, on the assumption that $\gamma_X^4 = \gamma_{H_2PO_4^-}^4 = \gamma_{HPO_4^{2-}}$, curves 2b, 2c, and 2i were obtained. Two different modes of application of a salt-effect correction $b_i I$ to these pa_H values led to the curves marked 3 and 4.

The NBS pH(S) values for the equimolal phosphate buffer, represented in Fig. 4–4 by dots, were obtained from e.m.f. measurements of chloride cells and activity coefficients for chloride ion calculated by equation 15. The mixture that contains potassium dihydrogen phosphate and disodium hydrogen phosphate, both at a molality of 0.025 mol kg^{-1}, is recommended for use as a standard reference solution.

Three important conclusions can be drawn from Fig. 4–4: first, that different conventions yield pa_H values that differ appreciably at ionic strengths above 0.1 ($m = 0.025$); second, that a change in the type of cell has no greater effect on the value of pa_H than a change in the convention; and third, that the results furnished by all three cells and all five conventions are in agreement within ± 0.01 pH unit below an ionic strength of 0.1.

This last observation led to the choice of reference solutions that are rather dilute and for which pa_H values could therefore be assigned that are consistent to ± 0.01 unit with several of the most reasonable conventional definitions of the individual activity coefficient. A *single* convention was later selected because (a) it is often desirable to evaluate pH changes with a precision greater than 0.01 unit, (b) the reproducibility of some modern pH instruments makes it possible to do this, and (c) under favorable circumstances the pa_H values assigned independently to different primary standards display an internal consistency better than 0.01 unit.

Phthalate Buffer Solutions

The 0.05m solution of potassium hydrogen o-phthalate is probably employed more widely for the standardization of pH assemblies than is any other reference solution. The NBS pH(S) values for solutions of potassium hydrogen phthalate from 0 to 60 °C are based on the work of Hamer, Pinching, and Acree,[37] who determined the first and second dissociation constants of phthalic acid by e.m.f. methods. In connection with the determination of the second constant, the quantity γ_r and the parameters $\mathring{a}^*$ and b^* were obtained (compare equation 38, Chapter 3). These data have been supplemented by later measurements from 60 to 95 °C.[38]

The value of $\mathring{a}^*$ was found to be 3.73 Å and to be uninfluenced by changes of composition, but b^* changed both with the buffer ratio and with the ratio of potassium chloride to phthalate. On the basis of certain assumptions regarding their mode of combination, the individual parameters $\mathring{a}$ and b_i for the chloride ion and the two phthalate anions were evaluated. For chloride ion in phthalate solutions, these two parameters were assigned the values 3.76 and 0.10 respectively, and the ionic activity coefficients could then be computed. The pa_H values of phthalate solutions were derived by the mass law from the thermodynamic dissociation constants, the compositions of the solutions, and the activity coefficients of the phthalate anions. The values of pH(S) given in Table 4–6, however, were derived from the $p(a_H\gamma_{Cl})°$ values for $0.05m$ potassium hydrogen phthalate,[37] with the convention set forth in equation 15. The difference between the pa_H found by the two methods is, on the average, only 0.004 unit at the 13 temperatures from 0 to 60 °C.

In Fig. 4–5 the pa_H at 25 °C of phthalate buffer solutions is plotted as a function of molality.[39] The dots represent the NBS values, obtained as described before, and the separation of the four curves illustrates the dependence of the pa_H on the chosen convention. Curves 1 and 4 resulted when γ_{Cl} in equation 14 was computed by the Debye-Hückel formula (equation 18) with extreme values of 8 and 3 Å, respectively, for $\mathring{a}$. The other two curves were obtained by assuming that the activity coefficient of chloride ion in solutions of potassium hydrogen phthalate is equal to the mean activity coefficient of hydrochloric acid (curve 2) or of potassium chloride (curve 3) in a solution of potassium chloride alone of the same ionic strength as the phthalate buffer.

Borax Buffer Solutions

The NBS pH(S) values from 0 to 60 °C for solutions of sodium tetraborate decahydrate (borax, $Na_2B_4O_7 \cdot 10H_2O$) were derived from the e.m.f. data of

---

[37] W. J. Hamer and S. F. Acree, *J. Res. Nat. Bur. Stand.*, **32**, 215 (1944); **35**, 381 (1945); W. J. Hamer, G. D. Pinching, and S. F. Acree, *ibid.*, **35**, 539 (1945); **36**, 47 (1946).

[38] V. E. Bower and R. G. Bates, *J. Res. Nat. Bur. Stand.*, **59**, 261 (1957).

[39] R. G. Bates, *Analyst*, **77**, 653 (1952).

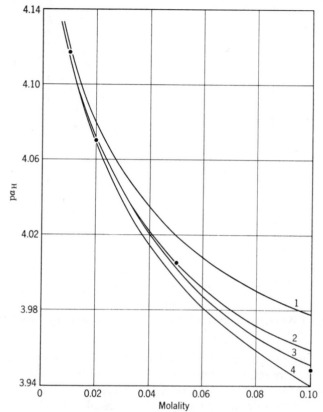

FIG 4–5 pa_H of phthalate buffer solutions as a function of molality. Dots represent NBS values.

Manov, DeLollis, Lindvall, and Acree[40] for aqueous mixtures of borax and sodium chloride. The activity coefficient of chloride ion, needed to calculate pa_H by equation 14, was computed by equation 15.

The effect of various conventional definitions of the individual activity coefficient on the pa_H obtained from these e.m.f. measurements is demonstrated in Fig. 4–6. The dot represents the pa_H of the $0.01m$ solution calculated as described before, whereas curves 1 and 2 locate the pa_H based on values of γ_{Cl} defined by equation 18 with $\mathring{a}$ parameters of 8 and 3 Å respectively.

[40] G. G. Manov, N. J. DeLollis, P. W. Lindvall, and S. F. Acree, *J. Res. Nat. Bur. Stand.*, **36**, 543 (1946). The measurements were extended to 95 °C by V. E. Bower and R. G. Bates, *ibid.*, **59**, 261 (1957).

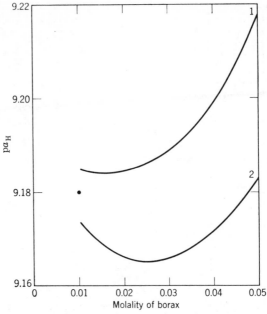

FIG 4–6. pa_H of borax solutions as a function of molality. Dot represents NBS value.

As the figure shows, the pH of solutions of borax is almost independent of changes in concentration between molalities of 0.01 and 0.05 mol kg^{-1}. The unusual shape of the pa_H-concentration curves doubtless reflects the tendency of metaboric acid to polymerize,[41] causing the buffer ratio to depart from unity. Consequently, the pa_H computed by a simple application of the mass law to the dissociation of metaboric acid, ignoring other equilibria, is in error. However, the value derived from equation 14 presumably is approximately correct, for the secondary equilibria influence this computation only through a change in the ionic strength; furthermore, little polymerization is to be expected in borax solutions of low concentration.

Tartrate Buffer Solutions

The pa_H values of solutions of potassium hydrogen tartrate from 0 to 95 °C were obtained from e.m.f. measurements of aqueous mixtures of the tartrate and sodium chloride by the procedures outlined in earlier sections.[42]

---

[41] See, for example, I. M. Kolthoff and W. Bosch, *Rec. Trav. Chim.*, **46**, 180 (1927).

[42] R. G. Bates, V. E. Bower, R. G. Miller, and E. R. Smith, *J. Res. Nat. Bur. Stand.*, **47**, 433 (1951); V. E. Bower and R. G. Bates, *ibid.*, **59**, 261 (1957).

The pa_H at 25 °C is plotted as a function of molality in Fig. 4–7. The activity coefficient of chloride ion with which the three curves were obtained was defined by equation 18. An $\mathring{a}$ parameter of 6 Å gave the upper line, 4 the middle line, and 2 the lower line. If the ionic activity coefficient were identified with the mean activity coefficient of hydrochloric acid in a solution of potassium chloride of the same ionic strength as the tartrate solution, the pa_H-molality line would lie between the upper and middle curves of Fig. 4–7

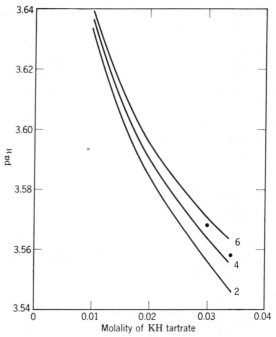

FIG 4–7. pa_H of potassium hydrogen tartrate solutions as a function of molality, based on three different conventions. The number beside each curve denotes the value of $\mathring{a}$ in Å (equation 18) used to compute the activity coefficient of chloride ion. Dots represent NBS values.

(compare Table 3–4), close to the NBS values based on the convention of equation 15.

In spite of a pronounced tendency to molding,[43] saturated solutions of potassium hydrogen tartrate are often convenient and useful standards.

---

[43] R. G. Bates, *Anal. Chem.*, **23**, 813 (1951).

The solution saturated at 25 °C is about 0.034 M and has an assigned pH(S) of 3.557 at 25 °C, in good agreement with pH 3.57 found by Lingane[44] and with 3.57 assigned to the 0.03 M solution by Hitchcock and Taylor.[8]

Citrate and Carbonate Buffer Solutions

Two new primary standards, supplementing the original five, were established by Staples and Bates.[45] These are a solution of potassium dihydrogen citrate (molality = 0.05 mol kg^{-1}) and a mixture of sodium bicarbonate and sodium carbonate (each with a molality of 0.025 mol kg^{-1}). The carbonate solution extends the standard scale to higher alkalinities, while the citrate solution is more stable than the tartrate solution which it closely matches in pH.

The conventional pa_H values of the citrate and carbonate standards from 0 to 50 °C were obtained from e.m.f. measurements of the buffer solutions with added potassium chloride by the procedures outlined earlier and embodied in equations 13 to 16. The ionic strengths were calculated with the aid of the dissociation constants of citric and carbonic acids. The values at 0, 25, and 50 °C are as follows: citrate solution, 0.0527, 0.0526, and 0.0524; carbonate solution, 0.1, 0.0999, 0.0995.

Standards for pH Measurements of Blood

Blood is a well-buffered medium. In order to correlate blood pH measurements in a useful way with physiological processes and with pathological conditions, the discrimination of small changes of pH is often necessary. It is desirable not only that the results of different workers have a common basis of reference but also that this reference solution have a pH value close to the pH of blood itself.

Two different standard solutions, both consisting of a mixture of phosphate salts, have been proposed for this purpose. The standard established at the National Bureau of Standards[46] has an ionic strength of 0.1 and contains potassium dihydrogen phosphate and disodium hydrogen phosphate in the molal ratio 1 : 3.5. The values of pH(S) are given in the sixth column of Table 4–6.

Semple, Mattock, and Uncles[47] have described another reference solution containing the same salts in the molar ratio 1 : 4; the exact composition is

---

[44] J. J. Lingane, *Anal. Chem.*, **19**, 810 (1947).

[45] B. R. Staples and R. G. Bates, *J. Res. Nat. Bur. Stand.*, **73A**, 37 (1969).

[46] V. E. Bower, M. Paabo, and R. G. Bates, *J. Res. Nat. Bur. Stand.*, **65A**, 267 (1961); *Clin. Chem.*, **7**, 292 (1961).

[47] S. J .G. Semple, G. Mattock, and R. Uncles, *J. Biol. Chem.*, **237**, 963 (1962).

KH$_2$PO$_4$ (0.01 M), Na$_2$HPO$_4$ (0.04 M). This solution, which has an ionic strength of 0.13, is prepared by dissolving 1.360 g of KH$_2$PO$_4$ and 5.677 g of Na$_2$HPO$_4$ (air weights) in sufficient carbon dioxide-free water to make a liter of solution. It was found to be stable for two to four weeks when stored in polyethylene containers at room temperature. The pH, found by careful comparison with the NBS primary standards using both glass and hydrogen electrodes, is 7.416 ± 0.004 at 37.5 and 38 °C. For the blood standard proposed by the NBS, Semple, Mattock, and Uncles found 7.388 at 38 °C, in good agreement with 7.384 given in Table 4–6. Additional secondary standards for blood pH measurements are described on pages 89 and 90.

Internal Consistency of the Standard Scale

The seven NBS primary standards fix a standard scale which shows satisfactory internal consistency from pH 3.5 to 9.2. This means that the pH furnished by a cell of type 1 with properly designed liquid junction is nearly the same regardless of which of these seven solutions is chosen as a standard. The same is true if a glass electrode with perfect pH response (electromotive efficiency of 1.0) is substituted for the hydrogen electrode.

Some data for the intercomparison of the primary standards are summarized in Table 4–8.[48] Two types of liquid junction and three different concentrations of the potassium chloride bridge solution were used. One junction was small, area about 1.0 mm^2, and was formed within a capillary tube. The other was a broad junction, about 3.5 cm^2 in area formed just above a sintered-glass disk sealed into a glass tube. The potassium chloride solutions which constituted the bridge solutions and were part of the calomel reference electrodes were 3.0 M, 3.5 M, or saturated. No significant differences were observed among the data obtained with the different bridge solutions and junctions, although the capillary junction appeared to retain its stability for a longer period of time than did the junction of larger dimensions. At 25 °C the five standards of pH(S) 3.7 to 9.2 appear to be consistent among themselves to ±0.006 unit in the pH cell with liquid junction. However, an inconsistency slightly greater than 0.01 unit is apparent for the carbonate standard and for the borax standard at elevated temperatures.

The assignment of uncertainties to the primary standards of pH may now be considered. In the range 0 to 50 °C, it should easily be possible with refined techniques to determine $E - E°$ with an error less than ±0.2 mV. Thus p($a_H \gamma_{Cl}$) and pa_H (equations 13 and 14) should have an uncertainty of about ±0.003 unit. The accuracy of pH(S), however, involves the added question of internal consistency, that is, the departure of each individual pH(S) value

---

[48] M. Paabo and R. G. Bates, unpublished data; Staples and Bates (reference 45).

TABLE 4–8. INTERNAL CONSISTENCY OF THE NBS STANDARD pH SCALE FROM 10 TO 40 °C

Standard Solution	10 °C	20 °C	25° C	30 °C	38 °C	40 °C
1 : 1 Phosphate (reference values) pH(S)	6.923	6.881	6.865	6.853	6.840	6.838
Citrate						
pH(S)	3.820	—	3.776	—	—	3.750[a]
pH	3.832	—	3.782	—	—	3.757[a]
Phthalate						
pH(S)	3.998	4.002	4.008	4.015	4.030	4.035
pH	4.008	4.009	4.012 ± 0.002	4.019	4.034 ± 0.002	4.040
n[b]	8	7	20	10	11	7
1 : 3.5 Phosphate						
pH(S)	7.472	7.429	7.413	7.400	7.384	7.380
pH	7.473	7.430	7.414 ± 0.001	7.403	7.389 ± 0.001[c]	7.388
n	4	3	17	5	11	6
Borax						
pH(S)	9.332	9.225	9.180	9.139	9.081	9.068
pH	9.327	9.227	9.184 ± 0.001	9.146	9.092 ± 0.001	9.080
n	4	4	11	6	9	6
Carbonate						
pH(S)	10.179	—	10.012	—	—	9.856[a]
pH	10.166	—	10.001	—	—	9.848[a]

[a] At 45°C.
[b] Number of measurements.
[c] A value of 7.388 was found by Semple, Mattock, and Uncles.[47]

from the standard scale fixed by the primary standards as a group. The uncertainty of pH(S) appears then to be about ±0.006 at 25 °C and somewhat larger at other temperatures.

Secondary Standards

Although the standard scale represented by pH(S) possesses a satisfactory degree of internal consistency in the intermediate region of the pH scale, the Henderson equation[49] predicts, and experiment confirms,[50] that this

---

[49] See equations 16, 17, and 18 of Chapter 3.
[50] R. G. Bates, G. D. Pinching, and E. R. Smith, *J. Res. Nat. Bur. Stand.*, **45**, 418 (1950).

would not be the case at either end of the scale. As the result of a variable liquid-junction potential the measured pH may be expected to differ seriously from the pa_H determined from cells without a liquid junction in solutions of high acidity or high alkalinity. The deviations are not a function of pH alone, unfortunately, and accordingly cannot be corrected by a suitable standardization of the pH cell. For this reason, *primary* standardization of the pH cell is limited to the intermediate region of the pH scale.

To affirm the proper functioning of the glass electrode, additional *secondary* standards of pH outside the intermediate region are desirable. No glass electrode has the ideal pH response over the entire useful range of the pH scale. When glass electrodes are used, the safest procedure is to assume that the e.m.f. of the cell varies linearly with the pH but that the proportionality factor is not necessarily exactly $(RT \ln 10)/F$ volts per pH unit. Two standards S_1 and S_2 are therefore needed, and these should preferably bracket the pH of the test solution X. The pH of the latter is then determined by

$$\text{pH(X)} - \text{pH(S}_1) = \frac{E_X - E_{S(1)}}{E_{S(2)} - E_{S(1)}} [\text{pH(S}_2) - \text{pH(S}_1)] \tag{19}$$

Even within the intermediate pH range, secondary standards prove convenient because they are more easily prepared or more stable than the primary standards. A good example is 0.05 M borax, which, although more stable than the 0.01 M solution of borax, could not be chosen as a primary standard because of its complex composition and uncertain ionic strength. The pH of secondary standards is assigned by careful comparison with the primary standards by means of cell 1 or a similar cell with a glass electrode.

A secondary standard might likewise be chosen to match closely the composition and pH of a group of unknowns on which repeated measurements are made, especially where reproducibility is the prime consideration. An excellent example of the utility of this type of secondary standard is given by Tuddenham and Anderson.[51] These authors recommend a solution 0.2 M in calcium chloride, 2.0 M in sodium chloride, and saturated with calcium hydroxide for checking pH meters used for the analytical control of highly alkaline photographic processing solutions. They assign a pH of 11.82 at 25 °C to this reference solution. It goes without saying that the pH values of concentrated alkaline solutions based on a reference standard of this sort have no precise meaning in terms of the conventional pa_H, but none is needed.

Secondary standards of pa_H outside the intermediate region of the pH scale are useful under certain circumstances. There is evidence, for example, that errors due to the residual liquid-junction potential in dilute hydrochloric

---

[51] W. M. Tuddenham and D. H. Anderson, *Color Control Dept. Report A-49*, Eastman Kodak Co., Rochester, N.Y., 1949.

acid solutions and in solutions of potassium tetroxalate of about the same pH are of the same sign and of approximately the same magnitude. Similarly, the error in dilute solutions of sodium hydroxide appears to match fairly well the error in a solution of calcium hydroxide of about the same pH. When the pa_H values of solutions of hydrochloric acid or sodium hydroxide are needed, therefore, it is sometimes useful to adjust the pH instrument to read correctly the known pa_H of a tetroxalate solution or a calcium hydroxide solution. This procedure should only be used with a full recognition of its limitations. The values obtained are not correctly called pH, as this term should be reserved for the numbers furnished by cell 1 (or its equivalent), adjusted to read correctly the pH(S) of the primary standards, in accordance with the operational definition, equation 3.

The pa_H of the solution of potassium tetroxalate, $KH_3(C_2O_4)_2 \cdot 2H_2O$, molality $= 0.05$ mol kg^{-1}, and the solution of calcium hydroxide saturated at 25 °C are summarized in Table 4–9. These values were derived in the manner described earlier in this chapter from e.m.f. measurements of cells without

TABLE 4–9. pa_H of Potassium Tetroxalate (Molality $=$ 0.05 mol kg^{-1}) and the Saturated Solution of Calcium Hydroxide

t, °C	0.05m $KH_3(C_2O_4)_2 \cdot 2H_2O$	Ca(OH)$_2$ Satd. at 25 °C
0	1.666	13.423
5	1.668	13.207
10	1.670	13.003
15	1.672	12.810
20	1.675	12.627
25	1.679	12.454
30	1.683	12.289
35	1.688	12.133
38	1.691	12.043
40	1.694	11.984
45	1.700	11.841
50	1.707	11.705
55	1.715	11.574
60	1.723	11.449
70	1.743	—
80	1.766	—
90	1.792	—
95	1.806	—

liquid junction.[52] The change of pa_H with absolute temperature T is expressed by the equations

0.05m *Potassium tetroxalate*:

$$pa_H = -\frac{362.76}{T} + 6.1765 - 0.018710T + 2.5847 \times 10^5 T^2 \qquad (20)$$

Calcium hydroxide, saturated at 25 °C:

$$pa_H = \frac{7613.65}{T} - 38.5892 + 0.119217T - 11.2918 \times 10^5 T^2 \qquad (21)$$

The standard deviations of "observed" values from these equations are 0.0019 (tetroxalate) and 0.0028 (calcium hydroxide).

The pa_H of solutions of hydrochloric acid is only slightly affected by changes of temperature. The 0.1m solution has a pa_H of 1.09 (see Table 3–5) from 0 to 50 °C. It has been suggested that this solution might serve as the basis for an operational extension of the pH scale up to 275 °C.[53] The pH of 0.1m HCl, defined by $-\log(0.1\gamma_\pm)$ is 1.2 at 275 °C.

Measurements made at the National Bureau of Standards have provided conventional pa_H values for buffer solutions composed of tris(hydroxymethyl)-aminomethane ("tris") and tris hydrochloride. These values are listed in Table 4–10. These buffers should be useful for pH control in the physiological range, where phosphates cause undesirable side reactions. They should probably be regarded as secondary standards for pH until consistency with the primary standard scale has been demonstrated. Tris and its hydrochloride are supplied by the NBS as standard reference materials.

A tris buffer for pH measurements in media of ionic strength 0.16 is based on e.m.f. measurements made by Durst and Staples.[54] This medium approximates "isotonic saline" in ionic strength, and sodium chloride is the predominant electrolyte in both. The buffer solutions studied had the compositions: tris·HCl, molality m; tris (molality m or $m/3$); NaCl $(0.16 - m)$. Conventional pa_H values were derived from the $p(a_H \gamma_{Cl})$ given by equation 13 by extending the Bates-Guggenheim pH convention (equation

---

[52] R. G. Bates, *J. Res. Nat. Bur. Stand.*, **66A**, 179 (1962). Sources of the primary data: tetroxalate, 0 to 60 °C, V. E. Bower, R. G. Bates, and E. R. Smith, *ibid.*, **51**, 189 (1953); tetroxalate, 60 to 95 °C, V. E. Bower and R. G. Bates, *ibid.*, **59**, 261 (1957); calcium hydroxide, R. G. Bates, V. E. Bower, and E. R. Smith, *ibid.*, **56**, 305 (1956).

[53] R. S. Greeley, *Anal. Chem.*, **32**, 1717 (1960).

[54] R. A. Durst and B. R. Staples, *Clin. Chem.*, **18**, 206 (1972). Tris buffers are compatible with synthetic sea waters and make useful secondary standards in these media; see W. H. Smith, Jr., and D. W. Hood, in *Recent Researches in the Fields of Hydrosphere, Atmosphere, and Nuclear Geochemistry*, Maruzen Co. Ltd., Tokyo, 1964.

TABLE 4–10. pa_H OF BUFFER SOLUTIONS COMPOSED OF TRIS-
(HYDROXYMETHYL) AMINOMETHANE ("TRIS") AND ITS HYDRO-
CHLORIDE, WITH AND WITHOUT ADDED SODIUM CHLORIDE
(m = molality)

t, °C	Tris ($m = 0.05$), tris · HCl ($m = 0.05$)[a]	Tris ($m = 0.01667$), tris · HCl ($m = 0.05$)[b]
0	8.946	8.471
5	8.774	8.303
10	8.614	8.142
15	8.461	7.988
20	8.313	7.840
25	8.173	7.699
30	8.036	7.563
35	7.904	7.433
37	7.851	7.382
40	7.777	7.307
45	7.654	7.186
50	7.537	7.070

| | Molality | | pa_H[b] | |
Tris	Tris · HCl	NaCl	25 °C	37 °C
0.02	0.02	0.14	8.220	7.904
0.05	0.05	0.11	8.225	7.908
0.006667	0.02	0.14	7.745	7.428
0.01667	0.05	0.11	7.745	7.427

[a] Calculated from data given by R. G. Bates and H. B.
Hetzer, *J. Phys. Chem.*, **65**, 667 (1961).
[b] Durst and Staples[54]

15) to $I = 0.16$. Their values, which should be useful as secondary standards
for pH in isotonic saline and in many biological media, are included in
Table 4–10.

Properties of the NBS pH Standards

The primary standard pH scale is defined in terms of the conventional
pa_H of seven solutions. The pH(S) values assigned to these reference solutions
are listed in Table 4–6. These solutions were selected to give a satisfactory stability with respect to changes of concentration and accidental contamination

and to have an ionic strength not in excess of 0.1. The seven solutions cover adequately the intermediate region of the pH scale, where the residual liquid-junction potential may be expected to be small. Because of the tendency of boric acid to polymerize at higher concentrations, presumably altering the ionic strength and the pH in a complex manner, a rather dilute solution of borax was chosen.

In addition to the seven primary standards for the operational pH scale, two supplementary standards of pa_H have been established. One of these, $0.05m$ potassium tetroxalate, is highly acidic ($pa_H = 1.68$ at 25 °C), and the other, a solution of calcium hydroxide saturated at 25 °C, is highly alkaline ($pa_H = 12.45$ at 25 °C).

The pH(S) of six of the primary standards is plotted as a function of temperature in Figs. 4–8 and 4–9. The pH(S)-temperature curve for the phosphate solution of buffer ratio 1 : 3.5 (not shown) parallels that for the equimolal phosphate solution but is displaced in the alkaline direction by about 0.54 pH unit. Similar plots for the pa_H of the tetroxalate solution and the solution of calcium hydroxide are shown in Fig. 4–10. The compositions, density, dilution value, buffer value, and temperature coefficient of pa_H for the seven primary standards and the two supplementary standards are summarized in Table 4–11.[31]

PREPARATION OF THE STANDARDS

The compositions of the seven primary standard solutions are given in Table 4–12. The citrate, phthalate, phosphate, borax, and carbonate solutions are prepared simply by dissolving the indicated weights of the pure materials in pure water and diluting to 1 dm³ (1 liter). The saturated solution of potassium hydrogen tartrate is made by shaking vigorously an excess (5–10g) of the pure crystalline salt with 100–300 ml of water (temperature 25 ± 3 °C) contained in a glass-stoppered bottle. The excess of salt should be removed by filtration or decantation. This solution is about 0.034 M. Certified materials for the preparation of these standard solutions, with the exception of potassium dihydrogen citrate, are available as NBS standard samples from the National Bureau of Standards. The citrate is expected to be added to the list at some future date.

The $0.05m$ solution of potassium tetroxalate [$KH_3(C_2O_4)_2 \cdot 2H_2O$] is prepared by dissolving 12.61 g (air weight) of the salt in water and diluting to 1 dm³. To prepare the calcium hydroxide solution, a considerable excess of the pure, finely granular hydroxide is shaken vigorously in a stoppered bottle with water at a temperature near 25 °C. The gross excess of solid is allowed to settle, the temperature recorded to the nearest degree, and the suspended material removed by filtration with suction on a sintered-glass funnel of

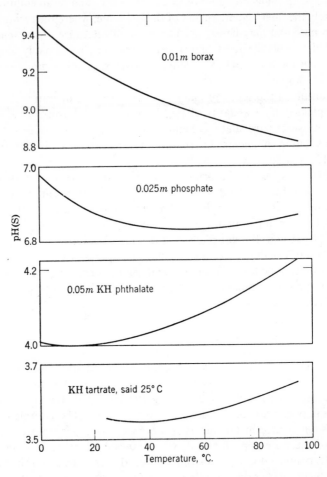

FIG. 4–8. Values of pH(S) for four primary standards as a function of temperature.

medium porosity. A polyethylene bottle is very satisfactory for both the preparation and the preservation of this solution.

The solubility of calcium hydroxide decreases with increase of the temperature, but the solution supersaturates readily and no precipitation of solid is ordinarily observed at 60 °C. Unfortunately, the change of the solubility with temperature is sufficiently large to require the application of a correction to the values given in Table 4–9 when the temperature of saturation

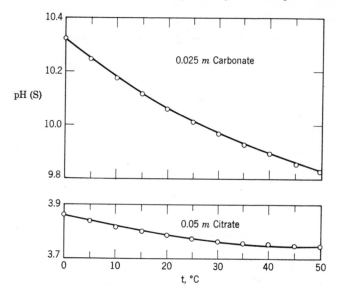

FIG. 4–9. pH(S) for the citrate and carbonate primary standards plotted as a function of temperature.

differs from 25 °C. If t_s, the saturation temperature, lies between 20 and 30 °C, the pa_H is given by[55]

$$pa_H = pa_H \text{ (Table 4–9)} - 0.003(t_s - 25) \qquad (22)$$

Of the buffer salts, only disodium hydrogen phosphate and sodium carbonate are appreciably hygroscopic. The tartrate, phthalate, and phosphates should be dried at 110 °C for 1 to 2 hours before use. Potassium tetroxalate dihydrate and potassium dihydrogen citrate should not be dried at temperatures above 60 °C, nor should borax and sodium bicarbonate be heated above normal room temperatures. Sodium carbonate should be dried for two hours at 275 °C.

The conductivity of the water used to prepare the buffer solutions should be less than 2×10^{-6} ohm^{-1} cm^{-1} at 25 °C. Higher conductances may indicate the presence of acidic or alkaline impurities that have a relatively large effect on the pH. For the borax, phosphate, and carbonate solutions, water purged with carbon dioxide-free air or freshly boiled water, pH 6.7 to 7.3, should be used. Boiled water should be protected from contamination by atmospheric carbon dioxide while cooling. Air-equilibrium water, pH 5.6 to

[55] R. G. Bates, V. E. Bower, and E. R. Smith, *J. Res. Nat. Bur. Stand.*, **56**, 305 (1956).

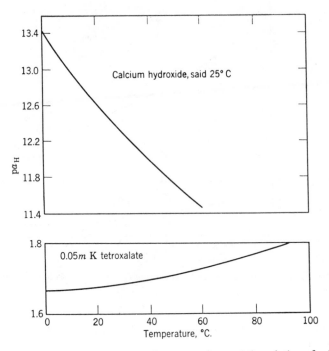

FIG. 4–10. Values of pa_H for 0.05m potassium tetroxalate and the solution of calcium hydroxide (saturated at 25 °C) as a function of temperature.

6.0, is satisfactory for the preparation of the citrate, phthalate, and tartrate standards, as well as for the solutions of potassium tetroxalate and calcium hydroxide.

PRESERVATION OF THE SOLUTIONS

The buffer solutions can be stored in bottles of resistant glass or polyethylene. The tartrate solution is subject to mold growth accompanied by an increase of 0.01 to 0.10 unit of pH[65]. Consequently, this solution must be replaced every few days unless a preservative is added. Thymol has been found to preserve tartrate solutions effectively for 2 months or more.[57] A crystal about 8 mm in diameter appears to be sufficient to preserve 200 ml of

[56] R. G. Bates, *Anal. Chem.*, **23**, 813 (1951).

[57] R. G. Bates, V. E. Bower, R. G. Miller, and E. R. Smith, *J. Res. Nat. Bur. Stand.*, **47**, 433 (1951). Nevertheless, for the highest accuracy the use of preservatives should probably be avoided.

TABLE 4–11. PROPERTIES OF NINE STANDARD BUFFER SOLUTIONS AT 25 °C

Solution	Molality	Density g/ml	Molar Conc.	Dilution Value $\Delta pH_{1/2}$	Buffer Value β equiv/pa_H	Temperature Coefficient $\partial pa_H/\partial t$ units/K
Tetroxalate	0.05	1.0032	0.04962	+0.186	0.070	+0.001
Tartrate	0.0341	1.0036	0.034	+0.049	0.027	−0.0014
Citrate	0.05	1.0029	0.04958	+0.024	0.034	−0.0022
Phthalate	0.05	1.0017	0.04958	+0.052	0.016	+0.0012
Phosphate 1:1	0.025[a]	1.0028	0.02490[a]	+0.080	0.029	−0.0028
Phosphate 1:3.5	0.008695[b] 0.03043[c]	1.0020	0.008665[b] 0.03032[c]	+0.07	0.016	−0.0028
Borax	0.01	0.9996	0.009971	+0.01	0.020	−0.0082
Carbonate	0.025[a]	1.0013	0.02492[a]	+0.079	0.029	−0.0096
Calcium hydroxide	0.0203	0.9991	0.02025	−0.28	0.09	−0.033

[a] Of each buffer salt.
[b] KH_2PO_4.
[c] Na_2HPO_4.

the solution. The solubility of thymol is about 0.006 M in aqueous solutions at room temperature, and the preservative has been found not to alter pH of tartrate buffer solutions by more than 0.01 unit.

Molds may also appear in the citrate and phthalate solutions after several weeks or months, and the phosphate solution may be found to contain a light sediment. In view of the contamination that will normally occur during use of a standard buffer solution, sterilization of the bottles and the solutions is not usually warranted. Instead, it is recommended that the buffer solutions be replaced about 4 weeks after preparation. With the exception of the tartrate standard, they will usually show good stability during this period. In 28 months of storage, initially sterile phthalate, phosphate, and borax solutions have been found to change less than 0.007 unit in pa_H, although some mold or sediment was evident in each.[58]

[58] G. G. Manov, *Symposium on pH Measurement*, p. 31, ASTM Tech. Publ. 73, Philadelphia, Pa., 1946.

TABLE 4–12. COMPOSITIONS OF PRIMARY STANDARD BUFFER SOLUTIONS

(Weight of buffer substance, in air near sea level,
per liter of buffer solution at 25 °C)

Solution	Buffer Substance	Weight in Air g
Tartrate, about 0.034 M	$KHC_4H_4O_6$	Satd. at 25 °C
Citrate, 0.05m	$KH_2C_6H_5O_7$	11.41
Phthalate, 0.05m	$KHC_8H_4O_4$	10.12
Phosphate, 1 : 1		
0.025m	KH_2PO_4	3.388
0.025m	Na_2HPO_4	3.533
Phosphate, 1 : 3.5		
0.008695m	KH_2PO_4	1.179
0.03043m	Na_2HPO_4	4.302
Borax, 0.01m	$Na_2B_4O_7 \cdot 10H_2O$	3.80
Carbonate		
0.025m	$NaHCO_3$	2.092
0.025m	Na_2CO_3	2.640

TETROXALATE BUFFER, 0.05m

Potassium tetroxalate of the theoretical composition is somewhat difficult to prepare, but it can be obtained by recrystallization from water if no
crystals are permitted to separate from the mother liquor at temperatures
above 50 °C.[59] Although the pa_H of the tetroxalate solution shows a more
pronounced change with concentration than that of the primary standards,
no appreciable error is likely to result from differences in the water content of
different lots of the salt. The buffer capacity is high. The pa_H of this solution
is 1.679 at 25 °C, but the pH on the operational scale fixed by the primary
standards is usually found to be about 1.65.

TARTRATE BUFFER, SATURATED AT 25 °C

A 10 per cent error in the concentration of this solution changes the pa_H by
only 0.01 unit; hence elaborate precautions to insure saturation are unnecessary. Furthermore, the temperature at which the saturated solution is prepared may vary within the limits 22 to 28 °C. It is suggested that the excess of
powdered salt be removed by filtration or decantation, for the presence of
finely divided solids at the liquid junction sometimes causes erratic and

[59] R. G. Bates, G. D. Pinching, and E. R. Smith, *J. Res. Nat. Bur. Stand.*, **45**, 418 (1950)

unstable boundary potentials. Because of its limited stability, the tartrate standard is the least satisfactory of the primary standards. It is recommended that the citrate solution be used instead of the tartrate solution whenever it is possible to do so.

CITRATE BUFFER, 0.05*m*

The solution of potassium dihydrogen citrate (molality 0.05 mol kg^{-1}) is an excellent buffer. As the data of Table 4–11 show, the citrate solution has a higher buffer value than any of the other primary standards. Furthermore, its change of pH on dilution with an equal volume of water is smaller than that of any other primary standard except the borax solution. The citrate buffer is less subject to molding than is the tartrate solution. Unlike the phthalate solution, it is stable in contact with the platinum-hydrogen electrode.

PHTHALATE BUFFER, 0.05*m*

This solution has a rather low buffer value and should be guarded carefully from accidental contamination with strong acid or alkali. The platinized hydrogen electrode reduces phthalates, and stable potentials cannot be obtained. However, the difficulty can be remedied by coating the electrode with palladium black instead of with platinum, a more active catalyst.[60] The density of the 0.05*m* solution is 1.0030 g/ml at 20 °C and 1.0017 g/ml at 25 °C. Potassium hydrogen phthalate can be dried safely at temperatures below 135 °C.[61]

PHOSPHATE BUFFERS

Anhydrous disodium hydrogen phosphate absorbs moisture when the relative humidity exceeds 41 per cent at 25 °C. Inasmuch as the two phosphates are weighed separately, it is a wise precaution always to dry the disodium hydrogen phosphate for two hours at 110 to 130 °C. For equimolal buffers at 25 °C, the ratio of the molar concentration (*c*) to the molality (*m*) of each phosphate is given by

$$\frac{c}{m} = 0.99707 - 0.0477m - 0.024m^2$$

when *c* and *m* are less than 0.1.

---

[60] W. J. Hamer and S. F. Acree, *J. Res. Nat. Bur. Stand.*, **33**, 87 (1944).
[61] E. R. Caley and R. H. Brundin, *Anal. Chem.*, **25**, 142 (1953).

TRIS BUFFERS

The buffer solution containing tris and tris hydrochloride, each at a molality of 0.05 mol kg^{-1}, is prepared by dissolving 5.970 g of tris and 7.768 g of the hydrochloride in water and diluting the resulting solution to 1 dm^3 at 25 °C. The buffer solution 0.01667m in tris and 0.0500m in tris hydrochloride is prepared by dissolving 1.999 g of tris and 7.800 g of tris hydrochloride in water and diluting it to 1 dm^3 with distilled water at 25 °C. The figures given are air weights. Tris and tris.HCl should not be heated above 80 and 40 °C, respectively. Carbon dioxide-free water should be used in preparing the solution. Reference electrodes containing linen-fiber junctions have been found unsatisfactory in tris buffers. Large unstable liquid-junction potentials have been observed when they are used.

BORAX BUFFER, 0.01m

When borax is stored for a year or longer, its water content is usually found to have fallen from 10 moles of water per mole of borax to 9.0, 8.5, or even lower, unless the container is tightly sealed. The dilution value of the borax solution is so small, however, that this change in water content is not serious. However, the change of composition precludes the use of the sample as an acidimetric standard. The pH of the 0.01m solution is altered 0.001 unit when 0.2 per cent of carbon dioxide is absorbed. Hence it is advisable to keep the solution well stoppered whenever a portion is not actually being removed.

CARBONATE BUFFER, 0.025m

The equimolal mixture of sodium bicarbonate ($m = 0.025$) and sodium carbonate ($m = 0.025$) has a high buffer capacity but suffers a rather large change of pH on dilution. The equilibrium tension of carbon dioxide is so low that no difficulty is experienced in using the hydrogen electrode in the carbonate buffer solution. The sodium bicarbonate should not be dried by heating; the sodium carbonate should be ignited for two hours at 275 °C before use in order to dry it and to convert any bicarbonate present into the normal carbonate.

CALCIUM HYDROXIDE, SATURATED AT 25° C

Calcium hydroxide should be prepared from well-washed calcium carbonate of low-alkali grade. The carbonate is heated to 1000 °C and ignited at that temperature for at least 45 minutes. The oxide formed is allowed to cool and is added slowly with stirring to water. The suspension is heated to boiling, cooled, and filtered on a sintered-glass funnel of medium porosity. The solid is dried and crushed to a uniformly finely granular state for use.

It is advisable to determine the concentration of a saturated solution of each lot of calcium hydroxide prepared. This is easily done by titration with a standard solution of a strong acid. If the molar concentration of the solution saturated at 25 °C exceeds 0.0206, the calcium carbonate from which the hydroxide was prepared probably contained soluble alkalies. If a higher grade cannot be obtained, these impurities can be extracted with water before the carbonate is used. The pa_H of the calcium hydroxide solution is 12.454 at 25 °C, but the pH on the operational scale fixed by the primary standards is usually found to be about 12.42.[62]

The British Standard pH Scale

The British Standard for the pH scale,[63] which is intended to apply strictly to aqueous solutions at temperatures between 0 and 95 °C, is consistent in nearly all respects with the NBS conventional scale described in the foregoing sections. The operational definition of pH (equation 3) is adopted, and the 0.05 M solution of potassium hydrogen phthalate is chosen as the single primary standard. The pH of this primary standard is *defined* by the equation

$$pH = 4.000 + \frac{1}{2}\left(\frac{t-15}{100}\right)^2 \tag{23}$$

between temperatures t of 0 and 55 °C, and by

$$pH = 4.000 + \frac{1}{2}\left(\frac{t-15}{100}\right)^2 - \frac{t-55}{500} \tag{24}$$

between 55 and 95 °C. The standard values for the British primary standard, although based on NBS data,[64] are not identical with the pH(S) given in Table 4–6. The differences exceed 0.005 unit only at four temperatures: 0 °C (0.008 unit), 5 °C (0.006 unit), 90 °C (0.006 unit), and 95 °C (0.013 unit). The pH values of the secondary standards recommended in the British standard pH method are given in Table 4–13.

The British standard emphasizes that the pH measured in accordance with the recommendations set forth therein has no precise simple fundamental meaning and that the standard definition is a practical one. Nevertheless, it is suggested that the measured pH of dilute solutions (less than 0.1 M) that are neither strongly acidic nor strongly alkaline (pH between 2

[62] R. G. Bates, V. E. Bower, and E. R. Smith, *J. Res. Nat. Bur. Stand.*, **56**, 305 (1956).

[63] *Specification for pH Scale*, British Standard 1647 (1961), British Standards Institution, 2 Park St., London, W.1.

[64] W. J. Hamer, G. D. Pinching, and S. F. Acree, *J. Res. Nat. Bur. Stand.*, **36**, 47 (1946); V. E. Bower and R. G. Bates, *ibid.*, **59**, 261 (1957).

TABLE 4–13. Standards for the Calibration of Glass Electrodes Recommended by the British Standard Method

	pH Values at		
Reference Solution	12 °C	25 °C	38 °C
Primary Standard			
0.05 M KH phthalate	4.000	4.005	4.026
Secondary Standards			
0.1 M $KH_3(C_2O_4)_2 \cdot 2H_2O$	—	1.48	1.50
0.01 M HCl + 0.09 M KCl	—	2.07	2.08
0.1 M CH_3COOH + 0.1 M CH_3COONa	4.65	4.64	4.65
0.01 M CH_3COOH + 0.01 M CH_3COONa	4.71	4.70	4.72
0.025 M KH_2PO_4 + 0.025 M Na_2HPO_4	—	6.85	6.84
0.05 M $Na_2B_4O_7 \cdot 10H_2O$	—	9.18	9.07
0.025 M $NaHCO_3$ + 0.025 M Na_2CO_3	—	10.00	—

Sources of data: Hitchcock and Taylor;[8] MacInnes, Belcher, and Shedlovsky;[17] Bates, Pinching, and Smith.[59]

and 12) may be considered to conform with the equation

$$pH = -\log c_H y_1 \pm 0.02 \qquad (25)$$

Here c_H is the concentration of hydrogen ion in moles per liter and y_1 denotes the mean activity coefficient (molar concentration scale) of a typical uni-univalent electrolyte in the solution. Numerically this formulation is substantially equivalent to defining the activity coefficient of hydrogen ion by equation 18 with any value of $å$ between 3.5 and 6.5 Å. Hence it is generally consistent with the interpretation outlined in the last section.

A careful study of ten secondary reference buffer solutions from 0 to 60 °C has been made by Alner, Greczek, and Smeeth.[65] The cell contained platinum or palladium hydrogen electrodes and a saturated calomel reference electrode. It was standardized, in close but not exact accordance with the British standard, by measuring the e.m.f. for the primary standard solution, potassium hydrogen phthalate, molality 0.0500 mol kg^{-1}. A very stable liquid junction was formed in a precision capillary tube of 1 mm bore.

---

[65] D. J. Alner, J. J. Greczek, and A. G. Smeeth, *J. Chem. Soc. (A)*, 1205 (1967). See also G. Mattock and D. M. Band, Chapter 2 in *Glass Electrodes for Hydrogen and Other Cations*, G. Eisenman, ed., Marcel Dekker, New York, 1967.

The results of this study make it feasible to assign another significant figure to the pH(S) of the secondary standards of the British standard scale. They also provide an illuminating comparison between the pH determined by a cell with liquid junction in reference to a single primary standard and the conventional pa_H determined by cells without liquid junction. Such a comparison of data at 25 °C is shown in Table 4–14. The very satisfactory agreement in the pH range 3.5 to 10 confirms the belief that the residual liquid-junction potential for dilute buffer solutions causes no concern in this range of pH. At extreme values of pH, however, agreement is not expected, for the data furnished by the British standard include a liquid-junction potential that no longer matches that present in the standardization process.

TABLE 4–14. COMPARISON OF pH VALUES ON THE BRITISH STANDARD AND NBS SCALES AT 25 °C

Solution (molality)	British Standard	NBS
K tetroxalate (0.05)	1.650	1.679
KH tartrate (satd. 25 °C)	3.559	3.557
KH phthalate (0.05)	4.008[a]	4.008
Acetic acid (0.01), Na acetate (0.01)	4.716	4.718
KH_2PO_4 (0.025), Na_2HPO_4 (0.025)	6.858	6.865
$Na_2B_4O_7$ (0.01)	9.185	9.180
$NaHCO_3$ (0.025), Na_2CO_3 (0.025)	10.001	10.012
$Ca(OH)_2$ (satd. 20 °C)	12.435	12.47

[a] Reference value.

Other National pH Methods

In the United States, the NBS scale is an integral part of the ASTM standard method for determining the pH of aqueous solutions.[66] This document deals with the definition of pH and sets forth procedures for the electrometric measurement of the pH of aqueous solutions or extracts.

The Japanese standard pH method[67] is consistent with both the NBS

---

[66] ASTM Method E70–68, *Standard Method of Test for pH of Aqueous Solutions With the Glass Electrode*, American Society for Testing Materials, Philadelphia, Pa., 1968.

[67] *Method for Determination of pH*, Japanese Industrial Standard Z8802, Japanese Standards Assn., Tokyo, 1958.

procedures and the British standard, incorporating elements of each. The standard applies to the temperature range 0 to 60 °C. The 0.05 M phthalate solution is designated the primary standard and other recommended standard solutions are the 0.05 M solution of potassium tetroxalate; the 0.025 M (equimolal) phosphate solution; 0.01 M borax; and a solution 0.025 M with respect to both sodium bicarbonate and sodium carbonate. The assigned pH values correspond to the pa_H determined by the NBS method.

Although selected as the only primary standard, the phthalate solution is apparently not so used. For the standardization of the pH meter the Japanese method prescribes an initial adjustment with the equimolal phosphate buffer. If the test solution has a pH between 2 and 7, the phthalate standard is also used and the temperature compensator of the instrument adjusted to bring these two standards into harmony.[68] In each case the selection of the second standard buffer solution is in fact determined by the pH of the test solution: pH less than 2, tetroxalate standard; pH 7 to 10, borax standard; pH 10 to 11, carbonate standard. When measurements at a pH greater than 11 are to be made, it is recommended that the meter be standardized with a 0.1 M solution of sodium hydroxide or with the saturated calcium hydroxide solution.

Standard methods for pH measurements have also been formulated in Germany and in the U.S.S.R. Both prescribe the use of standard pH values based on the NBS scale. The German method specifies all the primary standards of Table 4–6 except the citrate and carbonate solutions, and, in addition, the supplementary standards, tetroxalate and calcium hydroxide. The five primary standards of the Russian method are the tetroxalate, tartrate, phthalate, phosphate (1 : 1), and borax buffer solutions of Table 4–6.

IUPAC Recommendations

Essentially, the differences among national standards for pH measurement are minor, and variations are only of detail. In 1960, the International Union of Pure and Applied Chemistry set forth guidelines to which an international pH scale should conform, recommending that methods for pH measurements in the member nations should be made consistent with the approaches already formulated by the BSI and the NBS.[69] Specific recommendations for a

---

[68] This procedure was suggested by the author, *Symposium on pH Measurement*, p. 1, ASTM Tech. Publ. 190, Philadelphia, Pa., 1957, as a simple means of adjusting the meter to read pH in accordance with equation 19.

[69] R. G. Bates and E. A. Guggenheim, *Pure Appl. Chem.*, **1**, 163 (1960).

IUPAC pH method were contained in the IUPAC Manual of 1970.[70] This document suggests that the operational definition of pH be adopted universally and that the standard scale be fixed by five primary standards of the NBS scale, namely the tartrate, phthalate, phosphate (1 : 1), phosphate (1 : 3.5), and borax solutions. The values assigned to these solutions are given in Table 4–6. Preliminary endorsement by IUPAC has also been given to a proposal for the measurement of pH in nonaqueous and mixed solvents.[71]

Interpretation of pH

The operational definition of the practical pH value (equation 3) is an expression of the fact that the experimental pH can never be an exact measure of either the concentration or the activity of hydrogen ion. The purpose of establishing pH standards is twofold: first, to provide a common basis for all pH measurements, in order that the pH numbers obtained will be truly reproducible from time to time and from place to place; and second, through assignment of suitable standard pH values, to permit the measured pH to assume as much quantitative meaning in terms of chemical equilibria as possible.

It is safe to say that no quantitative interpretation of measured pH values should be attempted unless the medium can be classified as a dilute aqueous solution of simple solutes. This requirement excludes all nonaqueous media, suspensions, colloids, and aqueous solutions of ionic strengths greater than 0.2. From this point of view, the "ideal" solutions are those that match the standards of reference, namely aqueous solutions of buffers and simple salts with ionic strengths between 0.01 and 0.1. Under these very restricted conditions, the measured pH may be expected to approach an experimental $-\log m_H \gamma_H$, where γ_H is defined in a conventional manner consistent with the assignment of pa_H values to the standards with which the instrument was adjusted. For most practical purposes, the value of γ_i in this dilute range can be estimated by equation 18 with $B\mathring{a} = 1.5 \text{ kg}^{1/2} \text{ mol}^{-1/2}$.

The manner in which the measured pH value can be interpreted and applied is illustrated by the following examples.

EXAMPLE 1.   The pH of a mixture of acetic acid (0.025 M) and sodium acetate (0.05 M) is 4.98 at 25 °C. What is the thermodynamic dissociation constant of acetic acid at this temperature? What is the concentration of hydrogen ion in the solution?

---

[70] *Manual of Symbols and Terminology for Physicochemical Quantities and Units*, International Union of Pure and Applied Chemistry, Butterworths, London, 1970.
[71] R. G. Bates, *Pure Appl. Chem.*, **18**, 421 (1969).

The ionic strength is $0.05 + m_H$ or, as a first approximation, 0.05. Hence,

$$-\log m_H = pH - \frac{0.51\sqrt{0.05}}{1 + 1.5\sqrt{0.05}}$$

or 4.89. If necessary, a more accurate value of $-\log m_H$ can be obtained by a second approximation. From the mass law and equation 18,

$$-\log K = pH - \log\frac{0.05 + m_H}{0.025 - m_H} + \frac{0.51\sqrt{0.05}}{1 + 1.5\sqrt{0.05}}$$

or 4.76, and $K = 1.74 \times 10^{-5}$.

EXAMPLE 2. What is the pH at the equivalence point in the titration of potassium hydrogen phthalate with strong alkali if the titration is performed under such conditions that the final concentration of phthalate ion is 0.05 M?

The ionic strength is 0.15. From the mass law equation for the hydrolysis of phthalate ion, equation 18, and the definition pH $\approx -\log m_H \gamma_H$, one finds

$$pH = -\tfrac{1}{2}\log K_2 K_w + \tfrac{1}{2}\log 0.05 - \frac{1.02\sqrt{0.15}}{1 + 1.5\sqrt{0.15}}$$

$$= 9.70 - 0.65 - 0.25 = 8.80$$

In this equation K_2 and K_w are, respectively, the second thermodynamic dissociation constant of phthalic acid and the ion product constant for water.

CHAPTER 5

properties of
buffer solutions

The *total acidity* or neutralizing power of equal amounts of 0.1 M solutions of hydrochloric acid, acetic acid, and boric acid is found to be identical by titration of these solutions to their respective equivalence points with a solution of strong alkali. Nevertheless, the concentrations of hydronium (hydrogen) ion in 0.1 M aqueous solutions of these three acids are approximately 0.1, 0.0013, and 0.000004, respectively. Inasmuch as the hydronium or hydrogen ion is largely responsible for the properties characteristic of aqueous acids, the *degree of acidity* differs greatly from one of these solutions to another.

These differences are explained quite simply in terms of the Brønsted–Lowry, or proton-transfer, concept of acid-base interaction. A strong base like hydroxide ion causes complete transfer of protons from all but the very weakest acids, whereas the weak base water effects only partial transfer from all but the strongest acids. The extent to which this reaction occurs depends on the relative strengths of the two bases concerned, namely water and the anion of the acid.

Buffer Action

When 1 ml of a 0.1 M solution of a strong acid or base is added to a liter of pure water, the concentration of hydrogen or hydroxyl ion in the water is altered about a thousandfold. If the liter of water contains 0.1 mole of acetic acid and 0.1 mole of sodium acetate when the hydrochloric acid is added, however, the change is scarcely detectable, for the acetate mixture is a good *buffer*.

Buffers have been defined by Van Slyke[1] as "substances which by their presence in solution increase the amount of acid or alkali that must be added to cause unit change in pH." Buffers make possible the accurate regulation of hydrogen and hydroxyl ion concentrations and thus the control of reactions dependent on the level of acidity. *Buffer action* is a consequence of the equilibria between water, weak acids (HA), bases (B), and ampholytes ($Z^{\mp}$), and the ions into which these several species are partially converted in aqueous solution:

$$HA^{z+1} + H_2O \rightleftharpoons A^z + H_3O^+ \tag{1}$$

$$B^z + H_2O \rightleftharpoons BH^{z+1} + OH^- \tag{2}$$

$$Z^{\mp} + H_3O^+ \rightleftharpoons Z^+ + H_2O \tag{3}$$

and

$$Z^{\mp} + OH^- \rightleftharpoons Z^- + H_2O \tag{4}$$

Because of the reversibility of reaction 1, for example, the addition of strong acid (furnishing H_3O^+) to an acetate buffer solution fails to disturb the hydrogen ion concentration seriously, for the greater part of the added acid is neutralized by the base, acetate ion (A^-), which is present in much larger concentration than is H_3O^+. Similarly, the principal end result of the addition of strong alkali is the conversion of acetic acid molecules to acetate.

In succeeding chapters we have occasion to characterize the charge type of the equilibrium between an acid and its conjugate base. This will be done by affixing appropriate charges to the letters A and B. Thus A^0B^- denotes an acid-base pair of the type HA, A^-; other common examples are A^+B^0, $A^-B^=$, and $A^{\mp}B^-$.

The properties of buffer solutions can be explained satisfactorily in terms of equilibria similar to those represented by equations 1 to 4. The effect of changes in concentration (molality) and activity of the participating species is expressed quantitatively by the mass law. Thus for a monobasic acid HA with dissociation constant K_a

$$K_a = \frac{a_{H^+} a_{A^-}}{a_{HA}} = \frac{m_{H^+} m_{A^-}}{m_{HA}} \cdot \frac{\gamma_{H^+} \gamma_{A^-}}{\gamma_{HA}} \tag{5}$$

and for a monoacidic base B whose dissociation constant is K_b

$$K_b = \frac{a_{OH^-} a_{BH^+}}{a_B} = \frac{K_w a_{BH^+}}{a_{H^+} a_B} = \frac{K_w}{m_{H^+}} \cdot \frac{m_{BH^+}}{m_B} \cdot \frac{\gamma_{BH^+}}{\gamma_{H^+} \gamma_B} \tag{6}$$

---

[1] D. D. Van Slyke, *J. Biol. Chem.*, **52**, 525 (1922).

where K_w is the thermodynamic ion product constant for water. It is evident from these equations that (a) the hydrogen ion concentration of a buffer solution composed of equal molal amounts of HA and A^- is approximately equal to K_a, if the activity coefficient term does not depart greatly from unity, and (b) the hydrogen ion concentration of an equimolal buffer solution of B and BH^+ is about equal to K_w/K_b under the same conditions.

The intensity of buffer action, that is, the effectiveness with which a solution resists changes in hydrogen ion concentration, is called the *buffer capacity*. Equations 5 and 6 suggest that the buffer capacity is determined, in the first approximation, by the change in the *buffer ratio*, or ratio of m_{A^-} to m_{HA} (or m_{BH^+} to m_B), brought about by addition of small amounts of other acids or bases. If these ratios are markedly changed the buffer capacity is said to be low; if the ratios are altered only slightly, the solutions have high buffer capacity.

The primary effect of adding an increment Δm of strong acid or base is to increase one member (numerator or denominator) of the buffer ratio and to decrease the other by a like amount. Hence the necessary condition for a high buffer capacity is that

$$\frac{m_1 + \Delta m}{m_2 - \Delta m}$$

shall not be very different from m_1/m_2, the buffer ratio before the addition of acid or alkali. Evidently a given Δm will affect the ratio least when the concentrations of both components of the buffer are high. Furthermore, if the total concentration of weak electrolyte $m_1 + m_2$ is fixed, the buffer capacity will be at its maximum value when m_1 and m_2 are equal, that is, when half of the weak acid or base has been neutralized. The condition for maximum buffer efficiency at a particular concentration is, then, that pH shall be equal to $-\log K$. The following sections will be devoted to a consideration of the measurement of the buffer capacity and the effect of dilution, addition of neutral salt, and changes of temperature and pressure on the pH of a buffer solution.

The Buffer Value

A unit for the quantitative measurement and numerical expression of buffer effects, through which the buffer capacity can be related to the dissociation constant and the pH, was proposed by Van Slyke in 1922.[1] expressed mathematically, the buffer unit, which Van Slyke called the *buffer value* and designated β, is a differential ratio,

$$\beta \equiv \frac{db}{dpH} \tag{7}$$

where db is an increment of strong base expressed in moles of hydroxide ion per cubic decimeter (liter).

An increment of strong acid, equivalent in effect to a negative increment $-$db of strong base, produces a decrease in pH. The ratio db/dpH is therefore always positive. For the estimation of buffer values, measured increments Δb are added and the corresponding changes of pH, ΔpH, determined. If the increments are not too great, this approximation serves the purpose quite well.

STRONG ACIDS AND BASES

Let us examine first the buffer capacity of aqueous solutions of strong acids and bases, in which the solute is completely dissociated. Such solutions are not ordinarily classified as "buffer solutions"; however, their buffer capacity is often larger than that of "buffer solutions."

Considering first a solution of strong base in which the hydroxyl ion concentration is m_{OH}, one can write db = dm_{OH} and dpH = $-$d log m_H = d log m_{OH}, and, consequently,

$$\frac{db}{dpH} = \frac{dm_{OH}}{d \log m_{OH}} = 2.303 m_{OH} \tag{8}$$

where pH is defined as $-$log m_H or $-$log c_H. Similarly, for a solution of strong acid (for which dm_H = $-$db),

$$\frac{db}{dpH} = \frac{-dm_H}{-d \log m_H} = 2.303 m_H \tag{9}$$

The total buffer value of water plus completely dissociated acid or alkali is, at all pH values,

$$\frac{db}{dpH} = 2.303(m_H + m_{OH}) \tag{10}$$

SOLUTIONS OF WEAK ELECTROLYTES

In order to relate the buffer value to pH and to the concentration of the weak electrolyte in a buffer solution, it is convenient and sufficiently accurate to employ the "classical" dissociation constants k written in terms of concentrations, rather than the thermodynamic constants K expressed in activities.[2] For a weak acid HA the constant k is written

$$k \equiv \frac{m_H m_A}{m_{HA}} = K \frac{\gamma_{HA}}{\gamma_H \gamma_A} \tag{11}$$

---

[2] This treatment is essentially that of Van Slyke. Equations for the points of maximum and minimum buffer capacity in acid-base titrations have been developed by S. Kilpi, Z. Physik. Chem., A173, 223 (1935).

Let c be the initial concentration of weak acid and b the moles of hydroxide ion (strong alkali) added, the total volume remaining constant.[3] Inasmuch as the solution must be electrically neutral,

$$b = m_A - m_H + m_{OH} = m_A - m_H + \frac{k_w}{m_H} \qquad (12)$$

where k_w is the ion product constant for water, $m_H m_{OH}$. From equation 11,

$$m_A = \frac{k(c - m_A)}{m_H} = \frac{kc}{k + m_H} \qquad (13)$$

Substitution of equation 13 into equation 12 and differentiation with respect to pH gives

$$\beta \equiv \frac{db}{dpH} = 2.303 \left[\frac{kcm_H}{(k + m_H)^2} + m_H + m_{OH} \right] \qquad (14)$$

MAXIMUM BUFFER VALUE

When the concentration c of the weak acid is 0.01 to 0.1 M and the pH is between 3 and 11, the buffer value is determined almost wholly by the first term within the brackets. At low and high pH, however, the buffer capacity of the free hydrogen and hydroxyl ions is large (compare equation 10). As has been stated, the maximum buffer effect in a weak acid system is found, for a given value of c, when $m_H = k$. Hence from equation 14

$$\beta_{max} = \frac{2.303c}{4} = 0.576c \qquad (15)$$

As a first approximation, the maximum value is independent of the dissociation constant of the weak acid. Similar conclusions can be deduced for buffer systems composed of weak bases and their salts.

It must be remembered that these equations were derived without taking into account the changes in activity coefficients that occur when the buffer concentration c is increased. Kortüm[4] has observed that interionic effects impose a practical upper limit of about 0.2 on the buffer value.

The buffer action of strong and weak acids and strong bases is illustrated in Fig. 5–1, where the buffer values β of solutions of hydrochloric acid and sodium hydroxide are compared with those of acetate, phthalate, phosphate,

---

[3] If the buffer solution is formed from HA and its salt MA, $b = m_M$ and $c = m_{HA} + m_A$. For the present purpose, concentrations may be expressed either in moles per cubic decimeter or in molality.

[4] G. Kortüm, *Lehrbuch der Elektrochemie*, 2nd ed., p. 313, Verlag Chemie, Weinheim, 1957.

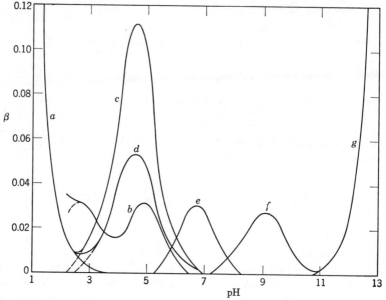

FIG. 5–1. Buffer values β for solutions of strong acid and alkali and for buffer solutions composed of weak acids and strong alkali, as a function of pH.

a. HCl.

b. 0.05M phthalic acid + NaOH.

c. 0.02M acetic acid + NaOH.

d. 0.1M acetic acid + NaOH.

e. 0.05M KH_2PO_4 + NaOH.

f. 0.05M H_3BO_3 + NaOH.

g. NaOH.

and borate buffer solutions.[5] It is evident, first of all, that the buffer capacities of the strong acid and strong base exceed those of weak acid buffers of the same concentration. Furthermore, the maximum buffer values of the acetate, phthalate, phosphate, and borate systems are approximately those predicted by equation 15, namely 0.028 for the 0.05 M systems, 0.057 for the 0.1 M system, and 0.115 for the 0.2 M system. Below pH 3 the hydrogen ion contribution (the second term within the brackets of equation 14) becomes apparent and the buffer value of the acetate and phthalate solutions exceeds the limit for a weak acid system set forth in equation 15. Finally, the

---

[5] The data for hydrochloric acid and sodium hydroxide were calculated by equation 10. The data for acetate solutions were taken from Van Slyke's paper, and those for the other buffer solutions were computed from the pH values of W. M. Clark and H. A. Lubs, *J. Biol. Chem.*, **25**, 479 (1916); see also W. M. Clark, *The Determination of Hydrogen Ions*, 3rd ed., Chapter 9, The Williams and Wilkins Co., Baltimore, Md., 1928.

figure demonstrates that the useful buffer action associated with the neutralization of a single acid group has a range of 2 to 3 pH units.

Dilution Effects

If a buffer solution is to be most effective in regulating the acidity, its pH should be not only little affected by the addition of small amounts of strong acid or strong base but also relatively insensitive to changes in the total concentration of the buffer components. It is advantageous to express the dilution effect, like the buffer effect, in a quantitative manner. This can be accomplished by defining a quantity $\Delta pH_{1/2}$, which has been called the *dilution value*,[6] as the increase of pH suffered by a solution of initial concentration c_i on dilution with an equal volume of pure water:

$$\Delta pH_{1/2} \equiv (pH)_{c_i/2} - (pH)_{c_i} \tag{16}$$

The dilution value is positive when the pH increases with dilution and negative when it decreases.

The dilution value can be determined experimentally, and, as with the buffer value, its approximate value can also be calculated from theoretical considerations. If v (in liters per mole) is defined as $1/c$, we have for the interval from c_i to $c_i/2$:

$$\Delta pH_{1/2} = \Delta v \left(\frac{dpH}{dv}\right)_{c=0.75c_i} = \frac{1}{c_i}\left(\frac{dpH}{dv}\right)_{c=0.75c_i} \tag{17}$$

where the value of the differential ratio dpH/dv corresponds to the mean concentration $0.75c_i$. By differentiation of the various expressions relating pH to concentration or volume, equations for dpH/dv as a function of c have been obtained for four common types of solutions. For these derivations pH was defined as $-\log a_H = -\log c_H \gamma_H$, no distinction being made between molar concentration and molality.

SOLUTIONS OF STRONG ACIDS AND BASES

The change of pH on dilution is caused both by concentration changes, expressed in terms of the quantity D,

$$D \equiv c_H - c_{OH} \tag{18}$$

---

[6] R. G. Bates, *Anal. Chem.*, **26**, 871 (1954). A thorough treatment of dilution effects in buffer solutions has also been given by I. M. Kolthoff and C. Rosenblum, *Acid-Base Indicators*, p. 269, The Macmillan Co., New York, 1937. The change of pH on dilution is also useful in elucidating the constitution of species with acidic or basic properties, as, for example, basic salts; see R. Schaal and J. Faucherre, *Bull. Soc. Chim. France*, Ser. 5, **14**, 927 (1947).

and by changes in the activity coefficients brought about by the alteration of the ionic strength I. When only strong acids, bases, and salts are present in the solution,

$$g = D = \frac{1}{\gamma}\left(a_H - \frac{K_w}{a_H}\right) \tag{19}$$

where γ is the activity coefficient of a "normal" or average univalent ion, K_w is the thermodynamic ion product constant for water, and g is defined by

$$g \equiv c_a - c_b \tag{20}$$

In equation 20, c_a and c_b are, respectively, the stoichiometric molar concentration of uni-univalent strong acid and base.

By differentiation of equation 19 with respect to v, one obtains

$$\frac{dpH}{dv} = -\frac{1}{\sqrt{g^2 + 4K_w/\gamma^2}}\left(0.4343\frac{dg}{dv} + g\frac{d\log\gamma}{dv}\right) \tag{21}$$

The absolute value of dg/dv is g^2, but the sign is negative for positive g (acid solutions) and positive for negative g (alkaline solutions). To express $d\log\gamma/dv$ in terms of $d\log\gamma/dI$, v is set equal to $1/I$. When these substitutions are made, one finds

$$\frac{dpH}{dv} = \frac{\pm 0.4343g^2}{\sqrt{g^2 + 4K_w/\gamma^2}} + \frac{gI^2}{\sqrt{g^2 + 4K_w/\gamma^2}}\left(\frac{d\log\gamma}{dI}\right) \tag{22}$$

where the first term has a positive sign when g is positive and a negative sign when g is negative.

The differential ratio $d\log\gamma/dI$ can be estimated by differentiation of the Debye–Hückel equation

$$\log\gamma = \frac{-A\sqrt{I}}{1 + \rho\sqrt{I}} \tag{23}$$

In equation 23, A is the Debye–Hückel slope and a value of 1.3 was chosen for ρ. From equation 23,

$$\frac{d\log\gamma}{dI} = \frac{-0.5A}{\sqrt{I}(1 + 1.3\sqrt{I})^2} \tag{24}$$

If the concentration of the solution of strong acid or base is greater than 10^{-5} M and no added salts are present, $4K_w/\gamma^2$ is negligible with respect to g^2. Therefore equation 22 can be simplified and combined with equation 24 to obtain (for 25 °C)

Monobasic strong acids:

$$\frac{dpH}{dv} = 0.4343c_a\left[1 - \frac{0.587\sqrt{c_a}}{(1 + 1.3\sqrt{c_a})^2}\right] \tag{25}$$

and

Monoacidic strong bases:

$$\frac{dpH}{dv} = -0.4343c_b\left[1 - \frac{0.587\sqrt{c_b}}{(1 + 1.3\sqrt{c_b})^2}\right] \tag{26}$$

SOLUTIONS CONTAINING A WEAK ACID OR BASE
AND ITS STRONG SALT IN EQUAL MOLAR CONCENTRATIONS

Buffer solutions composed of (1) weak acid HB and its strong salt MB, both at concentration c, or (2) weak base B and its strong salt HBX, both at concentration c, belong in this class. The weak acid and base may bear a charge. The ionic equilibrium for both types is written

$$HB^{z+1} \rightleftharpoons B^z + H^+ \tag{27}$$

for which the equilibrium constant is K_a.

If the hydrogen ion concentration of the solution depends only on equilibrium 27 and the ionization of water,

$$pH = -\log K_a + \log\frac{1 + vD}{1 - vD} - (2z + 1)\log\gamma \tag{28}$$

where D is again given by

$$D = \frac{1}{\gamma}\left(a_H - \frac{K_w}{a_H}\right) \tag{29}$$

By differentiation of equations 28 and 29 with respect to v and combination with equation 24 for $d\log\gamma/dI$, one obtains

$$\frac{dpH}{dv} = \frac{\gamma c^2}{\gamma(c^2 - D^2) + 2c(a_H + a_{OH})}$$
$$\left\{0.8686D - \left(\frac{I^2}{c^2}\right)\frac{0.254}{\sqrt{I}(1 + 1.3\sqrt{I})^2}[(2z + 1)(c^2 - D^2) + 2Dc]\right\} \tag{30}$$

Both z and I/c are determined by the changes borne by the buffer ions; the values of these coefficients for five types of acid-base buffer systems without added neutral salt are given in Table 5-1.

TABLE 5–1. VALUES OF z AND I/c FOR ACID-BASE BUFFER SYSTEMS

Acid	Base	z	I/c
HB^{2+}	B^+	$+1$	4
HB^+	B	0	1
HB	B^-	-1	1
HB^-	B^{2-}	-2	4
HB^{2-}	B^{3-}	-3	9

SOLUTIONS OF THE ACID SALT OF A DIBASIC ACID

The pH of a c molar solution of the acid salt MHB of the dibasic acid H_2B, whose dissociation constants are K_1 and K_2, is given by the approximate equation[7]

$$pH = -\tfrac{1}{2} \log K_1 K_2 + \tfrac{1}{2} \log \frac{xc + D}{xc - D} + 2 \log \gamma \qquad (31)$$

where x is defined by

$$x \equiv \frac{\dfrac{2(K_2/K_1)^{1/2}}{\gamma}}{1 + \dfrac{2(K_2/K_1)^{1/2}}{\gamma}} \qquad (32)$$

When D^2 is negligible compared with $x^2 c^2$, as is usually the case between pH 4 and 9 if c is no smaller than 0.005 M, the equation for $d\text{pH}/dv$ becomes

$$\frac{d\text{pH}}{dv} = \frac{\dfrac{0.4343 D}{xc}\left(c - \dfrac{d \ln x}{dv}\right) - I^2 \dfrac{d \log \gamma}{dI}\left(2 - \dfrac{D}{xc}\right)}{1 + \dfrac{a_H + a_{OH}}{\gamma xc}} \qquad (33)$$

Equation 24 can be used for the estimation of $d \log \gamma/dI$. It has been shown that the approximate value of I/c is 1.11 ± 0.1 and that $d \ln x/dv$ amounts to only a few per cent of c when the latter is less than 0.1.[8]

---

[7] R. G. Bates, *J. Amer. Chem. Soc.*, **70**, 1579 (1948). This equation does not yield accurate pH values below a concentration of about 0.02 for $D = \pm 0.001$. Between pH 4.5 and 9.5, however, it is valid to lower concentrations.

[8] R. G. Bates, *Anal. Chem.*, **26**, 871 (1954).

DILUTION PHENOMENA

The observed changes of pa_H on dilution of solutions of hydrochloric acid and phthalate and phosphate buffer solutions are compared in Table 5–2 with the values of $\Delta pH_{1/2}$ computed by equation 17 with the aid of the expressions for dpH/dv in the special cases discussed in the foregoing sections. The calculated dilution effects for hydrochloric acid and for four concentrations of a buffer solution composed of a monobasic weak acid and its salt (buffer charge type A^0B^-) are plotted in Fig. 5–2. Figure 5–3 is a similar plot for sodium hydroxide and for buffer solutions composed of a monoacidic weak base and its salt (charge type A^+B^0) at four different concentrations. The dilution effect for several values of the acidic dissociation constant K_a and of the basic dissociation constant $K_b = K_w/K_a$ is indicated.

Inasmuch as the dilution value of an equimolal buffer solution of the acid type changes sign at high pH, the pH will pass through a maximum on dilution. Similarly, the dilution value of a weak base buffer changes sign at low pH, and the pH may pass through a minimum on dilution of the solution.

In the region of intermediate pH, the curves for each buffer concentration become straight lines, and the dilution value does not depend on the dissociation constant of the weak acid or base. It changes only with concentration, type of buffer (acid or base), and valence type of the buffer components. The dilution values $\Delta pH_{1/2}$ of buffer solutions of four types in this intermediate region are listed in Table 5–3 for seven values of c, the molar concentration of each component.

TABLE 5–2. DILUTION VALUES $\Delta pH_{1/2}$ OF HYDROCHLORIC ACID AND BUFFER SOLUTIONS

Solution	Molar Concentration of Each Solute Species				
	0.1	0.05	0.02	0.01	0.005
Hydrochloric acid					
Observed	0.28	0.28	0.29	0.29	0.29
Calculated	0.30	0.30	0.31	0.31	0.32
Potassium hydrogen phthalate					
Observed	0.057	0.052	0.047	0.050	0.066
Calculated	0.061	0.054	0.049	0.055	0.072
Potassium dihydrogen phosphate, disodium hydrogen phosphate					
Observed	0.101	0.088	0.073	0.059	0.046
Calculated	0.105	0.096	0.079	0.065	0.047

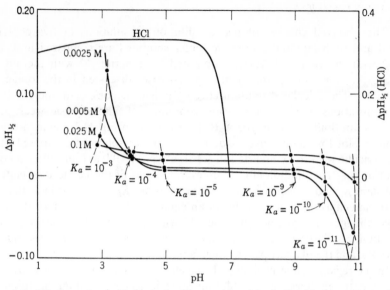

FIG. 5-2. Dilution value ΔpH_2 for hydrochloric acid and for buffer solutions composed of a monobasic weak acid and its strong salt in equal molar concentration, as a function of pH.

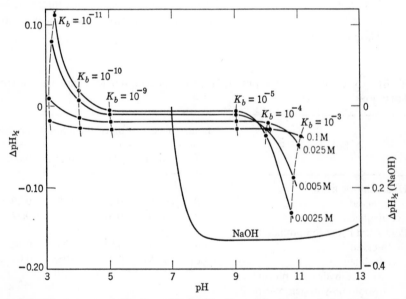

FIG. 5-3. Dilution value ΔpH_2 for sodium hydroxide and for buffer solutions composed of a monoacidic uncharged weak base and its strong salt in equal molar concentration, as a function of pH.

TABLE 5-3. DILUTION VALUES (IN pH UNITS) FOR FOUR
TYPES OF BUFFER SOLUTIONS BETWEEN pH 4.5 AND 9.5

c	A^0B^-	$A^-B^=$	A^+B^0	Acid salt MHB
0.1	0.028	0.105	−0.028	0.059
0.05	0.023	0.096	−0.023	0.050
0.025	0.019	0.082	−0.019	0.039
0.02	0.017	0.079	−0.017	0.037
0.01	0.013	0.065	−0.013	0.028
0.005	0.010	0.047	−0.010	0.022
0.0025	0.007	0.040	−0.007	0.016

The dilution value of monobasic weak acid buffers decreases with decrease
of concentration between pH 4.5 and 9.5. An inversion occurs near pH 4, and
at low pH the dilution value increases with decreasing concentration. Monoacidic weak base buffers display an analogous behavior, with an inversion near
pH 10, but the dilution value is negative at intermediate pH. The dilution
value of solutions of potassium hydrogen phthalate is at a minimum near
$c = 0.02$. It is interesting to note that solutions of strong acids and bases,
although having a higher buffer capacity than dilute buffer solutions composed of weak acids or bases, display a considerably larger change of pH
on dilution than do the systems of incompletely dissociated solutes.

There are two effects of dilution, and these may profitably be considered
separately. The first is the effect of the water, which is both a weak acid and a
weak base, on the buffer equilibria. It causes a shift toward neutrality. The
second is the effect of dilution on the activity coefficients. In dilute solutions,
where equation 23 is valid, d log γ/dI is always negative. The changes in
interionic forces on dilution of the solution tend to increase the activity
coefficient and to raise the pH of solutions of strong bases, weak acid buffer
solutions ($2z + 1$ negative), and acid salts (compare equations 19, 28, and 31).
The pH of solutions of strong acids (equation 19) and buffer solutions composed of weak bases and their salts (equation 28, $2z + 1$ positive) tends to
decrease. The net result of these two effects is to cause the pH of weak acid
buffer systems with pH greater than 7 to pass through a maximum at some
low concentration. In the same way, the pH of a buffer solution composed
of a weak base and its salt may pass through a minimum if the base is so
weak that the pH of the solution is less than 7.

The pH of a buffer solution containing a weak uncharged or anion acid
and its salt in equal concentrations usually increases on dilution when c

is in the range 0.1 to 0.01. This is true even for solutions as alkaline as 1 : 1 bicarbonate-carbonate mixtures (pH about 10). As the curves of Fig. 5–2 show, however, $\Delta pH_{1/2}$ eventually becomes negative when the acid is largely dissociated, and the pH drops toward the neutral point. At still higher pH, the acid-base reaction with the added water completely obscures the effect of dilution on the activity coefficients. For such solutions $\Delta pH_{1/2}$ is negative at relatively high concentrations.

Salt Effects

The effect on the pa_H of adding neutral salt to a buffer solution is readily predicted in a qualitative manner from a consideration of the buffer equilibria, if one bears in mind that the primary effect of salt addition in dilute solutions is to lower the activity coefficients of the ions. The activity coefficients of uncharged species are affected much less strongly. The effect of the addition of neutral salt is opposite in sign to the effect of dilution. Accordingly, the pa_H of solutions of strong acids and of weak base-salt buffer mixtures is raised, whereas that of solutions of strong bases, weak acid-salt buffer mixtures and solutions of acid salts is lowered. In the absence of direct interactions between the added salt and the buffer species, which alter the buffer ratio in an abnormal manner, the change of pa_H is presumably determined primarily by the effect of the salt on the ratio of the activity coefficients of the buffer. A smaller secondary change in the buffer ratio is to be expected. Nevertheless, the magnitude of the salt effect will depend first on the increase in ionic strength and second on the specific properties of the added ions.

Salt effects can be estimated from the mass law expression for the buffer equilibrium together with a formula for the change of the activity coefficients with increase of ionic strength. For a buffer solution composed of a monobasic uncharged weak acid HA and its completely dissociated salt, we have

$$pa_H = -\log K_a + \log \frac{m_{A^-}}{m_{HA}} + \log \frac{\gamma_{A^-}}{\gamma_{HA}} \tag{34}$$

and for one composed of a monoacidic uncharged weak base B and its cation BH^+,

$$pa_H = -\log K_w + \log K_b - \log \frac{m_{BH^+}}{m_B} - \log \frac{\gamma_{BH^+}}{\gamma_B} \tag{35}$$

In these equations m is molality, K_a is the dissociation constant of the weak acid, and K_b is the dissociation constant of the weak base, that is, the equilibrium constant of equation 2.

Kortüm[9] has used the Debye-Hückel limiting law to compute the approximate effect of adding neutral salt (0.1 M) to dilute solutions of both strong and weak acids and bases and to buffer solutions. The results of his calculations are given in Table 5-4. The changes in each quantity are indicated by placing Δ before the appropriate symbol. A plus sign signifies an increase on addition of the salt, a negative sign a decrease. The changes of pa_H caused by the addition of sodium or potassium chloride to three standard buffer solutions at 25 °C are given in Table 5-5. These values were determined by e.m.f. measurements of cells without liquid junction in the manner described in Chapter 4. The sources are indicated at the foot of the table.

TABLE 5-4. SALT EFFECT CAUSED BY ADDITION OF NEUTRAL SALT (0.1 M)

Solution	(After Kortüm) $(100 \Delta c_H)/c_H$ per cent	$(100 \Delta a_H)/a_H$ per cent	$\Delta(-\log a_H)$ $= \Delta pa_H$
HCl, 0.005 M	0	−20	+0.10
NaOH, 0.005 M	+56	+25	−0.10
CH₃COOH, 0.005 M	+29	0	0
NH₃, 0.005 M	+32	0	0
CH₃COOH, 0.005 M; CH₃COONa, 0.005 M	+56	+25	−0.10
NH₃, 0.005 M; NH₄Cl, 0.005 M	0	−20	+0.10

SALT EFFECT VS. DILUTION EFFECT

Bates, Diamond, Eden, and Acree[10] have investigated the effect of salts on the activity coefficients in buffer solutions formed by half-neutralization of the weak acid potassium *p*-phenolsulfonate (KHPs) with sodium hydroxide.

[9] G. Kortüm, *Z. Elektrochem.*, **48**, 145 (1942). The effect of concentrated salt media on acid-base equilibria has been investigated by D. Rosenthal and J. S. Dwyer, *Anal. Chem.*, **35**, 161 (1963); *J. Phys. Chem.*, **66**, 2687 (1962). Acid-base constants in 1 M NaCl have been determined by S. Bruckenstein and D. C. Nelson, *J. Chem. Eng. Data*, **6**, 605 (1961). Neutral salt effects on the pH of dilute solutions of strong acids have been studied by Schwabe and his coworkers; see K. Schwabe, *Z. Physik. Chem.*, **41**, 368 (1964); *Electrochim. Acta.* **12**, 67 (1967); G. Beinroth, K. Schwabe, and H. D. Suschke, *Z. Physik. Chem.*, **235**, 133 (1967).
[10] R. G. Bates, P. T. Diamond, M. Eden, and S. F. Acree, *J. Res. Nat. Bur. Stand.*, **37**, 251 (1946).

TABLE 5–5. Salt Effects (Δpa_H) on Three Standard Buffer Solutions

Buffer solution	Molality of Added Salt		
	0.01	0.02	0.05
Potassium hydrogen phthalate, $0.05m$ (added KCl)	−0.008	−0.019	−0.044
Potassium dihydrogen phosphate, $0.025m$; disodium hydrogen phosphate, $0.025m$ (added NaCl)	−0.012	−0.022	−0.051
Borax, $0.01m$ (added NaCl)	−0.007	−0.014	−0.035

Sources: Phthalate buffer: W. J. Hamer and S. F. Acree, *J. Res. Nat. Bur. Stand.* **32**, 215 (1944); phosphate buffer: R. G. Bates and S. F. Acree, *ibid.*, **34**, 373 (1945); borax buffer: G. G. Manov, N. J. DeLollis. P. W. Lindvall, and S. F. Acree, *ibid.*, **36**, 553 (1946).

The pa_H of these buffer solutions is given in terms of the mass law by

$$pa_H = -\log K_2 - \log \frac{m_{HPs^-}}{m_{Ps^{2-}}} - \log \frac{\gamma_{HPs^-}}{\gamma_{Ps^{2-}}} \tag{36}$$

where K_2 is the second dissociation constant of p-phenolsulfonic acid. Inasmuch as the last term of equation 36, like the activity coefficient of an individual ionic species, is physically undefined and not measurable, the authors studied the character of the salt effect on the quantity $\gamma_{HPs^-}\gamma_{Cl^-}^0/\gamma_{Ps^{2-}}$, in which $\gamma_{Cl^-}^0$ represents the activity coefficient in the chloride-free buffer solution. The activity coefficient term was obtained from e.m.f. measurements of cells with hydrogen and silver-silver chloride electrodes containing the buffer-chloride mixtures with and without another added salt:

$$-\log \frac{\gamma_{HPs^-}\gamma_{Cl^-}}{\gamma_{Ps^{2-}}} = \frac{(E - E^\circ)F}{RT \ln 10} + \log K_2 + \log \frac{m_{HPs^-}m_{Cl^-}}{m_{Ps^{2-}}} \tag{37}$$

The salts studied were potassium nitrate, sodium sulfate, and sodium citrate. The effect of the sodium chloride, added to assure proper functioning of the silver chloride electrode, was eliminated by extrapolation.

The values of $\log(\gamma_{HPs^-}\gamma_{Cl^-}^0/\gamma_{Ps^{2-}})$ at 25 °C for phenolsulfonate buffers with sufficient added salt to contribute one-fifth of the total ionic strength are listed in Table 5–6, together with the values of the same quantity in buffer solutions without added salt. No specific effects larger than the

TABLE 5–6. SALT EFFECTS ON $\log (\gamma_{HPs} - \gamma^0_{Cl^-}/\gamma_{Ps^{2-}})$ AT 25 °C

(The added salt contributes one-fifth of the ionic strength)

Ionic Strength	$\log(\gamma_{HPs} - \gamma^0_{Cl^-}/\gamma_{Ps^{2-}})$ for Buffer Solutions Containing			
	No Added Salt	KNO_3	Na_2SO_4	$Na_3C_6H_5O_7$
0.02	0.104	0.104	0.103	0.103
0.05	0.140	0.140	0.140	0.139
0.1	0.169	0.170	0.169	0.168
0.2	0.196	0.198	0.197	0.195
0.3	0.209	0.211	0.210	0.208
0.4	0.215	0.219	0.217	0.215
0.5	0.219	0.223	0.221	0.219

uncertainty of the measurement are apparent. The effectiveness of salts of the 1-1, 1-2, and 1-3 valence types at equal ionic strengths in altering the activity coefficients appears to be nearly the same, and no marked influence of the increased ionic charge is evident. A parallel experiment indicated that the salt effect of barium chloride is no larger than that of sodium sulfate.

Unfortunately, this evidence bears only indirectly on the problem of the salt effect on pa_H. If the change of $\log(\gamma_{HPs} - /\gamma_{Ps^{2-}})$ with addition of neutral salt approximates the behavior of $\log(\gamma_{HPs} - \gamma^0_{Cl^-}/\gamma_{Ps^{2-}})$, however, it may be judged that the salt effect on the pa_H of phenolsulfonate buffer solutions is governed almost exclusively by the increase in ionic strength. The pa_H of the solution after addition of salt would then be nearly the same as that of a more concentrated solution of the salt-free buffer with the same total ionic strength.

It seems evident that the salt effect may be regarded, in approximate fashion, as a "concentration effect," opposite in sign to the dilution effect discussed in the foregoing section. This is strictly true, however, only for buffer solutions near the neutral pH, where the dilution effect, as well as the salt effect, is substantially the result of changes in activity coefficients alone. In Table 5–7 the salt effect of potassium chloride on the 0.05 M solution of potassium hydrogen phthalate and of sodium chloride on three phosphate buffer solutions and the 0.01 M solution of borax is compared with the pa_H change $\Delta pH_{1/2}$ on dilution of the pure buffer solution from the higher ionic strength to the lower. The dilution effect calculated by the equations of the previous section is given in the last column. Except for the borax buffer solution, the agreement is good. The difference can doubtless be ascribed in this case to the well-known tendency for complex polyboric acids to form when the concentration of boric acid is increased.

TABLE 5-7. COMPARISON OF SALT EFFECTS AND DILUTION EFFECTS

Buffer Solution	Initial I	Final I	Salt Effect, Δpa_H	$\Delta pH_{1/2}$ Observed	$\Delta pH_{1/2}$ Calculated
KH phthalate, 0.05 M	0.053	0.106	−0.044	0.057	0.061
KH$_2$PO$_4$, 0.025 M;					
Na$_2$HPO$_4$, 0.025 M	0.1	0.2	−0.088	0.088	0.096
KH$_2$PO$_4$, 0.01 M;					
Na$_2$HPO$_4$, 0.01 M	0.04	0.08	−0.074	0.073	0.079
KH$_2$PO$_4$, 0.005 M;					
Na$_2$HPO$_4$, 0.005 M	0.02	0.04	−0.057	0.059	0.065
Na$_2$B$_4$O$_7$, 0.01 M	0.02	0.04	−0.014	0.004	0.018

pa_H and Temperature

STRONG ACIDS AND BASES

The pa_H of solutions of strong acids and bases of the 1-1 valence type at molalities m_a and m_b exceeding 10^{-6} mol kg^{-1} is approximately $-\log m_a \gamma$ and $-\log K_w/m_b\gamma$, respectively, where γ is the activity coefficient of an average univalent ion. It should be remembered that concentrations on the molal scale do not change with temperature. At constant pressure and concentration the change of pa_H with T, the thermodynamic temperature, is, for solutions of strong acids,

$$\left(\frac{\partial pa_H}{\partial T}\right)_{P,m} = -\left(\frac{\partial \log \gamma}{\partial T}\right)_{P,m} \tag{38}$$

and for solutions of strong bases,

$$\left(\frac{\partial pa_H}{\partial T}\right)_{P,m} = -\left(\frac{\partial \log K_w}{\partial T}\right)_{P} + \left(\frac{\partial \log \gamma}{\partial T}\right)_{P,m} \tag{39}$$

BUFFER SOLUTIONS

Expressions for the effect of temperature changes on the pa_H of buffer solutions are obtained by differentiating the appropriate mass law equations with respect to temperature. Changes in the concentration terms with temperature, although of secondary importance, must be considered in the exact treatment, for the extent to which water reacts with (that is, hydrolyzes) the components of the buffer is a function of the temperature. The temperature coefficient obtained by neglecting the influence of temperature on the concentration term is an excellent approximation for solutions of intermediate pH and will fulfill adequately the needs of the present discussion.

The change of the pa_H of a buffer solution composed of a single weak acid and its salt can be derived from equations similar to equation 34. The general expression is[11]

$$\frac{\partial pa_H}{\partial T} \approx -\frac{\partial \log K}{\partial T} - (2z + 1)\frac{\partial \log \gamma}{\partial T} \tag{40}$$

For a buffer solution consisting of a weak base and its salt (compare equation 35),

$$\frac{\partial pa_H}{\partial T} \approx -\frac{\partial \log K_w}{\partial T} + \frac{\partial \log K_b}{\partial T} - (2z + 1)\frac{\partial \log \gamma}{\partial T} \tag{41}$$

Similarly, for solutions of the acid salt of a dibasic acid, we find from equation 31

$$\frac{\partial pa_H}{\partial T} \approx -\frac{1}{2}\frac{\partial \log K_1}{\partial T} - \frac{1}{2}\frac{\partial \log K_2}{\partial T} + 2\frac{\partial \log \gamma}{\partial T} \tag{42}$$

The values of z for buffers of several valence types are given in Table 5–1.

ESTIMATION OF $\partial \log K/\partial T$ AND $\partial \log \gamma/\partial T$

The temperature coefficients of $\log K$ and $\log \gamma$, in terms of which the temperature coefficient of pa_H is expressed, can readily be obtained or estimated for many buffer systems from data in the literature. At constant pressure, the van't Hoff equation gives

$$\frac{\partial \ln K}{\partial T} = \frac{\Delta H°}{RT^2} \tag{43}$$

where R is the gas constant and $\Delta H°$ is the change of heat content for the dissociation of 1 mole of the acid or base in the standard state. Furthermore, the temperature coefficient of $\log \gamma$ can be estimated from the relative partial molal heat content $\bar{L}_2$ of uni-univalent electrolytes, for γ, the activity coefficient of the "average" ion is, to the extent of our knowledge, identical with $\gamma_\pm$, the mean activity coefficient of an "average" electrolyte of this valence type. Hence

$$\frac{\partial \ln \gamma}{\partial T} = -\frac{\bar{L}_2}{2RT^2} \tag{44}$$

---

[11] In dilute solutions the activity coefficient of an uncharged species is nearly unity at all temperatures, and the valence relationships of the Debye-Hückel theory serve to relate the activity coefficients of ions of different charges. Hence the activity coefficient of a bivalent ion is approximately γ^4 and that of a trivalent ion γ^9, where γ is the activity coefficient of a univalent ion.

It is somewhat more convenient to estimate $\partial \log \gamma / \partial T$ by this equation than by differentiating the Debye-Hückel equation with respect to temperature.

The values of $\bar{L}_2$ for the halogen acids, alkali halides, and sodium hydroxide in their aqueous solutions at 25 °C vary from 200 to 300 joules per mole at an ionic strength of 0.01, from 300 to 550 J mol⁻¹ at $I = 0.04$, and from 300 to 800 J mol⁻¹ at $I = 0.1$.[12] The values are positive in this range of concentrations. Hence $\partial \log \gamma_\pm / \partial T$ appears to be of the order of -0.00007 per kelvin at $I = 0.01$, -0.00012 K⁻¹ at $I = 0.04$, and -0.0002 K⁻¹ at $I = 0.1$. The values for hydrochloric acid, sodium and potassium chlorides, and sodium hydroxide, all at 25 °C, are given in Table 5–8.

TABLE 5–8. $\partial \log \gamma_\pm / \partial T$ FOR STRONG ELECTROLYTES AT 25 °C

| | Ionic Strength (molality) | | |
	0.01	0.05	0.1
HCl	−0.00009	−0.00018	−0.00025
NaCl	−0.00007	−0.00012	−0.00013
KCl	−0.00007	−0.00010	−0.00011
NaOH	−0.00008	−0.00014	−0.00016

The plots of $-\log K$ (pK) as a function of temperature are roughly parabolic in form, and pK for many weak electrolytes has a minimum value in the range 0 to 60 °C. Hence, $\partial \log K / \partial T$ may be either positive or negative at room temperature. In general, the characteristic minimum is displaced toward higher temperatures as the acid strength decreases, so that pK increases with temperature for the stronger acids and higher temperatures and decreases with temperature for weaker acids and lower temperatures. For weak bases, pK_b usually decreases with rising temperature. This variation is illustrated in Table 5–9, where the temperature coefficients of $-\log K$ for water and a number of weak acids and bases are listed. The temperature coefficient of pa_H for several solutions at 25 °C is given in Table 5–10. The solutions are arranged in order of increasing pH.

GENERAL CONSIDERATIONS

We are now in a position to draw some general conclusions regarding the effect of temperature on pH values. In dilute solutions the change of the activity coefficient with temperature is negative and uniform over the whole

---

[12] H. S. Harned and B. B. Owen, *The Physical Chemistry of Electrolytic Solutions*, 3rd ed., Reinhold Publishing Corp., New York, 1958. 1J = (1/4.184) cal.

TABLE 5-9. $\partial \log K/\partial T$ FOR WEAK ELECTROLYTES AT 25 °C

Weak Electrolyte	Constant	$-\partial \log K/\partial T$
Water	K_w	-0.0330
Formic acid	K_a	$+0.0001$
Acetic acid	K_a	$+0.0002$
Propionic acid	K_a	$+0.0004$
Lactic acid	K_a	$+0.0002$
Glycolic acid	K_a	-0.0004
Diethylbarbituric acid	K_a	-0.0143
Boric acid	K_a	-0.0082
Sulfuric acid	K_{2a}	$+0.0128$
Phosphoric acid	K_{1a}	$+0.0045$
	K_{2a}	-0.0024
Carbonic acid	K_{1a}	-0.0054
	K_{2a}	-0.0086
Oxalic acid	K_{2a}	$+0.0041$
Malonic acid	K_{2a}	$+0.0029$
Succinic acid	K_{1a}	-0.0019
	K_{2a}	$+0.0003$
d-Tartaric acid	K_{1a}	-0.0018
	K_{2a}	-0.0006
o-Phthalic acid	K_{1a}	$+0.0016$
	K_{2a}	$+0.0012$
Glycine	K_a	-0.0028
Ammonia	K_b	-0.0026
Methylamine	K_b	-0.0010
Ethanolamine	K_b	-0.0035
Tris(hydroxymethyl)aminomethane	K_b	-0.0054
Glycine	K_b	-0.0067

pH range. This change tends to lower $\partial pa_H/\partial T$ for strong bases, acid salts, and weak acid buffers (negative z) and to raise it for strong acids and buffers containing weak bases (positive z). The effect of the altered activity coefficient is, however, so small that it is often completely overshadowed by $\partial \log K/\partial T$. Changes of concentration are frequently without noticeable effect on the temperature coefficient.[13] For these reasons pa_H-temperature curves are of much the same form as the curves of pK vs. temperature. Like pK, the pa_H often passes through a minimum in the experimental range; the effect of the

[13] W. J. Hamer, *J. Wash. Acad. Sci.*, **29**, 551 (1939).

TABLE 5–10. Change of pa_H with Temperature at 25 °C

Solution	$\partial pa_H/\partial T$ pa_H units K^{-1}
Hydrochloric acid, 0.1 M	+0.0003
Potassium tetroxalate, 0.05 M	+0.001
Potassium hydrogen tartrate, 0.03 M	−0.0014
Potassium hydrogen phthalate, 0.05 M	+0.0012
Acetic acid, 0.1 M; sodium acetate, 0.1 M	+0.0001
Acetic acid, 0.01 M; sodium acetate, 0.01 M	+0.0002
Potassium dihydrogen phosphate, 0.025 M; disodium hydrogen phosphate, 0.025 M	−0.0028
Diethylbarbituric acid, 0.01 M; sodium diethylbarbiturate, 0.01 M	−0.0144
Ammonia, 0.1 M; ammonium chloride, 0.1 M	−0.0303
Borax, 0.01 M	−0.0082
Sodium bicarbonate, 0.025 M; sodium carbonate, 0.025 M	−0.0090
Trisodium phosphate, 0.01 M	−0.026
Sodium hydroxide, 0.1 M	−0.0332

term $\partial \log \gamma/\partial T$ is to shift the minimum to a temperature higher or lower than that characteristic of the dissociation constants alone. It is evident that the temperature coefficient of pH may appear deceptively small at temperatures in the vicinity of the minimum.

Most striking, perhaps, is the contrast between the effect of temperature changes on the pa_H of strong acids and on that of strong bases. The pa_H of the former increases slightly with increase of temperature through the influence of the small $\partial \log \gamma/\partial T$ term, whereas the latter reflects the large negative temperature coefficient of pK_w, the water constant. Indeed, as the pa_H value rises above and progressively beyond the neutral point, the ionization of water with its large heat effect plays an increasingly important part in the equilibria that fix the pa_H value. Hence the pa_H of alkaline solutions usually falls with rising temperature, whereas that of acid solutions usually increases.

Effect of Pressure Changes

Ionic equilibria are much less sensitive to pressure changes than to changes of temperature, and the alteration of pa_H caused by the normal fluctuations in barometric pressure is completely negligible. Indeed, Gibson and Loeffler[14]

[14] R. E. Gibson and O. H. Loeffler, *Trans. Am. Geophys. Union*, 503 (1941).

have pointed out that important effects of very high pressures are to be expected only when the solutions contain slightly dissociated acids and bases. They determined the change in the pH of phosphate, borate, carbonate, and ammonia buffer solutions with increase of pressure from 1 to 1200 atm by observing the changes in the absorption spectra of the indicators cresol red and bromphenol blue. The hydrogen ion concentration in the weak acid buffer solutions was found to increase by approximately 100 per cent when the pressure was raised from 1 to 1000 atm, whereas that in the ammonia solution decreased in about the same ratio. These results are in general agreement with the conductance measurements of Brander.[15]

The magnitude of pressure effects is governed by the volume changes resulting from electrostriction of the solvent, which increase as the ionic charge increases and as the radius of the ions decreases. Hence pressure effects can be estimated from electrostatic theory. The change of the ionization constant of water with increase of pressure was calculated by Owen and Brinkley[16] from the partial molal volumes and compressibilities of ions. The relative increase in K_w with pressure is greater the lower the temperature. At 25 °C an increase of pressure from 1 to 1000 atm increases K_w in the ratio 1 : 2.36; at 0 °C the ratio is 1 : 2.8. Current research in oceanography promises to shed further light on the effect of high pressures on ionic equilibria and on pH values.

Buffers for pH Control

The yield of a chemical process, the rates of growth of organisms, and the electrodeposition of metals are among the many reactions that are frequently profoundly influenced by the pH of the medium. In the study of reaction kinetics as well as chemical equilibria it is often necessary to adjust the acidity to a definite, perhaps predetermined, level and to nullify changes of pH that would otherwise occur during the course of the reaction. Both of these objectives can be accomplished by the addition of buffer substances.

In choosing a suitable buffer system, one should not only consider the pH required but should also take into account the nature of the reaction to be regulated. A high buffer capacity may be essential for the control of an acid-base reaction; if dilution of the medium is likely to occur, a low dilution value is desirable. Salt effects and the possibility of temperature changes will occasionally influence the selection. The chemical nature of the buffer materials may be a factor, for the added substances must not form insoluble compounds or complexes or enter into other undesired side reactions with

[15] E. Brander, *Soc. Sci. Fennica, Commentationes Phys.-Math.*, **6**, 3 (1932); No. 17, 1 (1936).
[16] B. B. Owen and S. R. Brinkley, Jr., *Chem. Rev.*, **29**, 461 (1941).

the medium. In the first part of this chapter we considered the influence of the buffer type and the buffer concentration on the buffer value, the dilution value, the salt effect, and the temperature effect. The discussion will serve as a guide to the choice of a suitable buffer system.

The range of the buffering effect of a single weak acid group or weak base group is approximately 2 pH units, or 1 unit on either side of $-\log K$. When there are two or more "overlapping" groups per molecule, the range is broader. The useful ranges of some buffer systems are collected in Table 5–11. When a buffer system is used for control purposes, the components must be added to the medium under study. In precise work, the pH of the buffered medium should be determined when all the reactants are present. Directions for the preparation of buffers of a particular pH can, however, be found in the sources indicated in the table. The compositions, buffer values, and pH of a number of useful buffer systems are listed in Tables 8 and 9 of the Appendix.

UNIVERSAL BUFFER MIXTURES

Universal buffer solutions, which have a relatively high buffer capacity over a wide pH range, have been proposed by Prideaux and Ward[17] and by Britton and Robinson.[18] The Prideaux and Ward buffer mixture is composed of three acids, namely phosphoric, phenylacetic, and boric. The titration curves of the five acid groups overlap to such an extent that the pH varies from 2 to 12 in approximately linear fashion with the amount of strong alkali added.

The Britton-Robinson universal buffer mixture also has a range from pH 2 to 12. It is 0.02857 M with respect to each of the four components: diethylbarbituric acid, citric acid, potassium dihydrogen phosphate, and boric acid. There are thus seven acid groups, and the neutralization of these with strong alkali provides an exact linear relationship between pH and volume of alkali within the range pH 4 to 8.4. Britton and Welford[19] have established the neutralization curves for this mixture at seven temperatures from 12.5 to 91 °C.

THE pH RANGE 6 TO 9

There are many weak acids with dissociation constants in the range 10^{-3} to 10^{-6}; hence buffer solutions for pH control in the acid range are

---

[17] E. B. R. Prideaux and A. T. Ward, *J. Chem. Soc.*, **125**, 426 (1924).
[18] H. T. S. Britton and R. A. Robinson, *J. Chem. Soc.*, 458 (1931); H. T. S. Britton, *Hydrogen Ions*, 4th ed., Vol. I, D. Van Nostrand Co., Princeton, N.J., 1956. Another universal buffer mixture has been proposed by D. A. Ellis, *Nature*, **191**, 1099 (1961). It is composed of sodium carbonate, sodium dihydrogen phosphate, citric acid, and 2-amino-2-methyl-1,3-propanediol.
[19] H. T. S. Britton and G. Welford, *J. Chem. Soc.*, 1848 (1937).

TABLE 5-11. pH Ranges of Buffer Solutions

Acidic Component	Basic Component	pH Range	Ref.
HCl	Glycine	1.0–3.7	a
HCl	Na_2H citrate	1.0–5.0	a
p-Toluenesulfonic acid	Na p-toluenesulfonate	1.1–3.3	d
KH sulfosalicylate	NaOH	2.0–4.0	
HCl	KH phthalate	2.2–4.0	b
Citric acid	NaOH	2.2–6.5	e
Citric acid	Na_2HPO_4	2.2–8.0	e
Furoic acid	Na furoate	2.8–4.4	d
Formic acid	NaOH	2.8–4.6	
Succinic acid	Borax	3.0–5.8	e
Phenylacetic acid	Na phenylacetate	3.4–5.1	d
Acetic acid	Na acetate	3.7–5.6	e
KH phthalate	NaOH	4.0–6.2	b
NaH succinate	Na_2 succinate	4.8–6.3	d
Na_2H citrate	NaOH	5.0–6.3	a
NaH maleate	NaOH	5.2–6.8	
KH_2PO_4	NaOH	5.8–8.0	b
KH_2PO_4	Borax	5.8–9.2	c
NaH_2PO_4	Na_2HPO_4	5.9–8.0	a
HCl	Triethanolamine	6.7–8.7	
HCl	Na diethylbarbiturate	7.0–9.0	e
Diethylbarbituric acid	Na diethylbarbiturate	7.0–9.0	
H_3BO_3 or HCl	Borax	7.0–9.2	a, e
HCl	Tris(hydroxymethyl)-aminomethane	7.2–9.0	
H_3BO_3	NaOH	8.0–10.0	b
K p-phenolsulfonate	NaOH	8.2–9.8	
Glycine	NaOH	8.2–10.1	a
NH_4Cl	NH_4OH	8.3–9.2	
Glycine, Na_2HPO_4	NaOH	8.3–11.9	
HCl	Ethanolamine	8.6–10.4	
Borax	NaOH	9.2–11.0	a
$NaHCO_3$ or HCl	Na_2CO_3	9.2–11.0	c, e
Borax	Na_2CO_3	9.2–11.0	c
Na_2HPO_4	NaOH	11.0–12.0	e

* S. P. L. Sørensen, *Biochem. Z.*, **21**, 131 (1909); **22**, 352 (1909); *Ergeb. Physiol.*, **19**, 393 (1912).
* W. M. Clark and H. A. Lubs, *J. Biol. Chem.*, **25**, 479 (1916); W. M. Clark, *The Determination of Hydrogen Ions*, 3rd ed., Chapter 9, The Williams and Wilkins Co., Baltimore, Md., 1928.
* I. M. Kolthoff and J. J. Vleeschhouwer, *Biochem. Z.*, **189**, 191 (1927).
* W. L. German and A. I. Vogel, *Analyst*, **62**, 271 (1937).
* H. T. S. Britton, *Hydrogen Ions*, 4th ed., Vol. I, Chapter 17, D. Van Nostrand Co., Princeton, N.J., 1956.

rather plentiful. This is not the case, however, in the physiologically important range pH 6 to 9. Furthermore, phosphate, carbonate and borate buffers,
which might otherwise be suitable, precipitate calcium or enter into other
disturbing side reactions with many media of biochemical interest. Diethylbarbiturate buffers inhibit certain enzyme systems, and some phenols have
undesirable antiseptic properties.

Léonis[20] has found that buffer solutions for this important range can be
prepared from N-dimethylleucylglycine. The pH values are said to change
but little with temperature. Buffers containing diethanolamine, triethanolamine, or imidazole (glyoxaline, $C_3H_4N_2$)[21] are sometimes useful. In addition
Gomori[22] has proposed three other base-salt systems suitable for pH control
in the range pH 6.5 to 9.7. One of the bases, collidine (2,4,6-trimethyl-
pyridine), is a liquid, and the other two are solids at room temperature. They
are 2-amino-2-(hydroxymethyl)-1,3-propanediol, more commonly known as
the tris(hydroxymethyl)aminomethane, THAM, or simply "tris", and the
closely related base 2-amino-2-methyl-1,3-propanediol.

Tris is perhaps the most useful of these buffer substances.[23] Both the free
base and its hydrochloride are available commercially in a state of high
purity.[24] The pH, compositions, and properties of tris buffers at 25 °C are
given in Table 9 of the Appendix and on pages 90 and 98.

Good and his coworkers[25] selected 12 buffer substances with pK_a ranging
from 6.15 to 8.35 and tested them for their suitability for biological pH
control. Ten were amino acids and two were primary aliphatic amines. The
amino acids were either N-substituted glycines or N-substituted taurines. It
was shown that the stabilities of the complexes formed between these ligands
and the metal cations Mg^{2+}, Ca^{2+}, Mn^{2+}, and Cu^{2+} are, in general, small.
Approximate pK_a values at 20 °C were given, and trivial names for the buffer
substances were proposed. Several of these buffers are commercially available
for example, five of the most promising, namely Bicine, Tricine, HEPES, TES,
and MES are supplied by Calbiochem of Los Angeles, California.

Buffer substances useful for pH control in biological media, pH 6 to 9, are
listed in Table 5–12. The buffers recommended by Good are included. The

---

[20] J. Léonis, *Compt. Rend. Trav. Lab. Carlsberg*, *Sér. Chim.*, **26**, 357 (1948).

[21] A. H. M. Kirby and A. Neuberger, *Biochem. J.*, **32**, 1146 (1938); E. T. Mertz and
C. A. Owen, *Proc. Soc. Exptl. Biol. Med.*, **43**, 204 (1940). According to H. Thies and
G. Kallinich, *Biochem. Z.*, **324**, 485 (1953), a mixture of the phosphate salts of mono
ethanolamine and triethanolamine is a useful buffer from pH 5.5 to 12.8.

[22] G. Gomori, *Proc. Soc. Exptl. Biol. Med.*, **62**, 33 (1946).

[23] R. G. Bates, *Ann. N.Y. Acad. Sci.*, **92**, 341 (1961). A detailed summary of the properties
of amine buffers, including pH ranges, temperature coefficients of pH, buffer capacity, salt
effects, and dilution effects, is given in this paper.

[24] J. A. Riddick, *Ann. N. Y. Acad. Sci.*, **92**, 357 (1961).

[25] N. E. Good, G. D. Winget, W. Winter, T. N. Connolly, S. Izawa, and R. M. M. Singh
Biochem., **5**, 467 (1966).

TABLE 5–12. BUFFER SUBSTANCES FOR pH CONTROL IN BIOLOGICAL MEDIA, pH 6 TO 9[a]

Substance	Trivial Name	pK_a at 20 °C
1. 2-(*N*-Morpholino)ethanesulfonic acid	MES	6.15
2. 2,2-Bis(hydroxymethyl)-2,2′,2″-nitrilotriethanol	Bis-tris	6.567[b]
3. *N*-(2-Acetamido)iminodiacetic acid	ADA	6.6
4. Piperazine-*N*,*N*′-bis(2-ethanesulfonic acid)	PIPES	6.8
5. *N*-(2-Acetamido)-2-aminoethanesulfonic acid	ACES	6.9
6. (2-Aminoethyl)trimethylammonium chloride hydrochloride	Cholamine chloride	7.1
7. *N*,*N*-Bis(2-hydroxyethyl)-2-aminoethanesulfonic acid	BES	7.15
8. *N*-Tris(hydroxymethyl)methyl-2-aminoethanesulfonic acid	TES	7.5
9. *N*-2-Hydroxyethylpiperazine-*N*′-2-ethanesulfonic acid	HEPES	7.55
10. *N*-(2-Acetamido)glycine	Acetamidoglycine	7.7
11. *N*-Tris(hydroxymethyl)methylglycine	Tricine	8.15
12. Glycinamide hydrochloride	Glycinamide	8.2
13. Tris(hydroxymethyl)aminomethane	Tris	8.214[c]
14. Glycylglycine	Glycylglycine	8.4
15. *N*,*N*-Bis(2-hydroxyethyl)glycine	Bicine	8.413[c]
16. 2-Amino-2-methyl-1,3-propanediol	AMP	8.951[d]

[a] From Good *et al.* (reference 25) unless otherwise noted.
[b] Reference 29.
[c] Reference 26.
[d] H. B. Hetzer and R. G. Bates, *J. Phys. Chem.*, **66**, 308 (1962).

pK_a of tris.H^+ has been found to range from 8.850 at 0 °C to 7.437 at 50 °C and that of Bicine to range from 8.753 at 0 °C to 7.979 at 55 °C.[26] A series of buffer combinations useful in mammalian cell culture, including several from the list of Good, has been recommended.[27]

Another useful buffer substance with pK_a about 6.5 at 25 °C was proposed by Lewis.[28] Its acid-base properties from 0 to 50 °C were studied in detail by

[26] Tris: R. G. Bates and H. B. Hetzer, *J. Phys. Chem.*, **65**, 667 (1961); S. P. Datta, A. K. Grzybowski, and B. A. Weston, *J. Chem. Soc.*, 792 (1963). Bicine: S. P. Datta, A. K. Grzybowski, and R. G. Bates, *J. Phys. Chem.*, **68**, 275 (1964).
[27] H. Eagle, *Science*, **174**, 500 (1971).
[28] J. C. Lewis, *Anal. Biochem.*, **14**, 495 (1966).

Paabo and Bates.[29] This compound, termed Bis-tris, is a di-N-substituted derivative of tris, in which the two hydrogens of the primary amino group are replaced by $-C_2H_4OH$. The free base is named 2,2-bis(hydroxymethyl)-2,2',2''-nitriloethanol, and it is not difficult to prepare. The pK of protonated Bis-tris ranges from 6.931 at 0° C to 6.096 at 50 °C. Conventional pa_H values for buffer solutions containing Bis-tris and its hydrochloride in equal molalities (m from 0.02 to 0.1) at 11 temperatures from 0 to 50 °C have been determined.[2⁹]

BUFFER SOLUTIONS OF A DEFINITE IONIC STRENGTH

Buffer solutions of varying pH but fixed ionic strength can be prepared readily in the monobasic acid and monoacidic base systems. It is necessary, however, that the acid or base be so weak that the contribution of its ionization to the total ionic strength can be ignored.

A solution of the salt NaA of the weak monobasic acid is prepared, and sufficient neutral salt, for example potassium chloride, is added to give the desired ionic strength. The hydrochloric acid solution that is added to form the buffers of the type HA,A⁻ will dilute the stock solution and lower the ionic strength; consequently, neutral salt must be added to the acid reagent in sufficient quantity to compensate exactly for this dilution. The concentration of NaA need only be great enough to yield mixtures of adequate buffer capacity. Weak base buffer mixtures can be prepared in a similar manner from a stock solution of the hydrochloride of the base (B·HCl) with added neutral salt by addition of varying amounts of a reagent containing sodium hydroxide and neutral salt.

As long as the number of moles of hydrochloric acid (or sodium hydroxide) added does not exceed the amount of NaA (or B·HCl) present, the ionic strength remains constant and the pH may be shifted over the entire useful range of the buffer system. Representative compositions of buffer mixtures of these two types are given for five ionic strengths in Table 5–13. The useful ranges of buffer systems suited to this method are indicated.

Buffer solutions of varying pH but constant ionic strength (I from 0.5 to 1.0) have been described by Elving, Markowitz, and Rosenthal.[30] The buffer compositions were selected from the McIlvaine formulas,[31] and the desired ionic strength is maintained by the addition of potassium chloride.

---

[29] M. Paabo and R. G. Bates, *J. Phys. Chem.*, **74**, 702 (1970).

[30] P. J. Elving, J. M. Markowitz, and I. Rosenthal, *Anal. Chem.*, **28**, 1179 (1956). The preparation of Britton-Robinson buffer solutions of constant ionic strength has been described by J. A. C. Frugoni, *Gazz. Chim. Ital.*, **87**, 403 (1957).

[31] T. C. McIlvaine, *J. Biol. Chem.*, **49**, 183 (1921). These buffer solutions are prepared from citric acid and disodium hydrogen phosphate.

TABLE 5-13. PREPARATION OF BUFFER MIXTURES OF A DEFINITE
IONIC STRENGTH

	Monobasic Weak Acid System			
	Base Stock Solution		Acid Reagent	
Ionic Strength	NaA	KCl	HCl	KCl
0.05	0.05 M	—	0.2 M	0.05 M
0.1	0.05 M	0.05 M	0.2 M	0.1 M
0.15	0.05 M	0.1 M	0.2 M	0.15 M
0.2	0.05 M	0.15 M	0.2 M	0.2 M
0.25	0.05 M	0.2 M	0.2 M	0.25 M

Useful Buffer Ranges	
A	pH
Formate	3.0–4.5
Phenylacetate	3.5–5.0
Acetate	4.0–5.5
Barbiturate	7.0–9.0
Borate	9.0–10.0

	Monoacidic Weak Base System			
	Acid Stock Solution		Alkaline Reagent	
Ionic Strength	B.HCl	KCl	NaOH	KCl
0.05	0.05 M	—	0.2 M	0.05 M
0.1	0.05 M	0.05 M	0.2 M	0.1 M
0.15	0.05 M	0.1 M	0.2 M	0.15 M
0.2	0.05 M	0.15 M	0.2 M	0.2 M
0.25	0.05 M	0.2 M	0.2 M	0.25 M

Useful Buffer Ranges	
B	pH
Triethanolamine	7.0–8.5
Tris(hydroxymethyl)-aminomethane	7.2–9.0
Ammonia	8.2–9.2
Ethanolamine	8.6–10.4

CHAPTER 6

measurements of acidity
with indicators

As we have seen, the hydrogen electrode is the fundamental standard by which reference values of pH are assigned. Consequently, the dissociation constants of many (but not all) acid-base indicators and the color tone of indicator solutions are likewise dependent on electrometric standards. Therefore colorimetry with acid-base indicators must be regarded as a secondary means of determining pH values.

Although the indicator properties of naturally occurring coloring matters have been recognized for centuries, the determination of acidity by optical means is of more than historical interest. Measurements of this sort can be performed rapidly and reproducibly, and the technique may be so simple that it is easily mastered by untrained personnel. The apparatus required for visual colorimetry is inexpensive and portable. Photometric titrations[1] are amenable to automation, and colorimetry can sometimes be used to advantage in industrial process control.[2] Differential spectrophotometry with indicators may well provide the most precise detection of the equivalence point of acid-base titrations that has yet been devised.[3]

Under optimum conditions, indicator pH measurements in aqueous media can be made as precise as electrometric determinations. Colorimetry and photometry can likewise yield highly accurate values of the dissociation constants of moderately strong acids and bases whose equilibria are not well

---

[1] A. Ringbom and F. Sundman, Z. Anal. Chem., 116, 104 (1939).
[2] K. Schwabe, Fortschritte der pH-Messtechnik, Chapter 2, VEB Verlag Technik, Berlin, 1958.
[3] See S. Bruckenstein and M. M. T. K. Gracias, Anal. Chem., 34, 975 (1962).

suited for accurate analysis by electrometric methods.[4,5] In addition, an important part of our information on the effect of solvent character on the interactions of acids and bases has been derived from indicator studies,[6] and the useful Hammett acidity function is based on colorimetry.[7]

A full discussion of colorimetric measurements is beyond the scope of this chapter. For a detailed consideration of indicators, their behavior and their applications in pH measurements, the reader is referred to the excellent monograph by Kolthoff and Rosenblum[8] and to the earlier but no less authoritative book by Clark.[9]

Behavior of Acid-Base Indicators

MECHANISM OF THE COLOR CHANGE

Acid-base indicators are essentially weak acids or bases that exhibit a change of color on the conversion of the acidic form to the basic form, or vice versa. Although these colors are not necessarily, as Ostwald believed,[10] characteristic of the undissociated and ionic forms of the indicator molecule, the gradual removal of protons from the acidic form is accompanied by a progressive change in the optical absorption characteristics of the solution, that is, in the *color tone*. The measurement of pH by means of indicators is usually based on the presumption that two solutions of the same color tone have equal pH, provided the same indicator is present at the same concentration in each of them and the two solutions are at the same temperature. As we shall see, this is not always true.

The first useful indicators were of natural origin. New and improved acid-base indicators were later synthesized, and the correlation of color with structure has been the subject of intensive investigations. These studies led Kolthoff to formulate a new definition of an indicator, differing somewhat from that of Ostwald: "Indicators are (apparent) weak acids or bases of

[4] C. R. Singleterry, Ph.D. thesis, University of Chicago, 1940.

[5] The advantages of a combination of e.m.f. and optical data for this purpose have been set forth by R. G. Bates and G. Schwarzenbach, *Helv. Chim. Acta*, **37**, 1069 (1954).

[6] See, for example, the extensive work of Hantzsch, of which the following papers are representative: A. Hantzsch, *Z. Elektrochem.*, **29**, 221 (1923); **30**, 194 (1924); **31**, 167 (1925); *Ber.*, **58**, 612, 941 (1925); *Z. Physik. Chem.*, **125**, 251 (1927).

[7] L. P. Hammett and A. J. Deyrup, *J. Amer. Chem. Soc.*, **54**, 2721, 4239 (1932); L. P. Hammett, *Chem. Rev.*, **13**, 61 (1933).

[8] I. M. Kolthoff, *Acid-Base Indicators* (translated by C. Rosenblum), The Macmillan Co., New York, 1937.

[9] W. M. Clark, *The Determination of Hydrogen Ions*, 3rd ed., Chapters 3 to 8, The Williams and Wilkins Co., Baltimore, Md., 1928.

[10] See W. M. Clark, *ibid.*, Chapter 5.

which the ionogenic (aci- or baso-respectively) form possesses a color and constitution different from the color and structure of the pseudo- or normal compound."[11]

Detailed examination of the behavior and constitution of indicator dyes shows that the structural changes which accompany or bring about the change of color are often quite complex. Most phthaleins, for example, apparently exist largely in a colorless lactone form in moderately acidic media, whereas the colored alkaline form has a quinone-phenolate structure. The structures of these two forms of phenophthalein are

Phenolphthalein—colorless lactone   Phenolphthalein—red quinone-phenolate
(acid form)   (alkaline form)

The lactone form behaves as a dibasic acid. Removal of one proton from the hydrated lactone produces a univalent carboxylate ion which is colorless. The colored form results from removal of a second proton and tautomeric rearrangement. The color fades slowly in strong alkali as a colorless carbinol is formed. In strongly acidic media a colored quinoid structure is also present.

In the sulfonphthaleins, perhaps the most useful series of acid-base indicators, the presence of a sulfonic acid group in the molecule makes the indicator more soluble in aqueous solutions than are the corresponding phthaleins. The stability in alkali is also increased, but at the same time the electric charge type is altered.

The sulfonphthalein indicators, which are listed in Table 6-1, have two color transformations. Although quinoid configurations may exist, the yellow form is usually represented as a hybrid ion (with both positive and negative charges) to which the sulfonate group imparts an extra negative charge. The charge type of the yellow form may be denoted $A^{\mp-}$ (or $B^{\mp-}$). The color change in strongly acidic media is believed to result from the conversion of this compound into a hybrid ion $A^{\mp}$, or even into a cationic

---

[11] Kolthoff and Rosenblum, *op. cit.*, p. 234.

species in which the sulfonic acid group is undissociated. The change of color in alkaline solutions is doubtless accompanied by dissociation of the phenol group to give a hybrid ion with two extra negative charges ($B^{\mp=}$).[12, 13] The probable structures of three forms of phenolsulfonphthalein (phenol red) are as follows:

| Phenol red–
orange form
(strong acids) | Phenol red–
yellow acid
form | Phenol red–
red form
(alkaline solutions) |

The azo indicators, on the other hand, are often yellow or orange in neutral or alkaline solutions and turn red in the presence of acids. The red acidic form of methyl orange is a hybrid ion that is converted into a singly charged anionic species (orange) by removal of a proton:[14]

| Methyl orange–
red form
(acid solutions) | Methyl orange–
orange form
(alkaline solutions) |

In spite of this complex behavior, it is nonetheless true that the equilibria that account for the color change of acid-base indicators can be treated successfully by the simple formulation

$$\text{HIn (color I)} \rightleftharpoons \text{In (color II)} + H^{+} \qquad (1)$$

The requirements of the simple Ostwald theory are thus formally met, although the overall dissociation constant of the indicator acid or base is a function of several equilibrium and ionization constants involving the several

---

[12] H. Lund, *J. Chem. Soc.*, 1844 (1930).
[13] G. Schwarzenbach, *Helv. Chim. Acta*, **20**, 490 (1937); I. M. Kolthoff and L. S. Guss, *J. Amer. Chem. Soc.*, **60**, 2516 (1938).
[14] A. Meretoja, *Ann. Acad. Sci. Fennicae, Ser. A. II*, No. 12 (1944).

TABLE 6–1. SULFONPHTHALEIN INDICATORS

Indicator	Trans-formation Range, pH	Color Change Acid → Alk.	Absorption Maximum	
			Acidic Form, nm	Basic Form, nm
Metacresol purple	1.2–2.8	Red–yellow	533	—
Thymol blue	1.2–2.8	Red–yellow	544	—
Bromophenol blue	3.0–4.6	Yellow-blue	—	592
Bromocresol green	3.8–5.4	Yellow-blue	—	617
Chlorophenol red	5.2–6.8	Yellow–red	—	573
Bromophenol red	5.2–6.8	Yellow–red	—	574
Bromocresol purple	5.2–6.8	Yellow–purple	—	591
Bromothymol blue	6.0–7.6	Yellow–blue	—	617
Phenol red	6.8–8.4	Yellow–red	—	558
Cresol red	7.2–8.8	Yellow–red	—	572
Metacresol purple	7.6–9.2	Yellow–purple	—	580
Thymol blue	8.0–9.6	Yellow–blue	—	596

Sources: W. M. Clark and H. A. Lubs, *J. Bact.*, **2**, 1, 109, 191 (1917); Clark, *op. cit.*, p. 94; B. Cohen, *U.S. Pub. Health Rpts.*, **41**, 3051 (1927).

tautomeric forms.[15] One need know little, in fact, about the possible resonant configurations that produce the change of color when the structure of the molecule is disturbed by the removal of a proton. In order to allow for salt effects and medium effects, however, the electric charges of the species HIn and In must be known, so that activity coefficients can be estimated. Calculations of this sort are notoriously unsatisfactory for large complex molecules and ions, and, indeed, it is still not completely clear how such a calculation should best be performed for hybrid ions. These uncertainties constitute the major barrier to highly accurate indicator pH measurements.

TRANSFORMATION RANGE

If, for the moment, activity coefficients are ignored and use is made of the concentration dissociation constant K'_{HIn} of the indicator, we can write

$$pm_H = pK'_{HIn} + \log \frac{m_{In}}{m_{HIn}} = pK'_{HIn} + \log \frac{\alpha}{1 - \alpha} \qquad (2)$$

---

[15] Clark, *op. cit.*, Chapter 5.

where α is the fraction of the indicator existing in the form In and pm_H is $-\log m_H$. It is evident that the logarithm of the ratio of concentrations of the colored forms varies linearly with pm_H and that the colored forms are present in equal concentrations when pm_H is equal to pK'_{HIn}.

Equation 2 shows further that the color change is not a sudden event, but instead the concentrations of the colored forms change continuously with the pm_H of the solution. As pm_H increases, the concentration of the colored form In increases steadily while that of the colored form HIn decreases steadily. The ratio of the concentrations of the two forms is fixed by the magnitude of the quantity $pm_H - pK'_{HIn}$. In Fig. 6–1 the fraction α of the indicator present in the alkaline form at various pm_H values is plotted for three common indicators. The logarithm of the ratio of the concentrations of the two forms, namely $\alpha/(1 - \alpha)$, is likewise shown.

It is clear from Fig. 6–1 and the related equation 2 that there exists for each indicator a region or interval of pH above and below which one or the other of the indicator forms is present in negligibly small concentration. For practical purposes, the width of this *transformation range* is determined by the sensitivity of the human eye or of the colorimeter used to detect one color in the presence of the other. A useful "rule of the thumb" for visual color

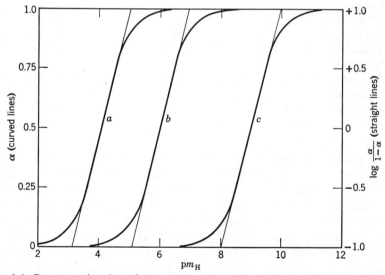

FIG. 6–1. Representative plots of the variation of α and $\log [\alpha/(1 - \alpha)]$ with pm_H. The intercepts of the straight lines with the top and bottom edges of the figure mark the approximate upper and lower limits, respectively, of the transformation range. (*a*) Bromophenol blue, (*b*) bromothymol blue, (*c*) thymol blue (alkaline range).

comparisons is that the lower limit of visual detection corresponds to about 10 per cent of one form in the presence of 90 per cent of the other. The limiting values of $\alpha/(1 - \alpha)$ are then approximately 0.1 and 10. The corresponding transformation range is accordingly two pH units wide, extending from a pm_H value of $(pK'_{HIn} - 1)$ on the low side to $(pK'_{HIn} + 1)$ on the high side. The intercepts of the straight lines in Fig. 6–1 with the top and bottom edges of the plot mark off this effective transformation interval for these three indicators.

There are, of course, many "one-color" indicators, only one form of which (usually the alkaline species) absorbs light in the visible region of the spectrum. Although the same equations hold for one-color indicators as for two-color indicators, the color of the mixture of two forms changes only in intensity as the pH of the solution is altered. The one-color indicators selected by Michaelis[16] are listed in Table 6–2, together with their colors and transformation ranges.

TABLE 6–2. ONE-COLOR INDICATORS RECOMMENDED BY MICHAELIS

Name	(Acid form colorless) Transformation Range, pH	Alkaline Color
Picric acid	0.0–1.3	Yellow
2,6-Dinitrophenol	1.7–4.4	Yellow
2,4-Dinitrophenol	2.4–4.0	Yellow
2,5-Dinitrophenol	4.0–5.8	Yellow
p-Nitrophenol	5.3–7.6	Yellow
m-Nitrophenol	6.4–8.8	Yellow
Phenolphthalein	8.2–9.8	Red
Salicyl yellow	10.0–12.0	Yellow

The practical transformation ranges of one-color indicators may be un symmetrical with respect to the value of pK'_{HIn}. This is because it is more difficult to detect the point at which the intensity of a color is substantially at its maximum than it is to determine the point at which the color is first perceptible. Likewise the human eye is more sensitive to some colors than to others. For these reasons the width of the effective transformation interval varies from one indicator to another and also with different total concentrations of the same indicator. The chief difficulties in the discrimination of color

---

[16] L. Michaelis and A. Gyemant, *Biochem. Z.*, **109**, 165 (1920); L. Michaelis and R. Krüger, *ibid.*, **119**, 307 (1921).

intensities can often be circumvented, however, by reducing the concentration of indicator when measurements are made near the colored end of the transformation range.

MIXED INDICATORS

For acid-base titrations and for certain other specialized applications indicators with a *transformation point* are often useful. The so-called "mixed indicators," consisting of an acid-base indicator and a suitable dye, have been developed to meet this need. A familiar example is the methyl orange-xylene cyanole mixture which has an easily detectable and sharp point of transformation at pH 3.8, near the equivalence point in the neutralization of carbonates with strong acid.[17] The improvement in the sharpness results from the superposition of the (pH-dependent) color of the indicator and the color of the dye.

A typical mixed indicator consists of a solution of 1 per cent phenolphthalein and 0.2 per cent methyl green. According to Kolthoff and Rosenblum,[18] this mixture imparts a green color to acidic solutions. The color turns gray when the color change of phenolphthalein commences, and at pH 8.8 the color is pale blue. At pH 9.0 and above, the red-violet form of phenolphthalein predominates and the color of the solution is violet. The blue color (pH 8.8) represents an easily detectable transformation point. Mixtures of two acid-base indicators extend the pH range that can be covered by a single indicator test solution. Apart from this, however, they are of limited application in the determination of pH values.

INDICATOR SOLUTIONS

The selection of indicators appearing in Table 6–3 is that of Kolthoff and Stenger.[19] The transformation range and the color change are given, and the electric charges borne by the acid form A and base form B have been added when it was possible to do so. The last column gives the "salt correction index" i_s of the indicator. This quantity is defined in a later section of this chapter.

In general, one to three drops of a 0.1 to 0.5 per cent aqueous indicator solution in each 10 ml of test solution (or reference buffer solution) is sufficient for pH measurements. When the indicator itself is only slightly soluble in water, it is often convenient to utilize the water-soluble salt form. If this is not available, the acid form can sometimes be dissolved in dilute alkali or a

[17] K. C. D. Hickman and R. P. Linstead, *J. Chem. Soc.*, **121**, 2502 (1922).
[18] *Op. cit.*, Chapter 5.
[19] I. M. Kolthoff and V. A. Stenger, *Volumetric Analysis*, 2nd ed., Vol. I, pp. 92–93, Interscience Publishers, New York, 1942.

solution of the acid form prepared in 50 per cent ethanol. An excess of alkali should be carefully avoided. Kolthoff suggests that the following solutions of the Michaelis one-color indicators (Table 6–2) are suitable: picric acid and the dinitrophenols, 0.04 per cent aqueous solutions of the sodium salts; mononitrophenols, 0.1 per cent aqueous solution; phenolphthalein, 0.04 per cent solution in 30 per cent ethanol; salicyl yellow, 0.1 per cent alcoholic solution for pH 10 to 11, 0.025 per cent alcoholic solution for pH 11 to 12.

Indicator solutions are available commercially, together with equipment (comparators, color charts, and artificial color standards) suitable for visual color matching with the aid of reference buffer solutions. The pH indicators sold by W. A. Taylor and Co. for use with the Taylor pH slide comparator are listed in Table 6–4.[20] Solutions of all of these indicators can be purchased; in addition, the first thirteen are also available in crystalline form.

TABLE 6–3. List of Recommended Indicators

Indicator	Transformation Range, pH	Color Change	Charge Type	i_s
o-Cresol red (acid range)	0.2–1.8	Red–yellow	$A^{\mp}B^{\mp-}$	1
Thymol blue (acid range)	1.2–2.8	Red–yellow	$A^{\mp}B^{\mp-}$	1
Pentamethoxy red	1.2–3.2	Red violet–colorless	$A^{+}B^{0}$	−1
Tropeolin 00	1.3–3.2	Red–yellow	$A^{\mp}B^{-}$	−1
2,4-Dinitrophenol	2.4–4.0	Colorless–yellow	$A^{\mp}B^{-}$	−1
Methyl yellow	2.9–4.0	Red–yellow	$A^{+}B^{0}$	−1
Methyl orange	3.1–4.4	Red–orange	$A^{\mp}B^{-}$	−1
Bromophenol blue	3.0–4.6	Yellow–blue violet	$A^{\mp-}B^{\mp=}$	3
Tetrabromophenol blue	3.0–4.6	Yellow–blue	$A^{\mp-}B^{\mp=}$	3
Alizarin sodium sulfonate	3.7–5.2	Yellow–violet	$A^{-}B^{=}$	3
α-Napthyl red	3.7–5.0	Red–yellow	$A^{+}B^{0}$	−1
p-Ethoxychrysoidine	3.5–5.5	Red–yellow		
Bromocresol green	4.0–5.6	Yellow–blue	$A^{\mp-}B^{\mp=}$	3
Methyl red	4.4–6.2	Red–yellow	$A^{0}B^{-}$	1
Bromocresol purple	5.2–6.8	Yellow–purple	$A^{\mp-}B^{\mp=}$	3
Chlorophenol red	5.4–6.8	Yellow–red	$A^{\mp-}B^{\mp=}$	3
Bromothymol blue	6.2–7.6	Yellow–blue	$A^{\mp-}B^{\mp=}$	3
p-Nitrophenol	5.0–7.0	Colorless–yellow	$A^{0}B^{-}$	1
Azolitmin	5.0–8.0	Red–blue		
Phenol red	6.4–8.0	Yellow–red	$A^{\mp-}B^{\mp=}$	3
Neutral red	6.8–8.0	Red–yellow	$A^{+}B^{0}$	−1

[20] *Modern pH and Chlorine Control*, 16th ed., p. 22, W. A. Taylor and Co., 7300 York Road, Baltimore 4, Md., 1962.

TABLE 6–3 (*Concluded*)

Indicator	Transformation Range, pH	Color Change	Charge Type	i_s
Rosolic acid	6.8–8.0	Yellow–red	A^0B^-	1
Cresol red (alk. range)	7.2–8.8	Yellow–red	$A^{\mp}-B^{\mp=}$	3
α-Naphtholphthalein	7.3–8.7	Rose–green	$A^-B^=$	3
Tropeolin 000	7.6–8.9	Yellow–rose red	$A^{\mp}B^-$	−1
Thymol blue (alk. range)	8.0–9.6	Yellow-blue	$A^{\mp}-B^{\mp=}$	3
Phenolphthalein	8.0–10.0	Colorless–red	$A^-B^=$	3
α-Naphtholbenzein	9.0–11.0	Yellow–blue	A^0B^-	1
Thymolphthalein	9.4–10.6	Colorless–blue	$A^-B^=$	3
Nile blue	10.1–11.1	Blue–red	A^+B^0	−1
Alizarin yellow	10.0–12.0	Yellow–lilac	$A^-B^=$	3
Salicyl yellow	10.0–12.0	Yellow–orange brown	$A^-B^=$	3
Diazo violet	10.1–12.0	Yellow–violet		
Tropeolin 0	11.0–13.0	Yellow–orange brown	$A^{\mp}B^-$	−1
Nitramine	11.0–13.0	Colorless–orange brown	A^+B^0	−1
Poirrier's blue	11.0–13.0	Blue violet–pink		
Trinitrobenzoic acid (indicator salt)	12.0–13.4	Colorless–orange red	A^0B^-	1

Detailed Treatment of Indicator Equilibria

SALT EFFECTS

The exact form of equation 2 contains the thermodynamic dissociation constant K_{HIn} of the acid form of the indicator and also a term representing the ratio of the activity coefficients of the two colored forms:

$$pa_H = pK_{HIn} + \log \frac{\alpha}{1 - \alpha} + \log \frac{\gamma_{In}}{\gamma_{HIn}} \qquad (3)$$

where pa_H is written for $-\log a_H$. Visual or instrumental color matching, if successful, assures that the ratio of the concentrations of the two colored forms (the second term on the right of equation 3) is the same in the unknown solution as in a standard of known pa_H or pH. More sophisticated colorimetric or photometric techniques evaluate this concentration term directly. They cannot, however, detect changes in the last term of the

TABLE 6-4. LIST OF INDICATORS OF W. A. TAYLOR AND CO.

Indicator	pH Range	Color Change
Acid cresol red	0.2–1.8	Red–yellow
Acid metacresol purple	1.2–2.8	Red–yellow
Acid thymol blue	1.2–2.8	Red–yellow
Bromophenol blue	3.0–4.6	Yellow–blue
Bromocresol green	3.8–5.4	Yellow–blue
Methyl red	4.4–6.0	Red–yellow
Chlorophenol red	5.2–6.8	Yellow–red
Bromocresol purple	5.2–6.8	Yellow–purple
Bromothymol blue	6.0–7.6	Yellow–blue
Phenol red	6.8–8.4	Yellow–red
Cresol red	7.2–8.8	Yellow–red
Metacresol purple	7.6–9.2	Yellow–purple
Thymol blue	8.0–9.6	Yellow–blue
Phthalein red	8.6–10.2	Yellow–red
Tolyl red	10.0–11.6	Red–yellow
Acyl red	10.0–11.6	Red–yellow
Parazo orange	11.0–12.6	Yellow–orange
Acyl blue	12.0–13.6	Red–blue
Benzo yellow	2.4–4.0	Red–yellow
Benzo red	4.4–7.6	Red–blue
Thymol red	8.0–11.2	Yellow–red
Long-range indicator	3.0–11.0	Red–violet

equation, the activity coefficient ratio, or deal with them. These changes arise from an alteration of the ionic strength of the medium or differences in the ionic strength of the two media of matching color. The magnitude of the activity coefficient term depends not only on the ionic strength but on the charge type of the indicator acid-base conjugate pair and, particularly at moderate and high ionic strengths, on specific effects not easily evaluated.

Neglect of these salt or electrolyte effects early led to the distrust of many indicators and to their classification as unreliable. The difficulty was recognized by Clark, who wrote in 1928,[21] ". . . the indiscriminate use of miscellaneous indicators may lead to gross errors or at least to such a diversity of data that their correlation will become complex during the coming period when the specific salt effects and general conduct of the individual indicators are still being determined."

---

[21] Op. cit., p. 70.

The salt error of an indicator can be determined by comparing the "indicator pH" for a series of solutions of different ionic strengths with the electrometric (hydrogen electrode) pH value for the same solutions, the latter being accepted as correct. Salt corrections have been determined in this way for many indicators.[22] For some purposes it is convenient to combine the first and third terms on the right of equation 3, as Sendroy and Hastings have done,[23] into an "apparent" or *formal* value of pK_{HIn} to be used for solutions of a given ionic strength. It must be remembered, however, that the formal indicator constant depends, at moderate and high ionic strengths, on the specific properties of the electrolytes present.

It is possible also to predict the salt error with some success in relatively dilute solutions, when the charge type of the indicator is known, by utilizing the Debye-Hückel equation or one of its modifications to compute the last term of equation 3. Unfortunately, calculated salt corrections are of limited usefulness for several reasons. First of all must be mentioned the inadequacy of the Debye-Hückel equation at ionic strengths above about 0.2, particularly when the magnitude of the ion-size parameter is unknown. Furthermore, indicator ions are large and complex, and the effective charge type is often unknown. It may be expected that the distribution of electrostatic potential in the vicinity of the large and irregular indicator ions does not correspond closely to the relatively simple models for which the theoretical equations were developed. Ionization of the indicator salts may indeed not be complete. Finally, the proper method of calculating the activity coefficient of hybrid ions is still not completely clear.

HYBRID IONS

Bjerrum[24] suggested that the charges on a hybrid ion are often so widely separated that the hybrid behaves as two ions. This supposition is in agreement with the results of Güntelberg and Schiödt and of Meretoja[25] on methyl orange and with that of Kolthoff and Guss[26] on the sulfonphthaleins.

The work of Kolthoff and Guss [26] sheds considerable light on the role of hybrid ions (that is, ions with two equal but opposite charges) in the manifestation of the salt effect and on the use of salt effects to identify the charge types of the acid and base forms of an indicator. In very dilute solutions, it

[22] See, for example, I. M. Kolthoff, *J. Phys. Chem.*, **32**, 1820 (1928).

[23] J. Sendroy, Jr., and A. B. Hastings, *J. Biol. Chem.*, **82**, 198 (1929). The formal dissociation constants of a number of indicators were determined in the presence of various salts over a range of ionic strengths.

[24] N. Bjerrum, *Z. Physik. Chem.*, **103**, 147 (1923).

[25] E. Güntelberg and E. Schiödt, *Z. Physik. Chem.*, **135**, 393 (1928); A. Meretoja, *Ann. Acad. Sci. Fennicae, Ser. A. II*, No. 12 (1944).

[26] I. M. Kolthoff and L. S. Guss, *J. Amer. Chem. Soc.*, **60**, 2516 (1938).

appears, the hybrid ion behaves much like an uncharged molecule,[27] whereas the hybrid ion with an extra negative charge behaves as a univalent anion. At higher ionic strengths, however, there is evidence that each of the charge centers behaves as an independent univalent ion, if they are sufficiently widely spaced.

When a small amount of acid-base indicator is added to a buffer solution in which the concentrations of buffer acid HA and its conjugate base A are known, a measurement of $\alpha/(1-\alpha)$ permits the apparent equilibrium constant K' for the reaction

$$HIn + A \ \rightleftharpoons \ HA + In \tag{4}$$

to be determined:

$$K' \equiv K \frac{\gamma_A \gamma_{HIn}}{\gamma_{HA} \gamma_{In}} = \frac{m_{HA}}{m_A} \frac{\alpha}{1-\alpha} \tag{5}$$

where K is the thermodynamic equilibrium constant, namely K_{HIn}/K_{HA}. If z_{In} and z_A are the numbers of charges borne by the indicator base In and the basic species A of the buffer, respectively, the activity coefficient of each species (charge z_i) in the second term of equation 5 can, for sufficiently dilute solutions, be expressed in terms of the activity coefficient γ_1 of an "average" univalent ion, as follows:

$$\log \gamma_i = z_i^2 \log \gamma_1 \tag{6}$$

By combining equations 5 and 6 one obtains

$$pK' - pK = 2(z_A - z_{In}) \log \gamma_1 \tag{7}$$

Values of $\log \gamma_1$ for ionic strengths from 0.01 to 0.5 in water are listed in Table 6–5; they were calculated by the equation proposed by Davies.[28]

If indicator and buffer are of the same charge type, the apparent value of pK should be constant at low ionic strengths.[29] If they are of different charge types, the plot of $pK' - pK$ as a function of ionic strength will be curved, and its slope at any point will depend on the respective charge types of the buffer and the indicator.

Kolthoff and Guss[30] showed that the behavior of methyl orange in tri-chloroacetate buffers reveals the hybrid character of the acid form of the

---

[27] E. J. Cohn and J. T. Edsall, *Proteins, Amino Acids and Peptides*, Chapters 4 and 12, Reinhold Publishing Corp., New York, 1943.

[28] C. W. Davies, *J. Chem. Soc.*, 2093 (1938). The Davies equation is $-\log \gamma_1 = A\sqrt{I}/(1 + \sqrt{I}) - 0.1I$, where A is the Debye-Hückel slope (see Appendix, Table 4).

[29] This principle has been used by R. G. Bates and G. Schwarzenbach, *Helv. Chim. Acta*, 37, 1069 (1954), to facilitate the determination of the dissociation constants of indicator acids.

[30] *Op. cit.*

TABLE 6–5. VALUES OF $\log \gamma_1$ IN WATER AT 25 °C FOR VARIOUS IONIC STRENGTHS I

	(Calculated by the equation of Davies[28])		
I	$-\log \gamma_1$	I	$-\log \gamma_1$
0.01	0.0454	0.14	0.1251
0.02	0.0612	0.16	0.1299
0.03	0.0724	0.18	0.1341
0.04	0.0811	0.20	0.1378
0.05	0.0883	0.25	0.1453
0.06	0.0944	0.30	0.1508
0.08	0.1046	0.35	0.1548
0.10	0.1127	0.40	0.1579
0.12	0.1194	0.45	0.1601
		0.50	0.1615

indicator. At low ionic strengths, the values of pK' are constant, but at higher concentrations the curve has the slope to be expected if the acidic form bears two independent single charges. Equations 6 and 7 no longer apply, for the hybrid species is behaving as a number (n) of separate univalent ions. Instead of the z^2 relationship of equation 6, one must write

$$\log \gamma(\text{hybrid}) = n \log \gamma_1 \qquad (8)$$

Accordingly, $\log \gamma_{\mp} = 2 \log \gamma_1$ for the hybrid bearing both a positive and a negative charge center rather widely separated.

In the same way, the sulfonphthaleins in their alkaline ranges yield slopes which suggest that the yellow form of the indicator, behaving as a univalent anion at low ionic strengths, is actually a species with two negative charges and one positive charge. The alkaline form accordingly bears one positive and three negative charges: hence $\log \gamma_{\mp =} = 4 \log \gamma_1$, whereas for the yellow form $\log \gamma_{\mp -} = 3 \log \gamma_1$.

SALT CORRECTIONS

Qualitatively, one would expect from equation 3 that the apparent pa_H found by indicator measurement will be too high in the presence of added neutral salt, if the acid form of the indicator is an uncharged molecule or an anion. If the acid form bears one or more excess positive charges, the apparent pa_H should be too low. When hybrid ions are involved, the salt effect may have either sign, depending on whether the proton added to the base neutralizes an existing negative charge center or creates a new center of positive charge.

In a practical consideration of salt errors, however, it must be remembered that the addition of a neutral salt to a buffer-indicator mixture not only produces a salt error attributable to a shift of the indicator equilibrium but also affects the equilibrium between buffer acid and salt in such a way that a real change of the pa_H is brought about. Table 6–6 illustrates these concomitant effects.[31, 32] The actual change in the pa_H ("true pH") of six solutions on addition of a neutral salt to a concentration of 0.1 M is given, together with the apparent change of pH as shown by indicators of three different charge types.

TABLE 6–6. CALCULATED EFFECTS OF ADDING NEUTRAL SALT (0.1 M)

(+ sign signifies an increase in true or apparent pH)

Solution	Δ(True pH)	Apparent Change of pH		
		Indicator A^0B^-	Indicator $A^-B^=$	Indicator A^+B^0
Hydrochloric acid, 0.005 M	+0.10	+0.19	+0.35	0
Sodium hydroxide, 0.005 M	+0.10	0	+0.15	−0.19
Acetic acid, 0.005 M	0	+0.12	+0.31	−0.11
Ammonia, 0.005 M	0	+0.11	+0.31	−0.12
Acetic acid, 0.005 M; sodium acetate, 0.005 M	−0.10	0	+0.15	−0.19
Ammonia, 0.005 M; ammonium chloride, 0.005 M	+0.10	+0.19	+0.35	0

The effect on the indicator pH value of adding neutral salt to a buffer system (HA,A) containing a small concentration of an indicator (HIn, In) can be treated in the following manner. By rearrangement of equation 5 and conversion to common logarithms one finds

$$pK_{HA} + \log \frac{m_A}{m_{HA}} + \log \frac{\gamma_A}{\gamma_{HA}} = pK_{HIn} + \log \frac{\alpha}{1 - \alpha} + \log \frac{\gamma_{In}}{\gamma_{HIn}} \qquad (9a)$$

Evidently this equation can also be written

$$\text{True pH} = \text{Indicator pH} + \log \frac{\gamma_{In}}{\gamma_{HIn}} \qquad (9b)$$

---

[31] G. Kortüm, *Z. Elektrochem.*, **48**, 145 (1942).
[32] R. G. Bates in *Treatise on Analytical Chemistry*, I. M. Kolthoff and P. J. Elving, eds., Part I, Vol. I, Chapter 10, Interscience Publishers, New York, 1959.

The change Δ(true pH) in the pH of the buffer solution on addition of neutral salt is

$$\Delta(\text{true pH}) = \Delta\left(\log \frac{\gamma_A}{\gamma_{HA}}\right) = -(2z_A + 1)\cdot\Delta(\log \gamma_1) \qquad (10)$$

Hence

$$\Delta(\text{indicator pH}) = \Delta(\text{true pH}) - \Delta\left(\log \frac{\gamma_{In}}{\gamma_{HIn}}\right)$$

$$= \Delta\left(\log \frac{\gamma_A\gamma_{HIn}}{\gamma_{HA}\gamma_{In}}\right) \qquad (11)$$

When no hybrid ions are present, we can therefore write, with the aid of equation 6,

$$\Delta(\text{indicator pH}) = 2(z_{In} - z_A)\cdot\Delta(\log \gamma_1) \qquad (12)$$

for the resultant effect on the indicator pH of adding neutral salt to a buffer-indicator system. The activity coefficients of hybrid ions, if present, should be evaluated by equation 8 while those of the simple ions are estimated by equation 6.

The *salt correction* which must be added to the indicator pH to obtain the true pH is, from equation 9b,

$$\text{Salt correction} \equiv (\text{true pH}) - (\text{indicator pH})$$

$$= \left[\log \frac{\gamma_{In}}{\gamma_{HIn}}\right]_I - \left[\log \frac{\gamma_{In}}{\gamma_{HIn}}\right]_{I_{ref}} \qquad (13)$$

The right side of equation 13 represents the difference between the logarithm of the ratio of the activity coefficients of the two indicator forms at the ionic strength of the actual measurement and at the ionic strength of the comparison buffer (where the salt error is "calibrated out").

Inasmuch as $\log (\gamma_{In}/\gamma_{HIn})$ is, as a first approximation, some multiple of $\log \gamma_1$ (values of which are collected in Table 6–5), the salt correction can be characterized by an index number i_s, which will be termed the *salt correction index*. This number is determined by the charges of the acid and base forms of the indicator in accordance with equation 6 for simple ions and equation 8 for hybrid ions. Values of i_s are given in the last column of Table 6–3 for each of the indicators of known charge type. Thus at a given ionic strength I,

$$\log \left(\frac{\gamma_{In}}{\gamma_{HIn}}\right)_I = i_s \log (\gamma_1)_I \qquad (14)$$

and consequently

$$\text{Salt correction} = i_s [\log(\gamma_1)_I - \log(\gamma_1)_{I_{ref}}] \qquad (15)$$

TABLE 6–7. Comparison of Calculated and Observed Salt Corrections for Six Indicators of Various Charge Types, in pH Units

| | (Reference ionic strength $= 0.1$) | | | | | |
| Ionic Strength | Thymol Blue (acid) $A^{\mp}B^{\mp-}$ | | Methyl Orange $A^{\mp}B^-$ | | p-Nitrophenol A^0B^- | |
	Calc.	Obs.	Calc.	Obs.	Calc.	Obs.
0.01	+0.07	0	−0.07	−0.02	+0.07	+0.03
0.02	+0.05	0	−0.05	0	+0.05	+0.02
0.05	+0.02	0	−0.02	0	+0.02	+0.01
0.1	0	0	0	0	0	0
0.5 (KCl)	−0.05	0	+0.05	0	−0.05	−0.18

| | Phenol Red $A^{\mp}-B^{\mp=}$ | | Neutral Red A^+B^0 | | Phenolphthalein $A^-B^=$ | |
| Ionic Strength | | | | | | |
	Calc.	Obs.	Calc.	Obs.	Calc.	Obs.
0.01	+0.20	+0.11	−0.07	−0.05	+0.20	+0.12
0.02	+0.16	+0.07	−0.05	−0.04	+0.16	+0.10
0.05	+0.07	+0.04	−0.02	−0.02	+0.07	+0.05
0.1	0	0	0	0	0	0
0.5 (KCl)	−0.15	−0.20	+0.05	+0.07	−0.15	−0.26

Salt corrections calculated for certain acid-base indicators by equation 15 are compared in Table 6–7 with the corrections determined experimentally by Kolthoff.[33] It is evident that calculated salt effects are of no more than qualitative significance. This conclusion emphasizes the fact that our knowledge of the structure of indicator molecules and of the behavior of these complex species in solution is far from complete. Nevertheless, the comparison of observed and calculated salt effects is sometimes helpful in elucidating structure. For example, the absence of a pronounced salt effect with acid thymol blue may mean that structural changes, possibly initiated by partial neutralization of the sulfonate group, convert the basic species into a univalent anion which, in strongly acidic media, becomes a cation by acceptance of two protons. The value of i_s would then be 0. Schwarzenbach[34] has

[33] I. M. Kolthoff, J. Phys. Chem., **32**, 1820 (1928).
[34] G. Schwarzenbach, Helv. Chim. Acta, **26**, 418 (1943).

described an indicator (5-pyridiniumglutacondialdehyde perchlorate) which appears to be of this charge type.

Before using the data for indicator salt errors tabulated in the literature, one should inquire whether the correction represents the effect of the salt on the indicator equilibrium alone or whether the salt effect on the pH of the buffer is included as well. Salt corrections, determined experimentally by comparing indicator results with electrometric data for the same solutions, usually combine the two effects, but unfortunately they are strictly valid only for the particular buffer system employed. Calculated salt corrections, on the other hand, can treat the two effects separately, but their limitations should always be recognized.

SPECIFIC BEHAVIOR

If pH values of unknowns are to be determined by colorimetric methods, it is evidently desirable that the reference buffer solutions of known pH have nearly the same ionic strength as the test solutions. In actual practice, the maintenance of constant ionic strength is not a sufficient condition for highly accurate results. The ratio of the activity coefficients of the two indicator forms is not solely dependent on the ionic strength but is influenced by the nature of the electrolytes of which the solution is composed. Specific effects of this sort become particularly pronounced at ionic strengths above 0.2. Thus Kolthoff found[35] that the salt correction for methyl red in a citrate buffer (0.01 M) with added potassium chloride (0.5 M) is $+0.05$ pH unit, whereas the correction is -0.08 unit when sodium chloride is substituted for the potassium chloride. The ionic strength was 0.52 in both instances.

Another type of salt error is rarely of concern when the concentration of electrolyte is less than 0.5 M. Large concentrations of salt sometimes bring about a modification of the color itself, that is, they cause a displacement of the absorption maximum to longer or shorter wavelengths.[36] It seems likely that this shift is caused by the enhanced electrostatic field of the added ions and its effect on the solvation equilibria. In some instances, the indicator may interact chemically with one of the salt ions.

SOLVENT EFFECTS

Changes in the composition of the solvent medium often influence chemical equilibria much more profoundly than do changes normally encountered in the total concentration of ionic solutes. Large changes of composition are produced more readily by the addition of nonelectrolytes than by the addition

---

[35] I. M. Kolthoff, *J. Phys. Chem.*, **32**, 1820 (1928).
[36] H. von Halban and L. Ebert, *Z. Physik. Chem.*, **112**, 321 (1924); H. von Halban and J. Eisenbrand, *ibid.*, **132**, 401, 433 (1928); **133**, 476 (1928); **134**, 334 (1928).

of electrolytes. Because of limited solubility, salt effects at concentrations greater than 2 M are infrequently studied. The same reduction in the mole fraction of water is produced by the addition of only 6 per cent of methanol, yet the study of solvent effects may extend over the entire range of methanol-water mixtures. The behavior of ionic solutions as influenced by a change of solvent character and composition is discussed more fully in Chapter 7.

Acid-base equilibria are, in general, rather sensitive to changes in the dielectric constant and in the basicity of the medium. As the dielectric constant is lowered, the formation of ion pairs and aggregates of even higher order assumes more importance. Kortüm and Andrussow[37] found, however, that the relatively labile ion pairs of the Bjerrum type present in methanolic solutions cannot be distinguished spectrocopically from free ions. In glacial acetic acid, whose dielectric constant (6.1) is considerably lower than that of methanol (32.6), the color reactions of indicators can only be successfully accounted for by recognizing both ionization and dissociation of ion aggregates.[38, 39] There is, indeed, no simple relationship in this solvent between the ratio of the concentrations of the alkaline and acid forms of the indicator and the hydrogen ion concentration such as exists in water.[39] In an aprotic solvent like benzene, free ions probably play a negligible role in the interaction of acids and bases, yet indicator color transformations persist.[40]

In water and partially aqueous solvents of relatively high dielectric constant, however, the change of color of an indicator involves a transfer of protons. In this process the acidic form of the indicator loses a proton to one of the basic solute species or to the protophilic solvent. Alternatively, the basic form of the indicator is partially neutralized through acceptance of protons from proton donors—acidic solutes or possibly the protogenic solvent. This reaction, called *salt formation*, is usually an ionic process. If ions are formed from neutral species, as in the reaction

$$HA + In \;\rightleftharpoons\; HIn^+ + A^- \tag{16}$$

a lowering of the dielectric constant (which increases the work required to separate the ions) will shift the point of equilibrium toward the left, favoring the basic form of the indicator. The effect will correspondingly be smaller when an indicator acid is concerned:

$$HA + In^- \;\rightleftharpoons\; HIn + A^- \tag{17}$$

for salt formation can take place without the creation of new electric charges.

[37] G. Kortüm and K. Andrussow, *Z. Physik. Chem. (Frankfurt)*, **25**, 321 (1960).
[38] I. M. Kolthoff and S. Bruckenstein, *J. Amer. Chem. Soc.*, **78**, 1 (1956).
[39] S. Bruckenstein and I. M. Kolthoff, *J. Amer. Chem. Soc.*, **78**, 10 (1956).
[40] See, for example, M. M. Davis, *Acid-Base Behavior in Aprotic Organic Solvents*, National Bureau of Standards Monograph 105, U.S. Government Printing Office, Washington, 1968.

Simple considerations such as these are helpful in understanding the observations of Hantzsch,[41] which sometimes seem to be at variance with the results of electrometric measurements in the same systems. As Hammett has pointed out,[42] we must be careful not to confuse the phenomena attendant on salt formation with changes in the true "acidity" or proton energy level of the solution.

The equilibrium expression for the indicator transformation in a solvent medium other than pure water can be written as in equation 3 if the standard state in water (w) is retained, or, alternatively, with reference to a standard state in the *particular solvent medium* concerned (s):

$$-\log(m_H \cdot {}_s\gamma_H) = p({}_sK_{HIn}) + \log \frac{\alpha}{1-\alpha} + \log \frac{{}_s\gamma_{In}}{{}_s\gamma_{HIn}} \quad (18)$$

where ${}_s\gamma_i$ is the activity coefficient of the species i referred to unity at infinite dilution in the particular solvent medium in question. Like γ_i (that is, ${}_w\gamma_i$) in the water medium, ${}_s\gamma_i$ embodies interionic and other effects dependent on solute concentration, or what we have called *salt effects*.

It is permissible to retain the standard state in water, if it is advantageous to do so, even though measurements in nonaqueous or partially aqueous media are under discussion. The value of pK_{HIn}, which can be written formally as $p({}_wK_{HIn})$, is then unchanged, of course, but each activity coefficient has a finite value different from unity at infinite dilution in the particular solvent (where interionic effects are absent). This limiting value is the *primary medium effect*,[43, 44] ${}_m\gamma_i$, which is related to ${}_s\gamma_i$ and ${}_w\gamma_i$ (the activity coefficient referred to the aqueous standard state) by

$$_w\gamma_i = {}_m\gamma_i \cdot {}_s\gamma_i \quad (19)$$

The difference between the values of pK_{HIn} referred to the two different standard states is therefore a medium effect term:

$$p({}_sK_{HIn}) - p({}_wK_{HIn}) = \log \frac{{}_m\gamma_H \cdot {}_m\gamma_{In}}{{}_m\gamma_{HIn}} \quad (20)$$

In dilute solutions, where salt effects can be overlooked, matched, or corrected for, the solvent effect is largely a question of the difference between $p({}_sK_{HIn})$ and $p({}_wK_{HIn})$ or of its equivalent, the medium effect term of equation 20. Individual values of ${}_m\gamma_i$ for single ionic species are indeterminate, as are

---

[41] A. Hantzsch, *Z. Elektrochem.*, **29**, 221 (1923); **30**, 194 (1924); **31**, 167 (1925); *Z. Physik. Chem.*, **125**, 251 (1927).

[42] L. P. Hammett, *J. Amer. Chem. Soc.*, **50**, 2666 (1928).

[43] B. B. Owen, *J. Amer. Chem. Soc.*, **54**, 1758 (1932).

[44] Medium effects are considered in greater detail in chapter 8.

those of $_w\gamma_i$ and $_s\gamma_i$ for single ions. Grunwald and his coworkers,[45] however, have based a calculation of $_m\gamma_H$ and $_m\gamma_A/_m\gamma_{HA}$ in ethanol-water mixtures on regularities observed in the change of the dissociation constants of acids HA of two charge types as the composition of the solvent was changed from pure water to pure ethanol. Their value for log $_m\gamma_H$ in pure ethanol is 4.707. This result is, of course, not subject to proof. Furthermore, it seems likely that values of $_m\gamma_A/_m\gamma_{HA}$, where HA is a carboxylic acid or substituted ammonium ion, do not constitute a useful estimate of the medium effect upon indicator species of complex structure. This conclusion is supported by the discussion of salt effects in the earlier sections of this chapter.

From studies of six aromatic acids in alcoholic media, Kolthoff, Lingane, and Larson[46] found that $_m\gamma_{HA}$ for the uncharged molecules varied from 0.005 to 0.01 in ethanol and from 0.005 to 0.02 in methanol. The corresponding values of $_m\gamma_H \cdot _m\gamma_{A^-}$ ranged from 13 to 30,000 in ethanol and from 7 to 1770 in methanol.[47] These figures are in general agreement with the increase of about 4 to 6 units often observed in the pK for an uncharged weak acid when the solvent composition is changed from pure water to pure alcohol.

In a purely qualitative way, then, the sign of the solvent effect, like that of the salt effect, can be predicted by equation 3. The activity coefficients of ions, for example, increase as alcohol is added to the medium, whereas those of uncharged species often (but not always) decrease. The activity coefficients of charged species (aqueous standard state) increase as alcohol is added to the solution in spite of a decrease in $_s\gamma_i$ (equation 19), which is not large in dilute solutions. In other words, the medium effect outweighs differences in the salt effect.

Medium effects are perhaps best characterized in terms of the change in pK_{HIn} produced by a change in the composition of the solvent. In accord with expectation, the value of pK_{HIn} for an indicator acid (charge type A^0B^- or $A^-B^=$, for example) is increased rather sharply when alcohol is added to the aqueous solvent, whereas that for an indicator base A^+B^0 is usually decreased to a less extent and often passes through a minimum. In Table 6-8 values of $p(_sK_{HIn})$ for several indicators in 33.4 per cent (w/w) methanol[48-51]

[45] B. Gutbezahl and E. Grunwald, *J. Amer. Chem. Soc.*, **75**, 559, 565 (1953); E. F. Sieckmann and E. Grunwald, *ibid.*, **76**, 3855 (1954).

[46] I. M. Kolthoff, J. J. Lingane, and W. D. Larson, *J. Amer. Chem. Soc.*, **60**, 2512 (1938).

[47] The increase in the activity coefficients of charged species as the dielectric constant of the medium decreases is in qualitative accord with the change in the electrostatic free energy, as predicted by the Born equation; M. Born, *Z. Physik*, **1**, 45 (1920).

[48] D. Rosenthal, H. B. Hetzer, and R. G. Bates, *J. Amer. Chem. Soc.*, **86**, 549 (1964).

[49] G. Kortüm and M. Buck, *Z. Elektrochem.*, **62**, 1083 (1958).

[50] C. L. de Ligny, H. Loriaux, and A. Ruiter, *Rec. Trav. Chim.*, **80**, 725 (1961).

[51] E. E. Sager and V. E. Bower, *J. Res. Nat. Bur. Stand.* **64A**, 351 (1960); E. E. Sager, R. A. Robinson, and R. G. Bates, *ibid.*, **68A**, 305 (1964).

TABLE 6-8. DISSOCIATION CONSTANTS OF INDICATORS IN WATER AND IN 33.4 WT. PER CENT METHANOL

Indicator	Charge Type	$(\Delta = \mathrm{p}(_sK_{HIn}) - \mathrm{p}(_wK_{HIn}))$		
		$\mathrm{p}(_wK_{HIn})$ (water)	$\mathrm{p}(_sK_{HIn})$ (33.4% MeOH)	Δ
Methylpicric acid[48]	A^0B^-	0.808	0.979	+0.171
Dimethylpicric acid[48]	A^0B^-	1.376	1.712	+0.336
2,6-Dinitrophenol[49]	A^0B^-	3.693	4.195	+0.502
p-Nitrophenol[50]	A^0B^-	7.15	7.51	+0.36
4-Chloro-2,6-dinitrophenol[51]	A^0B^-	2.969	3.176	+0.207
m-Nitroaniline[51]	A^+B^0	2.454	1.944	−0.510
p-Nitroaniline[48]	A^+B^0	1.003	0.408	−0.595
o-Chloroaniline[51]	A^+B^0	2.615	2.221	−0.394
Methyl orange[50]	$A^{\mp}B^-$	3.38	2.81	−0.57
m-Cresol purple (acid)[48]	$A^{\mp}B^-$ or $A^{\mp}B^{\mp-}$	1.703	1.599	−0.104

are compared with $\mathrm{p}(_wK_{HIn})$ in pure water. The dissociation constants of indicators in pure methanol, taken from the paper of Kolthoff and Guss,[52] are summarized in Table 6-9. The effect of changing solvent composition on the $\mathrm{p}(_sK_{HIn})$ of seven indicators in alcohol-water solvents is illustrated in Fig. 6-2.

We have already seen that salt effects in a given solvent medium may yield information on the structure of the indicator. Like salt effects, the change of $\mathrm{p}K_{HIn}$ when alcohol is added to the aqueous solvent may also yield valuable information on the charge type of the acid and base forms. The dissociation constants of tropeolin 00 and methyl orange, for example, are nearly the same regardless of whether the indicators are dissolved in pure water or in pure methanol (see Table 6-9), but the $\mathrm{p}K_{HIn}$ values for both indicators pass through a minimum when the water solvent is enriched with methanol. The $\mathrm{p}(_sK_{HIn})$ at 64 wt. per cent methanol [a composition near that at which $\mathrm{p}(_sK_{HIn})$ reaches its lowest value] is about 1.3 units lower than $\mathrm{p}(_wK_{HIn})$, the value in pure water.[50] In this respect, these indicators behave like m-nitroaniline and o-chloroaniline in methanol-water mixtures (see Fig. 6-2). Inasmuch as the charge type of the anilines is A^+B^0, while tropeolin 00 and methyl orange are of type $A^{\mp}B^-$, the close resemblance in behavior is

---

[52] I. M. Kolthoff and L. S. Guss, *J. Amer. Chem. Soc.*, **60**, 2516 (1938).

TABLE 6–9. DISSOCIATION CONSTANTS OF INDICATORS IN WATER AND IN METHANOL

Indicator	(After Kolthoff and Guss) $p(_wK_{HIn})$ (water)	$p(_sK_{HIn})$ (methanol)
Tropeolin 00	2.0	2.2
Methyl yellow	3.2	3.4
Thymolbenzein (acid)	—	3.5
Methyl orange	3.4	3.8
Methyl red (acid)	2.3	4.1
Thymol blue (acid)	1.6	4.7
Neutral red	7.4	8.2
Bromophenol blue	4.1	8.9
Methyl red (alk.)	5.0	9.2
Bromocresol green	4.9	9.8
Bromocresol purple	6.4	11.3
Bromothymol blue	7.3	12.4
Phenol red	8.0	12.8
Thymolbenzein (alk.)	—	13.1
Thymol blue (alk.)	9.2	14.0

strong support for the belief that the hybrid ion behaves as two independe univalent ions.

In the same way, the decrease of pK_{HIn} for m-cresol purple when methan (33.4 per cent) is added to the aqueous solvent (Table 6–8) suggests th $A^{\mp}B^{-}$ is a better representation of the charge of this indicator than is $A^{\mp}B^{\mp}$ The latter, like $A^{0}B^{-}$, would be expected to lead to an increase of pK_H Likewise, Bates and Schwarzenbach[53] were able to show that the behavi of 2,4-dinitrophenol in ethanol-water mixtures corresponds to the char type $A^{\mp}B^{-}$ instead of $A^{0}B^{-}$.

At the present time the prediction of solvent effects on indicator equilib in any general way is quite unsatisfactory. Although values of pK_{HIn} for a f indicators in nonaqueous media have been determined, only a few solvents solvent mixtures have been studied in any detail. A convenient and adequ correction of solvent errors on indicator measurements must await the esta lishment of pH scales and reference buffer solutions of known pH in nc aqueous and partially aqueous media.

[53] R. G. Bates and G. Schwarzenbach, Helv. Chim. Acta, 38, 699 (1955).

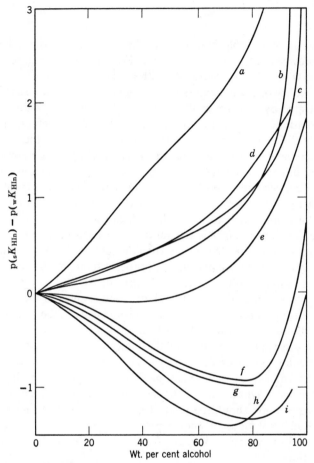

FIG. 6–2. Effect of changing solvent composition on pK_{HIn} for certain indicators in alcohol-water mixtures at 25 °C.

    a. Anisic acid in ethanol-water.[51]
    b. 2,6-Dinitrophenol in ethanol-water.[49]
    c. 2,6-Dinitrophenol in methanol-water.[49]
    d. *p*-Nitrophenol in methanol-water.[50]
    e. 4-Chloro-2,6-dinitrophenol in ethanol-water.[53]
    f. *o*-Chloroaniline in methanol-water.[51]
    g. *m*-Nitroaniline in methanol-water.[51]
    h. *m*-Nitroaniline in ethanol-water.[53]
    i. Methyl orange in methanol-water.[50]

Colorimetric Techniques for pH Measurement

PROCEDURES

Before the most suitable indicator can be selected, the approximate pH of the unknown solution must be found. For this purpose a preliminary examination with indicator papers or with a wide-range indicator will often suffice. The utility of the indicators most likely to prove suitable should then be demonstrated by treating small portions of the unknown with a drop of indicator solution. To be useful an indicator must show a definite transition color visibly different from the color of either the acid or the base form. Kolthoff recommends that pH measurements should always be made with two different indicators, unless only an approximate value is required.

It is often convenient to use the sets of indicators that are available commercially. A representative list is given in Table 6–4. If one-color indicators are chosen, it is important to know the indicator concentration accurately. With two-color indicators, it is the ratio of the concentrations of the two colored forms that is important rather than the total indicator concentration.

A *precision* of 0.05 to 0.1 pH unit in properly standardized routine measurements of buffered solutions can be expected. An *accuracy* of the same order can only be achieved when special attention is given to the application of salt corrections. Procedures for the measurement of the pH of poorly buffered media will be considered in a later section.

When the dissociation constant of the indicator is known, the measurement of the color tone, specifically $\alpha/(1 - \alpha)$, permits the pH of the solution to be determined. This procedure is sometimes called the Michaelis method. The *formal dissociation constant* of the indicator acid at the ionic strength of the unknown solution should be used, if possible, in order to minimize the salt error. Buffer solutions of known pH, determined electrometrically, are useful in establishing practical values for the formal dissociation constants of the indicators used.

The most common procedure for pH determination by visual color matching is the so-called Sørensen method. In the application of this procedure, it is not necessary to know the dissociation constant of the indicator. Instead of determining experimentally the ratio of concentrations of the colored forms, $\alpha/(1 - \alpha)$, the color of a portion of the unknown solution to which indicator has been added is compared with the colors of several solutions of known pH containing the same indicator in the same concentration. Slide comparators, available commercially, are provided with color standards which make the preparation of reference buffers unnecessary in many instances.

Buffer solutions covering a wide range of pH in steps of 0.2 pH unit or less are needed for the most accurate and effective color matching. The pH values of two series of buffer solutions useful for this purpose are given in Tables 8 and 9 of the Appendix. The first is the well-known series of Clark and Lubs,[54] the pH values of which have been redetermined[55] at 25 °C on the conventional activity pH scale in use today. The data in the second table were taken from the paper of Bates and Bower.[56] The pH is probably correct to ± 0.02 unit. It should be remembered that most of the pH numbers for reference buffer solutions found in the older monographs are based on the Sørensen scale. Consequently they are about 0.04 unit lower than the activity pH defined by the standards of the present day.

For the highest accuracy, the color of the unknown solution to which indicator has been added should be bracketed by two reference buffer-indicator mixtures, one of slightly higher pH and the other of slightly lower pH than the unknown itself. The ionic strength given in Table 9 of the Appendix is useful in matching the salt error or in the application of corrections for the salt effect. The Van Slyke buffer value[57] β is given for both series of buffer solutions, and Table 9 (Appendix) also lists the dilution value[58] $\Delta pH_{1/2}$ and gives an estimate of the temperature coefficient of the pH value.

PHOTOMETRY AND COLORIMETRY

The determination of the concentration c of a colored species in solution by optical means is based on the validity of the Bouguer-Lambert-Beer law:

$$A = \log \frac{I_i}{I_e} = cl\varepsilon_\lambda \tag{21}$$

where l is the length of the solution path traversed by light of initial intensity I_i, which emerges with an intensity I_e; A is the *absorbance*, and ε_λ is the *molar absorption coefficient* at the wavelength λ.

Equation 21, usually known simply as Beer's law, is generally valid for monochromatic light. The absorbance A can be measured directly by a photometer; the molar absorption coefficient ε_λ is a measure of the power of the solution to absorb light of wavelength λ. Inasmuch as a certain amount of light may be absorbed by the solvent and other species present in addition to the substance of predominant color, it is customary to apply a "blank" correction. The latter is the absorbance of a portion of the solution without indicator, measured at the same wavelength in a cell of the same length as that used for the measurement of the absorbance of the solution with indicator.

[54] W. M. Clark and H. A. Lubs, *J. Biol. Chem.*, **25**, 479 (1916).
[55] V. E. Bower and R. G. Bates, *J. Res. Nat. Bur. Stand.*, **55**, 197 (1955).
[56] R. G. Bates and V. E. Bower, *Anal. Chem.*, **28**, 1322 (1956).
[57] D. D. Van Slyke, *J. Biol. Chem.*, **52**, 525 (1922).
[58] R. G. Bates, *Anal. Chem.*, **26**, 871 (1954).

With the aid of a spectrophotometer, the absorption curve, that is, a plot of ε or A as a function of wavelength, λ, can be constructed for each of the forms of the indicator, when both forms absorb light in the spectral region studied. For this purpose solutions of the indicator of identical concentration are prepared in media sufficiently acidic or alkaline to assure that all of the indicator is in the colored acid form (a) or colored basic form (b), respectively. The concentrations of these forms are accordingly known, as is the cell length l. Plots of the absorbance A for the two forms of p-nitrophenol are shown as curves a and b in Fig. 6–3.[59]

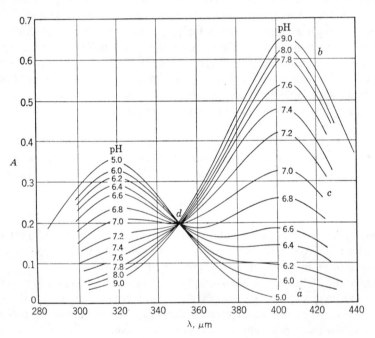

FIG. 6–3. Absorption curves for a solution of p-nitrophenol in aqueous buffer solutions of varying pH. Data of Biggs.[59] (1 cm cells, 3.6×10^{-5}M indicator).

In solutions of pH within the transformation range of the indicator however, both forms are present. The absorption curve consequently follows an intermediate path such as that of curve c. The absorbance at a given wavelength due to the presence of two absorbing species is the sum of the

[59] A. I. Biggs, Trans. Faraday Soc., 50, 800 (1954).

absorbances of each species; hence the measured A_x is given in terms of the molar absorption coefficients of the two colored species by

$$A_x = c_a l \varepsilon_a + c_b l \varepsilon_b \tag{22}$$

If c is the total molar concentration of indicator, the concentrations of the colored acid and base forms are $(1 - \alpha)c$ and αc, respectively. When the same cells or cells of equal length are used for all the measurements of absorbance, therefore,

$$A_x = (1 - \alpha)A_a + \alpha A_b \tag{23}$$

where a, b, and x refer respectively to the acid form, the base form, and the partially transformed indicator. For pH measurements one needs to know the ratio of the concentrations of the base and acid forms, namely $\alpha/(1 - \alpha)$. This quantity is evidently related to the measured absorbances at a given wavelength by

$$\frac{\alpha}{1 - \alpha} = \frac{A_x - A_a}{A_b - A_x} \tag{24}$$

It is desirable, for the greatest accuracy, to choose a wavelength at which the absorptions of the acidic and basic forms are widely different. When a suitable photoelectric spectrophotometer is used, it is not necessary that this absorption band be located in the visible region of the spectrum. In order to affirm that the system obeys Beer's law, however, the absorption curves should be plotted over a considerable range of wavelengths for both limiting forms as well as for solutions of intermediate pH. These curves should intersect at sharply defined *isosbestic points*, such as point *d* in Fig. 6–3.

The spectrophotometer provides an incident beam of almost monochromatic light, and the intensity of the emergent beam is measured by one of a variety of optical devices, usually with the aid of photoelectric detection. Most indicator measurements, however, are made with optical instruments of much simpler design. Visual colorimeters or photoelectric absorptiometers are satisfactory when only one of the indicator forms is colored in the visible region of the spectrum. These instruments commonly use white (polychromatic) light. If two beams of light of the same spectral composition are passed through two solutions containing the same colored substance in different concentrations c_u and c_r, Beer's law suggests that the path lengths necessary to give the same light absorption will be inversely proportional to the concentrations:

$$c_u = c_r \cdot \frac{l_r}{l_u} \tag{25}$$

The concentration of one solution being known, therefore, the measuremen of the ratio of path lengths l_r/l_u required for equal depth of color permits th other concentration to be determined. A measurement of this sort is easil made with the Duboscq colorimeter.

Direct measurements of $\alpha/(1 - \alpha)$ can also be made for two-color indicato by the use of a mixed color instrument such as the Gillespie bicolorimeter or the Thiel mixed colorimeter.[61] In these ingenious devices, the light tran. mitted through a colored solution prepared by adding indicator to th "unknown" solution is matched visually by the light passing through tw portions of solution in series, one of which contains the indicator in its aci form (path length l_a) and the other the indicator in its basic form (pat length l_b). The total indicator concentration and the length of the light pat are the same for unknown and comparison standard, but the ratio l_b/l_a in th reference standard can be varied at will. If l_u and l_r are the depths of unknow and reference solution through which the two light beams pass, this instru ment permits l_b/l_a to be varied while $l_a + l_b$ (that is, l_r) remains always equ to l_u. A similar synthesis of transition colors can be achieved by use of th Bjerrum color wedge.[62]

The reader is referred to the monographs of Brode[63] and Kortüm[64] for authoritative descriptions of colorimeters and photometers and discu sions of the theory of their operation.

Indicator Test Papers

Indicator papers covering various ranges of pH are available commerciall and are a great convenience in determining the approximate pH of a unknown solution. The accuracy obtainable is, however, never much bett than ± 0.5 pH unit. Consequently indicator test papers are a poor substitu for indicator solutions.

The complications attendant on the measurement of pH with indicat papers have been set forth by Kolthoff and Rosenblum.[65] Considerab difficulty attaches to the selection of a suitable grade of paper. Sized pap tends to give the most distinct color reactions because the liquid does n spread easily and the color is localized. The buffering effect of the sizin may, however, lead to erroneous pH readings, particularly for poorly buffere

[60] L. J. Gillespie, *J. Bacteriol.*, **6**, 399 (1921).

[61] A. Thiel, *Marburger Sitzungsber.*, **66**, 37 (1931).

[62] See Clark, *op. cit.*, p. 170.

[63] W. R. Brode, *Chemical Spectroscopy*, 2nd ed., John Wiley and Sons, New York, 194

[64] G. Kortüm, *Kolorimetrie, Photometrie, und Spektrometrie*, 4th ed., Springer-Verla Berlin-Göttingen-Heidelberg, 1962.

[65] *Op. cit.*, Chapter 11.

solutions. On the other hand, filter paper permits the solution to spread too readily, and as a result of absorption by the fibers of the paper changes of composition take place outward from the center of the drop. Total immersion of the paper may lead to an alteration of the color through dissolution of a part of the indicator in the solution whose pH is to be determined. Considerable trouble from absorption of carbon dioxide by the wet paper may be experienced when the test solution has a pH greater than 6.

The most common procedure for the use of indicator paper is to place a drop of the unknown solution on the paper strip and to compare the color produced with printed color charts. Alternatively, the end of the paper strip is dipped into the solution. The accuracy of the measurement can be improved somewhat with the aid of known reference buffer solutions of pH close to that of the unknown, but the severe limitations of indicator test papers rarely justify this refinement.

Limitations of Indicator Measurements

POORLY BUFFERED MEDIA

Inasmuch as indicators are themselves weak acids and bases, it is manifestly impossible to add them to a test solution without risk of a change in the pH. The concentration of indicator is, however, small (of the order of 10^{-5} M), and the alteration of the pH is often completely undetectable when the solution has a moderate or high buffer capacity. Nevertheless, poorly buffered media, notably distilled water, may undergo such a large change of pH when indicator is added that no reliable estimate of the initial pH can be obtained. Ideally the indicator solution should have the same pH as the solution under test.

These difficulties can be circumvented to a large degree by using an "isohydric" indicator solution.[66] This is a solution of indicator whose pH has been adjusted until it is so close to that of the test solution that no alteration of the pH of the latter occurs when the two solutions are combined in the necessary proportions.

Indicator solutions should, in fact, always be completely free of excess acid or alkali. If dilute alkali is added to effect solution of a difficulty soluble acid indicator, any excess should be carefully neutralized before the indicator is used. The indicator solutions supplied by W. A. Taylor and Co. for measurements between pH 3.0 and 9.6 (see Table 6–4) are adjusted to the midpoint of the transformation range.

---

[66] E. H. Fawcett and S. F. Acree, *J. Bacteriol.*, **17**, 163 (1929); *Ind. Eng. Chem., Anal. Ed.*, **2**, 78 (1930).

For the application of the isohydric technique many indicator solutior adjusted to different pH values are required. In principle, the measurement made by adding several successive increments of indicator to the test solutio and repeating the colorimetric pH determination after each addition. gradually decreasing measured pH signifies that the indicator solution more acidic than the test solution, and, conversely, an increase in pH mear that the indicator solution is too alkaline. This experience enables a mor suitable solution of the indicator to be selected, and the process is repeate until the truly isohydric indicator solution is found.

Isohydric indicator solutions are partially neutralized, that is, both colore forms are present in the proportion corresponding to a given intermediat pH. They can be prepared by partial titration of the indicator with dilut solutions of strong acid or base, and the pH of each solution can be deter mined electrometrically. Perhaps the simplest procedure is that of Koltho and Kameda,[67] who preferred to keep only two solutions of indicator, on containing solely the acid form and the other the alkaline form (without, c course, any excess of strong acid or alkali). The correct mixture of these tw indicator solutions for a particular measurement is readily found after tw trials.

Kolthoff and Kameda were able to measure the pH of pure water (cor tained in a closed cell through which carbon dioxide-free air was passec with either of two isohydric indicators: bromothymol blue (acid form: bas form = 100/55) or phenol red (acid form: base form = 100/10). An equa amount of indicator mixture was added to the reference buffer in a cell c similar dimensions, and the measurement was made by color comparison i the usual way.

SALT AND MEDIUM EFFECTS

Variations in the ionic strength and solvent composition, as we hav already seen, can produce large uncertainties in pH measurements wit indicators. The effects they produce can be understood in large part anc indeed, could be allowed for, if more information on the pertinent equilibri were available. The corresponding uncertainties arise from oversimplifica tions in the quantitative treatment of the acid-base interactions concerne and are not in any sense random errors. They constitute nonetheless a important limitation to the usefulness of the colorimetric pH method. In s far as possible, reference buffer solutions should match the unknown close in both ionic strength and solvent composition, if errors from these source are to be minimized.

---

[67] I. M. Kolthoff and T. Kameda, *J. Amer. Chem. Soc.*, **53**, 825 (1931).

PROTEIN (COLLOID) ERRORS

Specific chemical reactions between solutes (as, for example, ions of the heavy metals) and the indicator may result in appreciable pH errors. The interference caused by proteins and colloidal substances with the indicator pH measurement, however, appears to result from adsorption of one of the indicator forms by the amphoteric protein molecules or charged colloidal particles. It seems clear that the binding of indicator dyes is strongly influenced by the charge state of the protein; it is least at the isoelectric point. Protein errors, in pH units, for several indicators have been listed by Clark.[68]

The Hammett Acidity Function

The definition of an acidity function H_0 by Hammett and coworkers has provided a useful approach to the problem of comparing acidities in different media.[69] This function is defined in terms of the acidity constants K_{HIn^+} of uncharged indicator bases (In) as follows:

$$pK_{HIn^+} = -\log a_H \frac{m_{In}\gamma_{In}}{m_{HIn^+}\gamma_{HIn^+}} = H_0 - \log \frac{m_{In}}{m_{HIn^+}} \tag{26}$$

where m represents molality. From equation 26 it is seen that

$$H_0 \equiv -\log \frac{a_H\gamma_{In}}{\gamma_{HIn^+}} \tag{27}$$

The acidity constant and the activity coefficients relate to the standard state for aqueous solutions.

The subscript 0 signifies that the function is derived from measurements with uncharged indicator bases, that is, indicators of the charge type A^+B^0. The corresponding function for indicator bases with a single negative charge is designated H_-. If the acidity constant of a suitable indicator is known, a colorimetric determination of the ratio m_{In}/m_{HIn^+} permits H_0 in any medium in which the indicator base is measurably but not completely dissociated to be calculated by equation 26.

---

[58] *Op. cit.*, pp. 185–186.

[59] L. P. Hammett and A. J. Deyrup, *J. Amer. Chem. Soc.*, **54**, 2721, 4239 (1932); L. P. Hammett, *Chem. Rev.*, **13**, 61 (1933). G. Schwarzenbach and R. Sulzberger, *Helv. Chim. Acta*, **27**, 348 (1944), regard the acidity function as a practical means of extending the pH scale to other media. V. Gold and B. W. V. Hawes, *J. Chem. Soc.*, 2102 (1951), have defined an acidity function J_0 which plays for the triphenylcarbinol bases a role analogous to H_0 for an uncharged Brønsted base. The characteristics of the Hammett functions and related acidity functions have been thoroughly discussed by R. H. Boyd, Chapter 3 in *Solute-Solvent Interactions*, J. F. Coetzee and C. D. Ritchie, eds., Marcel Dekker, New York, 1969.

Evidently the practical determination of H_0 requires a series of indicators capable of covering a wide range of acidities, and K_{HIn^+} for each one must be known. These constants are determined in the following manner. An indicator whose strength in water is well known is compared colorimetrically with a slightly weaker indicator base in a suitable solvent. By repetition of such a stepwise procedure in solvents of increasing acidity, the acidity constants of even extremely weak bases can eventually be obtained. Some bases that are practically undissociated in water are extensively protonated in strong sulfuric acid.[70] Therefore it is possible to choose a solvent mixture of sulfuric acid and water of such a composition that two indicator bases of nearly equal strength are both protonated to a degree convenient for accurate colorimetric comparisons.

The difference between the two acidity constants is, from equation 26,

$$(pK_{HIn^+})_1 - (pK_{HIn^+})_2$$

$$= \log \left(\frac{m_{HIn^+}}{m_{In}}\right)_1 \left(\frac{m_{In}}{m_{HIn^+}}\right)_2 + \log \left(\frac{\gamma_{HIn^+}}{\gamma_{In}}\right)_1 \left(\frac{\gamma_{In}}{\gamma_{HIn^+}}\right)_2 . \quad (28)$$

The first term on the right of this expression can be determined colorimetrically. If $\gamma_{HIn^+}/\gamma_{In}$ has the same value for all indicator bases in the same solution, the last term is zero. This assumption is equivalent to the statement that the relative strengths of indicators of the same valence type are independent of the medium. Unfortunately, this is not true. Nevertheless, the requirement has been shown to be satisfactorily fulfilled for mixtures of sulfuric, perchloric, hydrochloric, and nitric acids with water, and in anhydrous formic acid.[71] The acidity constants K_{HIn^+} of seventeen indicators of charge type A^+B^0, as determined by Hammett and his coworkers,[72] are listed in Table 6–10.

Each series of indicator constants listed in Table 6–10 is based on known values in the water medium. Accordingly, the activity coefficients in the definition of H_0 (equation 27) are referred to unity in the infinitely dilute

[70] H. Lemaire and H. J. Lucas, *J. Amer. Chem. Soc.*, **73**, 5198 (1951), used the H_0 function to measure the strengths of weak bases in glacial acetic acid. Acidity functions for solutions of hydrochloric acid in mixtures of ethanol, acetone, dioxane, and water have been used to study the solvation of protons in oxygen-containing solvents: E. A. Braude and E. S. Stern *Nature*, **161**, 169 (1947); *J. Chem. Soc.*, 1976, 1982 (1948); E. A. Braude, *ibid.*, 1971 (1948). Similar data for solutions of hydrochloric acid in methanol-water solvents have been reported by P. Salomaa, *Acta Chem. Scand.*, **11**, 125 (1957).

[71] L. P. Hammett and A. J. Deyrup, *J. Amer. Chem. Soc.*, **54**, 2721, 4239 (1932); L. P. Hammett and M. A. Paul, *ibid.*, **56**, 827 (1934); M. A. Paul and F. A. Long, *Chem. Rev.* **57**, 1 (1957).

[72] L. P. Hammett, *Chem. Rev.*, **16**, 67 (1935); *Physical-Organic Chemistry*, Chapter 9 McGraw-Hill Book Co., New York, 1940.

TABLE 6–10. pK_{HIn^+} FOR HAMMETT INDICATORS IN ACIDIC SOLVENTS

Indicator	HCl, H_2O	HNO$_3$, H_2O	H_2SO_4, H_2O	HClO$_4$, H_2O	HCOOH
		Solvent Mixture			
Aminoazobenzene	2.80				
Benzeneazodiphenyl-amine	1.52				
p-Nitroaniline	1.11	1.11[a]	1.11[a]	1.11[a]	
o-Nitroaniline	−0.17	−0.20	−0.13	−0.19	−0.17[a]
p-Chloro-*o*-nitroaniline	−0.91	−0.97	−0.85	−0.91	−0.94
p-Nitrodiphenylamine			−2.38		−2.51
2,4-Dichloro-6-nitro-aniline			−3.22	−3.18	−3.31
p-Nitroazobenzene			−3.35	−3.35	−3.29
2,6-Dinitro-4-methyl-aniline			−4.32		
2,4-Dinitroaniline			−4.38	−4.43	
N,N-Dimethyl-2,4,6-trinitroaniline			−4.69		
Benzalacetophenone			−5.61		
β-Benzoylnaphthalene			−5.92		
p-Benzoyldiphenyl			−6.19		
6-Bromo-2,4-Dinitro-aniline			−6.59		
Anthraquinone			−8.15		
2,4,6-Trinitroaniline			−9.29		

[a] Reference value.

aqueous solution, and H_0 consequently becomes equal to the usual pH value in aqueous solutions so dilute that each activity coefficient is substantially unity. Nevertheless, H_0 is not a direct measure of the proton activity, for in media other than water $\gamma_{HIn^+}/\gamma_{In}$ will depart considerably from unity because of the medium effect and the altered interionic forces. If a series of negatively charged indicator bases is used, the function H_- differs from H_0 for similar reasons.

An unusual indicator, 5-pyridiniumglutacondialdehyde perchlorate, has been described by Schwarzenbach.[73] The univalent cation (colorless) is

[73] G. Schwarzenbach, *Helv. Chim. Acta*, **26**, 418 (1943).

transformed to a univalent anion (red) on addition of alkali. Inasmuch as γ_+/γ_- is presumably altered less than $\gamma_{HIn^+}/\gamma_{In}$ by a change of medium, acidity functions based on colorimetric comparisons with this indicator would be expected to indicate the proton activity more accurately than does either H_0 or H_-.

The Hammett acidity function has been extremely useful in evaluating the relative acidities of media of different composition and remarkably successful in correlating reaction rates in different solvents.[74] The validity of the H_0 function has been firmly established in aqueous solutions of strong acids.[75] One means of testing its validity in other media is to determine the degree of regularity shown by the medium effect term $_m\gamma_{HIn^+}/_m\gamma_{In}$ for different indicators of the same charge type (A^+B^0) when the composition of the solvent is altered.

From equation 20,

$$p(_sK_{HIn}) - p(_wK_{HIn}) = \log{} _m\gamma_H - \log\frac{_m\gamma_{HIn^+}}{_m\gamma_{In}} \tag{29}$$

If a series of indicators displays identical changes of pK_{HIn} when the solvent composition is altered, the Hammett H_0 function is experimentally justifiable for these indicators in the solvents examined. There is a growing belief that the regularities necessary for a valid scale of H_0 are to be found almost exclusively in solvents of high dielectric constant and moderate or high acidity.

It has been reported that practical scales of H_0 and H_- do not "exist" in ethanol-water mixtures[76] and that H_0 does not exist in methanol-water mixtures of high methanol content, at least as far as a series of azo indicators is concerned.[77] Paul and Long[75] regard these criteria of existence as too drastic and are of the opinion that a useful H_0 function can be established in mixtures of water and alcohols, acetone, or dioxane by the careful selection of indicators. Kolling and Stevens[78] have chosen indicators for binary solvent mixtures of acetic anhydride and acetic acid, yet Kolthoff and Bruckenstein[7] report that the Hammett function has no exact significance in glacial acetic acid. The H_0 concept appears to have utility in solvents containing ethylene glycol and water, and the H_- function has been studied in pure ethylene

[74] F. A. Long and M. A. Paul, *Chem. Rev.*, **57**, 935 (1957).

[75] M. A. Paul and F. A. Long, *Chem. Rev.*, **57**, 1 (1957).

[76] B. Gutbezahl and E. Grunwald, *J. Amer. Chem. Soc.*, **75**, 559 (1953). These authors doubt that H_- exists even in sulfuric acid-water mixtures.

[77] C. L. de Ligny, H. Loriaux, and A. Ruiter, *Rec. Trav. Chim.*, **80**, 725 (1961).

[78] O. W. Kolling and T. L. Stevens, *Anal. Chem.*, **34**, 1653 (1962).

[79] I. M. Kolthoff and S. Bruckenstein, *J. Amer. Chem. Soc.*, **78**, 1 (1956).

and propylene glycols.[80] It is evident that much work remains to be done in this difficult field.

The effect of added neutral salts on the H_0 value for aqueous solutions of hydrochloric acid has been studied by Paul.[81] When the concentration of acid was 0.01 M or 0.1 M, H_0 changed linearly with the concentration of added salt up to 4 M. The slopes of the lines, however, were quite different for different salts. It is clear from equation 27 that the change in H_0 results from the salt effect on $\gamma_H \gamma_{In}/\gamma_{mIn^+}$. Solubility measurements led Paul to the conclusion that the effect of neutral salts on H_0 is largely accounted for by the change in the activity coefficient of the uncharged indicator base In and that the ratio γ_H/γ_{HIn^+} is relatively unaffected. The establishment of a scale of D_0 in heavy water, analogous to H_0 in ordinary water, has been considered by Högfeldt and Bigeleisen.[82]

[80] C. Kalidas and S. R. Palit, *J. Chem. Soc.*, 3998 (1961); N. Chattanathan and C. Kalidas *Indian J. Chem.*, **9**, 169 (1971); *Australian J. Chem.*, **24**, 83, (1971).
[81] M. A. Paul, *J. Amer. Chem. Soc.*, **76**, 3236 (1954). See also the discussion of the work of I. I. Moiseev and R. M. Flid, *Zh. Prikl. Khim.*, **27**, 1110 (1954), in the review paper of Paul and Long, *Chem. Rev.*, **57**, 1 (1957). C. H. Brubaker, Jr., P. G. Rasmussen, and D. C. Luehrs, *J. Chem. Eng. Data*, **17**, 563 (1962), have determined H_0 in acid-salt mixtures (HCl, AlCl₃ and HCl, KCl) of constant total ionic strength, while D. Rosenthal and J. S. Dwyer, *Anal. Chem.*, **35**, 161 (1963), have tabulated values of $pm_H - H_0$ in a number of concentrated salt solutions.
[82] E. Högfeldt and J. Bigeleisen, *J. Amer. Chem. Soc.*, **82**, 15 (1960).

acidity and basicity
in nonaqueous solutions

The quantitative comparison of the acidities of nonaqueous or partially aqueous solutions is a problem of far greater complexity than the measurement of pH values in aqueous media. Strictly speaking, the interpretation of measured pH values is limited to water solutions, and, in fact, to dilute water solutions, if the pH is to retain its full significance in terms of the hydrogen ion. As a basis for an examination of the possible useful scales for media other than water, it will be necessary to look further into the meaning of acidity and basicity and into the role of the solvent in acid-base phenomena. In this chapter the term "nonaqueous" will include mixtures such as ethanol-water, dioxane-water, and strong aqueous solutions of sulfuric acid, as well as anhydrous solvents like benzene and acetic acid.

Concepts of Acids and Bases

ACID-BASE PHENOMENA

The Arrhenius theory of electrolytic dissociation ascribed the acidic properties of solutions to the presence of hydrogen ions and basic properties to hydroxyl ions. Neutralization was regarded as the formation of water and a salt. Inasmuch as the degree of ionization could be determined by measurements of electrolytic conductance, the strengths of acids and bases were compared by conductivity. This view is sometimes called the *water theory* or *ionization theory* of acids and bases.

It has long been known, however, that many substances (for example, tin tetrachloride) whose solutions do not contain appreciable concentrations of hydrogen ions can nevertheless be titrated with bases in organic solvents.[1] Furthermore, ionization has been found to be neither a necessary prerequisite nor a necessary accompaniment to the reaction of acids with bases.[2]

In a series of papers of great importance to the development of acid-base theory, Hantzsch[3] pointed out that differences among the strengths of acids that appear equally strong in water (perchloric, hydrochloric, hydrobromic, and nitric acids, for example) are apparent in solvents like ether and chloroform. Furthermore, he found that these strong acids are much more acidic in some nonaqueous solutions than in their highly ionized aqueous solutions. This high acidity is manifested by extensive salt formation with basic indicators and by the rapid rate of decomposition of diazoacetic ester or inversion of cane sugar in solutions of the acid. The basicity and the dielectric constant of the solvent appeared to be the principal factors in the manifestation of acidity. Hantzsch considered that chemical methods are a better index of relative acid strengths than are electrometric methods.[4]

BRØNSTED–LOWRY DEFINITION

The concept of acids and bases advanced independently in 1923 by Brønsted[5] and by Lowry[6] has done much to clarify the role of the solvent in acid-base reactions. It places particular emphasis on the proton, regarding acids as substances that can furnish a proton and bases as substances capable of combining with a proton. An acid accordingly consists of a base in combination with a proton:

$$A \;\rightleftharpoons\; B + H^+ \tag{1}$$
$$\text{Acid} \qquad\quad \text{Base} \;\; \text{Proton}$$

The acid and its corresponding base form a *conjugate pair*.

---

[1] See, for example, O. Folin and F. F. Flanders, *J. Amer. Chem. Soc.*, **34**, 774 (1912).

[2] The evidence against the ionization theory of acidity has been summarized by W. F. Luder and S. Zuffanti, *The Electronic Theory of Acids and Bases*, Chapters 1 and 3, John Wiley and Sons, New York, 1946.

[3] For example: A. Hantzsch, *Z. Elektrochem.*, **29**, 221 (1923); **30**, 194 (1924); **31**, 167 (1925); *Ber.*, **58**, 612, 941 (1925); **59**, 793, 1096 (1926); **60**, 1933 (1927); *Z. Physik. Chem.*, **125**, 251 (1927); A. Hantzsch and W. Voigt, *Ber.*, **62**, 975 (1929).

[4] The views of Hantzsch have been discussed by N. F. Hall and J. B. Conant, *J. Amer. Chem. Soc.*, **49**, 3047 (1927); L. P. Hammett, *ibid.*, **50**, 2666 (1928); M. M. Davis and H. B. Hetzer, *J. Res. Nat. Bur. Stand.*, **48**, 381 (1952); and A. P. Kreshkov, *Talanta*, **17**, 1029 (1970).

[5] J. N. Brønsted, *Rec. Trav. Chim.*, **42**, 718 (1923); *Chem. Rev.*, **5**, 231 (1928).

[6] T. M. Lowry, *Chem. Ind. (London)*, **42**, 43 (1923); *Trans. Faraday Soc.*, **20**, 13 (1924).

The idea that acids do not dissociate spontaneously as suggested by equation 1 but, instead, interact with other conjugate pairs in a transfer of protons is fundamental to the Brønsted–Lowry concept:

$$A_1 + B_2 \rightleftharpoons A_2 + B_1 \qquad (2)$$

The stronger the acid A_1 and the weaker A_2, the more complete will be reaction 2. The stronger acid releases its proton more readily than the weaker; similarly, the stronger base binds a proton more strongly than does the weaker base. It is evident from these considerations that the base or acid conjugate to a strong acid or a strong base is always weak, whereas that conjugate to a weak acid or weak base is always strong.

Brønsted has pointed out that a general and precise definition of acids and bases must (1) attribute characteristic acid-base properties to the molecules themselves, rather than to their solutions; (2) relate the ideas of acid and base logically to each other; (3) offer an illuminating explanation of the peculiar character of these substances; and (4) be formulated independently of the solvent.[7]

LEWIS DEFINITION

G. N. Lewis proposed his electronic theory of acids and bases[8] in 1923, the same year in which the proton-transfer concept was announced, but his interpretation of acid-base reactions did not receive wide attention for more than a decade. The Lewis definition of acids is more comprehensive than that of Brønsted and Lowry, and it is of wide significance in both inorganic and organic chemistry.[9] A base in the Lewis definition is a substance capable of furnishing an electron pair, whereas an acid is able to accept such a pair to form a coordinate covalent bond. Acids and bases correspond, respectively, to Sidgwick's acceptor and donor species.[10] Neutralization is regarded as an association of acid and base through the formation of a coordinate covalent bond. Subsequent ionization may or may not occur.

The scope of the electronic theory is sufficiently broad to include the proton-transfer definition as a special case. Inasmuch as electron-donor molecules are able to combine with protons, the Lewis bases are identical with those of the Brønsted–Lowry definition. On the other hand, the Lewis

---

[7] J. N. Brønsted, *Z. Angew. Chem.*, **43**, 229 (1930).

[8] G. N. Lewis, *Valence and the Structure of Atoms and Molecules*, Chemical Catalog Co., New York, 1923; *J. Franklin Inst.*, **226**, 293 (1938).

[9] W. F. Luder and S. Zuffanti, *The Electronic Theory of Acids and Bases*, John Wiley and Sons, New York, 1946; D. W. Meek, Chapter 1 in *The Chemistry of Non-Aqueous Solvents*, J. J. Lagowski, ed., Vol. I, Academic Press, New York, 1966.

[10] N. V. Sidgwick, *The Electronic Theory of Valency*, Oxford University Press, London, 1927.

definition of an acid embraces many substances (for example, silver ion and boron trichloride) that are excluded if the term acid is restricted to proton donors. The major disadvantage of the Lewis system is on the quantitative side. The "hydrogen acids" form a group displaying greater uniformity than do the nonhydrogen acids of the Lewis definition. In the opinion of Bell,[11] the hydrogen acids can be included only artificially in the electronic definition. Nevertheless, the concepts of both the Brønsted and the Lewis definitions are useful, and there is no reason why one system should entirely supplant the other.[12]

CLASSIFICATION OF SOLVENTS

The solvent plays an important part in many acid-base reactions because most of the common solvents are themselves acidic or basic. Solvents which act predominantly as proton donors (for example, glacial acetic acid and hydrofluoric acid) are termed acidic or *protogenic*; solvents which are proton acceptors (like liquid ammonia and ether) are called basic or *protophilic*. Liquids like water and the alcohols that are capable both of accepting and donating protons are called *amphiprotic*. Solvents that can neither gain nor lose protons (for example the hydrocarbons) are termed inert or *aprotic*.[13] The molecules of many solvents contain oxygen or nitrogen atoms which endow these liquids with basic properties. On the other hand, acidic solvents are relatively uncommon. Amphiprotic solvents are often subdivided, on the basis of their dominant property, into *acidic amphiprotic* and *basic amphiprotic* solvents.

The phenomenon termed acidic or basic dissociation is in reality a proton-transfer process or *protolysis* in which the solvent participates. The reaction is customarily represented formally by equation 2. The extent of this acid-base reaction or "salt formation" is governed by the acidic or basic properties of the solvent as well as by those of the acid or base itself. If ions are formed, the dielectric constant of the solvent also influences the extent of dissociation.

Brønsted[14] has characterized solvents on the basis of three properties which have an important bearing on the acid-base behavior of solutions in

[11] R. P. Bell, *Quart. Rev. (London)*, **1**, 113 (1947).
[12] I. M. Kolthoff, *J. Phys. Chem.*, **48**, 51 (1944).
[13] P. Walden, *Salts, Acids, and Bases*, McGraw-Hill Book Co., New York, 1929, has expressed doubt that any truly inert solvent exists. Many independent lines of evidence, for example, the results of S. J. O'Brien, C. L. Kenney, and R. A. Zuercher, *J. Amer. Chem. Soc.*, **61**, 2504 (1939), on the solubility of hydrogen chloride in organic media, have shown that benzene and other compounds of similar electronic configuration can function as weak bases.
[14] J. N. Brønsted, *Ber.*, **61**, 2049 (1928).

these media. These properties are proton-donating power (acidity), proton-accepting power (basicity), and ionizing power (high dielectric constant ε). This classification is illustrated in Table 7–1, where a plus sign signifies that the solvent has this property in marked degree and a minus sign means that the property is possessed only in a slight degree, if at all. An example of each of the eight classes of solvent is given in the last column. Type 4 solvents are often classed as "dipolar aprotic."

TABLE 7–1. CLASSIFICATION OF SOLVENTS

Type	Dielectric Constant ε	(After Brønsted) Acidic Strength	Basic Strength	Example (ε)
1	+	+	+	Water (78.3)
2	+	+	−	Sulfuric acid (101)
3	+	−	+	Ethanolamine (37.7)
4	+	−	−	Acetonitrile (36.0)
5	−	+	+	Ethanol (24.3)
6	−	+	−	Acetic acid (6.1)
7	−	−	+	Pyridine (12.3)
8	−	−	−	Benzene (2.3)

The so-called strong acids appear to be almost equally strong in water, for the protophilic nature of the solvent is sufficiently pronounced to convert them almost completely to hydrogen (hydronium) ion and the acid anion. This phenomenon, which has its analog in other protophilic solvents, was called by Hantzsch a *leveling effect*. Similarly, all strong bases are essentially completely converted by water to an equivalent quantity of hydroxide ion, OH^-. Differences among the strengths of the strong acids become apparent only in indifferent solvents or those of very low basicity (weak protophilic nature). Such solvents are sometimes termed *differentiating*. Solutions of the strong acids in glacial acetic acid, a solvent of very low basicity, are much more acidic than in water, where the strongest acid that can exist is hydronium ion, H_3O^+. Hall and Conant[15] found that a 0.064M solution of sulfuric acid in anhydrous acetic acid has about the same acidity as the 1M aqueous solution. Furthermore, from measurements of the conductivity of their solutions, the common mineral acids have been found to have the following

---

[15] N. F. Hall and J. B. Conant, *J. Amer. Chem. Soc.*, **49**, 3047 (1927).

strengths in anhydrous acetic acid, relative to nitric acid as unity:[16] perchloric 400, hydrobromic 160, sulfuric 30, and hydrochloric 9.

The Brønsted classification of solvents presented in Table 7–1 does not always serve to categorize the properties of nonaqueous solvents in a useful way. This is because it is very difficult to decide whether acidic and basic properties of a solvent are truly absent; it is much easier to recognize that acidic or basic properties predominate. To compound the difficulties in classification, it is virtually impossible to decide whether an extremely weak apparent autoprotolysis is real or an artifact. Trace amounts of impurities such as water may be responsible for a measurable autoprotolysis constant. For example, extreme precautions were required by Coetzee and Padmanabhan in order to obtain the autoprotolysis constant of acetonitrile, which is stated to be $10^{-28.5}$. Acetonitrile has an exceptionally weak basic property, and its structure and behavior make it doubtful that an acidic property exists.[17]

It is thus often convenient to identify "acidic amphiprotic" and "basic amphiprotic" solvents when the words "protogenic" and "protophilic" seem unduly restrictive. Similarly, in practice, solvents that are nearly inert in their acid-base behavior are classified as aprotic; these solvents of relatively high dielectric constant (nitrobenzene, acetonitrile, dimethylsulfoxide, etc.) are termed "dipolar aprotic" solvents to distinguish them from the hydrocarbon solvents of low dielectric constant and low capacity to promote ionization. Considerable attention has been devoted in recent years to acid-base behavior in dipolar aprotic media.[17–19]

The Hydrogen Ion

It is unlikely that free protons exist in solution any more than free electrons do. The proton is doubtless combined with at least one molecule of water in aqueous solutions at ordinary temperatures to form the H_3O^+ ion, called the *oxonium* or *hydronium* ion. Although there is evidence that a substantial fraction of the protons in aqueous solutions of acids are in the form of

---

[16] R. P. Bell, *Acids and Bases*, Chapter 3, Methuen and Co., London, 1961.

[17] J. F. Coetzee and G. R. Padmanabhan, *J. Phys. Chem.*, **69**, 3193 (1965). I. M. Kolthoff and M. K. Chantooni, Jr., *J. Phys. Chem.*, **72**, 2270 (1968), believe this value of pK_S (28.5) to be too small. Their estimate is 33.3, but they feel the true value may be still larger.

[18] I. M. Kolthoff and M. K. Chantooni, Jr., *J. Amer. Chem. Soc.*, **87**, 4428 (1965).

[19] J. F. Coetzee and G. R. Padmanabhan, *J. Phys. Chem.*, **66**, 1708 (1962); *J. Amer. Chem. Soc.*, **87**, 5005 (1965); J. F. Coetzee, G. R. Padmanabhan, and G. P. Cunningham, *Talanta*, **11**, 93 (1964); J. F. Coetzee, J. M. Simon, and R. J. Bertozzi, *Anal. Chem.*, **41**, 766 (1969); C. D. Ritchie, Chapter 4 in *Solute-Solvent Interactions*, J. F. Coetzee and C. D. Ritchie, eds., Marcel Dekker, New York, 1969.

tetrahydrates,[20] there is also justification for the belief that one solvent molecule is attached particularly firmly in aqueous solutions and, by analogy, that the same is true in many other solvents. The term hydrogen ion is applied generally to the solvated proton, whether it be H_3O^+ in water, ROH_2^+ in an alcohol, RNH_3^+ in an amine, $RCOOH_2^+$ in an acid solvent, or still another form.[21] Likewise, the symbol a_H will be used here for "proton activity" or "hydrogen ion activity" regardless of the nature of the solvent.

Strong support for the existence of the ion H_3O^+ in the solid monohydrates of nitric acid, perchloric acid, and sulfuric acid is provided by proton magnetic resonance studies,[22] which have shown that the three protons are all located at equal distances from the oxygen atom in the crystal lattice. Bell[23] has summarized the additional evidence, from infrared and Raman studies, for the existence of a stable hydronium ion. He has also set forth the reasons why it is difficult to demonstrate unambiguously the structure of the hydrogen ion in solutions. The chief reasons are the close resemblance between the spectral properties of H_3O^+ and those of H_2O and the rapid exchange of protons among water molecules, which limit the information that can be gained from nuclear magnetic resonance measurements in solutions. Furthermore, it is likely that additional water molecules are attached more or less firmly, on the average, to a given hydronium ion.

Studies of acidity functions in solutions of the strong acids also lend support to the structure $H^+(H_2O)_4$ for the hydrated proton. Wyatt[24] and Högfeldt[25] have pointed out that the Hammett H_0 functions for concentrated aqueous solutions of the strong acids fall on a single curve when plotted as a function of the water activity. It would therefore appear that H_0 is independent of the anion and that the hydration state of the proton is the same in all of the acids when the comparison is made at equal water activities. Bell has shown[26] that no integral hydration number other than 4 accounts

---

[20] E. Wicke, M. Eigen, and T. Ackermann, Z. Physik. Chem. (Frankfurt), 1, 340 (1954); E. Glueckauf, Trans. Faraday Soc., 51, 1235 (1955).

[21] J. N. Brønsted, Z. Angew. Chem., 43, 229 (1930).

[22] R. E. Richards and J. A. S. Smith, Trans. Faraday Soc., 47, 1261 (1951). See also C. C. Ferriso and D. F. Hornig, J. Amer. Chem. Soc., 75, 4113 (1953); J. Chem. Phys., 23, 1464 (1955).

[23] R. P. Bell, The Proton in Chemistry, Chapter 3, Cornell University Press, Ithaca, N.Y., 1959.

[24] P. A. H. Wyatt, Discussions Faraday Soc., 24, 162 (1957). See also R. P. Bell and K. N. Bascombe, ibid., 24, 158 (1957); K. N. Bascombe and R. P. Bell, J. Chem. Soc., 1096 (1959); M. A. Paul and F. A. Long, Chem. Rev., 57, 1 (1957); and R. H. Boyd, Chapter 3 in Solute-Solvent Interactions, J. F. Coetzee and C. D. Ritchie, eds., Marcel Dekker, New York, 1969.

[25] E. Högfeldt, Acta Chem. Scand., 14, 1627 (1960); Svensk Kem. Tidskr., 75, 63 (1963).

[26] R. P. Bell, op. cit., Chapter 6.

satisfactorily for the observed increase in the apparent acidity with increase in the concentration of the acid. In the work of Rosenthal and Dwyer,[27] the variation of H_0 with pm_H in 1–9M lithium chloride, 1–9M perchloric acid, and 1–11M hydrochloric acid was shown to be consistent with hydration numbers of 3 for both H_3O^+ and the indicator acid BH^+ and 1 or 0 for the indicator base B.

An attempt has been made by Högfeldt[28] to calculate an average degree of hydration for the ion pairs (H^+, HSO_4^-), (H^+, ClO_4^-), and (H^+, NO_3^-) in concentrated aqueous solutions of the three acids by applying the methods widely used to determine the formulas of coordination complexes. Although not highly reliable, the calculation suggests that the hydration number decreases steadily with decreasing water activity from a value of about 9 (7 for nitric acid) to a value close to unity for pure sulfuric acid.

Solvation Energies of Ions

The solvation energy of the proton is a quantity of considerable interest for its bearing on the intrinsic basicity of a solvent SH. This quantity is the negative of the free energy change $G_{s(H^+)}$ in the process

$$H^+(g) + \infty \, SH(l) = SH_2^+ \text{ (in solvent SH)} \tag{3}$$

Izmailov[29] has calculated the sum of the solvation energies of hydrogen and chloride ions in 12 solvents by a Born–Haber cyclic process starting from the standard e.m.f. of the cell

$$Pt; H_2, HCl \text{ (in solvent SH)}, AgCl; Ag$$

To make the calculation it is necessary to know (a) the dissociation energy of molecular hydrogen, (b) the free energy of sublimation of silver, (c) the lattice energy of silver chloride, and (d) the ionization energies of atomic hydrogen and silver in the gas phase. The values obtained are given in Table 7–2.

Unfortunately, there are no direct methods for separating these free energies into the terms representing the individual ionic solvation energies. Izmailov based a separation on the accepted fact that the solvation energy decreases sharply with increasing ion radius. To obtain $G_{s(H^+)}$, for example, he first plotted $G_{s(H^+)} + G_{s(X^-)}$, where X^- represents a halide ion, as a

[27] D. Rosenthal and J. S. Dwyer, *Can. J. Chem.*, **41**, 80 (1963).

[28] E. Högfeldt, *Acta Chem. Scand.*, **14**, 1597 (1960).

[29] N. A. Izmailov, *Zh. Fiz. Khim.*, **34**, 2414 (1960); *Dokl. Akad. Nauk SSSR*, **149**, 884 (1963). See also the calculation of E. C. Baughan, *J. Chem. Soc.*, 1403 (1940), and the theoretical discussion of N. E. Khomutov, *Zh. Fiz. Khim.*, **34**, 380 (1960).

TABLE 7–2. Sum of the Solvation Energies of Hydrogen and Chloride Ions

Solvent	$-[G_{s(H+)} + G_{s(Cl-)}]$ kcal mol^{-1}	Solvent	$-[G_{s(H+)} + G_{s(Cl-)}]$ kcal mol^{-1}
Ammonia	346.0	Isobutanol	322.8
Water	331.1	Isopentyl alcohol	322.8
Methanol	325.4	Benzyl alcohol	322.2
Ethanol	324.1	Acetone	318.7
Propanol	323.5	Formic acid	323.5
Butanol	322.9	Acetic acid	310.3

function of $1/r_{X-}$. Values of $G_{s(H+)} - G_{s(M+)}$ for various metal cations M^+, obtained from data for cells with transport, were likewise plotted as a function of $1/r_{M+}$. If the solvation energies of M^+ and X^- are zero when these ions are of infinite size, the two curves will extrapolate to the same value, namely $G_{s(H+)}$, at $1/r = 0$. The extrapolation was long and admittedly not very reliable, but one of the lines was usually straight, facilitating the location of the common limit. Procedures for deriving individual ionic solvation energies from experimental data have been reviewed by Popovych.[30]

The solvation energies of individual ions found by Izmailov are summarized in Table 7–3. It is of interest to note that in a given solvent the energy of solvating the proton exceeds by more than 125 kcal mol^{-1} the solvation energy of any other univalent ion listed. This large difference can doubtless be attributed to the small size and intense electric field of the proton. It is responsible for the special position occupied by the hydrogen ion.[31] Furthermore, for these solvents, the magnitude of the proton solvation energy is not strongly dependent on the structure and properties of the solvent.

It is also noteworthy that the solvation energies for the proton and other cations decrease from the most basic solvent (ammonia) to the least basic (formic acid). Orders of basicity are not easy to establish for solvents in which pronounced acidic or basic character is lacking. Equilibrium measurements are difficult to interpret, as an electrostatic effect caused by a change in dielectric constant is usually superimposed on the basicity effect.[32] Nevertheless, it is widely accepted that basicity in solution decreases in the order

[30] O. Popovych, *Crit. Rev. Anal. Chem.*, **1**, 73 (1970).

[31] G. Schwarzenbach, *Chimia*, **3**, 1 (1949).

[32] I. M. Kolthoff and S. Bruckenstein, Chapter 13 in *Treatise on Analytical Chemistry*, I. M. Kolthoff and P. J. Elving, eds., Part I, Vol. 1, Interscience Publishers, New York, 1959.

TABLE 7-3. SOLVATION FREE ENERGIES OF IONS (in kcal mol^{-1})

(After Izmailov)

Solvent

Ion	Ammonia	Water	Methanol	Ethanol	Formic Acid
H$^+$	281.0	258.0	253.0	252.0	246.0
Li$^+$	124.0	117.0	116.0	115.0	116.0
Na$^+$	99.0	96.0	93.0	90.0	99.5
K$^+$	79.4	78.0	76.0	73.2	73.9
Rb$^+$	73.3	74.4	—	—	73.2
Cs$^+$	65.6	64.0	60.4	—	65.0
Ag$^+$	132.0	112.0	108.0	108.0	120.8
Ca^{2+}	360.0	372.4	—	—	309.2
Zn^{2+}	536.0	492.0	481.0	473.5	488.3
Cd^{2+}	546.0	430.4	417.0	413.0	410.4
Cl$^-$	65.5	74.0	71.0	71.3	78.3
Br$^-$	62.8	68.0	67.0	66.2	—
I$^-$	57.0	59.4	59.6	58.5	—

$H_2O > CH_3OH > C_2H_5OH$. This conclusion is based in part on the kinetics of the reaction of ethylene oxide with aqueous methanol and ethanol[33] and on studies of the effect of water on the conductivity of hydrochloric acid in methanol.[34] It is also borne out by estimates of proton exchange constants.[35]

Complete agreement on this point, however, has not been reached. Braude and Stern[36] and others have attributed the initial decrease in the pK_a of protonated amines and substituted anilinium ions as alcohol is added to the aqueous solvent to an increase of basicity resulting from breakdown of the water structure. Other interpretations based on solute-solvent interactions have nonetheless been offered to explain this phenomenon.[37] Feakins

---

[33] C. E. Newall and A. M. Eastham, *Can. J. Chem.*, **39**, 1752 (1961). See also E. A. Braude and E. S. Stern, *J. Chem. Soc.*, 1976 (1948), and J. Koskikallio, *Acta Chem. Scand.*, **13**, 671 (1959).

[34] H. Strehlow, *Z. Physik. Chem.* (*Frankfurt*), **24**, 240 (1960).

[35] E. J. King, *Acid-Base Equilibria*, Chapter 11, Pergamon Press, Oxford, 1965; L. S. Levitt and B. W. Levitt, *J. Phys. Chem.*, **74**, 1812 (1970).

[36] E. A. Braude and E. S. Stern, *J. Chem. Soc.*, 1971, 1976, 1982 (1948). See also C. F. Wells, *Nature*, **196**, 770 (1962), and M. Paabo, R. G. Bates, and R. A. Robinson, *J. Phys. Chem.*, **70**, 247 (1966).

[37] R. Gaboriaud, *Ann. Chim.* (*Paris*), Ser. 14, **2**, 201 (1967); R. G. Bates, *J. Electroanal. Chem.*, **29**, 1 (1971); H. Ohtaki, *Bull. Chem. Soc. Japan*, **42**, 1573 (1969).

and Watson[38] have derived values for the Gibbs energy of the proton in 10 and 43.12 per cent methanol by an extrapolation of data for the energy of transfer of the hydrogen halides, somewhat in the manner of Izmailov.[29] They have concluded that the proton is in a lower Gibbs energy state in the methanolic solvents, and in pure methanol, than it is in pure water. Franks and Ives[39] have pointed out, however, that "solvent basicity" may not be a valid concept in hydrogen-bonded solvent mixtures. It may be noted here that the gas-phase acidities of the alcohols are greater than that of water, possibly because of charge stabilization by the alkyl groups.[40]

Acidity Potential and Proton Activity

POTENTIALS OF ACID-BASE REACTIONS

There is a close analogy between oxidation-reduction reactions and acid-base phenomena.[41] The former require electron transfer, whereas the latter, according to the Brønsted view, involve a transfer of protons.

The concept of the acidity potential (or acid-base potential) is accordingly similar to the concept of the oxidation-reduction potential. Inasmuch as protolytic reactions involve a transfer of protons, the reversible work of transfer $-\Delta G$ of the proton from a given system into an arbitrary standard state can be regarded as a measure of the acidity of a solution, whatever the solvent:

$$-\Delta G = FE = RT \ln (a_H)_1 \qquad (4)$$

where E is the *acidity potential*.[42] In theory, E represents the difference of potential between two standard hydrogen electrodes, one immersed in a given solution 1 and the other in a solution of unit proton activity. The numerical value of $(a_H)_1$ is fixed by the choice of this standard state. By combination of equation 4 with the mass-law expression for the dissociation of the acid HA, we obtain[43]

$$E = E_{ac}{}^\circ + \frac{RT}{F} \ln \frac{a_{HA}}{a_A} \qquad (5)$$

---

[38] D. Feakins and P. Watson, *Chem. Ind. (London)*, 2008 (1962); A. L. Andrews, H. P. Bennetto, D. Feakins, K. G. Lawrence, and R. P. T. Tomkins, *J. Chem. Soc., A*, 1486 (1968).

[39] F. Franks and D. J. G. Ives, *Quart. Rev.*, **20**, 1 (1966).

[40] J. I. Brauman and L. K. Blair, *J. Amer. Chem. Soc.*, **90**, 6561 (1968).

[41] J. N. Brønsted, *Rec. Trav. Chim.*, **42**, 718 (1923); G. Schwarzenbach, *Helv. Chim. Acta*, **13**, 870 (1930); N. F. Hall, *Chem. Rev.*, **8**, 191 (1931); T. H. Hazlehurst, *J. Chem. Educ.*, **17**, 466 (1940).

[42] J. N. Brønsted, *Rec. Trav. Chim.*, **42**, 718 (1923); *Z. Physik. Chem.*, **A169**, 52 (1934).

[43] G. Schwarzenbach, *Helv. Chim. Acta*, **13**, 870 (1930); G. Kortüm, *Elektrolytlösungen*, Chapter 14, Becker und Erler Kom.-Ges., Leipzig, 1941.

where E_{ac}° is a constant at a particular temperature. When the activities of acid and base (HA and A, respectively) are equal, E becomes E_{ac}°, the *normal acidity potential.*

A high acidity potential and high proton activity mean that the protons are loosely bound, a low potential that they are firmly bound. In reactions of type 2, we are concerned with two single potentials; the proton is transferred from the higher potential to the lower. Either the free energy change or the proton activity can serve as a measure of acidity.

In order to place acidity potentials on a common unitary basis in all solvent media, Gurney[44] has calculated proton energy levels, J, in a manner that allows for differences in solvent concentration. Thus

$$J = \frac{RT}{F} \left(-\ln K + n \ln \frac{1000}{M} \right) \tag{6}$$

where M is the molecular weight of the solvent, K is the molal equilibrium constant for the acid-base reaction in question,[45] and n is the number of reactant solvent molecules SH. For the dissociation reactions

$$HA + SH \rightleftharpoons SH_2^+ + A \tag{7a}$$

and

$$A + SH \rightleftharpoons S^- + HA \tag{7b}$$

the value of n is 1, whereas for the autoprotolysis of the solvent SH

$$2SH \rightleftharpoons SH_2^+ + S^- \tag{8}$$

n is 2. If RT/F is in volts, as is usual, J is an energy in electron volts.

The numerical value of J calculated by equation 6 is perhaps best regarded as a measure of the energy of the *vacant* proton level in the base A, that is, of the affinity of the base for protons. It is proportional to $-\Delta G^{\circ}$ for the transfer of protons from SH_2^+ (in the standard state) to the base A in the system: $HA(a = 1), A(a = 1)$.[46]

RANGE OF PROTON ACTIVITY

The range of proton levels capable of stable existence in a given amphiprotic solvent SH is the difference in the energy of the (occupied) proton level of SH_2^+, the strongest acid (weakest base, SH) that can exist in that

---

[44] R. W. Gurney, *Ionic Processes in Solution*, Chapters 7 and 14, McGraw-Hill Book Co., New York, 1953.

[45] Formulated in the conventional way by taking the solvent activity as unity.

[46] On the contrary, the acidity potential E_{ac}° is proportional to the free energy change for a transfer of protons in the opposite direction.

solvent, and that of the occupied level of SH, the weakest acid (strongest base, S^-) that can exist unchanged therein. The value of J for the first limit is, by our convention, 0 eV; for the second it is given by equation 6 with the use of the autoprotolysis constant K_S. Data for the autoprotolysis of several pure solvents are given in Table 7–4. Dielectric constants of these solvents are listed in Table 5 of the Appendix.

In order to preserve the parallelism between J and the proton activity or escaping tendency, it is helpful to give J a numerical value expressing the tendency of the system HA,A to transfer protons to the base S^- in the system: $SH(a = 1),S^-(a = 1)$. This can be accomplished by subtracting each value of J calculated by equation 6 from the highest value for that solvent, namely J_S (Table 7–4). The numbers so obtained represent the proton energies in the *occupied* levels of the species HA. The proton energy levels obtained in this way for several acid-base pairs in water, methanol, and formic acid are shown in Fig. 7–1.

It is evident that the range of strengths of the acids and bases that can exist in formic acid is much more limited than in methanol and water. The stable band has a width of only 0.52 eV, whereas in water the range is 1.03 eV and in methanol 1.16 eV. Unfortunately, the *positions* of the stable bands in these solvents with respect to one another cannot be located exactly. This is indeed the reason why a single scale of acidity for all solvents has not been devised. In Fig. 7–1, for example, the potentials of the occupied proton levels in H_2O, CH_3OH, and $HCOOH$ have all been taken arbitrarily to be 0 in the solvents water, methanol, and formic acid, respectively. All experimental attempts to interrelate acidity potentials in different solvents have been thwarted thus far by the indeterminate liquid-junction potential. Qualitatively, however, it is safe to say that the proton level for $HCOOH_2^+,HCOOH$ in formic acid is higher than that for H_3O^+,H_2O in water, and that the proton level of the pair NH_3,NH_2^- in liquid ammonia doubtless lies considerably lower than $H_2O.OH^-$ in water. These concepts shed some light on the range of hydrogen ion activity with which we would have to deal if a general pH scale embracing all solvents could be realized. This matter will be considered further in the next chapter.

The solvation free energies of hydrogen ion calculated by Izmailov[29] and listed in Table 7–3 provide a possible means of interrelating scales of proton potential J in different amphiprotic solvents. From the values of $G_{s(H)^+}$ in liquid ammonia, formic acid, and water, together with the best estimates of J_S, it may be shown, for example, that the most acidic solutions in liquid ammonia are more alkaline than any aqueous solutions, while the most alkaline solutions in formic acid are more acidic than the most acidic aqueous solutions.

Because of differences in their chemical structures, some acids bind the

TABLE 7-4. AUTOPROTOLYSIS CONSTANTS K_S AND RANGES OF PROTON ENERGIES J_S FOR PURE SOLVENTS AT 25 °C

Solvent	$-\log K_S$	J_S (eV)
Water	14.0	1.03
Deuterium oxide	14.8[a]	1.08
Hydrogen peroxide	About 13.0[b]	0.94
Methanol	16.7	1.16
Ethanol	19.1	1.29
Sulfuric acid	3.6[c]	0.38
Formic acid	6.2[d]	0.52
Acetic acid	14.5[d]	1.00
Acetonitrile	28.5[e]	1.85
Ethylenediamine	15.3[f]	1.05
Ethanolamine	5.1(20 °C)[g]	0.44(20 °C)
Dimethylformamide	18.0(20 °C)[g]	1.18(20 °C)
Dimethylsulfoxide	33.3[h]	2.10
Ammonia	32(−60 °C)[i]	1.50
Hydrofluoric acid	10.7(0 °C)[j]	0.76
Formamide	16.8[k]	1.15
Hydrazine	13[l]	0.95
n-Propanol	19.4[m]	1.29
Isopropanol	20.8[n]	1.38
Ethylene glycol	15.8[o]	1.08
Propylene glycol	17.2[o]	1.15

[a] E. Abel, E. Bratu, and O. Redlich, Z. Physik. Chem., **A173**, 353 (1935).
[b] A. G. Mitchell and W. F. K. Wynne-Jones, Trans. Faraday Soc., **52**, 824 (1956).
[c] R. L. Flowers, R. J. Gillespie, and E. A. Robinson, Can. J. Chem., **38**, 1363 (1960).
[d] I. M. Kolthoff and S. Bruckenstein, Chapter 13 in Treatise on Analytical Chemistry, I. M. Kolthoff and P. J. Elving, eds., Part I, Vol. 1, Interscience Publishers, New York, 1959.
[e] J. F. Coetzee and G. R. Padmanabhan, J. Phys. Chem., **69**, 3193 (1965).
[f] S. Bruckenstein and L. M. Mukherjee, J. Phys. Chem., **66**, 2228 (1962).
[g] C. Jacquinot-Vermesse and R. Schaal, Compt. Rend., **254**, 3679 (1962); M. Tézé and R. Schaal, Bull. Soc. Chim. France, 1372 (1962).
[h] J. Courtot-Coupez and M. Le Demezet, Bull. Soc. Chim. France, 1033 (1969).
[i] M. Herlem, ibid., 1687 (1967).
[j] Data of M. Kilpatrick. Value quoted by B. Trémillon, La Chimie en Solvants Non-aqueux, Presses Universitaires de France, Paris, 1971.
[k] F. M. Verhoek, J. Amer. Chem. Soc., **58**, 2577 (1936).
[l] L. J. Vieland and R. P. Seward, J. Phys. Chem., **59**, 466 (1955).
[m] R. Schaal and A. Tézé, Bull. Soc. Chim. France, 1783 (1961).
[n] E. A. Braude and E. S. Stern, J. Chem. Soc., 1976 (1948).
[o] K. K. Kundu, P. K. Chattopadhyay, D. Jana, and M. N. Das, J. Phys. Chem., **74**, 2633 (1970).

proton more firmly than do others, and intrinsic acid strengths accordingly differ widely. If the proton activity or level of proton availability could be measured accurately in solutions of these acids in indifferent solvents, it would presumably reflect these intrinsic differences. In a basic solvent, how-ever, protolysis will occur and the proton potential (and activity) will fall to a new level corresponding to the lowered escaping tendency of the proton.

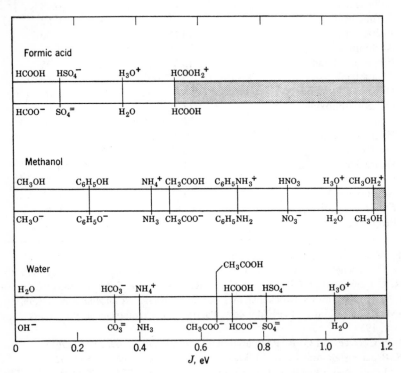

FIG. 7-1. Proton energy levels for acid-base pairs in water, methanol, and formic acid. Occupied levels above line, vacant levels below.

Solute acids are unable to raise the proton potential above the level character-istic of the solvated proton, which is accordingly the strongest acid that can exist unaltered in that particular solvent. Similarly, in protogenic solvents bases will be unable to reduce the proton potential below the level fixed by the acidic character of the solvent itself; strong bases are converted by re-action with the solvent into the base conjugate to the solvent acid. In acetic acid, for example, no base stronger than acetate ion can exist.

Only amphiprotic solvents impose both an upper and a lower limit to the proton activity. When the solution of acid or base is no more concentrated than 1M, the approximate range of the pH scale in water is from 0 to 14. This range is determined by the autoprotolysis constant for water, as we have already seen. The ends of the scale are fixed, then, primarily by the acidic and basic character of the solvent.[47] Figure 7–2 illustrates in a qualitative way the range of the scale of proton activity in solvents of different types. The activity is defined in such a way that $-\log a_H$ is equal to pH when the medium is water. The three bands on the right of the figure demonstrate in a qualitative fashion the activity range in a solvent that possesses only basic properties (*b*), in one with only acidic properties (*c*), and finally in one with little or no acidic or basic character (*d*).

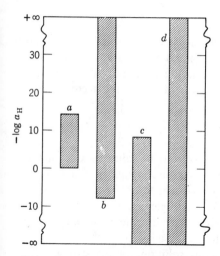

FIG. 7–2. Limits on the proton activity in solvents of different types. (*a*) Water (amphiprotic solvent), (*b*) protophilic solvent (only basic properties), (*c*) protogenic solvent (only acidic properties), (*d*) aprotic solvent (neither acidic nor basic properties).

It is clear that the concept of "neutrality" has meaning only in amphiprotic solvents. The characteristic acidic or basic properties of these media can be expressed in terms of the relative concentrations of the solvent acid and solvent base, that is, the acid formed by adding a proton to the solvent molecule and the base resulting from removal of a proton from the same molecule. Solutions in aprotic solvents cannot be neutral in this sense.

Kolthoff and Bruckenstein have shown how these principles apply to acidimetric titrations in nonaqueous media.[48] The "pH jump" at the

---

[47] N. Bjerrum, *Chem. Rev.*, **16**, 287 (1935).

equivalence point is largest with inert solvents, which do not limit the proton potential achieved by addition of the first drop of excess titrant. If amphiprotic solvents must be used for reasons of solubility, acidic amphiprotic solvents give the best results in the titration of bases, and basic amphiprotic solvents are preferred for the titration of acids.

The range of proton potentials that can be achieved is clearly an important consideration in choosing a solvent for nonaqueous acid-base titrations. This is especially true when it is desired to minimize overlapping and to differentiate two acids or bases of different strengths in a single titration. Although the magnitude of pK_S is a qualitative guide, the range of proton potentials is further restricted when a low solvent dielectric constant permits extensive ion association. Kreshkov[49] proposes that the differentiating capability of a solvent be characterized by a quantity K_{ES} which he terms the "relative acidity constant:"

$$K_{ES} \equiv \frac{E_{1/2}(HClO_4) - E_{1/2}(Et_4NOH)}{pK_S} \tag{9}$$

The numerator of the fraction is the difference between the half-neutralization potentials ($E_{1/2}$) for a strong acid ($HClO_4$) and a strong base (Et_4NOH), in mV. The solvents with greatest differentiating capability have the largest values of K_{ES}; values of 39 are found for alcohols, 50 for acetonitrile, and 64 for dimethylformamide.

MEASUREMENT OF PROTON ACTIVITY

The negligible concentration of free electrons in solutions does not preclude the measurement of oxidation-reduction potentials. Similarly, one need not know whether free protons are present to use proton activity as a definition of acidity. The fact that the hydrogen electrode, which is presumed to be reversible to protons, gives reproducible potentials in many media means that a_H has a definite value in these solutions.[50]

When the pH of concentrated aqueous solutions of the strong acids is measured with the hydrogen electrode and a calomel reference electrode,

[48] I. M. Kolthoff and S. Bruckenstein, Chapter 13 in Treatise on Analytical Chemistry, I. M. Kolthoff and P. J. Elving, eds., Part I, Vol. 1, Interscience Publishers, New York, 1959. See also J. P. Wolff, Chim. Anal., 28, 87 (1946); Anal. Chim. Acta, 2, 90 (1947); G. A. Harlow and G. E. A. Wyld, Anal. Chem., 30, 69, 73 (1958); G. A. Harlow and D. B. Bruss, ibid., 30, 1833 (1958); D. B. Bruss and G. A. Harlow, ibid., 30, 1836 (1958); and A. P. Kreshkov, Zh. Analit. Khim., 17, 6 (1962).

[49] A. P. Kreshkov, Talanta, 17, 1029 (1970); A. P. Kreshkov, N. S. Aldarova, and B. B. Tanganov, Zh. Analit. Khim., 25, 362 (1970).

[50] G. Kortüm, Z. Elektrochem., 48, 145 (1942).

large negative values are obtained. Concentrated sulfuric acid, for example, has a pH near -10. Similarly, the pH of a saturated aqueous solution of sodium hydroxide is about 19 at 20 °C. It appears likely that the hydrogen electrode potential is an index primarily of the availability of protons and only an indirect measure of the concentrations of protonated species.

We can compare, although only inexactly, the proton activity in different solvents by means of e.m.f. measurements of cells of the type

$$\text{Pt}; H_2(g, 1 \text{ atm}), \text{Acid soln. } 2 \mid \text{Acid soln. } 1, H_2(g, 1 \text{ atm}); \text{Pt} \qquad (10)$$

where the vertical line marks a boundary between the two different liquid phases. If E represents the e.m.f. of cell 10 and the phase boundary potential is negligible,

$$-\log (a_H)_2 = -\log (a_H)_1 + \frac{E}{(RT \ln 10)/F} \qquad (11)$$

The similarity between equation 11 and the operational definition of pH (Chapter 2) is evident. The measurement of *relative* proton activity will be considered in some detail in the next chapter.

We should like, of course, to have a method of measuring the activity of protons in different media, preserving the standard state which makes the activity of protons numerically equal to the concentration of hydrogen ion in very dilute *aqueous* solutions. The proton activity, if it could be accurately established, would be a real measure of the availability of protons in the system, as compared with the ability of hydronium ions at unit activity to release protons.[51]

There is nothing in the formulation of equation 11 to prevent the choice of a single standard state to which a_H in different solvents could be referred. One might even choose the aqueous solution as the reference state for activity, making $-\log (a_H)_1 = \text{pH}$ in the aqueous medium,[52] and compute corresponding pH values in other media.[53] The scales of pH and activity would, of course, have no meaning in terms of hydrogen ion concentrations except in the purely aqueous medium. The potentials across the phase boundary (vertical line in cell 10) are often so large and uncertain that the pH values obtained experimentally would have very little significance.[54]

---

[51] See D. S. McKinney, P. Fugassi, and J. C. Warner, *Symposium on pH Measurement*, p. 19, ASTM Tech. Publ. 73, Philadelphia, Pa., 1947.

[52] By choice of a standard state in which $(a_H)_1 / m_{acid} = 1$ when $m_{acid} = 0$.

[53] L. Michaelis and M. Mizutani, *Z. Physik. Chem.*, **116**, 135 (1925).

[54] G. Schwarzenbach, *Helv. Chim. Acta*, **13**, 896 (1930). The potential across the junction between ethanol and saturated aqueous potassium chloride was estimated to be 139 mV (2.3 pH units) by B. Gutbezahl and E. Grunwald, *J. Amer. Chem. Soc.*, **75**, 565 (1953).

"ABSOLUTE" HYDROGEN ELECTRODE POTENTIALS

A reference electrode whose standard potential is independent of the solvent would provide a means of bridging the gap between different solvent media without recourse to liquid junctions of indeterminate potential. The solvation of electrode components is such a general phenomenon, however, that the discovery of a reference electrode meeting this requirement would seem to be a very unlikely event.

The selection of the hydrogen electrode as the zero point for the measurement of electrode potentials in different solvents is purely conventional, and there may be better choices. Pleskov[55] was of the opinion that the potential of the rubidium electrode is more suitable as a reference point than is the potential of the hydrogen electrode, inasmuch as the rubidium ion is large, singly charged, and not very polarizable.[56] Nevertheless, a free energy change on transfer of rubidium ion from one solvent to another of different dielectric constant may be expected, even in the absence of appreciable solvation. This free energy term is inversely proportional to the ionic radius, as a first approximation, and presumably is negligible for very large ions. Strehlow[57, 58] has allowed for the differences in standard potential to be expected from this source by a calculation based on a modification of the Born equation.

In the opinion of Koepp, Wendt, and Strehlow,[58] however, oxidation-reduction systems better suited than rubidium as a reference electrode can be found. In a careful study of eighteen redox systems, they chose the ferrocene, ferricinium couple[59] and the cobaltocene analogue as the most suitable. In both the charge is presumably distributed evenly over the surface of the large complex, so that the charge density and solvation are minimal. Furthermore, the metal ion is buried at the center of a nearly spherical

---

[55] V. A. Pleskov, *Advan. Chem. (USSR)*, **16**, 254 (1947).

[56] Cesium might have been a more logical choice, but the standard potential of this element in water was still unknown at that time. Nevertheless, it has been found more recently that $E_{Cs}^{\circ} - E_{Rb}^{\circ}$ is nearly constant regardless of the solvent studied; hence either element could serve as well as the other as a basis for a scale of potential.

[57] H. Strehlow, *Z. Elektrochem.*, **56**, 827 (1952); Chapter 4 in *The Chemistry of Non-Aqueous Solvents*, J. J. Lagowski, ed., Vol. I, Academic Press, New York, 1966. Two other experimental approaches to absolute electrode potentials deserve mention. One is the cell with "air junction" described by J. E. B. Randles, *Ann. Reports*, **56**, 32 (1959), and studied further by Yu. F. Rybkin and M. G. Karpenko, *Electrochim. Acta*, **11**, 1135 (1966), and by B. Case and R. Parsons, *Trans. Faraday Soc.*, **63**, 1224 (1967). The other is the suggestion of I. Oppenheim, *J. Phys. Chem.*, **68**, 2959 (1964), that the quadrupole radiation emitted by an electrode in harmonic motion might be utilized for this purpose. Neither approach has yet been refined to the point where reliable results are obtainable.

[58] H. M. Koepp, H. Wendt, and H. Strehlow, *Z. Elektrochem.*, **64**, 483 (1960).

[59] Ferrocene is the dicyclopentadienyl ferrous complex. Oxidation to the corresponding ferric complex (ferricinium) is a simple one-electron process.

complex; hence these substances undergo negligible chemical and steric changes when oxidation or reduction occurs.

The standard potentials of the ferrocene and cobaltocene couples differ by a constant amount in water, methanol, acetonitrile, and formamide. The ions are, unfortunately, still not large enough to eliminate completely the electrostatic free energy of transfer, but allowance can be made for the remaining differences at the time the zero of potential for each solvent is defined.

Nonthermodynamic procedures for deriving the medium effect for the proton,[60] if successful, will also furnish the standard potential of the hydrogen electrode in other solvents relative to the value in water. These procedures will be examined further in the next chapter. They would make possible a universal scale of acidity for all solvents, defined in terms of the differences in hydrogen electrode potential as determined in cells without liquid junction. Nevertheless, it is too early to say how much reduction in the uncertainty of establishing the points of reference in the different solvents (now of the order of 5 mV or 0.08 pH unit under the most favorable circumstances) can be achieved. The e.m.f. of the hydrogen electrode relative to the ferrocene-ferricinium couple in sulfuric acid-water mixtures has already been determined and the corresponding acidities compared with H_0 for these mixtures.[61] The data are given on page 199.

ACTIVITY COEFFICIENTS

Changes in the nature and composition of the solvent often produce large changes in the activity coefficients of chemical species and consequently in the proton potential. These may be especially large if the aqueous standard state is retained for the mixed solvents formed by adding appreciable amounts of an organic constituent to water. In these instances, the activity coefficient reflects changes in both interionic attractions and medium effects. A change in the character of the solvent medium is usually accompanied by an alteration of the dielectric constant and, hence, an alteration in the effective interionic forces. Moreover, a large effect persists even in the limit of zero molality. This effect can be attributed to alterations in the electrostatic Gibbs energies of the ions and to changes in the pattern of solute–solvent interactions.

---

[60] O. Popovych, *Crit. Rev. Anal. Chem.*, **1**, 73 (1970); C. Barraqué, J. Vedel, and B. Trémillon, *Bull. Soc. Chim. France*, 3421 (1968); A. J. Parker and R. Alexander, *J. Amer. Chem. Soc.*, **90**, 3313 (1968); J. F. Coetzee, J. M. Simon, and R. J. Bertozzi, *Anal. Chem.*, **41**, 766 (1969); R. G. Bates, Chapter 2 in *Solute-Solvent Interactions*, J. F. Coetzee and C. D. Ritchie, eds., Marcel Dekker, New York, 1969.

[61] H. Strehlow and H. Wendt, *Z. Physik. Chem. (Frankfurt)*, **30**, 141 (1961).

These interactions are often revealed by changes in solubility as the solvent medium is altered. If $S(w)$ and $S(s)$ are the solubilities of a given substance in water and in a nonaqueous or mixed solvent, respectively, and the aqueous standard state is applied to both solutions, the activity coefficient $\gamma(s)$ is given by $[S(w)/S(s)]\gamma(w)$. If the substance is not very soluble in water, $\gamma(w)$ is close to unity and $\gamma(s)$ is given, in close approximation, by $S(w)/S(s)$. An increase in solubility in medium s over that in water (salting in) corresponds to a lowered activity coefficient, while a decrease (salting out) corresponds to an increase in activity coefficient. Many weak organic acids and bases are more soluble in organic solvents such as alcohol than in water; hence, addition of alcohol to the water solvent produces a "salting in" or a decrease in the activity coefficient of these substances.

The opposite effect is often observed with electrolytes. Solubility is decreased by addition of alcohol, unless the electrolyte is strongly hydrophobic, and the activity coefficient is increased. Here the increase in Gibbs energy of the ions is doubtless the result of the decreased dielectric constant, the effect of which is expressed, qualitatively at least, by the $1/\varepsilon$ coefficient in the Born equation. This effect can be overcome only by a strong stabilization of the ions by the added solvent component. Since most organic solvents are less polar than water, this stabilization of simple ions by an ion-dipole mechanism is unlikely.

At low ionic strengths, $\gamma(s)$ is primarily a measure of the medium effect $(_m\gamma)$, as the contribution of interionic forces is negligible. The nature of the medium effect is discussed in the next chapter. Experimental evidence indicates that the activity coefficient is increased (that is, $_m\gamma_i > 1$) when an organic solvent (for example, alcohol) is added to an aqueous solution of an electrolyte, lowering the dielectric constant. This evidence is to be found in the decreased solubility of "normal" salts like the alkali halides. Furthermore, the data of many authors show that the e.m.f. of cells with hydrogen and silver–silver chloride electrodes in solutions of hydrochloric acid is often lower, for a given acid concentration, in nonaqueous and partially aqueous solvents than in water. This observation suggests that the activity coefficient is often higher in the nonaqueous medium.

EFFECT OF SOLVENT BASICITY

It is appropriate at this point to consider the acidity of a solution in terms of changes in the basicity of the solvent and solute, considered separately from the changes in activity coefficients of the ions brought about by alteration of the dielectric constant of the medium. These effects can be accounted for by introduction of the appropriate acidity constants. The proton activity a_H is chosen as a measure of acidity, and the customary standard state for aqueous solutions is used. The following treatment is essentially that of Hammett.[62]

Consider an acid HA at molality m in an amphiprotic solvent SH. It is assumed that the dielectric constant of SH is high enough to permit a certain amount of ionization. The following equilibrium is then established:

$$HA + SH \rightleftharpoons SH_2^+ + A^- \tag{12}$$

The molality of SH can be considered constant, and a_H can be expressed in terms of K_{HA} and K_{SH}, which are, respectively, the acidity constants of HA and SH_2^+:

$$a_H = K_{HA} \frac{m_{HA}\gamma_{HA}}{m_{A^-}\gamma_{A^-}} = K_{SH}m_{SH_2^+} \frac{\gamma_{SH_2^+}}{\gamma_{SH}} \tag{13}$$

Inasmuch as the aqueous standard state is retained, the values of K_{HA} and K_{SH} remain unaltered when the medium is changed. The concentration of free protons is negligible; hence

$$m = m_{HA} + m_{A^-} = m_{HA} + m_{SH_2^+} \tag{14}$$

Combination of equations 13 and 14 gives

$$a_H = \frac{1}{2} K_{HA} \frac{\gamma_{HA}}{\gamma_{A^-}} \left[-1 + \left(1 + 4m \frac{K_{SH}\gamma_{SH_2^+}\gamma_{A^-}}{K_{HA}\gamma_{SH}\gamma_{HA}} \right)^{1/2} \right] \tag{15}$$

According to equation 15, decreased basicity of the solvent (greater K_{SH}) results in increased acidity, even though the extent of dissociation of the acid is lessened. This conclusion is in agreement with the discovery of Hall and Conant and their coworkers[63] that solutions of some of the common acids in glacial acetic acid are so acidic that they can be used to titrate very weak bases. Likewise, Hardman and Lapworth[64] found that acidity potentials in alcoholic solutions of hydrochloric acid are lowered by the addition of small amounts of water. The same conclusion can be drawn from determinations of the activity of hydrochloric acid[65] and from the Hammett acidity functions[66] in these solutions, as well as from rate studies. Figure 7–3, taken from a paper of Hall and Werner,[67] shows the change of electromotive force in the titration of acetoxime with perchloric acid in water, methyl alcohol, and

[62] L. P. Hammett, *J. Amer. Chem. Soc.*, **50**, 2666 (1928).

[63] N. F. Hall and J. B. Conant, *J. Amer. Chem. Soc.*, **49**, 3047 (1927); N. F. Hall and T. H. Werner, *ibid.*, **50**, 2367 (1928); N. F. Hall, *ibid.*, **52**, 5115 (1930).

[64] R. T. Hardman and A. Lapworth, *J. Chem. Soc.*, **99**, 2242 (1911).

[65] P. S. Danner, *J. Amer. Chem. Soc.*, **44**, 2832 (1922); H. Millet, *Trans. Faraday Soc.*, **23** 515 (1927); J. W. Woolcock and H. Hartley, *Phil. Mag.*, [7] **5**, 1133 (1928); B. O. Heston and N. F. Hall, *J. Amer. Chem. Soc.*, **56**, 1462 (1934).

[66] E. A. Braude and E. S. Stern, *Nature*, **161**, 169 (1948).

[67] N. F. Hall and T. H. Werner, *J. Amer. Chem. Soc.*, **50**, 2367 (1928).

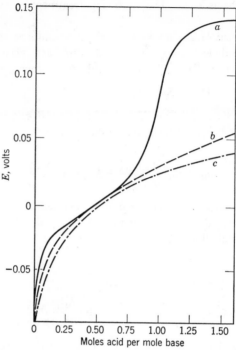

FIG 7–3. Titration of acetoxime with perchloric acid in three different solvents (after Hall and Werner). (*a*) acetic acid, (*b*) ethanol, (*c*) water.

glacial acetic acid. The inflection that occurs at the end point is much sharper in acetic acid than in the other more protophilic solvents.

Two special cases are of interest. If the acid is weak and the solvent only weakly basic, the last term of equation 15 is much greater than unity, and one obtains

$$a_H = \sqrt{mK_{HA}K_{SH}\frac{\gamma_{SH_2^+}\gamma_{HA}}{\gamma_{SH}\gamma_{A^-}}} \tag{16}$$

The reaction between HA and the solvent is incomplete under these conditions, and the acidity depends on both K_{HA} and K_{SH}. When, on the other hand, the solvent has pronounced basic properties and the acid is strong reaction 12 is substantially complete and HA is almost entirely converted to SH_2^+. Under these *leveling* conditions,

$$a_H = K_{SH}m\frac{\gamma_{SH_2^+}}{\gamma_{SH}} \tag{17}$$

The acidity of the solution is that of an m molal solution of the acid $SH_2{}^+$, weaker than HA. It is dependent only on the magnitude of K_{SH} and not at all on K_{HA}.

We have been mainly concerned here with the effect of the acid-base character of the solute and solvent on the proton activity. When the solvent is changed, however, the change in dielectric constant has a large effect on the activity coefficients in equations 15, 16, and 17 and, consequently, also on the proton activity. A decrease in the dielectric constant usually causes $\gamma_{SH_2^+}/\gamma_{SH}$ and γ_{A^-}/γ_{HA} to increase, and it is possible that both of these ratios will increase by about the same amount.[68] Hence, equation 16 suggests that a lowering of the dielectric constant will have little effect on the acidity of an unbuffered solution of a weak acid. But the acidity of a solution of a strong acid in a leveling medium of lowered dielectric constant is expected to be greater than that of an equally concentrated solution of the acid in a solution of high dielectric constant (compare equation 17).

These ideas may be extended to buffer solutions in the following manner. The ratio of the concentrations of the weak acid HA and its conjugate base A is fixed and equal to the buffer ratio R, and a_H is approximately equal to $RK_{HA}\gamma_{HA}/\gamma_A$. A decrease in the dielectric constant alters a_H through its effect on the ratio of the activity coefficients, which is, in turn, determined by the charges borne by HA and A. Hence the acidity of an aqueous buffer system of the type HA, A^- is expected to decrease as the dielectric constant is lowered, whereas that of type HA^+, A is expected to increase.

IONIC ASSOCIATION

Solvent effects on acid-base phenomena in amphiprotic media of intermediate and high dielectric constant are therefore often successfully interpreted in terms of changes in the dielectric constant (electrostatic effects) and in the basicity (nonelectrostatic effects). When the dielectric constant is lowered, however, an increasing fraction of the ions becomes associated to form Bjerrum ion pairs[69] and higher aggregates such as triple ions and dipole aggregates.[70] The stability of these associated products may be expressed by an equilibrium constant which can be determined by conductance measurements or by thermodynamic means. The magnitude of ion pair formation

[68] L. P. Hammett, *J. Amer. Chem. Soc.*, **50**, 2666 (1928).
[69] N. Bjerrum, *Kgl. Danske Videnskab. Selskab., Mat.-Fys. Medd.*, [7], 9 (1926).
[70] R. M. Fuoss and C. A. Kraus, *J. Amer. Chem. Soc.*, **55**, 2387 (1933); **57**, 1 (1935); R. M. Fuoss, *ibid.*, **56**, 1027 (1934); A. A. Maryott, *J. Res. Nat. Bur. Stand.*, **41**, 1 (1948). See also R. P. Bell, *The Proton in Chemistry*, Chapter 4, Cornell University Press, Ithaca, N.Y., 1959.

constants at a given dielectric constant can be estimated by an equation proposed by Denison and Ramsey.[71]

Studies of the dissociation of picrates in methanol solutions suggest that the total association found by conductance and diffusion studies, for example, should be regarded as the sum of two processes, namely the Bjerrum association of solvated ions and the formation of ion dipoles from unsolvated ions in contact.[72] Measurements of optical absorption cannot distinguish between free ions and solvent-separated ion pairs.

In solvents of low to moderate dielectric constant, acidity phenomena are governed largely by ionic association reactions, as Kolthoff and Bruckenstein have shown so convincingly.[73] Even in the intermediate range of dielectric constants ($25 < \varepsilon < 45$), ion pair formation can explain some puzzling anomalies in analytical processes.[74]

Extensive dimerization of carboxylic acids occurs in hydrocarbon solvents. In these aprotic nonionizing media, the concentration of free ions is negligible yet acid-base reactions take place and can be studied quantitatively with the aid of suitable indicators.[75] The primary product of the combination of a hydrogen acid HA with a base B in these solvents is probably a polar addition compound (BH^+A^-) or a hydrogen-bonded complex $B \cdots HA$.[76]

Hammett Acidity Functions

The Hammett acidity functions H_0 and H_- discussed in Chapter 6 provide a useful means of measuring the acidity of media that are not wholly aqueous

[71] J. T. Denison and J. B. Ramsey, *J. Amer. Chem. Soc.*, **77**, 2615 (1955). See also R. M. Fuoss and F. Accascina, *Electrolytic Conductance*, Chapter 16, Interscience Publishers, New York, 1959.

[72] G. Kortüm and K. Andrussow, *Z. Physik. Chem. (Frankfurt)*, **25**, 21 (1960).

[73] Glacial acetic acid ($\varepsilon = 6.1$): I. M. Kolthoff and S. Bruckenstein, *J. Amer. Chem. Soc.*, **78**, 1 (1956); **79**, 1 (1957); S. Bruckenstein and I. M. Kolthoff, *ibid.*, **78**, 2974 (1956). Ethylenediamine ($\varepsilon = 14$): S. Bruckenstein and L. M. Mukherjee, *J. Phys. Chem.*, **66**, 2228 (1962). See also O. Popovych, *ibid.*, **66**, 915 (1962), and B. Trémillon, *La Chimie e Solvants Non-aqueux*, Chapter 2, Presses Universitaires de France, Paris, 1971.

[74] E. Grunwald, *Anal. Chem.*, **26**, 1696 (1954); J. F. Coetzee, G. R. Padmanabhan, an G. P. Cunningham, *Talanta*, **11**, 93 (1964).

[75] M. M. Davis and P. J. Schuhmann, *J. Res. Nat. Bur. Stand.*, **39**, 221 (1947); M. M Davis, P. J. Schuhmann, and M. E. Lovelace, *ibid.*, **41**, 27 (1948); M. M. Davis and H. B Hetzer, *ibid.*, **48**, 381 (1952).

[76] A. F. Trotman-Dickenson, *J. Chem. Soc.*, 1293 (1949); M. M. Davis and E. A. McDonald *J. Res. Nat. Bur. Stand.*, **42**, 595 (1949); M. M. Davis, *J. Amer. Chem. Soc.*, **71**, 3544 (1949) M. M. Davis and H. B. Hetzer, *J. Res. Nat. Bur. Stand.*, **46**, 496 (1951); O. Popovyc *J. Phys. Chem.*, **66**, 915 (1962); M. M. Davis, *Acid-Base Behavior in Aprotic Organi Solvents*, National Bureau of Standards Monograph 105, U.S. Government Printing Offic Washington, 1968.

Nevertheless, these functions do not meet the requirements of an ideal general scale of acidity. In the first place, the Hammett function is based on the *extent* of the reaction of protons with an indicator base, either uncharged (H_0) or a univalent anion (H_-), rather than on proton activity alone:

$$H_0 = -\log a_H \frac{\gamma_{In}}{\gamma_{HIn^+}} \; ; \quad H_- = -\log a_H \frac{\gamma_{In^-}}{\gamma_{HIn}} \tag{18}$$

Neither acidity function is strictly comparable with $-\log a_H$, for salt effects and medium effects may make γ_{In}/γ_{HIn} very different from unity.

Furthermore, these acidity functions do not "exist" in many media. In other words, a series of indicators for all of which $_m\gamma_{In}/_m\gamma_{HIn}$ has a single identical value in each solvent medium cannot always be found. Such a series of indicators of the same charge type, but with a gradation of acidic strengths, is needed to establish a useful acidity function. In general it appears that the necessary regularity in the medium effect is most likely to occur for uncharged indicator bases in acidic solvents of rather high dielectric constant.[77, 78] The regularity in $_m\gamma_{In}/_m\gamma_{HIn}$ needed for a useful scale of H_- is still less commonly found. Hammett functions are of limited application in organic media of low dielectric constant. The characteristics of acidity functions and their applications have been reviewed thoroughly by Boyd.[78]

It is sometimes useful to distinguish between the *proton potential* and the *extent of reaction*, for example, with an indicator. The proton potential or activity, which Brønsted regarded as the only logically consistent measure of acidity,[5] denotes the degree of looseness with which the protons are held. According to this view, two solutions that have the same proton potential have the same true acidity, whatever the solvent.

On the other hand, the extent of a protolytic reaction depends not only on the strengths of the acid and base but is usually strongly dependent on the dielectric constant of the solvent and other factors as well. Nevertheless, measurements of acidity are often of practical importance for the very reason that they furnish a means of predicting the extent of a protolytic reaction. For this reason, Hantzsch[3] regarded a chemical method like an indicator measurement as providing a more useful index of acidity than proton activity. There is as yet no agreement on a basis for the establishment of an acidity scale valid in more than one medium. The effect of changes in the medium on the point of equilibrium is reflected in the changes the activity coefficient undergoes. Hammett[68] showed that the effect of decreased

[77] K. N. Bascombe and R. P. Bell, *J. Chem. Soc.*, 1096 (1959); M. A. Paul and F. A. Long, *Chem. Rev.*, **57**, 1 (1957).

[78] R. H. Boyd, Chapter 3 in *Solute-Solvent Interactions*, J. F. Coetzee and C. D. Ritchie, eds., Marcel Dekker, New York, 1969.

dielectric constant on the proton activity often lies between the effect shown by an indicator acid and that shown by an indicator base. It was for reasons such as this that Brønsted did not favor the use of protolytic reactions as a basis for establishing a general acidity scale.

Oxidation-Reduction Potentials

The residual liquid-junction potential in highly acidic media makes it impossible to obtain meaningful values for the pH of concentrated solutions of the strong acids by measurement of the usual pH cell with hydrogen (or glass) and calomel electrodes. Clark[79] has suggested a possible means for the extension of the pH scale to concentrated acid solutions and mixed solvent media. The scale he proposes would be based on the e.m.f. of a cell without liquid junction, consisting of an oxidation-reduction electrode coupled with the hydrogen electrode. To be suitable, the redox electrode would have to be reversible to hydrogen ions (as is the common quinhydrone electrode), but the magnitude of the change of its electrode potential with pH would have to be different from that of the hydrogen electrode (as that of the quinhydrone electrode is not).

The compilation of Clark[79] contains several examples of redox systems for which the change of potential per pH unit at 25 °C is 29 mV, 88 mV, or even 0 (as compared with 59 mV for the hydrogen electrode) when, at 50 per cent reduction of the system, the activities of the oxidized and reduced species are equal. The e.m.f. of a cell without liquid junction consisting of a hydrogen electrode in combination with one of these selected redox systems (half reduced) would vary with the pH. It would be desirable, of course, to have assurance that selective solvation in the various solvent media did not alter the electrode reaction in such a way that the response of the cell to proton activity was altered. Clark's proposal merits further study.

Michaelis and Granick[80] suggested a procedure for the empirical extension of the pH scale to strong solutions of sulfuric acid. This method is based on the measurement of the standard potentials of the two steps of oxidation in a bivalent reversible oxidation-reduction system. Three levels can be distinguished, namely, the reduced form, the semioxidized form, and the oxidized form. The *difference* between the standard potentials of the lower and higher steps of oxidation of certain dyestuffs such as 3-hydroxythiazine was found to be a function of the pH.

The relationship between this "spread" and pH was established in the

[79] W. M. Clark, *Oxidation-Reduction Potentials of Organic Systems*, pp. 201–203, The Williams and Wilkins Co., Baltimore, Md., 1960.

[80] L. Michaelis and S. Granick, *J. Amer. Chem. Soc.*, **64**, 1861 (1942).

region between pH 2 and 0, where the acidity is still measurable with the hydrogen-calomel cell. The curves, which were all linear, could be extended into the region of higher acidity. A change in the acidic ionization of the dye causes the slopes of the lines to be altered, but it is possible to identify these discontinuities and to make the appropriate modifications in the calibration plots.

By this procedure, Michaelis and Granick obtained pH values of solutions of sulfuric acid as concentrated as 11 M. Like H_0, the pH was found to be a linear function of the concentration above 1 M, as shown by the two upper curves of Fig. 7–4. The pH is in good agreement with H_0 at 1 M and somewhat lower at higher concentrations. The difference at 10 M is about 0.5 unit.

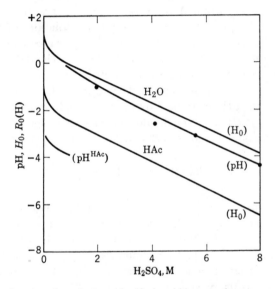

FIG 7–4. Acidity functions for solutions of sulfuric acid in water (upper curves) and in anhydrous acetic acid (lower curves). The dots represents values of $R_0(H)$.

The ferrocene, ferricinium redox system, proposed by Koepp, Wendt, and Strehlow[81] as a basis for intercomparing electrode potentials in different solvents, has been applied to the determination of acidity functions for sulfuric acid-water mixtures.[82] The half-wave potentials of the ferrocene,ferricinium couple in sulfuric acid solutions (17.8 to 100 wt. per cent)

[81] H. M. Koepp, H. Wendt, and H. Strehlow, *Z. Elektrochem.*, **64**, 483 (1960).
[82] H. Strehlow and H. Wendt, *Z. Physik. Chem.* (*Frankfurt*), **30**, 141 (1961).

were measured polarographically with respect to the mercury-mercurous sulfate electrode. These data were then combined with the e.m.f. of hydrogen-mercurous sulfate cells for the same sulfuric acid concentrations[83] to give the e.m.f. E_X of the cell

$$Pt;H_2(g, 1\ atm),H_2SO_4(aq.),Ferrocene,Ferricinium(1:1);Pt$$

Strehlow and Wendt defined an acidity function $R_0(H)$, termed the "redox function," as

$$R_0(H) = \frac{E_X - E_1}{(RT \ln 10)/F} \quad (19)$$

where E_1 is the e.m.f. of the cell at unit proton activity. This function is therefore a true (logarithmic) measure of relative proton activity in the solvents studied; it approaches pH for very dilute solutions of sulfuric acid. The values of $R_0(H)$ calculated from E_X and E_1[84] are compared with H_0 for the same solutions of sulfuric acid[85] in Table 7–5. The data at the lowest concentrations of acid are also shown as dots in the upper part of Fig. 7–4 (the value at 8 M was interpolated). The $R_0(H)$ values are in acceptable agreement with the pH determined by Michaelis and Granick.

The values of $R_0(H)$ and H_0 should not, of course, agree. If $R_0(H)$ is an approximate measure of $-\log a_H$, as it appears to be, equation 18 shows that

$$R_0(H) - H_0 = \log \frac{\gamma_{In}}{\gamma_{HIn^+}} \quad (20)$$

Except at concentrations of sulfuric acid approaching 100 per cent, $R_0(H)$ is lower than H_0 and $\gamma_{In}/\gamma_{HIn^+}$ is less than unity. Selective solvation of the acid form of the indicator by the water could explain the observed result.[82] Solvation of the ferrocene and ferricinium species should be relatively unimportant. The negative value of H_0 increases by 1.4 units when the last 1 per cent of water is removed.

[83] W. H. Beck, J. V. Dobson, and W. F. K. Wynne-Jones, *Trans. Faraday Soc.*, **56**, 1172 (1960).

[84] It appears that E_X for 65 per cent sulfuric acid should be $+ 80$ mV instead of $+ 180$ mV as given in Table 1 of the paper of Strehlow and Wendt.

[85] M. A. Paul and F. A. Long, *Chem. Rev.*, **57**, 1 (1957). M. J. Jorgenson and D. R. Hartter, *J. Amer. Chem. Soc.*, **85**, 878 (1963), have derived a new set of H_0 values from measurements with a series of primary anilines. Below 60 per cent sulfuric acid, agreement with the values of Paul and Long is complete, but differences appear in concentrated solutions of the acid. Comparisons of Hammett functions with $R_0(H)$ have also been made by J. Vedel, *Ann. Chim. (Paris)*, Ser. 14, **2**, 335 (1967).

TABLE 7-5. COMPARISON OF H_0 AND $R_0(H)$ FOR SULFURIC ACID-WATER MIXTURES AT 25 °C

Concentration of H_2SO_4

Wt. Per Cent	M/mol dm^{-3}	$H_0{}^a$	$R_0(H)^b$
10	1.08	−0.31	
15	1.68	−0.66	
17.8	1.96	−0.86	−1.01
20	2.32	−1.01	
30	3.73	−1.72	
32.6	4.12	−1.90	−2.6
40	5.3	−2.41	
40.8	5.5	−2.48	−3.0
50	7.1	−3.38	
60	9.1	−4.46 (−4.46)	
65	10.3	−5.04 (−5.07)	−5.4
70	11.5	−5.65 (−5.80)	
75	12.7	−6.30 (−6.56)	
76.5	13.2	−6.50 (−6.79)	−7.6
80	14.1	−6.97 (−7.34)	
88	16.1	−8.03 (−8.61)	−8.5
90	16.6	−8.27 (−8.92)	
100	18.6	−11.10	−9.7

[a] M. A. Paul and F. A. Long, *Chem. Rev.* **57**, 1 (1957). Values in parentheses are those of M. J. Jorgenson and D. R. Hartter, *J. Amer Chem. Soc.*, **85**, 878 (1963). See footnote 85.
[b] H. Strehlow and H. Wendt, *Z. Physik. Chem. (Frankfurt)*, **30**, 141 (1961).

Acidity in Nonaqueous Amphiprotic Media

Acid-base behavior has been studied in a number of pure amphiprotic solvents as well as in mixed solvents containing water as one component. We shall have more to say about water–alcohol mixtures in the next chapter. Careful e.m.f. measurements of hydrogen and silver–silver chloride electrodes in cells without liquid junction have been used to study the properties of hydrochloric acid and occasionally the dissociation constants of a few weak acids and bases in ethylene glycol,[86] propylene glycol,[86] glycerol,[87] n-propanol,[88] isopropanol,[89] and n-butanol,[90] as well as in their aqueous mixtures. In general the objective has not been the establishment of pH scales and little is known concerning the response of the glass electrode in these media.

Mitchell and Wynne-Jones[91] have established a standard scale of pc_H ($= - \log c_H$) in mixtures of hydrogen peroxide and water covering the entire range of solvent compositions. The cell

$$Ag;AgCl,HCl\,||\,Glass||\,Soln.\ in\ H_2O_2,H_2O\,|\,KCl,3.5\ M(aq.),AgCl;Ag$$

was used, and most of the measurements were made at a constant ionic strength, namely $I = 0.1 = c_{HClO_4} + c_{NaClO_4}$. The value of $E°$ in the equation

$$E = E° - 0.05916 \log c_{HClO_4} = E° + 0.05916\ pc_H \qquad (21)$$

was shown to be constant by the measurement of E (corrected for day-to-day variations in the potential of the glass electrode) for three different concentrations of perchloric acid at each solvent composition. Although this scale of pc_H applies strictly only to solutions of perchloric acid and sodium perchlorate of ionic strength 0.1 at 25 °C, the work illustrates one practical method of standardizing pH measurements in media of high dielectric constant.

Acidic Solvents

The earliest extensive electrometric investigation of acidity in an anhydrous medium was that of Hall and Conant[92] in glacial acetic acid. It had been observed that some bases which are very weak in water react extensively with strong acids in anhydrous acetic acid. Hall and Conant undertook to study quantitatively the hydrogen ion (proton) activity of these solutions,

[86] K. K. Kundu, P. K. Chattopadhyay, D. Jana, and M. N. Das, *J. Chem. Eng. Data*, **15**, 209 (1970).

[87] R. N. Roy, W. Vernon, J. J. Gibbons, and A. L. M. Bothwell, *J. Electroanal. Chem.*, **34**, 101 (1972).

[88] R. N. Roy, W. Vernon, J. J. Gibbons, and A. L. M. Bothwell, *J. Chem. Thermodynamics* **3**, 883 (1971); R. N. Roy, W. Vernon, and A. L. M. Bothwell, *Electrochim. Acta*, **17**, 5 (1972).

[89] R. N. Roy, W. Vernon, and A. L. M. Bothwell, *J. Chem. Thermodynamics*, **3**, 769 (1971) *J. Chem. Soc., Faraday I*, **68**, 2047 (1972).

[90] R. N. Roy, A. L. M. Bothwell, J. J. Gibbons, and W. Vernon, *J. Chem. Soc., A., Dalton* 530 (1972).

[91] A. G. Mitchell and W. F. K. Wynne-Jones, *Trans. Faraday Soc.*, **51**, 1690 (1955).

[92] N. F. Hall and J. B. Conant, *J. Amer. Chem. Soc.*, **49**, 3047 (1927).

which they termed "superacid."[93] As the most suitable electrode for e.m.f. measurements in acetic acid they chose the one formed of tetrachloroquinone (chloranil, $C_6Cl_4O_2$) and its hydroquinone, $C_6Cl_4(OH)_2$.[94] Their cell is represented by the following scheme:

$$
\begin{array}{c|c|c}
\begin{matrix} C_6Cl_4O_2 \text{ (satd.)} \\ Pt;C_6Cl_4(OH)_2 \text{ (satd.)} \\ HX \text{ (in acetic acid)} \end{matrix} & Bridge & KCl \text{ (satd. in } H_2O),Hg_2Cl_2;Hg
\end{array} \qquad (22)
$$

The bridge was a supersaturated solution of lithium chloride in acetic acid containing a small amount of gelatin. The arrangement provided good reproducibility. The liquid-junction potential was large (about 0.15 V) but appeared to be satisfactorily constant as the concentration of the acid HX was varied.

The differences in the electromotive force E are believed to be a direct measure of the relative proton activity of the acid solutions, in view of the apparent constancy of the potentials at the liquid–liquid boundaries. In order to express acidities in terms of a unit analogous to pH, rather than in volts, Hall and Conant defined pH^{HAc} at 25 °C by the equation

$$
pH^{HAc} \equiv \frac{E + 0.566}{0.0591} \qquad (23)
$$

The constant 0.566 was obtained in a partly arbitrary manner. This value was supported by a comparison of dissociation constants computed from pH^{HAc} with the known values in water.

Although pH^{HAc} is a conventional unit rather than a measure of the proton activity relative to the water standard, this unit is useful for comparisons of acidity in glacial acetic acid. The rate of inversion of sucrose in acetic acid solutions has been found to be approximately proportional to the proton activity (computed from pH^{HAc} on the assumption that this unit is $-\log a_H$) and in agreement with estimates of the acidity by means of the color changes of a series of triarylcarbinols and unsaturated ketones.[95] The high acidity of

---

[93] The concept of "superacidity" is considered misleading by G. Jander and H. Klaus, *J. Inorg. Nucl. Chem.*, **1**, 126 (1955), who feel that attention should be directed instead to the ability of the solvent to enhance the base-analog character of weak bases dissolved in it.

[94] Subsequent comparisons with the hydrogen electrode by B. O. Heston and N. F. Hall, *J. Amer. Chem. Soc.*, **55**, 4729 (1933), showed that the solid phase contains a solvate of the hydroquinone, both in glacial acetic acid and in aqueous dioxane. The standard potential of the cell is, of course, independent of the solvent only if the electrode substances are unsolvated. O. Tomicek and A. Heyrovsky, *Collection Czech. Chem. Commun.*, **15**, 984 (1950), found that the chloranil electrode cannot be used in acetic acid containing excess anhydride.

[95] J. B. Conant and N. F. Hall, *J. Amer. Chem. Soc.*, **49**, 3062 (1927).

solutions of the strong acids in acetic acid and of buffer solutions formed by partial neutralization of these acids with urea was found to be reduced in a regular manner by addition of the protophilic substances water and alcohol.[96] Titration curves and dilution curves for weak acids in anhydrous acetic acid have been obtained, and excellent inflection points were found for bases too weak to be titrated in water (compare Fig. 7–3). These bases resemble strong bases in their titration behavior, yet the changes of pH^{HAc} with dilution obey the law of concentration effect. The bases cannot therefore be regarded as completely dissociated in the sense in which this term is applied to aqueous electrolytes. This phenomenon, called "self-buffering," has also been noted by Schwarzenbach.[97] It is explained by the complex nature of the ionization-dissociation process in glacial acetic acid.[98]

A comparison of pH^{HAc} with the acidity function H_0 is of interest. In view of the partly arbitrary definition of the former unit, it is to be expected that these two quantities will not be equal numerically. Plots of pH^{HAc} and H_0 for solutions of sulfuric acid in anhydrous acetic acid, shown in the lower portion of Fig. 7–4, are approximately linear with respect to the logarithm of the concentration of sulfuric acid below 1 M. The values of pH^{HAc} are, however, about 1.7 to 2.0 units more negative than H_0.[99] Above 1 M, the values of H_0 both in acetic acid and in water vary linearly with the concentration of sulfuric acid. The curves are parallel and have slopes of about -0.5. Their separation indicates that sulfuric acid is over 400 times as acidic in acetic acid as in water at the same concentration.

An admirably thorough investigation by Kolthoff and Bruckenstein[100] has shown that acid-base behavior in glacial acetic acid can only be understood in terms of ionization of the solutes with association of the resulting ions to form ion pairs and often triplets and quadruplets as well. The "ionization constant" of an acid or base in acetic acid (to form ion pairs) gives a more useful expression of the acidic or basic strength than does the overall "dissociation constant." The dielectric constant of this solvent is low (6.13 at 25 °C) and the capacity for hydrogen bonding is high. The resulting interactions modify profoundly the course of Brønsted acid-base processes in this solvent.[101] Bruckenstein and Kolthoff found $pK = 4.87$ for perchloric acid,

[96] N. F. Hall and J. B. Conant, J. Amer. Chem. Soc., **49**, 3047 (1927).

[97] G. Schwarzenbach, Helv. Chim. Acta, **13**, 896 (1930).

[98] I. M. Kolthoff and S. Bruckenstein, J. Amer. Chem. Soc., **78**, 1 (1956).

[99] N. F. Hall and W. F. Spengeman, J. Amer. Chem. Soc., **62**, 2487 (1940); N. F. Hall and F. Meyer, ibid., **62**, 2493 (1940).

[100] I. M. Kolthoff and S. Bruckenstein, J. Amer. Chem. Soc., **78**, 1 (1956); **79**, 1 (1957) S. Bruckenstein and I. M. Kolthoff, ibid., **78**, 10, 2974 (1956); **79**, 5915 (1957).

[101] B. Trémillon, La Chimie en Solvants Non-aqueux, Chapter 2, Presses Universitaires de France, Paris, 1971.

the strongest acid in glacial acetic acid solvent.[102] The pK for hydrochloric acid is 8.55.

As there are very few ions in these solutions, ionic strength effects (salt effects) can be considered negligible. The complexity of the equilibria in glacial acetic acid is illustrated by the discovery that indicator bases respond colorimetrically to the activity of the ion pair form of an acid rather than to the proton activity. Apparent acid strengths are dependent on the indicator base chosen as a reference, and the order of strengths may differ from that found by potentiometric measurements.

In spite of the complexity of the equilibria in acetic acid, Kolthoff and Bruckenstein[100] found it possible to determine pH numbers that could serve as a useful index to acid-base interactions in this medium. Following the early work of Hall and Conant,[92] they utilized a cell with chloranil and saturated calomel electrodes, although the reference calomel electrode was prepared with acetic acid as the solvent. Using dissociation constants for the reference substances hydrochloric acid and pyridine, determined by spectrophotometric techniques, they outlined procedures by which the dissociation constants of other acids, bases, and salts could be determined in the acetic acid medium.

In a study comparable in many respects with an earlier one in ethanol–water mixtures,[103] Schwarzenbach and Stensby[104] determined the acidity functions $p(a_H\gamma_X)$ and H_0 for 0.01 M solutions of hydrochloric, hydrobromic, and hydriodic acids (where X is halide) in acetic acid–water solvents of compositions ranging from 0 to 99.9 per cent acetic acid. The activity coefficients of the three acids were determined relative to the aqueous standard state. Two H_0 indicators, *o*-nitroaniline and *p*-nitroaniline, were used. The acidity functions $p(a_H\gamma_X)$ and H_0 were also determined in 0.01 M perchloric acid, in mixtures of perchloric acid with hydrochloric acid and with tetramethylammonium chloride, and in a mixture of 0.01 M sodium hydroxide with tetramethylammonium chloride. The results demonstrate the differentiating action of acetic acid on the strengths of the strong acids.

Acid-base behavior in sulfuric acid and its mixtures with water has usually been studied with the aid of Hammett acidity functions. The pure solvent is amphiprotic and yields the ions $H_3SO_4^+$ and HSO_4^- when autoprotolysis takes place. The behavior of the acidic solvents sulfuric acid, hydrofluoric

[102] I. M. Kolthoff and A. Willman, *J. Amer. Chem. Soc.*, **56**, 1007 (1934); M. M. Jones and E. Griswold, *ibid.*, **76**, 3247 (1954); S. Bruckenstein and I. M. Kolthoff, *ibid.*, **78**, 2974 (1956); A. M. Shkodin and L. I. Karkuzaki, *Zh. Fiz. Khim.*, 33, 2795 (1959); A. M. Shkodin, *ibid.*, **34**, 1625 (1960).

[103] R. G. Bates and G. Schwarzenbach, *Helv. Chim. Acta*, **38**, 699 (1955).

[104] G. Schwarzenbach and P. Stensby, *Helv. Chim. Acta*, **42**, 2342 (1959).

acid, and nitric acid has been thoroughly reviewed.[105] Acid-base processes in other highly acidic media have also been studied.[106]

Basic Solvents

Among the well-known solvents in which the basic property is clearly dominant are found ethylenediamine, pyridine, tetramethylguanidine, hydrazine, and the ethanolamines. The analytical uses of some of these solvents have been studied in detail.[107] It is not impossible, when the dielectric constant is not too low, to establish useful scales of relative acidity in basic media, for these solvents may exhibit the same types of regularity found in aqueous media. For the practical realization of these scales, however, hydrogen ion electrodes of satisfactory response and reproducibility, as well as stable reference electrodes, must be available.

Acid-base behavior in ethylenediamine has been studied in considerable detail.[108, 109] The hydrogen electrode is capable of furnishing useful results in this solvent and has been used to obtain dissociation constants for weak acids. According to Deal and Wyld,[110] the glass electrode is also satisfactory, provided that sodium salts are not present. A reference electrode consisting of a saturated zinc amalgam in contact with solid zinc chloride functions satisfactorily in this solvent.[111] A suitable salt bridge contains lithium chloride (saturated or 0.25 M) in ethylenediamine. A corrosive sublimate electrode also makes a stable reference.[109] This electrode is formed from mercury in contact with a solution saturated with mercuric chloride and lithium chloride in ethylenediamine.

Weak acids can be titrated in 1,1,3,3-tetramethylguanidine and suitable end points established by either potentiometric or indicator methods.[112] The hydrogen electrode functions reversibly, and the pK of perchloric acid has been found to be about 3 in this solvent.

[105] The Chemistry of Non-Aqueous Solvents, J. J. Lagowski, ed., Vol. II, Academic Press, New York, 1967. H_2SO_4: W. H. Lee, Chapter 3. HF: M. Kilpatrick and J. G. Jones, Chapter 1. HNO_3: W. H. Lee, Chapter 4.

[106] Trifluoroacetic acid: M. G. Harriss and J. B. Milne, Can. J. Chem., 49, 1888, 2937 (1971). m-Cresol: M. Bos and E. A. M. F. Dahmen, Anal. Chim. Acta, 57, 361 (1971). $(CH_3COOH)_2BF_3$: G. Petit and J. Bessière, ibid., 57, 227 (1971).

[107] H. B. van der Heijde and E. A. M. F. Dahmen, Anal. Chim. Acta, 16, 378 (1957).

[108] W. B. Schaap, R. E. Bayer, J. R. Siefker, J. Y. Kim, P. W. Brewster, and F. C. Schmidt, Rec. Chem. Progr., 22, 197 (1961).

[109] S. Bruckenstein and L. M. Mukherjee, J. Phys. Chem., 64, 1601 (1960); 66, 2228 (1962). L. M. Mukherjee, S. Bruckenstein, and F. A. K. Badawi, ibid., 69, 2537 (1965).

[110] V. Z. Deal and G. E. A. Wyld, Anal. Chem., 27, 47 (1955).

[111] J. F. Coetzee, G. R. Padmanabhan, and G. P. Cunningham, Talanta, 11, 93 (1964).

[112] J. A. Caruso, P. G. Sears, and A. I. Popov, J. Phys. Chem., 71, 1756 (1967).

The chemistry of reactions in liquid ammonia has been reviewed.[113] Acid-base behavior in pyridine[114] and in water–pyridine mixtures[115] has been studied. Broadhead and Elving found that the Pt; Cu(I), Cu(II) redox electrode provides a stable and reproducible reference for use in pyridine. Their reference half cell contained 0.0025 M CuCl and 0.025 M $CuCl_2$ in 0.1 M tetraethylammonium chloride. Mukherjee used the ferrocene-ferricinium electrode successfully, while Bos and Dahmen standardized their glass electrode with perchloric acid in pyridine.

Bauer[116] has demonstrated that the hydrogen electrode behaves satisfactorily in water–hydrazine mixtures, and dissociation constants in these mixed solvents have been determined. The electrode Zn(Hg), $ZnSO_4$, $SO_4{}^{2-}$ is suitable as a reference in anhydrous hydrazine. It is unstable in mixtures of hydrazine and water, however, as zinc hydroxide precipitates slowly.

Dipolar Aprotic Solvents

Studies of acid-base behavior in dipolar aprotic solvents have centered in general on acetonitrile, acetone, dimethylformamide, and dimethylsulfoxide. Increased attention, however, is being given to ethylene and propylene carbonates; sulfolane (tetramethylenesulfone); hexamethylphosphoric triamide, $[(CH_3)_2N]_3PO$; and other amide solvents. The subject has been reviewed by Ritchie.[117]

There has been considerable interest in acid-base reactions in acetonitrile. This solvent is predominantly aprotic and has a rather high dielectric constant (36). Acids and bases with an extraordinarily wide range of proton energies can exist unleveled in it. Perchloric acid appears to be the only mineral acid that is completely dissociated in acetonitrile.[118] The others are relatively weak (the apparent dissociation constants of hydrochloric acid and nitric acid are about 10^{-9}), and the primary dissociation equilibrium appears[119] to be 2 HA $\rightleftharpoons$ $H^+ + AHA^-$. Unfortunately, solutions of acids in

[113] J. J. Lagowski and G. A. Moczygemba, Chapter 7 in *The Chemistry of Non-Aqueous Solvents*, J. J. Lagowski, ed., Vol. II, Academic Press, New York, 1967.

[114] J. Broadhead and P. J. Elving, *Anal. Chim. Acta*, **48**, 433 (1969); M. Bos and E. A. M. F. Dahmen, *Anal. Chim. Acta*, **53**, 39 (1971); **55**, 285 (1971); L. M. Mukherjee, *J. Phys. Chem.*, **76**, 243 (1972).

[115] J. L. Brisset, R. Gaboriaud, and R. Schaal, *J. Chim. Phys.*, **68**, 1506 (1971).

[116] D. Bauer, *Bull. Soc. Chim. France*, 3302 (1965); thesis, Paris. 1967.

[117] C. D. Ritchie, Chapter 4 in *Solute-Solvent Interactions*, J. F. Coetzee and C. D. Ritchie, eds., Marcel Dekker, New York, 1969.

[118] I. M. Kolthoff and J. F. Coetzee, *J. Amer. Chem. Soc.*, **79**, 6110 (1957); J. N. Butler, *J. Electroanal. Chem.*, **14**, 89 (1967).

[119] I. M. Kolthoff, S. Bruckenstein, and M. K. Chantooni, Jr., *J. Amer. Chem. Soc.*, **83**, 3927 (1961).

acetonitrile are somewhat unstable and display a time-dependent decrease of hydrogen ion concentration.[119] This behavior may be caused by an acid-catalyzed hydrolysis of the solvent, in the opinion of Coetzee and Ritchie, to form protonated acetamide.[120]

Kolthoff and Thomas[121] were successful in using both the hydrogen electrode and the ferrocene–ferricinium picrate couple in acetonitrile. Dissociation constants of weak electrolytes and pa_H values for certain reference solutions were determined.[122] It was also shown that the glass electrode, standardized with these reference solutions, would produce useful results for other acid-base systems. Selected as standards were solutions of sulfonic acids and their tetraalkylammonium salts, o-nitroaniline and its perchlorate salt, and picric acid and tetrabutylammonium picrate. Coetzee and Padmanabhan[123] favor a silver reference electrode in a 0.01M solution of silver nitrate in acetonitrile. A solution of tetraethylammonium perchlorate (0.1M) in acetonitrile makes a suitable salt bridge.

The amide solvents formamide and acetamide and their N-methyl derivatives, together with N-methylpropionamide, are sometimes classified as basic media. The basic property is extremely weak, however, and these solvents are properly considered to be dipolar aprotic media. Nevertheless, some of them, notably dimethylformamide, have been shown to have measurable autoprotolysis constants. The chemistry of amide solvents has been reviewed by Vaughn.[124]

In general, amide solvents have moderate to high dielectric constants. It might therefore be expected that the interpretation of experimental pH numbers in these solvents would be more straightforward than in media of lower polarity. These expectations are not always fulfilled, however, especially when media with widely different capabilities for hydrogen bonding are compared. Although N-methylpropionamide has a remarkably high dielectric constant (about 175 as 25 °C), conductivity measurements indicate that potassium chloride and hydrochloric acid are appreciably ion-paired in this solvent.[125] Likewise, the dissociation constants of acetic acid and

[120] See R. G. Bates, Chapter 2 in *Solute-Solvent Interactions*, J. F. Coetzee and C. D. Ritchie, eds., Marcel Dekker, New York, 1969.

[121] I. M. Kolthoff and F. G. Thomas, *J. Phys. Chem.*, **69**, 3049 (1965).

[122] I. M. Kolthoff and M. K. Chantooni, Jr., *J. Amer. Chem. Soc.*, **87**, 4428, (1965); **90**, 5961 (1968).

[123] J. F. Coetzee and G. R. Padmanabhan, *J. Phys. Chem.*, **66**, 1708 (1962).

[124] J. W. Vaughn, Chapter 5 in *The Chemistry of Non-Aqueous Solvents*, J. J. Lagowski, ed., Vol. II, Academic Press, New York, 1967.

[125] T. B. Hoover, *J. Phys. Chem.*, **68**, 876 (1964); W. C. Duer, unpublished data, University of Florida, 1971.

protonated tris(hydroxymethyl)aminomethane in *N*-methylpropionamide have been found to be smaller than in water.[126]

The platinum–hydrogen electrode and the silver–silver chloride electrode have been used in formamide (Mandel and Decroly) and in *N*-methylacetamide (Dawson and coworkers).[127] Likewise, Nayak and Sahu[127] have studied the cell with hydrogen electrode and calomel electrode in formamide. The glass electrode appears to function properly in anhydrous dimethylformamide.[128] When the solvent contains 0.01 to 0.3 per cent of water, the hydrogen electrode can also be used.[129] A saturated solution of potassium chloride in dimethylformamide can serve as a bridge solution between the nonaqueous solution and the reference electrode.

Buffers containing picric acid and potassium picrate were used by Juillard[128] to standardize the glass-calomel pH cell in dimethylformamide. Although Ritchie and Megerle believe that picric acid is completely dissociated in this solvent, conductivity measurements lead to pK = 1.2. The pH values for the buffer solution on these two bases differ by only 0.05 unit. The pH of the five reference solutions employed by Juillard is listed in Table 7–6.

TABLE 7–6. pa_H VALUES (MOLAR SCALE) OF ACID-SALT BUFFER SOLUTIONS IN DIMETHYLFORMAMIDE

(Solutions 0.02M with respect to both acid and salt)		
Acid (HA)	Salt	pc_H
Picric	KA	2.20
2,6-Dinitrophenol	KA	6.05
Salicylic	NaA	7.94
p-Nitrobenzoic	LiA	10.00
Benzoic	LiA	11.70

---

[126] Unpublished measurements by E. S. Etz, University of Florida, 1971. The hydrogen electrode and the silver-silver chloride electrode were found to behave satisfactorily in this amide solvent.

[127] M. Mandel and P. Decroly, *Trans. Faraday Soc.*, **56**, 29 (1960); L. R. Dawson, R. C. Sheridan, and H. C. Eckstrom, *J. Phys. Chem.*, **65**, 1829 (1961); L. R. Dawson, W. H. Zuber, Jr., and H. C. Eckstrom, *ibid.*, **69**, 1335 (1965); B. Nayak and D. K. Sahu, *Electrochim. Acta*, **16**, 1757 (1971).

[128] C. D. Ritchie and G. H. Megerle, *J. Amer. Chem. Soc.*, **89**, 1447 (1967); J. Juillard, thesis, Clermont-Ferrand, 1968. Although perchloric acid appears to be completely dissociated in this solvent, conductivity measurements reveal considerable ionic association in solutions of several other acids and salts. See P. G. Sears, R. K. Wolford, and L. R. Dawson, *J. Electrochem. Soc.*, **103**, 633 (1956).

[129] G. Demange-Guérin and J. Badoz-Lambling, *Bull. Soc. Chim. France*, 3277 (1964).

It has been reported[130] that the glass electrode responds well to changes in hydrogen ion activity extending over 25 powers of 10 in dimethylsulfoxide. Experimental measurements were made at pH values as high as 28. A salt bridge consisting of 0.1 M tetraethylammonium perchlorate in the nonaqueous solvent proved satisfactory when used with a reference electrode of silver dipping in a solution of silver perchlorate (0.05 M) in dimethylsulfoxide In spite of some increase in the solubility of silver chloride, the silver–silver chloride electrode has been used in water–dimethylsulfoxide solvents containing less than 80 wt. per cent of the organic component.[131] Butler[132] has found, however, that the thallium amalgam–thallous chloride electrode is the most stable reference for use in pure dimethylsulfoxide. Ritchie and Uschold[130] standardized the pH cell with solutions of p-toluenesulfonic acid, which they found to be completely dissociated in dimethylsulfoxide.

Sulfolane has been shown to be a useful solvent for the resolution of mixtures of acids that are leveled in more reactive solvents.[133,134] Coetzee and Bertozzi obtained good results with both the hydrogen electrode and the glass electrode in anhydrous sulfolane. A silver electrode in contact with 0.1 M silver perchlorate in sulfolane served as a reference. They reached the conclusion that perchloric acid has a pK of 2.7 in this solvent[135] and were able to determine pa_H values for three buffer solutions. Their reference values are given in Table 7-7. The picric acid buffer was particularly recommended for the standardization of glass electrodes in sulfolane.

Madic and Trémillon[136] have used the hydrogen electrode in hexamethylphosphoric triamide with good results. In this solvent (dielectric constant 30), nitric acid, perchloric acid, and p-toluenesulfonic acid are fully dissociated.

Nonpolar Aprotic Solvents

Aprotic or "inert" solvents possess almost no acid-base properties. Many of these media have low dielectric constants and, in general, they cannot be

[130] C. D. Ritchie and R. E. Uschold, *J. Amer. Chem. Soc.*, **89**, 1721 (1967); see also I. M. Kolthoff and T. B. Reddy, *Inorg. Chem.*, **1**, 189 (1962). Dissociation constants in dimethylsulfoxide have also been determined by I. M. Kolthoff, M. K. Chantooni, Jr., and S. Bhowmik, *J. Amer. Chem. Soc.*, **90**, 23 (1968), and by I. M. Kolthoff and M. K. Chantooni, Jr., *ibid.*, **90**, 5961 (1968).

[131] J. P. Morel, *Bull. Soc. Chim. France*, 1405 (1967); 896 (1968); K. H. Khoo, *J. Chem. Soc., A*, 2932 (1971).

[132] J. N. Butler, *J. Electroanal. Chem.*, **14**, 89 (1967).

[133] D. H. Morman and G. A. Harlow, *Anal. Chem.* **39**, 1869 (1967).

[134] J. F. Coetzee and R. J. Bertozzi, *Anal. Chem.*, **43**, 961 (1971). See also J. F. Coetzee, J. M. Simon, and R. J. Bertozzi, *ibid.*, **41**, 766 (1969).

[135] See also R. L. Benoit, C. Buisson, and G. Choux, *Can. J. Chem.*, **48**, 2353 (1970).

[136] C. Madic and B. Trémillon, *Bull. Soc. Chim. France*, 1634 (1968).

TABLE 7-7. pa_H VALUES (MOLAR SCALE) OF ACID-SALT
BUFFER SOLUTIONS IN SULFOLANE

(Solutions 0.02M with respect to both acid and salt)		
Acid (HA)	Salt	pa_H
Picric	(Et)$_4$NA	17.3
2,6-Dihydroxybenzoic	(Bu)$_4$NA	18.8
Salicylic	(Bu)$_4$NA	23.6

classified as ionizing solvents. Acid-base interaction differs in important respects from that characteristic of ionizing media. This subject has been treated thoroughly by Davis in two reviews.[137]

Spectrophotometric measurements with indicator acids are useful in revealing the nature of acid-base processes in benzene and other hydrocarbon and halogenated hydrocarbon solvents. By methods of this sort, Davis[138] showed that the primary process in the interaction of the base triethylamine (B) with six isomeric dinitrophenols (A) in benzene is an association

$$\begin{array}{ccc} \text{B} & + & \text{A} \rightleftharpoons \text{S} \\ \text{Base} & & \text{Acid} \quad\quad \text{Salt} \end{array} \qquad (24)$$

On introduction of the equilibrium constant K for process 24, one obtains

$$\log \frac{[\text{S}]}{[\text{A}]} = \log K - \log [\text{B}] \qquad (25)$$

In agreement with this formulation, plots of log ([S]/[A]) as a function of $-\log [\text{B}]$ were found to be straight lines with a slope of -1, as shown in Fig. 7-5.

In some aprotic solvents, notably those of higher dielectric constant termed dipolar aprotic, partial ionic dissociation of the salt may occur, although it is not observed in benzene. Thus, conductivity data[139] show that partial dissociation of anilinium and pyridinium picrates occurs in nitrobenzene (dielectric constant 35). In the view of Davis, acid-base interaction,

[137] M. M. Davis, *Acid-Base Behavior in Aprotic Organic Solvents*, National Bureau of Standards Monograph 105, U.S. Government Printing Office, Washington, 1968; Chapter 1 in *The Chemistry of Non-Aqueous Solvents*, J. J. Lagowski, ed., Vol. III, Academic Press, New York, 1970.

[138] M. M. Davis, *J. Amer. Chem. Soc.*, **84**, 3623 (1962).

[139] C. R. Witschonke and C. A. Kraus, *J. Amer. Chem. Soc.*, **69**, 2471 (1947).

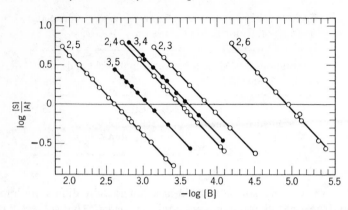

FIG. 7-5. Acid-base equilibrium relationships between triethylamine and six isomeric dini trophenols in benzene (after Davis).

even in polar media such as water, may be regarded to consist of a primary association of acid and base, followed by ionic dissociation. There is evi dence that the product (salt) formed in the first of these steps is often a hydrogen-bonded ion pair.

CHAPTER 8

medium effects and pH in
nonaqueous and mixed solvents

A few words concerning terminology are in order. The "solvent effect" refers to a change in a property of the system brought about by an alteration in the composition of the solvent. The "medium effect," on the other hand, is a property of a single solute species; it characterizes the difference in Gibbs energy of that species in two different selected standard states. The "medium effect" is a dimensionless quantity closely allied to the activity coefficient and the equilibrium constant for the transfer process it characterizes. In this chapter, the process of "determining the acidity" will mean the establishment of numbers on an operational scale that has some useful relationship to the ability of a system to react with a given base or to transfer protons to a standard hydrogen electrode.

At its biennial conference held in July 1971 at Washington, the International Union of Pure and Applied Chemistry initiated a study of symbols and terminology for the medium effect. Several terms are currently used in the literature in addition to "medium effect." These are "distribution coefficient" (Kolthoff and Bruckenstein[1]), "degenerate activity coefficient" (Grunwald and coworkers[2]), "medium activity coefficient" (Coetzee and Campion[3]), "solvent activity coefficient" (Parker and Alexander[4]), and, in

[1] I. M. Kolthoff and S. Bruckenstein, Chapter 13 in *Treatise on Analytical Chemistry*, I. M. Kolthoff and P. J. Elving, eds., Part 1, Vol. I, Interscience Publishers, New York, 1959.
[2] E. Grunwald and B. J. Berkowitz, *J. Amer. Chem. Soc.*, **73**, 4939 (1951); B. Gutbezahl and E. Grunwald, *ibid.*, **75**, 559, 565 (1953).
[3] J. F. Coetzee and J. J. Campion, *J. Amer. Chem. Soc.*, **89**, 2517 (1967).
[4] A. J. Parker and R. Alexander, *J. Amer. Chem. Soc.*, **90**, 3313 (1968).

French, "coefficient d'activité de solvatation" (Charlot and Trémillon[5]). There is also some variation in the symbol used for this quantity; we shall use $_m\gamma_i$ to denote the medium effect for the species i.

Nature of the Medium Effect

The relative activity of a species i is defined in terms of the chemical potential or partial Gibbs energy of that species in the particular state (μ_i) and in a selected standard state (μ_i°):

$$\mu_i = \mu_i^\circ + RT \ln a_i \tag{1}$$

By definition, the relative activity is unity in the standard state. Both the relative activity and the activity coefficient are regarded as dimensionless.[6] Consequently, to preserve dimensional consistency in the relationship $\gamma_i = a_i/m_i$, the molality m_i is often replaced by a relative molality, m_i/m°, where m° is the hypothetical unit molality. This device obviously does not alter the numerical value of m_i and will not be employed in this chapter.

The customary standard for aqueous solutions was chosen in such a way that activity approaches molality (and the activity coefficient approaches unity) as the solution becomes very dilute. Thus the approximate equality of activity and molality or concentration is preserved. We have already seen, however, that activities based on this standard state differ profoundly from the molality when the composition of the solvent is altered by addition of an organic constituent to the aqueous medium. For many purposes, therefore, it is convenient to alter the standard state in such a way that the activity and molality are nearly equal (and exactly so at infinite dilution) in each particular solvent medium.

The chemical potential of a given substance i in a nonaqueous or mixed solvent s can be compared with that in any number of standard states, chosen at will, with the result that different values of the activity and activity coefficient will be obtained (compare equation 1). In practice, however, it is usually not necessary to utilize more than two scales of activity; one is, of course, that referred to the aqueous standard state and the other is that based on an activity coefficient of unity at infinite dilution in the particular solvent or solvent mixture of concern.[7] If these standard states are identified by the

[5] G. Charlot and B. Trémillon, *Les Réactions Chimiques dans les Solvants et les Sels Fondus*, Gauthier-Villars, Paris, 1963.

[6] *Manual of Symbols and Terminology for Physicochemical Quantities and Units*, International Union of Pure and Applied Chemistry, Butterworths, London, 1970.

[7] R. G. Bates, Chapter 2 in *Solute-Solvent Interactions*, J. F. Coetzee and C. D. Ritchie, eds., Marcel Dekker, New York, 1969.

letters w and s, respectively, the chemical potential of a species i in a non-aqueous or mixed solvent can be expressed in two ways:

$$\mu_i = \mu_i^{\circ}(w) + RT \ln {}_wa_i = \mu_i^{\circ}(w) + RT \ln (m_i \cdot {}_w\gamma_i) \tag{2}$$

or

$$\mu_i = \mu_i^{\circ}(s) + RT \ln {}_sa_i = \mu_i^{\circ}(s) + RT \ln (m_i \cdot {}_s\gamma_i) \tag{3}$$

By subtraction of equation 3 from equation 2, one obtains

$$\Delta G_t^{\circ}(i) \equiv \mu_i^{\circ}(s) - \mu_i^{\circ}(w) = RT \ln \frac{{}_w\gamma_i}{{}_s\gamma_i} = RT \ln {}_m\gamma_i \tag{4}$$

Thus it is seen that the relative numerical value of the activity coefficients in a particular nonaqueous medium is dependent only on the Gibbs energy of that species in the two reference states. Equation 4 also defines the standard Gibbs energy of transfer, ΔG_t°, as well as the medium effect ${}_m\gamma_i$. The former is the change in Gibbs energy for the process

$$i \text{ (standard state, } w) = i \text{ (standard state, } s) \tag{5}$$

The medium effect is thus seen to be a measure of the work of transferring 1 mole of species i from the standard state in water to the standard state in solvent s. Medium effects for neutral species and for neutral ion combinations can be derived by rigorous thermodynamic methods. Formally, the transfer energy for an electrolyte is the sum of the transfer energies of the separate ions; the "mean ionic medium effect" ${}_m\gamma_{\pm}$ for an electrolyte is thus related to the transfer energy for 1 mole of electrolyte by

$$\Delta G_t^{\circ}(\pm) = \upsilon RT \ln {}_m\gamma_{\pm} \tag{6}$$

where υ is the number of moles of ions formed from 1 mole of the electrolyte. At 25 °C, the medium effect for a single ionic species i is readily calculated from the standard transfer energy (in kcal mol^{-1}) by

$$\log {}_m\gamma_i = 0.733 \Delta G_t^{\circ} \tag{7}$$

When equation 4 is applied to single ionic species, its limitations must be recognized. As Popovych[8] has pointed out, equation 4 does not adequately represent the energy change for process 5 when i is a single ionic species. This is because the medium effect is determined by differences in the chemical energies of solvation, whereas the actual transfer of a mole of a single ionic species would also involve an energy term for the transfer of charge. The term "transfer energy" should therefore be restricted to neutral species or to neutral ion combinations. It is, however, proper to consider the medium

---

[8] O. Popovych, *Crit. Rev. Anal. Chem.*, **1**, 73 (1970).

effect for a single ionic species to be a measure of the difference in "chemical solvation energy" in the two states identified in equation 5.

The experimental determination of medium effects is likewise restricted to the same ion combinations for which activity coefficients in water can be determined. Thus $_m\gamma_H \cdot {_m\gamma_{X^-}}$ (where X is a univalent anion) is thermodynamically defined, as are also $_m\gamma_H / _m\gamma_{M^+}$ (where M is a univalent cation), $_m\gamma_{Cl} / _m\gamma_X$, $_m\gamma_H \cdot {_m\gamma_A} / _m\gamma_{HA}$ for weak acids HA of all charge types, and $_m\gamma_{HA}$ or $_m\gamma_{MX}$ for uncharged species.

A knowledge of medium effects or their equivalent, chemical transfer energies, for individual ionic species would have many useful consequences. It would then be possible to determine liquid-junction potentials between solutions in media of different composition. Electrode potentials in different solvents could be related unambiguously to the hydrogen scale in water, and the same would be true for ion activities, including the pH value. Conversely, the estimation (or elimination) of liquid-junction potentials between solvent media and the selection of a reference electrode the potential of which is the same in all solvent media, are means for determining medium effects for ionic species.

These relationships may be made clear in the following manner. If one were to set up a cell consisting of two hydrogen electrodes immersed in different media, w and s:

$$\text{Pt};H_2(g, 1 \text{ atm}), H^+(m_1) \text{ in } w \,|\, H^+(m_2) \text{ in } s, H_2(g, 1 \text{ atm});\text{Pt} \qquad (8)$$

a liquid-junction potential E_j would arise at the point of contact of the two solutions. The e.m.f. E of this cell can be expressed by

$$E = {_w^s E^\circ} - {_w^w E^\circ} + k \log {_m\gamma_H} + E_j + k \log \frac{m_2 \, {_s^s\gamma_H}}{m_1 \, {_w^w\gamma_H}} \qquad (9)$$

where k is written for $(RT \ln 10)/F$.

As usual, the left-hand subscript in equation 9 denotes the standard state to which the quantity is referred. When needed for clarity, a left-hand superscript can also be used to denote the solvent medium. Thus $_w^s E^\circ$ is to be interpreted as "the standard potential in solvent s referred to the standard state in water" and $_s^s\gamma_H$ as "the activity coefficient of hydrogen ion in solvent s referred to the standard state in solvent s". A little reflection will convince the reader that $_w^s E^\circ$ and $_w^w E^\circ$ are not equal. In the latter case, $_w^w a_H$ is unity, and, by convention, $_w^w E^\circ = 0$ at all temperatures; in the case of the former, $_s^s a_H$ is unity and $_w^s a_H = {_m\gamma_H}$.

Returning now to equation 9, the difference of standard potentials on the right of the equation is given by [7,8]

$$_w^s E^\circ - {_w^w E^\circ} = k \log {_m\gamma_H} \qquad (10)$$

By combination of equations 9 and 10, therefore,

$$E_j = E - 2k \log {}_m\gamma_H - k \log \frac{m_2 \, {}_s^s\gamma_H}{m_1 \, {}_w^w\gamma_H} \tag{11}$$

In dilute solutions, ${}_s^s\gamma_H$ and ${}_w^w\gamma_H$ are near unity and can be estimated by the Debye–Hückel equation with reasonable accuracy.

With a knowledge of ${}_m\gamma_H$, equation 10 permits electrode potentials based on the hydrogen scale in a solvent system other than water to be related to the conventional hydrogen scale in water. Similarly, equation 11 offers a means of deriving liquid-junction potentials in cells of type 8 if ${}_m\gamma_H$ is known. Moreover, medium effects for other single ionic species can be derived, if ${}_m\gamma_H$ is known, from measurable mean medium effects. Thus,

$$_m\gamma_{Cl^-} = \frac{{}_m\gamma_\pm^{\,2}(HCl)}{{}_m\gamma_H} \tag{12}$$

and

$$_m\gamma_{Na^+} = \frac{{}_m\gamma_\pm^{\,2}(NaCl)}{{}_m\gamma_{Cl^-}} \tag{13}$$

and so on. It becomes possible, therefore, to determine liquid-junction potentials for cells similar to cell 8 but reversible to a variety of ions other than hydrogen.

Medium Effect and Salt Effect

The numerical values of activity coefficients are commonly assigned with reference to a value of unity at infinite dilution of the solute in the particular solvent under study. Unfortunately, this convention has obscured the change that the activity coefficient on the water scale undergoes when the composition of the solvent is altered by mixing dioxane, alcohol, and other organic liquids with the water. If the activity coefficient of a species i in different solvents is referred to the standard state in water, it will approach a value different from unity in solvents other than water as the concentration of solute is decreased. The limit is a measure of the *medium effect* ${}_m\gamma_i$, and this may differ greatly from unity. The activity coefficient ${}_w\gamma_i$ (or simply γ_i) in any given solvent, referred to the water basis, can be written as a product

$$\gamma_i = {}_m\gamma_i \cdot {}_s\gamma_i \tag{14}$$

The activity coefficient ${}_s\gamma_i$, which becomes unity at zero concentration in the particular solvent under consideration, essentially characterizes the interionic and ion–molecule forces (solute–solute interactions) in that solvent and other effects dependent on the concentration of solute in that medium. It may be termed the *concentration effect* or *salt effect*.[9]

The salt effect and the Debye–Hückel equation by which it can be estimated characterize the departure from ideal behavior brought about largely by the electrostatic interactions among the ions. The medium effect, on the other hand, is a measure of the free energy change on transfer of one mole of substance i from the standard state in water to the standard state in the solvent medium SH. It reflects differences in the electrostatic and chemical interactions of the substance i with the molecules of the two solvents (solute–solvent interactions). Of these interactions, solvation is probably the most important when ions are transferred from one medium to another. Indeed, $\Delta G_t^\circ(i)$ in equation 4 is the difference between Izmailov's "solvation free energy" for the ion i in the two solvents H_2O and SH (see Table 7–3).

The Debye–Hückel equation allows for the effect of changes of dielectric constant ε on the functional relationship between the activity coefficient $_s\gamma_i$ of the ionic species i (molal scale) and the ionic strength I, as follows:

$$-\log {}_s\gamma_i = \frac{(1.825 \times 10^6)z_i^2\sqrt{Id^\circ}}{(\varepsilon T)^{3/2}[1 + 50.29(\varepsilon T)^{-1/2}\mathring{a}\sqrt{Id^\circ}]} \qquad (15)$$

where $\mathring{a}$ is the ion-size parameter in Å, d° is the density of the solvent, and T is the thermodynamic temperature. When a solution containing ions is diluted, work must be done to separate the ions of opposite charge. Equation 15 reflects the increased coulombic forces in media of lowered dielectric constant. Lowering the value of ε causes the activity coefficient to decrease more rapidly with increasing ionic strength than it does at higher values of the dielectric constant. The activity coefficient $_s\gamma_i$ expresses the departure from ideal behavior; it becomes unity when a solution of ions is diluted without limit. For an uncharged species i in salt solutions (ion-molecule interactions), $\log {}_s\gamma_i$ usually varies almost linearly with ionic strength.[10]

Activity coefficients of hydrochloric acid have been determined in many mixed solvent media from e.m.f. measurements of cells with hydrogen and silver–silver chloride electrodes. The data serve well to illustrate the differences between $_s\gamma_\pm$ and $_w\gamma_\pm$. Figure 8–1 is a plot of the mean activity coefficient of hydrochloric acid in water and in five nonaqueous solvents as a

[9] The medium effect was called "primary medium effect" by B. B. Owen, *J. Amer. Chem. Soc.*, **54**, 1758 (1932). Owen's "secondary medium effect" represents the difference between the salt effects in two media of different but fixed solvent composition. The author prefers to regard $_m\gamma_i$ as the only true "medium effect."

[10] As shown, for example, by the activity coefficient of urea in sodium chloride solutions: V. E. Bower and R. A. Robinson, *J. Phys. Chem.*, **67**, 1524 (1963).

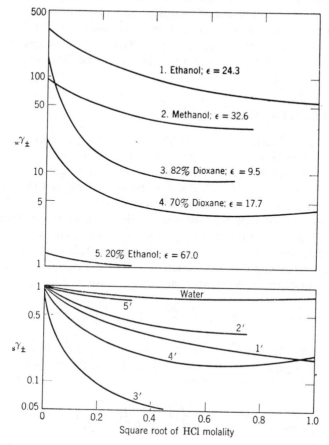

FIG. 8-1. Mean activity coefficients of hydrochloric acid in different solvent media at 25 °C.

function of the molality.[11] All the values plotted in the upper portion of the figure are referred to the standard state for aqueous solutions; hence the activity coefficient becomes unity only at zero concentration of the ions in pure water. The activity coefficient $_s\gamma_\pm$, referred to unity at infinite dilution in each particular solvent, is plotted in the lower portion of the figure. The

---

[11] Data for the standard e.m.f. of this cell in nonaqueous media have been summarized by H. D. Crockford in *Electrochemical Constants*, p. 153, National Bureau of Standards Circular 524, 1953; D. Feakins and C. M. French, *J. Chem. Soc.*, 3168 (1956); 2284, 2581 (1957); G. J. Janz, Chapter 4 in *Reference Electrodes*, D. J. G. Ives and G. J. Janz, eds., Academic Press, New York, 1961; and R. G. Bates, Chapter in *Hydrogen-Bonded Solvent Systems*, A. K. Covington and P. Jones, eds., p. 49, Taylor and Francis, London, 1968.

composition and dielectric constant of each solvent are indicated on the figure. The activity coefficient of hydrochloric acid at low concentrations in ethanol is seen to be more than 200 times its value in water at the same molality. The differences in the activity coefficients of uncharged molecules from solvent to solvent are usually smaller than the differences for ionic solutes.

Medium Effects from the Born Equation

As we have already seen, the Gibbs transfer energies for uni-univalent electrolytes may be regarded as the sum of the transfer (chemical) energies for the cation and the anion. Although we shall designate these terms for single ionic species i as $\Delta G_t^{\circ}(i)$, it should be understood that they represent chemical energies alone, and that no allowance for the transfer of charge has been made. The value of ΔG_t° for a single ionic species might therefore be represented by the difference $\Delta G_t^{\circ}(\text{el})$ between the electrostatic energy of charging a mole of ions in solvent s and the energy of charging the same ions in water. Introducing the Born equation[12] for these charging effects in the two media of dielectric constants ε_s and ε_w gives, for a single ionic species,

$$\Delta G_t^{\circ}(\text{el}) = \frac{Nz_i^2 e^2}{2r} \left(\frac{1}{\varepsilon_s} - \frac{1}{\varepsilon_w} \right) \tag{16}$$

where N is the Avogadro number, z_i is the valence of the species i, e is the electronic charge, and r is the radius of the ion in the two solvents. For a uni-univalent electrolyte, therefore,

$$\Delta G_t^{\circ}(\text{el}) = \frac{Ne^2}{2} \left(\frac{1}{\varepsilon_s} - \frac{1}{\varepsilon_w} \right) \left(\frac{1}{r_c} + \frac{1}{r_a} \right) \tag{17}$$

If the ionic radii for the cation (r_c) and the anion (r_a) are expressed in angstroms and the value of ε_w at 25 °C (78.30) is introduced, the transfer energy for the electrolyte in kcal mol^{-1} is given by

$$\Delta G_t^{\circ}(\text{el}) = 166(\varepsilon_s^{-1} - 0.0128)(r_c^{-1} + r_a^{-1}) \tag{18}$$

If the radii r_c and r_a remain unchanged in solvents of different compositions equation 17 assumes the form

$$\Delta G_t^{\circ}(\text{el}) = A\varepsilon_s^{-1} + B \tag{19}$$

where A and B are constants characteristic of the particular electrolyte and independent of the solvent or, at least, of the composition in a particular

---

[12] M. Born, Z. Physik, **1**, 45 (1920).

mixed solvent system. The changes in Gibbs energy on transfer of a mole of hydrochloric acid or sodium chloride from water to water–methanol mixtures are plotted in Fig. 8–2 as a function of the reciprocal of the dielectric constant of the nonaqueous medium. It is evident that a straight line is not obtained, as would be expected if equation 17 is an adequate representation of the changes in energy during the transfer process.

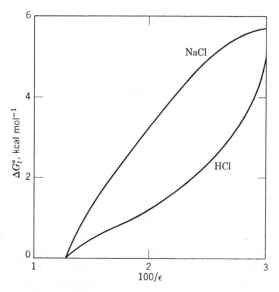

FIG. 8–2. Gibbs transfer energy for hydrochloric acid and sodium chloride from water to water-methanol solvents as a fuction of the reciprocal of the dielectric constant of the mixed solvent.

The simple Born model of the ionic transfer process thus does not account for the energies of transfer of electrolytes from water to the mixed solvents for which data are available. There are many reasons for its inadequacy. In the first place, most calculations with the Born equation are made with ionic radii from crystal measurements, in the absence of reliable information concerning the radii of ions in solution. Moreover, the macroscopic dielectric constants of the media s and w are usually employed. Although it seems likely that dipolar solvent molecules are oriented in the electric fields of the ions, producing a considerable amount of dielectric saturation near the ionic surface, the manner in which the effective dielectric constant varies with distance from the ion is unknown. There have been attempts, however, to improve the Born treatment by making allowance for variations in dielectric constant.[13,14]

In one of the earliest attempts to account for the observed transfer energies of electrolytes, Latimer, Pitzer, and Slansky[15] retained the form of the Born equation, utilizing the macroscopic dielectric constants of the solvent but adding a constant amount to the crystal radii of cations and another amount to those of the anions. In this way, approximate agreement with experiment could be obtained. This approach has recently been reexamined and refined by Strehlow.[16,17]

Ion-Solvent Interaction Energy

Possibly the most serious defect in the Born treatment of the transfer energy is that it greatly oversimplifies the nature of the transfer process. This suspicion is borne out by the observation that equation 16 predicts that the transfer chemical energies of cations and anions should be of the same sign and should differ only as $1/r_c$ differs from $1/r_a$. There is, however, mounting evidence that the chemical solvation energy change on transfer of a cation from water to a mixed solvent may have a sign different from that for the corresponding transfer of an anion. It would appear, for example, that cations are stabilized by transfer from water to water–methanol mixtures (at least, to those containing less than about 90 wt. per cent methanol), whereas anions are in a state of higher energy in the mixed solvent than in water. Tomkins, Feakins, and their coworkers [18-20] have offered an explanation for this phenomenon in terms of inductive effects of the alkyl groups and the transmission of these effects through hydrogen bonding to the water molecules in the primary solvation layer (see Fig. 8-3). The resulting increase in electron density may tend to strengthen the interaction between the cationic charge and the dipolar solvent molecules in the primary layer but to weaken the interaction with anions.

Drastic differences in the solvation energies of cations and anions are difficult to account for, and it is clear that electrostatic charging effects alone

[13] L. G. Hepler, *Australian J. Chem.*, **17**, 587 (1964).

[14] R. H. Stokes, *J. Amer. Chem. Soc.*, **86**, 979 (1964).

[15] W. M. Latimer, K. S. Pitzer, and C. M. Slansky, *J. Chem. Phys.*, **7**, 108 (1939).

[16] H. Strehlow, Chapter 4 in *The Chemistry of Non-Aqueous Solvents*, J. J. Lagowski, ed., Vol. I, Academic Press, New York, 1966.

[17] See also J. F. Coetzee and J. J. Campion, *J. Amer. Chem. Soc.*, **89**, 2513, 2517 (1967).

[18] R. P. T. Tomkins, thesis, Birkbeck College, London, 1966.

[19] H. P. Bennetto, D. Feakins, and D. J. Turner, *J. Chem. Soc. (A)*, 1211 (1966).

[20] A. L. Andrews, H. P. Bennetto, D. Feakins, K. G. Lawrence, and R. P. T. Tomkins, *J. Chem. Soc. (A)*, 1486 (1968).

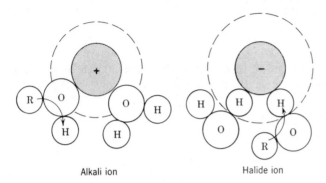

Alkali ion   Halide ion

FIG. 8–3. Structure of the primary solvation shell of spherical cations and anions in water-alcohol solvents (after Tomkins). (Arrows indicate the direction of inductive effects).

cannot explain them. Charged spheres in a medium of uniform dielectric constant provide a very crude model for ions in solution. In acid-base processes, charging effects may even be of secondary importance to selective interactions between solutes and preferred species of solvent molecules.[21-23]

A reasonable model for solvated ions incorporates a primary solvation shell of tightly bound dipolar solvent molecules in a region where dielectric saturation is more or less complete. In addition, there is a secondary solvation region characterized by partial orientation of solvent dipoles in the ionic field, with an effective dielectric constant that varies in an unspecified manner from the saturation value near the ion up to its macroscopic value at a distance of perhaps 10 Å from the center of the ion. By regarding the water molecule as a sphere with the dipole slightly off center, reasonable values for the hydration energies of 17 ions have been calculated.[24]

Medium Effect and the Dissociation Constant

In spite of the fact that it is incapable of allowing for the important ion–solvent interaction energy, the simple Born treatment is often useful in

[21] R. Gaboriaud, *Ann. Chim. (Paris)*, Ser. 14, **2**, 201 (1967).

[22] A. S. Quist and W. L. Marshall, *J. Phys. Chem.*, **72**, 684, 1536 (1968), D. J. Glover, *J. Amer. Chem. Soc.*, **87**, 5275, 5279 (1965).

[23] R. G. Bates, *J. Electroanal. Chem.*, **29**, 1 (1971). The subject of selective solvation has been reviewed by H. Schneider, Chapter 5 in *Solute-Solvent Interactions*, J. F. Coetzee and C. D. Ritchie, eds., Marcel Dekker, New York, 1969, and by H. Strehlow and H. Schneider, *Pure Appl. Chem.*, **25**, 327 (1971).

[24] S. Goldman and R. G. Bates, *J. Amer. Chem. Soc.*, **94**, 1476 (1972).

predicting solvent effects on proton transfer reactions in a qualitative fashion. The general acidic dissociation reaction in the solvent SH is formulated

$$HA + SH \rightleftharpoons SH_2^+ + A \qquad (20)$$

The ratio of the values of K_{HA} in the two solvents can be expressed in terms of equations 4 and 16 as follows:

$$\log \frac{{}_wK_{HA}}{{}_sK_{HA}} = \log \frac{{}_m\gamma_H \cdot {}_m\gamma_A}{{}_m\gamma_{HA}}$$

$$= 122\left(\frac{1}{r_{SH_2^+}} + \frac{z_A^2}{r_A} - \frac{z_{HA}^2}{r_{HA}}\right)\left(\frac{1}{\varepsilon_2} - 0.0128\right) \qquad (21)$$

As a first approximation one may assume that the radii of all the ions are equal. If this common radius is $\bar{r}$, we have

$$p({}_sK_{HA}) - p({}_wK_{HA}) = \frac{122n}{\bar{r}}\left(\frac{1}{\varepsilon_2} - 0.0128\right) \qquad (22)$$

where $n = 2$ for HA,A pairs of the charge type A^0B^-, $n = 4$ for the charge type $A^-B^=$, and $n = 0$ for the charge type A^+B^0. The dissociation constants in ethanol–water mixtures given in Table 8–1, taken from the papers of

TABLE 8–1. pK VALUES OF WEAK ACIDS IN ETHANOL-WATER SOLVENTS AT 25°C

	Wt. Per Cent Ethanol					
Acid	0 ($\varepsilon=78.3$)	20 ($\varepsilon=67.0$)	35 ($\varepsilon=58.0$)	65 ($\varepsilon=40.6$)	80 ($\varepsilon=32.8$)	100 ($\varepsilon=24.3$)
Formic	3.75	4.01	4.24	5.01	5.64	9.15
Acetic	4.76	5.12	5.43	6.29	6.87	10.32
Propionic	4.87	5.32	5.68	6.63	7.17	—
Malonic 1	2.75	3.13	3.38	3.94	4.41	—
Succinic 1	4.13	4.53	4.86	5.64	6.16	—
Benzoic	4.20	4.76	5.24	6.19	6.79	10.25
Ammonium	9.24	9.00	8.78	8.45	—	—
Methylammonium	10.70	10.41	10.11	9.58	—	—
Anilinium	4.636	4.416	4.161	3.799	3.750	5.70
N-Methylanilinium	4.839	4.62	4.276	3.636	3.417	4.86
p-Toluidinium	5.098	4.928	4.642	4.312	4.316	6.24

Grunwald and his associates,[25] show that pK for monobasic acids does indeed increase markedly as the dielectric constant decreases. The pK for cation acids, however, is not, as might be expected from equation 22, unaffected by a change in the dielectric constant. It usually decreases somewhat, passes through a minimum, and then increases at high concentrations of organic solvent. These two types of behavior are also characteristic of indicator acids, as has been seen in Chapter 6.

The form of equation 22 suggests that a plot of p($_sK_{HA}$) as a function of $1/\varepsilon_2$ should be a straight line. In practice, this expectation is usually fulfilled only by the dissociation constants determined with respect to a single reference acid,[26] doubtless because the ion–solvent interaction terms in the free energy, omitted in the Born treatment, are of about the same magnitude for each acid. For partially aqueous solvents it has been found that the linearity of the plot is improved when allowance is made for differences of water activity in the several solvents.[27]

The fact that the apparent relative strengths of acids depend not only on the solvent but also on the reference acid chosen[28] serves to emphasize the limited usefulness of the concept of intrinsic acid strengths. In some cases, however, the inconsistencies among the results in different solvents disappear when data extrapolated to infinite dielectric constant are used.[29]

Despite these limited successes, however, it has become clear that a purely electrostatic treatment is inadequate to account for solvent effects on the dissociation constants of weak acids and bases. A detailed examination of the dissociation equilibria for BH^+ acids in mixed solvents serves to emphasize the major role of solute–solvent interactions (selective solvation).[23] The solvent effect on pK_a, namely p($_sK_a$) − p($_wK_a$), is a function of the transfer chemical energies of the proton and of the acid and its conjugate base:

$$(RT \ln 10)[p(_sK_a) - p(_wK_a)] = \Delta G_t^\circ(B) - \Delta G_t^\circ(BH^+) + \Delta G_t^\circ(H^+) \quad (23)$$

When the conjugate base is uncharged (ammonia, nitroanilines, tris(hydroxymethyl)aminomethane, etc.), it is often possible to determine its chemical energy change on transfer from water to a solvent s by measurements of

[25] E. Grunwald and B. J. Berkowitz, *J. Amer. Chem. Soc.*, **73**, 4939 (1951); B. Gutbezahl and E. Grunwald, *ibid.*, **75**, 559 (1953).

[26] M. Kilpatrick, *Trans. Electrochem. Soc.*, **72**, 95 (1937); S. Glasstone, *Introduction to Electrochemistry*, Chapter 9, D. Van Nostrand Co., New York, 1942.

[27] M. Yasuda, *Bull. Chem. Soc. Japan*, **32**, 429 (1959).

[28] G. N. Lewis, *J. Franklin Inst.*, **226**, 293 (1938). See also F. H. Verhoek, *J. Amer. Chem. Soc.*, **58**, 2577 (1936).

[29] W. F. K. Wynne-Jones, *Proc. Roy. Soc. (London)*, **A140**, 440 (1933); L. J. Minnick and M. Kilpatrick, *J. Phys. Chem.*, **43**, 259 (1939).

solubility or solute vapor pressure in the two media.[21, 30] Consequently, one is able to determine from equation 23 the differences between the transfer chemical energies of the cation acid and the hydrogen ion, namely

$$\Delta G_t^{\circ}(BH^+) - \Delta G_t^{\circ}(H^+)$$

Some values obtained in this way are plotted in Fig. 8–4 as a function of the mole fraction x_2 of the organic component, and are compared with similar data for the differences in the energies of transfer of the alkali metal ions and hydrogen ion.[23]

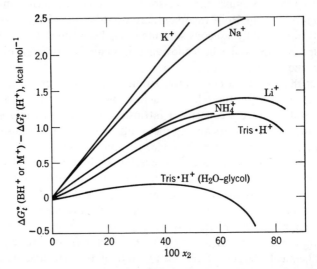

FIG. 8–4. Differences between the transfer chemical energies of cations in water-methanol and water-glycol mixed solvents, as a function of the mole fraction (x_2) of the methanol or glycol.

If the reasonable assumption that the radius of the hydrogen ion is smaller than that of any of the alkali metal ions or the cation acids BH^+ is made, the initial positive slope of all of the curves in the figure is understandable in terms of the Born equation (equation 16). The downward trend of all the curves when the mole fraction of the organic constituent exceeds 0.5, however, can only be satisfactorily explained by differences in the solvent stabilization of the two types of ions, brought about by changing the composition of the mixed solvent. A major determinant seems to be the marked preference of

[30] P. Schindler, R. A. Robinson, and R. G. Bates, *J. Res. Nat. Bur. Stand.*, **72A**, 141 (1968).

he proton for water and the consequent large increase in the transfer chemical :nergy of this ion as the activity of water in the solvent mixture becomes ,mall.

This preference of the proton for water tends to produce a sharp increase n the pK_a of acids of all charge types in water–organic solvent mixtures as he concentration of water becomes low. An exception is found for those mixtures where the organic constituent is more basic than water (ethylene-diamine,[31] pyridine,[32] etc.). From an examination of solvent effects both for markedly hydrophilic bases such as tris(hydroxymethyl)aminomethane and or markedly hydrophobic bases (nitroanilines), it may be concluded that $\Delta G_t°(BH^+)$ is always larger than $\Delta G_t°(B)$ for each conjugate acid-base pair, regardless of the preference of the base for one solvent component or the)ther.[23] Inasmuch as $\Delta G_t°(B) - \Delta G_t°(BH^+)$ decreases rather regularly with change of composition while $\Delta G_t°(H^+)$ increases at a highly accelerated rate as the mole fraction of water becomes low, the pK_a value of the BH^+ acids)ften passes through a minimum, as the data in Table 8–1 show.

Determination of the Medium Effect for the Proton

The search for a means of relating practical scales of proton activity a_H in nonaqueous and partially aqueous solvents to the conventional scale in water is in essence an attempt to determine medium effects $_m\gamma_H$ for the hydrogen ion. This is evident if one considers two dilute solutions of hydro-chloric acid (molality m) in media water (w) and nonaqueous solvent (s). The molality is the number of moles of solute per kilogram of pure or mixed solvent. The proton activity in both media relative to the aqueous standard state is expressed formally by $-\log(m \cdot _w\gamma_H)$; that in medium s can also be written $-\log(m \cdot _m\gamma_H \cdot _s\gamma_H)$. For dilute solutions, $_w\gamma_H$ in water and $_s\gamma_H$ in the solvent s can be estimated with fair success by the Debye–Hückel equation, but the estimation of $_m\gamma_H$ by the Born equation, as we have already seen, is not very useful. The difference in the zero points of the two scales of activity or)a_H is fixed by the transfer chemical energy of a mole of hydrogen ion from the standard state in water to the standard state in solvent s.

LIQUID-JUNCTION POTENTIALS

Bjerrum and Larsson[33] attempted to determine medium effects for in-dividual ionic species by e.m.f. measurements of cells with liquid junction

---

[31] D. Bauer, *Bull. Soc. Chim. France*, 3302 (1965); thesis, Paris, 1967.

[32] H. Wilski and G. Kortüm, *Z. Physik. Chem. (Frankfurt)*, **5**, 333 (1955).

[33] N. Bjerrum and E. Larsson, *Z. Physik. Chem.*, **127**, 358 (1927).

which utilized concentrated salt bridges. More recently, Oiwa[34] has made extensive measurements of hydrochloric acid solutions in aqueous methanol at 20, 25, and 30 °C, using cells with a hydrogen electrode and a saturated aqueous calomel reference electrode. He determined the difference between the standard potential of the cell in water and in the methanolic media by applying corrections for the liquid-junction potential. The activity coefficient $_s\gamma_H$ was estimated by the extended form of the Debye–Hückel equation, and the "apparent" standard potentials were found to be independent of the concentration of the acid, affording some indirect evidence of the validity of the corrections for the liquid-junction potential.

E.M.F. AND SOLUBILITY DATA

Izmailov and his coworkers[35] have determined medium effects for a considerable number of ion combinations by both e.m.f. methods and solubility measurements.[36] The medium effect term $\log (_m\gamma_H \cdot _m\gamma_{A^-})$ for the ions of a strong acid HA can be obtained by adding similar terms for KA and HBr and subtracting $\log (_m\gamma_{K^+} + _m\gamma_{Br^-})$ determined from measurements on potassium bromide. The solvent effect on the standard e.m.f. of cells with hydrogen electrodes or dilute potassium amalgam electrodes combined with the electrodes AgA; Ag or AgBr; Ag provide the primary data. Alternatively, the solvent effect on the solubility of silver salts, together with the known data for hydrochloric acid, can yield the same quantity by the algebraic combination

$$HCl + AgA - AgCl = HA$$

The regularities shown by these combinations of ionic medium effects may be expected to shed some light indirectly on the relative solvation energies of the ions in the solvents studied. It was in fact noted some years ago[37] that $_m\gamma_H \cdot _m\gamma_{A^-}$ (that is, $_m\gamma_\pm{}^2$) is practically the same for all monobasic strong acids in the same solvent medium, suggesting that the transfer free energies represent primarily the change in the energy of solvation of the proton. This observation was thought to be sufficient justification for using $_m\gamma_H = _m\gamma_\pm$ as a conventional means of relating a scale of proton activity in

---

[34] I. T. Oiwa, *Sci. Rpt. Tohoku Univ., First Ser.*, **41**, 129 (1957).

[35] Examples of the later work are: N. A. Izmailov and V. N. Izmailova, *Zh. Fiz. Khim.*, **29**, 1050 (1955); N. A. Izmailov and V. S. Chernyi, *ibid.*, **34**, 319 (1960); E. F. Ivanova and N. A. Izmailov, *ibid.*, **34**, 1021 (1960); N. A. Izmailov and T.V. Mozharova, *ibid.*, **34**, 1543, 1709 (1960); and N. A. Izmailov and L. L. Spivak, *ibid.*, **36**, 757 (1962).

[36] See also I. M. Kolthoff, J. J. Lingane, and W. D. Larson, *J. Amer. Chem. Soc.*, **60**, 2512 (1938).

[37] N. A. Izmailov, *Zh. Fiz. Khim.*, **23**, 639, 647 (1949).

nonaqueous and partially aqueous solvents to the aqueous pH scale.[37] Later refinements were based on a calculation of the chemical energy of proton solvation in water and nonaqueous solvents[38] and also on a calculation of $_m\gamma_H$ from the equilibrium constant for the transfer of a proton from H_3O^+ to SH, combined with a term that allowed for the different electrostatic environments of the two media[39] (as in the Born treatment). These two approaches gave results that agreed reasonably well. In pure methanol, they furnished values of 3.5 and 3.1, respectively, for $\log\ _m\gamma_H$; in pure ethanol the corresponding values of $\log\ _m\gamma_H$ were 4.3 and 4.1.[40]

The calculations of Izmailov[41] lead to the following estimates of the medium effect for hydrogen ion in amphiprotic solvents:

Solvent	$\log\ _m\gamma_H$
(Water	0)
Ammonia	−16.6
Methanol	3.3
Ethanol	4.2
Propanol	4.2
n-Butanol	4.7
Isobutanol	4.5
Benzyl alcohol	3.1
Formic acid	8.6

It is of interest to examine the relative positions of the scales of proton activity (pH) in these solvents as indicated by these figures for the medium effect. The length of the pH scale in each medium, in pH units, is equal to $-\log K_s$, where K_s is the autoprotolysis constant of the solvent (see Table 7-4). Furthermore, the pH or pa_H in any medium, based on the water scale, is $-\log (m_H \cdot {_s\gamma_H}) - \log\ _m\gamma_H$; hence the numerical value of the pH at the acidic end of the scale in each solvent is $-\log\ _m\gamma_H$, that is, 16.6 in liquid ammonia, −8.6 in formic acid, etc.[42] The relative positions of the pH scales in water, ethanol, ammonia, and formic acid, based on Izmailov's values of

---

[38] N. A. Izmailov, *Dokl. Akad. Nauk SSSR*, **126**, 1033 (1959).

[39] V. V. Aleksandrov and N. A. Izmailov, *Zh. Fiz. Khim.*, **32**, 404 (1958). See also N. A. Izmailov and V. V. Aleksandrov, *ibid.*, **31**, 2619 (1957).

[40] N. A. Izmailov, *Dokl. Akad. Nauk SSSR*, **127**, 104 (1959).

[41] N. A. Izmailov, reference 40 and earlier sources cited.

[42] This is the value of pa_H when $m_H \cdot {_s\gamma_H} = 1$, corresponding to $pa_H = 0$ in water. Higher proton activities may, of course, be achieved under conditions similar to those which produce negative pH values in aqueous solutions.

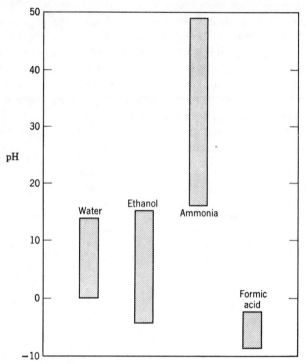

FIG. 8–5. Range and relative position of the pH scale in four different solvents.

$\log {}_m\gamma_H$, are shown in Fig. 8–5. It is apparent that the pH scales in liquid ammonia and formic acid do not overlap at all the pH scale in water, if these estimates of the medium effect on the hydrogen ion are correct.

LINEAR FREE-ENERGY RELATIONSHIPS

Grunwald and his coworkers have collected and analyzed data for the dissociation constants of weak uncharged acids HA (acetic acid type) and of cation acids HA^+ (ammonium type) in ethanol-water mixtures.[43] Values for the individual medium effects of hydrogen ion in these media were derived from the regularities observed in the solvent effect on pK by a method which presupposes the existence of certain linear free-energy relationships.[44] The

[43] E. Grunwald and B. J. Berkowitz, J. Amer. Chem. Soc., 73, 4939 (1951); B. Gutbezah and E. Grunwald, ibid., 75, 559 (1953).
[44] B. Gutbezahl and E. Grunwald, J. Amer. Chem. Soc., 75, 565 (1953).

procedure was patterned after an earlier successful treatment of specific solvolysis rates.[45]

An activity postulate, formulated as follows

$$\log \frac{{}_m\gamma_{A^-}}{{}_m\gamma_{HA}} = \overline{m}_{HA} Y_-$$

(24)

and

$$\log \frac{{}_m\gamma_A}{{}_m\gamma_{HA^+}} = \overline{m}_{HA^+} Y_0$$

(25)

was proposed. It separates $\log ({}_m\gamma_A/{}_m\gamma_{HA})$ into two factors; one $(\overline{m}_{HA})$ is independent of the solvent and characteristic only of the structure of HA, whereas the other (Y) is a solvent parameter, independent of the structure of the acid, though not of its charge type. By analysis of the data for the change of pK with composition of the ethanol-water solvent for acids of the two charge types at 25 °C, it was possible to derive consistent values for both the solvent parameter and the structural parameter and, therefore, to obtain ${}_m\gamma_H$ from the thermodynamically defined medium effect term, ${}_m\gamma_H \cdot {}_m\gamma_A/{}_m\gamma_{HA}$.[46]

EXTRAPOLATION PROCEDURES

A wide variety of other extrathermodynamic assumptions has been used to evaluate medium effects for single ionic species. The results of these have been reviewed by Parker and Alexander[47] and by Popovych.[48] Parker and Alexander have compared the medium effect (relative to methanol) for silver ion in 11 pure solvents, as derived by 16 different nonthermodynamic procedures.

Several recent attempts to derive the medium effect for the proton as well as that for other single ionic species have been based on procedures for separating experimental values of the medium effects for neutral ion combinations into the individual ionic contributions. The transfer energies and medium effects of the alkali halides, for example, can be determined by thermodynamic methods. Thus the standard potentials of cells without liquid junction of the types

$$Pt;H_2,H^+,X^-,AgX;Ag$$

(26a)

or

$$M;M^+,X^-,AgX;Ag$$

(26b)

[45] E. Grunwald and S. Winstein, *J. Amer. Chem. Soc.*, **70**, 846 (1948).

[46] B. Gutbezahl and E. Grunwald, *J. Amer. Chem. Soc.*, **75**, 565 (1953); E. F. Sieckmann and E. Grunwald, *ibid.*, **76**, 3855 (1954).

[47] A. J. Parker and R. Alexander, *J. Amer. Chem. Soc.*, **90**, 3313 (1968). See also A. J. Parker, *Pure Appl. Chem.*, **25**, 345 (1971), and R. Alexander, A. J. Parker, J. H. Sharp, and W. E. Waghorne, *J. Amer. Chem. Soc.*, **94**, 1148 (1972).

[48] O. Popovych, *Crit. Rev. Anal. Chem.*, **1**, 73 (1970).

or

$$M(Hg);M^+,X^-,AgX;Ag \qquad (26c)$$

where M^+ represents an alkali metal cation and X^- represents a halide ion, will furnish the primary data needed. The difference between the standard e.m.f. $E°$ in water and in solvent s is a direct measure of the medium effect in the transfer process

HX or MX (standard state, w) = HX or MX (standard state, s)

as shown by combining equation 6 with the relationship $\Delta G° = -nFE°$:

$$\log {}_m\gamma_\pm = \frac{{}_w^s E° - {}_s^s E°}{(2RT \ln 10)/F} \qquad (27)$$

Feakins and his coworkers[18-20] have collected and critically examined the data for the transfer energies of the halogen acids and the alkali halides in water–methanol solvents. Their values are summarized in Table 8-2.

TABLE 8-2. TRANSFER ENERGIES (MOLAR SCALE) FOR UNI-UNIVALENT ELECTROLYTES FROM WATER TO WATER-METHANOL MIXTURES (After Feakins and coworkers)

% Methanol (w/w)	$\Delta G_t°/\text{cal/mol}^{-1}$					
	HCl	HBr	HI	LiCl	NaCl	KCl
10	180	125	58	340	450	480
20.22	351	219	138	700	930	990
33.4	558	357	54	1170	1570	1700
43.12	701	438	12	1520	2070	2260
50	837	507	45	1800	2430	2650
68.33	1329	914	233	2640	3560	4130
90	2748	2273	1460	3970	5340	6530
100	5640	5120	4090	4800	5700	6600

As we have seen, the theoretical calculation of transfer chemical energies and medium effects for single ionic species is a matter of considerable difficulty and is not feasible at the present time. Although the Born electrostatic treatment is inadequate, it may still be capable of serving as a guide in formulating a nonthermodynamic procedure for separating transfer energies into their ionic components. If instead of assuming constant r_c and r_a in equation 17, one chooses to examine data for a series of electrolytes with a common cation in the same nonaqueous medium, r_c and ε_s will be constant, and equation 17 takes the form

$$\Delta G_t^{\,\circ}(\text{el}) = A + Br_a^{\,-1} \tag{28}$$

Plots of $\Delta G_t^{\,\circ}(\text{MX})$ as a function of the reciprocal of the radius of the anion should then be a straight line. When $1/r_a$ approaches 0, that is, when the anion is very large, ion-dipole interactions should decrease markedly in both water and solvent s. The contribution of $\Delta G_t^{\,\circ}(\text{X}^-)$ to $\Delta G_t^{\,\circ}(\text{HX})$ or $\Delta G_t^{\,\circ}(\text{MX})$ should fall correspondingly. Thus the transfer chemical energy of the proton could be obtained as the intercept of the plot of $\Delta G_t^{\,\circ}(\text{HX})$ as a function of $1/r_a$. In a similar fashion, $\Delta G_t^{\,\circ}(\text{Na}^+)$ would be obtained as the intercept of plots of $\Delta G_t^{\,\circ}(\text{NaX})$ as a function of $1/r_a$, while $\Delta G_t^{\,\circ}(\text{Cl}^-)$ could be obtained from plots of $\Delta G_t^{\,\circ}(\text{MCl})$ as a function of $1/r_c$.

This extrapolation procedure was devised by Izmailov[49] and was refined and applied extensively by Feakins and his coworkers.[50] Figure 8–6 is a graph of the data given in Table 8–2 for the three halogen acids and for the three alkali chlorides in 50 wt. per cent methanol as a function of the reciprocal of the crystal radius of the counter ion. The lines appear reasonably straight, but there is a long and uncertain extrapolation to $1/r = 0$. The accuracy of the extrapolation is somewhat improved when data for

$$\Delta G_t^{\,\circ}(\text{M}^+) - \Delta G_t^{\,\circ}(\text{H}^+)$$

are plotted.[49]

A more elaborate extrapolation procedure was described by Alfenaar and deLigny.[51] These authors regarded the transfer chemical energy of each ion to consist of two terms, one electrostatic and the other neutral:

$$\Delta G_t^{\,\circ}(i) = \Delta G_t^{\,\circ}(\text{el}) + \Delta G_t^{\,\circ}(\text{neut}) \tag{29}$$

The neutral part is related to the energy required to create a cavity in the solvent to receive the ion on its transfer from the gas phase to the liquid phase. This part of the energy does not disappear but indeed becomes larger as the radius of the ion increases. As others[52] have done, Alfenaar and deLigny identified the neutral part of the transfer energy of an ion with the measurable transfer energy for the isoelectronic noble gases, the atoms of which have radii close to those for the corresponding ions. After allowance had been made for the neutral transfer quantities, the residual portion of $\Delta G_t^{\,\circ}(\pm)$ for the same electrolytes considered by Feakins *et al.* was

[49] N. A. Izmailov, *Zh. Fiz. Khim.*, **34**, 2414 (1960); *Dokl. Akad. Nauk SSSR*, **149**, 884 (1963).

[50] See D. Feakins and P. Watson, *Chem. Ind.* (*London*), 2008 (1962), *J. Chem. Soc.*, 4734 (1963), and references 18 to 20.

[51] M. Alfenaar and C. L. deLigny, *Rec. Trav. Chim.*, **86**, 929 (1967).

[52] G. R. Haugen and H. L. Friedman, *J. Phys. Chem.*, **72**, 4549 (1968).

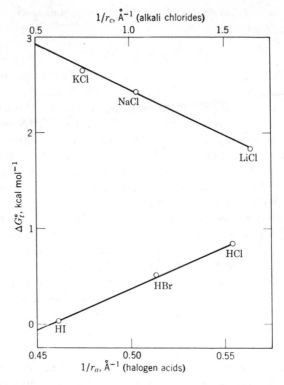

FIG. 8–6. Standard Gibbs energy change for the transfer of alkali chlorides and halogen acids from water to 50 wt per cent methanol, plotted as a function of the reciprocal of the radius of the alkali metal ion (top curve) or of the halide ion (bottom curve).

extrapolated to $1/r = 0$, on the assumption that the Born slope must be maintained when the radius of the ion exceeds 10 Å.

As Popovych[48] has stated, the chief difficulty in these extrapolation methods for determining single ionic medium effects is the dilemma of the neutral part of the Gibbs energy change. The approach of Izmailov and Feakins ignores the neutral part altogether, while the refinement of Alfenaar and deLigny probably overestimates it.

REFERENCE ELECTROLYTE METHOD

The concept, embodied in the $1/r$ relationship of the Born equation, that the electrostatic contribution to the transfer chemical energy is inversely proportional to the ionic radius has been applied in other ways. Pleskov's rubidium scale of electrode potentials[53] was based on the belief that solva-

---

[53] V. A. Pleskov, *Usp. Khim.*, **16**, 254 (1947).

tion of such a large ion as rubidium would be minimal in all solvent media and, hence, the transfer energy of rubidium ions from water to these media would approach a uniform value. Likewise, the ferrocene–ferricinium and cobaltocene–cobalticinium redox couples (see page 188) represent systems whose standard potentials can reasonably be expected to be nearly independent of the nature of the solvent. The reduced and oxidized species are presumably nearly spherical in conformation, large and of equal size, with the metal buried in the center of the molecule. Because of the large diameter, the density of charge at the surface is expected to be low, and specific interactions with solvent molecules are probably minimal. It is also likely that the ferrocene molecule represents an excellent neutral analog of the ferricinium ion and therefore provides a "built-in counterbalance to the neutral part of the solvation energy change."[48]

The same view of the factors contributing to an equalization of the chemical solvation energy as the solvent composition is changed can be utilized in another way. These concepts lead one to expect that a univalent cation and a univalent anion, both large, spherical, and of equal radius, will suffer the same change in Gibbs energy when moved from water to another solvent (s). If this is the case, the transfer chemical energy for each of the ions will be one-half of the measured transfer energy of the uni-univalent electrolyte of which they are a part. Once the transfer chemical energies of this pair of reference ions has been established, the additivity of ionic transfer energies makes it possible to derive medium effects for other ions from the energies of transfer of other electrolytes from water to the same solvents.

Selection of the most suitable reference electrolytes is evidently a matter of considerable concern. Solubility measurements are, in general, a favorite means of obtaining the Gibbs energies for the transfer of these electrolytes from water to another solvent or solvent mixture. In order to determine the medium effect most accurately, the solubility should be low in both solvents. Furthermore, it is essential that the solid phase with which the saturated solution is in equilibrium be of the same composition in each solvent. The ideal reference electrolyte, however, is so chosen as to have a minimal tendency toward solvation; hence, this latter consideration is often of little concern.

Among the reference electrolytes that have been proposed are tetraphenylphosphonium tetraphenylborate,[54] triisoamyl-*n*-butylammonium tetraphenylborate,[55] and tetraphenylarsonium tetraphenylborate.[56] These

---

[54] E. Grunwald, G. Baughman, and G. Kohnstam, *J. Amer. Chem. Soc.*, **82**, 5801 (1960).

[55] O. Popovych, *Anal. Chem.*, **38**, 556 (1966); O. Popovych and A. J. Dill, *ibid.*, **41**, 456 (1969).

[56] R. Alexander and A. J. Parker, *J. Amer. Chem. Soc.*, **89**, 5539 (1967).

electrolytes have been studied in dioxane, methanol, ethanol, water, aceto-nitrile, and in some mixtures of these solvents. The reader is referred to the original papers and to the reviews of Parker and Alexander[47] and Popovych[48] for a summary and a comparison of the results.

A recent nuclear magnetic resonance study[57] nonetheless suggests that tetraphenylarsonium, tetraphenylphosphonium, and tetraphenylborate ions are subject to a variety of specific solvation effects in some of these solvents. Although the magnitude of the energy changes involved in these inter-actions has not been evaluated, Coetzee and Sharpe[57] advise caution in the interpretation of transfer chemical energies based on the assumed equality of the behavior of these reference ions. Further work now under way will doubtless shed light on the validity of measurements made with these sub-stances.

pH Values on the Water Scale

Medium effects for the proton in water–methanol and water–ethanol solvents have been determined by the various methods discussed above. Some of the results are compared in Tables 8-3 and 8-4. It is evident that the several procedures do not yet yield results that agree well. There is, indeed, sometimes a disagreement on the sign of the effect. These discrepancies make it impossible at the present time to base the hydrogen ion activity in these mixed solvents on the standard pH scale in water. When $-\log a_H$ is related to the aqueous standard state it is designated pa_H and is defined by

$$pa_H = -\log (m_H \cdot {}_s\gamma_H \cdot {}_m\gamma_H) \tag{30}$$

Although one can estimate the value of ${}_s\gamma_H$ in dilute solutions from the Debye–Hückel equation when the ionic strength is known, values of ${}_m\gamma_H$ the medium effect, are needed but unknown.

As an illustration of the uncertainties now involved in any attempt to determine pa_H values in water-alcohol solvents, the pa_H values for acetate buffer solutions and for mixtures of hydrochloric acid with sodium or potassium chloride in water-methanol and water-ethanol mixed solvents are plotted in Figs. 8-7 and 8-8 as a function of the composition of the solvent. The values of $-\log m_H \cdot {}_s\gamma_H$ were derived from data in the literature,[58, 59] and the medium effects for the proton were taken from the four sources indicated on the figures. The practical realization of a single pH scale related

[57] J. F. Coetzee and W. R. Sharpe, J. Phys. Chem., 75, 3141 (1971).
[58] R. G. Bates, M. Paabo, and R. A. Robinson, J. Phys. Chem., 67, 1833 (1963).
[59] R. G. Bates and G. Schwarzenbach, Helv. Chim. Acta, 38, 699 (1955).

TABLE 8-3. MEDIUM EFFECTS FOR THE PROTON (MOLAL SCALE) IN WATER-METHANOL SOLVENTS AT 25 °C, DETERMINED BY EXTRAPOLATION METHODS[a]

	log $_m\gamma_H$		
% Methanol (w/w)	Izmailov	Feakins	Alfenaar and deLigny
10	—	−0.38	−0.21
20.22	0.14	−0.69	—
33.4	0.25	−1.44	—
43.12	0.35	−2.08	−1.28
50	0.43	−2.41	—
68.33	0.70	−3.44	—
87.7	—	—	−2.70
90	1.45	−4.00	—
100	3.21	−2.25	−1.45

[a] More detailed comparisons are given by R. G. Bates, Chapter 2 in *Solute-Solvent Interactions*, J. F. Coetzee and C. D. Ritchie, eds., Marcel Dekker, New York, 1969, and by O. Popovych, *Crit. Rev. Anal. Chem.*, **1**, 73 (1970).

TABLE 8-4. MEDIUM EFFECTS FOR THE PROTON (MOLAL SCALE) IN WATER-ETHANOL SOLVENTS AT 25 °C, DETERMINED BY THREE DIFFERENT METHODS

	log $_m\gamma_H$		
% Ethanol (w/w)	Izmailov[a]	Grunwald[b]	Popovych[c]
10	0.09	—	−0.06
20	0.18	−0.01	−0.26
50	—	+0.21	−1.03
72	0.87	—	—
80	—	+1.08	−0.74
90	1.61	—	−0.51
100	4.05	+4.60	+1.85

[a] Extrapolation method.
[b] Linear free-energy method applied to pK data.
[c] Reference electrolyte: triisoamyl-*n*-butylammonium tetraphenylborate.

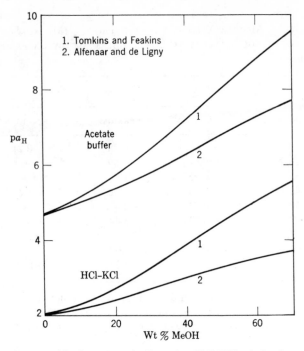

FIG. 8–7. pa_H (water scale) of an acetate buffer and an HCl-KCl solution in water-methanol solvents, based on two different estimates of the medium effect of the proton.

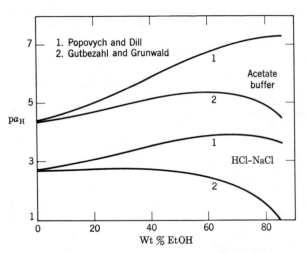

FIG. 8–8. pa_H (water scale) of an acetate buffer and an HCl-NaCl solution in water-ethanol solvents, based on two different estimates of the medium effect of the proton.

to the standard state in water evidently must await the refinement of methods for determining the medium effect of the proton.

Acidity Functions in Water-Alcohol Mixtures

A comparative study of acidity functions in water-ethanol mixtures was made by Bates and Schwarzenbach.[59, 60] The objective was to ascertain the behavior of these functions, to study the regularities in the activity coefficient ratios on which the Hammett function rests (see Chapter 6), and to select, if possible, reference points for the establishment of a practical acidity scale in these solvent media.

Three series of solutions were selected for study in seven solvent mixtures with ethanol content from 0 to 100 per cent. The first, containing only strong electrolytes, had the composition 0.002 M HCl, 0.008 M NaCl; the other two were buffer solutions of the charge types A^0B^- and A^+B^0, namely 0.02 M CH_3COOH, 0.01 M CH_3COONa, 0.005 M NaCl and 0.004 M triethanolammonium chloride, 0.002 M triethanolamine.

Four types of measurements were made for each solution in each solvent mixture at 25 °C: (1) The operational pH value was determined by measurement of a cell with hydrogen electrode and aqueous calomel reference electrode, standardized with aqueous buffer solutions; (2) the function $p(a_H\gamma_{Cl})$ was determined from the e.m.f. of the cell without liquid junction: Pt; H_2,Soln.X,AgCl; Ag (compare Chapter 4); (3) The Hammett function H_0 and (4) the Hammett H_- function were determined with the aid of selected indicators. The aqueous standard state was retained throughout. The indicators chosen for the determination of H_0 and H_- were as much alike structurally as possible, in the hope that the terms $\gamma_{In}/\gamma_{HIn^+}$ and $\gamma_{HIn}/\gamma_{In^-}$ would thereby be equalized,[61] so that

$$-\log a_H \approx \frac{H_0 + H_-}{2} \tag{31}$$

The results of the measurements on the three systems are shown in Figs. 8-9, 8-10, and 8-11. The general trend of these acidity functions is, with one or two exceptions, in qualitative agreement with the predictions of the Born and Hammett treatments of the effect on the proton activity of adding a solvent

---

[60] A comparison of the usefulness of H_0 and pH as measures of the acidity of nonaqueous media has been made by K. Schwabe, *Chimia*, **13**, 385 (1959).

[61] The results obtained by B. Gutbezahl and E. Grunwald, *J. Amer. Chem. Soc.*, **75**, 559, 565 (1953), by their empirical treatment of dissociation constants nonetheless suggest that $_m\gamma_A/_m\gamma_{HA^+}$ in pure ethanol is usually considerably smaller than $_m\gamma_{HA}/_m\gamma_{A^-}$ (10^{-4} to 10^{-3} as compared with 0.03 to 0.2), even when the acids are of similar structures.

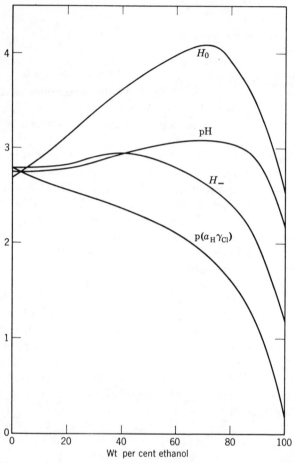

FIG. 8–9. Acidity functions in ethanol-water solvents: 0.002M HCl, 0.008M NaCl (Bates and Schwarzenbach). H_0 indicator, m-nitroaniline; H_- indicator, 4-chloro-2, 6-dinitrophenol.

of lower basicity and lower dielectric constant to the aqueous medium. Thus the acidity of the strong acid is higher in ethanol than in water, that of the acetate buffer decreases as alcohol is added, and that of the amine buffer increases.[62]

These data likewise serve to emphasize the highly specific nature of the medium effects on species of like charge. If $_m\gamma_A/_m\gamma_{HA}$ for the buffer acids

---

[62] The operational pH includes a large liquid-junction potential.

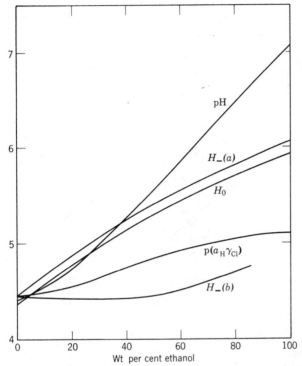

FIG. 8–10. Acidity functions in ethanol-water solvents: 0.02M CH₃COOH, 0.01M
CH₃COONa, 0.005M NaCl (Bates and Schwarzenbach). H_0 indicator, p-chloroaniline;
H_- indicators, (a) 2, 4-dinitrophenol, (b) dimedone.

HA^+ and HA in ethanol-water media were closely matched by $_m\gamma_{In}/_m\gamma_{HIn}$
for the Hammett indicator of the same charge type, H_0 for triethanolamine
buffer solutions would be independent of the composition of the solvent; the
same constancy would likewise be found for H_- in acetate buffer solutions.
Dimedone indicator appears to behave as expected in acetate solutions, but
the sharp (nearly linear) rise of H_- found with 2,4-dinitrophenol suggests
that the acid form of this indicator has the hybrid ion structure ∓ (see
Chapter 6). Likewise, H_0 in the amine solutions is far from constant.

The maximum in H_0 for the solution containing hydrochloric acid and
sodium chloride corresponds to the well-known minimum in $p(_sK_{HIn})$ for
weak cation acids in water-alcohol solvents (compare Fig. 6–2). Braude and

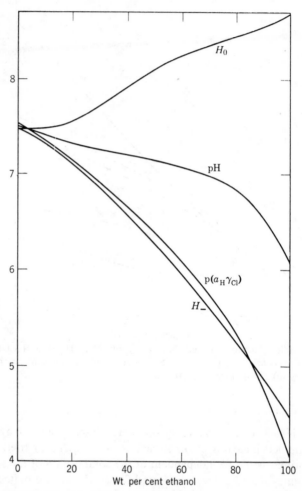

FIG. 8–11 Acidity functions in ethanol-water solvents: 0.004M triethanolammonium chloride, 0.002M triethanolamine (Bates and Schwarzenbach). H_0 indicator, N,N'-diethyltoluidine; H_- indicator, p-nitrophenol.

Stern[63] have pointed out that this maximum is not found in ethanol-acetone solvents. They attribute it to a gradual breakdown of the structure of the water by the organic solvent (increasing the total basicity of the medium), together with selective solvation of the proton by the water. When the water activity falls sufficiently, ethanol replaces water molecules in the hydration shell and the acidity rises once more. Inasmuch as $\log(_m\gamma_H \cdot _m\gamma_{Cl})$ for these media does not reveal a similar behavior,[64] it is evident that the maximum must be attributed to the medium effect on the indicator equilibrium.

Acidity phenomena in water-methanol solvents have been studied extensively by deLigny and his coworkers. In a determination of the dissociation constants of basic indicators in these media, they concluded[65] that an unambiguous scale of H_0 cannot be established at high methanol concentrations.

Percampus and Köhler[66] have determined the pH values of the Sørensen and McIlvaine buffer solutions in 10, 30, and 50 vol per cent methanol, using the hydrogen-aqueous calomel cell. The solvent effects on pH appear to parallel those on the pK of the buffer acid or base, as might be expected, but the results include, it appears, a variable residual liquid-junction error.

In reporting a careful study of the second dissociation constant of phosphoric acid in water and in 10 and 20 wt per cent methanol by means of hydrogen-silver chloride cells without liquid junction, Ender, Teltschik, and Schäfer[67] expressed the opinion that the hydrogen ion activity would be a suitable general measure of the acidity. No proposal for its evaluation in nonaqueous and partially aqueous media was made.

Relative Hydrogen Ion Activity

The most useful application of the pH cell

$$
\left.\begin{array}{l} \text{Pt}; \text{H}_2 \\ \text{or} \\ \text{glass electrode,} \end{array}\right| \begin{array}{l} \text{Soln. X in} \\ \text{nonaqueous} \\ \text{mixed solvent} \end{array} \left| \begin{array}{ll} \text{Salt} & \text{Reference} \\ \text{bridge,} & \text{electrode} \end{array} \right. \tag{32}
$$

outside the purely aqueous medium would appear to be in the determination

[63] E. A. Braude and E. S. Stern, *J. Chem. Soc.*, 1971, 1976, 1982 (1948); *Nature*, **161**, 169 (1948).

[64] See, for example, the curve of p($a_H\gamma_{Cl}$) in Fig. 8–9.

[65] C. L. deLigny, H. Loriaux, and A. Ruiter, *Rec. Trav. Chim.*, **80**, 725 (1961).

[66] H. H. Percampus and H. Köhler, *Z. Elektrochem.*, **61**, 836, 841 (1957). Acid-base properties in pure methanol have been investigated by R. Schaal and G. Lambert, *J. Chim. Phys.*, **59**, 1151, 1164 (1962); G. Lambert and R. Schaal, *ibid.*, **59**, 1170 (1962).

[67] F. Ender, W. Teltschik, and K. Schäfer, *Z. Elektrochem.*, **61**, 775 (1957).

of *relative* acidities in a solvent of fixed composition.[68] There is a rather wide belief that the residual liquid-junction error incurred in measurements of cell 32 is substantially constant in a given solvent medium and largely independent of the nature of the buffer electrolytes and of the pH.[69]

Likewise, the potential across the junction Soln. X in EtOH, H_2O | KCl (satd. aq.), calculated with the aid of values for $_m\gamma_H$, appears to be primarily dependent on the composition of the solvent.[70] Solutions of hydrochloric acid give about the same results for the liquid-junction potential as do carboxylic acid buffers and anilinium buffers, provided the composition of the solvent is unchanged.

The constancy of the liquid-junction potential between aqueous potassium chloride (3.5 M) and water-alcohol solvents has been examined in somewhat greater detail by Bates, Paabo, and Robinson.[71] For ethanol-water solvents, the data shown in Figs. 8–9, 8–10, and 8–11 were used.[72] In addition, values of pH and $p(a_H\gamma_{Cl})$ for twelve solutions in methanol-water mixtures were determined.[73] To avoid the complications due to appreciable ion pair formation, only solvents containing less than 68 wt per cent methanol were studied.

From the equation for the e.m.f. of the cell with liquid junction[74] we may write

$$pH = -\log a_H + \frac{E_j}{(RT \ln 10)/F} = -\log(m_H\gamma_H) + \bar{E}_j \qquad (33)$$

where $\bar{E}_j$ is the residual liquid-junction potential expressed in pH units.

---

[68] See D. S. McKinney, P. Fugassi, and J. C. Warner, *Symposium on pH Measurement*, p. 19, ASTM Tech. Publ. 73, Philadelphia, Pa., 1947.

[69] From the theory of the liquid-junction potential, it is evident that this supposition cannot be strictly true, any more than it is in water. Nevertheless, Wynne-Jones (private communication) has found that the mobility of the hydrogen ion in certain mixed solvents (including hydrogen peroxide-water solvents) is not very different from that of other cations. Similar conclusions were reached by H. Strehlow, *Z. Physik. Chem. (Frankfurt)*, **24**, 240 (1960), and by G. Kortüm and K. W. Koch, *Ber. Bunsenges. Phys. Chem.*, **69**, 677 (1965), in studies of conductivity in water-methanol solvents.

[70] B. Gutbezahl and E. Grunwald, *J. Amer. Chem. Soc.*, **75**, 565 (1953). It was later suggested by A. L. Bacarella, E. Grunwald, H. P. Marshall, and E. L. Purlee, *J. Org. Chem.*, **20**, 757 (1955); *J. Phys. Chem.*, **62**, 856 (1958), that alterations in the liquid-junction potential with the molar concentration c of solute could be allowed for by expressing the e.m.f. E of the pH cell at 25 °C by an equation of the form $E = A - 0.0591 \log a_H + B \log c$, where A and B are independent of the nature of the solute.

[71] R. G. Bates, M. Paabo, and R. A. Robinson, *J. Phys. Chem.*, **67**, 1833 (1963).

[72] R. G. Bates and G. Schwarzenbach, *Helv. Chim. Acta*, **38**, 699 (1955).

[73] Here again, pH is the operational pH number furnished by the hydrogen-aqueous calomel cell standardized with aqueous buffer solutions; $p(a_H\gamma_{Cl})$ is derived from the e.m.f. of the cell Pt; H_2,Soln. X,AgCl;Ag without liquid junction. The aqueous standard state is retained. For reproducible measurements, Soln. X must contain chloride ion.

[74] See Chapter 3.

Hence

$$pH - p(a_H\gamma_{Cl}) = \bar{E}_j + \log {}_m\gamma_{Cl} + \log {}_s\gamma_{Cl} \tag{34}$$

and if $\bar{E}_j$, like ${}_m\gamma_{Cl}$, is dependent only on the solvent composition,

$$pH - p(a_H\gamma_{Cl}) - \log {}_s\gamma_{Cl} = \text{Constant} = \bar{E}_j + \log {}_m\gamma_{Cl} \tag{35}$$

The chloride ion salt effect ${}_s\gamma_{Cl}$ in dilute solutions can be evaluated with sufficient accuracy for the present purpose by means of the Debye-Hückel formula (equation 15).

The values of the left side of equation 35 for twelve solutions (pH from 1.8 to 10.1 in water) were found to be about as constant in methanolic media as in pure water. The expected alteration of the liquid-junction potential in strongly acidic solutions (pH < 2) was observed in the mixed solvents, as in water. In addition, the values of $\bar{E}_j + \log {}_m\gamma_{Cl}$ in solutions of type A^+B^0 buffers fall slightly below those for buffers of types A^0B^- and $A^-B^=$. The difference is 0.02 pH unit in water and 0.05 unit in 68.1 per cent methanol. It could be accounted for by incomplete dissociation of the ammonium and substituted ammonium chlorides or attributed to a specific alteration in $\bar{E}_j$ not heretofore suspected.

The results obtained in ethanol-water mixtures are comparable with those found for methanol-water solvents. At 100 per cent ethanol, the constant of equation 35 amounted to 2.29 pH units in a dilute solution containing hydrochloric acid and sodium chloride, to 2.27 unit in a dilute acetate-chloride buffer, and to 2.18 unit in a dilute triethanolamine-chloride buffer.[75] Here also, the type A^+B^0 buffer solution appears to give a rather low result, but the sensitivity of the e.m.f. to traces of water makes the figures somewhat uncertain.

It appears from these findings that measurements of *relative* hydrogen ion activity in aqueous alcoholic solvents are possible. The usual pH meter with glass electrode is suitable for many measurements of this sort, as many glass electrodes display nearly the theoretical response to hydrogen ion, at least up to alcohol concentrations near 90 wt per cent. The choice of the most suitable pH scale and the establishment of reference points in alcoholic media will now be considered.

Standard pH Scale for Water-Alcohol Mixed Solvents

It will be recalled (see Chapter 2) that the replacement of the pc_H unit and the Sørensen unit by a practical pH unit representing a conventional activity of hydrogen ion was initiated by the recognition of activity as the quantity

---

[75] The compositions of the solutions are given on page 237.

formally most consistent with the experimental e.m.f. method by which most pH values were (and still are) determined. In the same way, we should examine, in terms of the experimental quantity (pH), the several possible units for expressing acidity in alcohol-water solvents.

The following relationships can be derived from equation 33:

$$pm_H = pH - \bar{E}_j + \log {}_m\gamma_H + \log {}_s\gamma_H = pH - \delta + \log {}_s\gamma_H \tag{36}$$

$$pa_H{}^* \equiv -\log(m_H \cdot {}_s\gamma_H) = pH - \bar{E}_j + \log {}_m\gamma_H = pH - \delta \tag{37}$$

and

$$pa_H = pH - \bar{E}_j \tag{38}$$

where δ is written for $\bar{E}_j - \log {}_m\gamma_H$. This quantity is substantially constant for a solvent medium of given composition; it is obtained by subtracting $\log({}_m\gamma_H \cdot {}_m\gamma_{Cl})$ from the constant $\bar{E}_j + \log {}_m\gamma_{Cl}$ of equation 35.[76]

The values of the constant δ for methanol-water mixtures and for ethanol-water mixtures from four investigations are compared in Table 8–5.[77, 78] It may be seen that δ is small up to about 80 wt per cent alcohol in both solvent series. This is apparently because $\bar{E}_j$ and $\log {}_m\gamma_H$ compensate to a considerable extent in this region of solvent composition.

From the practical point of view, the pm_H unit is unsatisfactory in water-alcohol media, inasmuch as ${}_s\gamma_H$ in each solution would not be known (see equation 36). Likewise, the liquid-junction potential $\bar{E}_j$ (needed for a scale of pa_H) can only be obtained by empirical extrathermodynamic methods the validity of which will always be open to question, convincing though the circumstantial evidence in their favor may be. The $pa_H{}^*$ is evidently related most directly to the experimental quantity pH (see equation 37); $a_H{}^*$ is the hydrogen ion activity referred to the standard state in the mixed solvent.

---

[76] The quantity $\log ({}_m\gamma_H \cdot {}_m\gamma_{Cl}) = 2 \log {}_m\gamma_\pm$ is readily obtained from the difference between the standard e.m.f. of the cell H_2;$HCl(m)$, $AgCl$;Ag in water and in alcohol-water solvents by equation 27. The standard e.m.f. of this cell in a number of partially aqueous media has been summarized by H. D. Crockford in *Electrochemical Constants*, p. 153, National Bureau of Standards Circular 524, 1953; I.T. Oiwa, *J. Phys. Chem.*, **61**, 1587 (1957); D. Feakins and C. M. French, *J. Chem. Soc.*, 2581 (1957); G. J. Janz in Chapter 4, *Reference Electrodes*, D. J. G. Ives and G. J. Janz, eds., Academic Press, New York, 1961; and R. G. Bates, in *Hydrogen-Bonded Solvent Systems*, A. K. Covington and P. Jones, eds., p. 49, Taylor and Francis, London, 1968.

[77] Methanol-water: C. L. deLigny and M. Rehbach, *Rec. Trav. Chim.*, **79**, 727 (1960). These authors refer to pH $-$ $pa_H{}^*$ (that is, δ) as the "liquid-junction potential," whereas it is actually a combination of liquid-junction potential and the medium effect for hydrogen ion. Ethanol-water: W. J. Gelsema, C. L. deLigny, A. G. Remijnse, and H. A. Blijleven, *Rec. Trav. Chim.*, **85**, 647 (1966).

[78] B. Gutbezahl and E. Grunwald, *J. Amer. Chem. Soc.*, **75**, 565 (1953). The values of δ were obtained by subtracting their $\log {}_m\gamma_H$ from their $\bar{E}_j$ (termed $\log l$).

TABLE 8–5. VALUES OF THE CONSTANT $\delta \equiv (\bar{E}_j - \log_m \gamma_H)$ IN METHANOL-WATER AND ETHANOL-WATER SOLVENTS AT 25 °C, IN pH UNITS

Wt Per Cent Alcohol	Methanol-Water Solvents		Ethanol-Water Solvents		
	Ref. 77	Ref. 71	Ref. 78	Ref. 71	Ref. 77
0	0	0	0	0	0
20	0.01	0.01	−0.03	0.02	—
30	—	—	—	—	0.11
35	0.05	0.06	+0.07	0.10	—
50	0.13	0.13	0.17	0.21	0.29
65	0.15	0.14	0.19	0.24	—
71.89	—	—	—	—	0.33
80	−0.06	—	0.11	0.11	—
90	−0.51	—	−0.40	−0.40	—
100	−2.34	—	−2.36	−2.91	—

Furthermore, pa_H^* is a useful quantity; that is, it plays a simple and unequivocal role in chemical equilibria in alcoholic media. For the dissociation equilibrium

$$HA + SH \rightleftharpoons A + SH_2^+ \tag{39}$$

for example,

$$p(_sK_{HA}) = pa_H^* - \log \frac{m_A}{m_{HA}} - \log \frac{_s\gamma_A}{_s\gamma_{HA}} \tag{40}$$

Thus pK values and other thermodynamic constants referred to the standard state s can be obtained, and pa_H^* meets many of the requirements of a useful acidity scale in these solvents.

The simplest way to determine approximate pa_H^* values experimentally would be to apply tabulated δ corrections (such as those given in Table 8–5) to the pH numbers furnished by a pH meter standardized with aqueous buffer solutions in the usual way. Because of the possibility of a shifting asymmetry potential when the glass electrode is transferred from an aqueous medium to an alcoholic medium, however, it would probably be preferable to utilize standard solutions of known pa_H^*, having the same solvent composition as the "unknowns."

It is convenient therefore to define an operational unit pH*, analogous to the operational quantity pH for aqueous solutions:

$$pH^*(X) = pH^*(S) + \frac{E_X - E_S}{(RT \ln 10)/F} \tag{41}$$

Under ideal conditions of measurement, the operational pH* approaches the dimensions of pa_H* (the scale on which pH*(S) is based), just as the pH in water approaches pa_H under restricted experimental conditions.

Standard values of pH*, namely pH*(S), can be assigned by the same method used to establish the NBS standards pH(S) for the aqueous scale (Chapter 4). By using $_sE°$ instead of $_wE°$, one obtains $p_s(a_H\gamma_{Cl})$ from the e.m.f. of hydrogen-silver chloride cells containing the standard solutions (with added chloride) in the water-alcohol medium:

$$p_s(a_H\gamma_{Cl}) = \frac{E - _sE°}{(RT\ln 10)/F} + \log m_{Cl} \tag{42}$$

and then pa_H* by the nonthermodynamic step

$$pa_H* = p_s(a_H\gamma_{Cl}) + \log {_s\gamma_{Cl}} \tag{43}$$

As in the water medium, the effect of the added chloride on $p_s(a_H\gamma_{Cl})$ can be removed by extrapolation and the pa_H* of the chloride-free buffer solution can be obtained. The values of pa_H* for selected buffer solutions are then identified with pH*(S) in the operational definition of pH* (equation 41).

The procedure set forth in equations 42 and 43 has been applied by Parks, Crockford, and Knight[79] to determine pa_H* for certain citrate and phosphate buffer solutions (with added chloride) in 10 and 20 wt per cent methanol. Likewise, deLigny and his associates[80] have determined pa_H* for an oxalate buffer solution and a succinate buffer solution in water-methanol and water-ethanol solvents. In addition, a salicylate buffer was studied in pure methanol and pure ethanol. For all three buffers in anhydrous methanol, the pa_H* values from the hydrogen-silver chloride cell agreed well with those derived from the e.m.f. of the hydrogen-silver bromide cell. The mean difference was, in fact, less than 0.01 pH unit.

In both investigations values of $\log {_s\gamma_{Cl}}$ for use in equation 43 were computed by a form of the Debye-Hückel equation. Parks, Crockford, and Knight chose the two-parameter form, estimating the ion size and "salting-out constant" from data in the literature.[81] Their values of pa_H* are based

[79] R. L. Parks, H. D. Crockford, and S. B. Knight, *J. Elisha Mitchell Sci. Soc.*, **73**, 289 (1957).

[80] C. L. deLigny, P. F. M. Luykx, M. Rehbach, and A. A. Wieneke, *Rec. Trav. Chim.*, **79**, 699, 713 (1960); C. L. deLigny and P. F. M. Luykx, *ibid.*, **77**, 154 (1958); W. J. Gelsema, thesis, Utrecht, 1964.

[81] See H. D. Crockford in *Electrochemical Constants*, p. 153, National Bureau of Standards Circular 524, 1953; J. F. Masi and S. B. Knight, *J. Amer. Chem. Soc.*, **67**, 1558 (1945). In his studies of acid-base behavior in water-acetone and water-ethanol solvents, R. Reynaud, *Bull. Soc. Chim. France*, 2686 (1967), standardized the glass-calomel cell with solutions of HCl on the scale of pa_H*. The activity coefficient $_s\gamma_H$ was calculated by a form of the Debye-Hückel equation.

TABLE 8–6. Reference Values of pH*(S) in Water-Methanol Solvents at 25 °C[a]

% Methanol (w/w)	Oxalate Buffer[b]	Succinate Buffer[c]
0	2.15	4.12
10	2.19	4.30
20	2.25	4.48
30	2.30	4.67
40	2.38	4.87
50	2.47	5.07
60	2.58	5.30
70	2.76	5.57
80	3.13	6.01
90	3.73	6.73
100	5.79	8.75

[a] Data of C. L. deLigny, P. F. M. Luykx, M. Rehbach, and A. A. Wieneke, *Rec. Trav. Chim.*, **79**, 713 (1960).
[b] Prepared from oxalic acid and ammonium hydrogen oxalate, each at a molality of 0.01 mol kg^{-1}.
[c] Prepared from succinic acid and lithium hydrogen succinate, each at a molality of 0.01 mol kg^{-1}.

on the molar scale of activity. DeLigny and his associates used the "extended terms" equation of Gronwall, LaMer, and Sandved.[82] Corrections were made for ion pair formation in the pure alcohols. Reference values of pH*(S) recommended by deLigny and his associates for use in water-methanol and water-ethanol solvents are summarized in Tables 8–6 and 8–7.[83]

Reference values of pH*(S) have also been assigned to three buffer solutions in 50 wt per cent methanol over the temperature range 10 to 40 °C.[84] In this work, the convention adopted for the evaluation of $_s\gamma_{Cl}$ in equation 43

[82] T. H. Gronwall, V. K. LaMer, and K. Sandved, *Physik. Z.*, **29**, 358 (1928).
[83] The values of $_sE°$ on which the pa$_H$* in methanol-water solvents given by deLigny et al. is based are those of J. M. Austin, A. H. Hunt, F. A. Johnson, and H. N. Parton (unpublished work cited by R. A. Robinson and R. H. Stokes, *Electrolyte Solutions*, 2nd ed. revised, Appendix 8.2, Butterworths, London, 1970). M. Paabo, R. G. Bates, and R. A. Robinson, *Anal. Chem.*, **37**, 462 (1965), interpolated $_sE°$ from a smooth plot of all available data as a function of weight per cent methanol. The two sets of standard e.m.f. values agree well at low methanol concentrations but differ by about 2.5 mV at 68 per cent methanol.
[84] M. Paabo, R. A. Robinson, and R. G. Bates, *J. Amer. Chem. Soc.*, **87**, 415 (1965).

TABLE 8-7. Reference Values of pH*(S) in Water-Ethanol Solvents at 25 °C[a]

% Ethanol (w/w)	Oxalate Buffer[b]	Succinate Buffer[c]
0	2.146	4.113
30	2.322	4.692
50	2.502	5.064
71.89	2.971	5.697

	Salicylate Buffer[d]	Diethylbarbiturate Buffer[e]
100	8.302	13.232

[a] Data of W. J. Gelsema, thesis, Utrecht, 1964.
[b] Prepared from oxalic acid and lithium hydrogen oxalate, each at a molality of 0.00998 mol kg^{-1} in water and 0.01 mol kg^{-1} in the mixed solvents.
[c] Prepared from succinic acid and lithium hydrogen succinate, each at a molality of 0.00998 mol kg^{-1} in water, 0.00994 mol kg^{-1} in 30% ethanol, and 0.01 mol kg^{-1} in the other two mixed solvents.
[d] Prepared from salicylic acid (molality 0.01006 mol kg^{-1}) and lithium salicylate (molality 0.00985 mol kg^{-1}).
[e] Prepared from diethylbarbituric acid (molality 0.00990 mol kg^{-1}) and lithium diethylbarbiturate (molality 0.01006 mol kg^{-1}).

was a modification of that employed in the establishment of pH standards in water:

$$-\log {}_s\gamma_{Cl} = \frac{AI^{1/2}}{1 + B\mathring{a}I^{1/2}} \tag{44}$$

In the pH convention, $B\mathring{a}$ is assigned the value 1.5 at all temperatures; thus $\mathring{a} = 4.56$ Å in water at 25 °C. This value of $\mathring{a}$ was retained in the convention for ${}_s\gamma_{Cl}$ in methanol-water solvents, but the constants A and B were given the values appropriate to the dielectric constant and density $d°$ of the solvent mixtures and the temperature, namely

$$A = \frac{1.825 \times 10^6}{(\varepsilon T)^{3/2}} d°^{1/2} \tag{45}$$

and

$$B = \frac{50.29}{(\varepsilon T)^{1/2}} d^{\circ 1/2} \tag{46}$$

The resulting standard reference values of pH*(S) are given in Table 8–8.

TABLE 8–8. REFERENCE VALUES OF pH*(S) IN THE 50% (w/w) MIXTURE OF WATER AND METHANOL FROM 10 TO 40 °C[a, b]

t, °C	Acetate Buffer	Succinate Buffer	Phosphate Buffer
10	5.518	5.720	7.937
15	5.506	5.697	7.916
20	5.498	5.680	7.898
25	5.493	5.666	7.884
30	5.493	5.656	7.872
35	5.496	5.650	7.863
40	5.502	5.648	7.858

[a] Data of M. Paabo, R. A. Robinson, and R. G. Bates, *J. Amer. Chem. Soc.*, **87**, 415 (1965).
[b] Compositions of the buffer solutions (in mol kg^{-1}):
Acetate: acetic acid (0.05), sodium acetate (0.05), sodium chloride (0.05).
Succinate: sodium hydrogen succinate (0.05), sodium chloride (0.05).
Phosphate: potassium dihydrogen phosphate (0.02), disodium hydrogen phosphate (0.02), sodium chloride (0.02).

Nature of the pH* Scale

The pH* is a practical, useful unit consistent in every way both with the experimental methods for pH measurement and with the thermodynamic equations for acid-base equilibria in the solvents concerned. Although it reduces to pH when the medium is diluted with a large amount of water, it does not relate the proton activity directly to the water scale; pH* is a succession of scales rather than a single universal scale of acidity. Consequently, two solutions in solvent media of different compositions may have the same pH* yet behave in a totally different manner in acid-base reactions.

The differences among three acidity scales are illustrated in Fig. 8–12, where pm_H, pa_H*, and pa_H of a dilute hydrochloric acid-sodium chloride solution, an acetate-chloride buffer, and a triethanolamine-chloride buffer

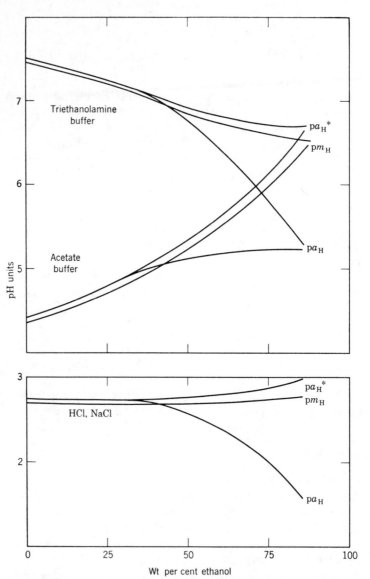

FIG. 8–12. Comparison of pa_H^*, pm_H, and pa_H for a strong acid-salt mixture and for two buffer solutions of different charge types in ethanol-water solvents.

are plotted as a function of the composition of the ethanol-water solvent.[85] The relative positions of the curves emphasize that pa_H^* and pH* are, in effect, pm_H values corrected for salt effects ($_s\gamma_H$). On the other hand, pa_H includes not only the salt effect but also the medium effect $_m\gamma_H$ which represents the change in the free energy of the proton on transfer from the aqueous medium to an ethanol-water solvent. This medium effect may be quite large at high ethanol concentrations. Evidently the pH* value fails completely to indicate the increased acidity (proton availability) that accompanies the replacement of water by a solvent of lower dielectric constant and probably somewhat lower basicity.

Nevertheless, it is unfortunately true that a single pH scale, referred to the aqueous standard state, cannot be realized at the present time. As we have seen, the establishment of standards for " universal pH " in a variety of media will require the adoption of a nonthermodynamic procedure permitting the independent evaluation of log $_m\gamma_H$ or $\bar{E}_j$ in all of the media of interest. At best this will be a very laborious procedure. Until it can be accomplished, the pH* scale can be used to advantage in the measurement of acidity in many nonaqueous and partially aqueous media.

Measurement of pD in Deuterium Oxide

A reversible deuterium gas electrode is easily set up by the same procedures which lead to a reversible hydrogen electrode (see Chapter 10). Likewise, the silver-silver halide electrodes function well in deuterium oxide; hence, the tools are available for the accurate investigation of acid-base equilibria in heavy water and for the establishment of a standard pD scale. It appears that the glass electrode develops a theoretical response to both hydrogen and deuterium ions;[86] therefore no barrier to practical measurements of pD remains.

Acidity functions $p(a_D\gamma_{Cl})$ for suitable standard reference solutions in deuterium oxide have been obtained from e.m.f. measurements of cells of the type

$$Pt; D_2(g, 1 \text{ atm}), buffer, Cl^- \text{ in } D_2O, AgCl; Ag \qquad (47)$$

---

[85] The calculations were based on the e.m.f. data of R. G. Bates and G. Schwarzenbach, *Helv. Chim. Acta*, **38**, 699 (1955). To derive pa_H, the values of log $_m\gamma_H$ given by B. Gutbezahl and E. Grunwald, *J. Amer. Chem. Soc.*, **75**, 565 (1953), were subtracted from pa_H^*. Likewise, pm_H was obtained from pa_H^* by adding log $_s\gamma_H$ estimated by the Debye-Hückel equation. The slight upward trend in pm_H for the mixtures rich in ethanol is doubtless indicative of ion pair formation.

[86] P. K. Glasoe and F. A. Long, *J. Phys. Chem.*, **64**, 188 (1960); P. R. Hammond, *Chem. Ind.*, (*London*), 311 (1962); A. K. Covington, M. Paabo, R. A. Robinson, and R. G. Bates, *Anal. Chem.*, **40**, 700 (1968).

for which the standard potential is known over a range of temperatures.[87] It was therefore possible to derive conventional pa_D values for the solutions by the relationship (compare equation 43)

$$pa_D = p(a_D\gamma_{Cl}) + \log \gamma_{Cl} \tag{48}$$

The convention chosen was the analog of that used in establishing the conventional pa_H^* in mixed solvents. The values of the Debye-Hückel constants A and B in the equation for $\log \gamma_{Cl}$ (see equation 44) were calculated by equations 45 and 46 to correspond with the temperature of measurement and with the dielectric constant and density of deuterium oxide. The value of 4.56 Å for å was retained.

TABLE 8–9. REFERENCE VALUES OF pD(S) IN DEUTERIUM OXIDE
FROM 5 TO 50 °C[a]

t, °C	Citrate Buffer[b]	Phosphate Buffer[b]	Carbonate Buffer[b]
5	4.378	7.539	10.998
10	4.352	7.504	10.924
15	4.329	7.475	10.855
20	4.310	7.449	10.793
25	4.293	7.428	10.736
30	4.279	7.411	10.685
35	4.268	7.397	10.638
40	4.260	7.387	10.597
45	4.253	7.381	10.560
50	4.250	7.377	10.527

[a] Data of M. Paabo and R. G. Bates, *Anal. Chem.*, **41**, 283 (1969).
[b] Compositions (in mol kg^{-1}):
Citrate buffer: $KD_2C_6H_5O_7$ (0.05)
Phosphate buffer: KD_2PO_4 (0.025), Na_2DPO_4 (0.025)
Carbonate buffer: $NaDCO_3$ (0.025), Na_2CO_3 (0.025)

The conventional pa_D values for a citrate buffer, a phosphate buffer, and a carbonate buffer in deuterium oxide have been determined in this way[88] and are listed in Table 8–9. The solutions can be prepared by dissolving the protiated salts in pure deuterium oxide; the effect of isotopic exchange on the

[87] R. Gary, R. G. Bates, and R. A. Robinson, *J. Phys. Chem.*, **68**, 1186 (1964).
[88] R. Gary, R. G. Bates, and R. A. Robinson, *J. Phys. Chem.*, **68**, 3806 (1964); **69**, 2750 (1965); M. Paabo and R. G. Bates, *Anal. Chem.*, **41**, 283 (1969).

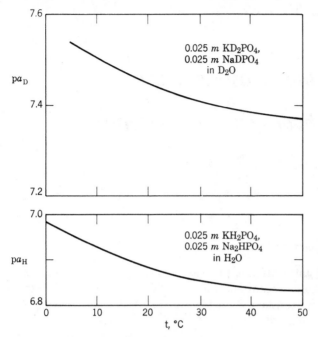

FIG. 8–13. Comparison of the pa_D for a phosphate buffer solution in deuterium oxide with the pa_H of a solution of the same molal composition in water.

pa_D of the reference solutions is less than the estimated uncertainty of 0.006 unit in the values themselves. By identification of pa_D with pD(S) in the operational formula

$$pD(X) = pD(S) + \frac{E_X - E_S}{(RT \ln 10)/F} \tag{49}$$

these standard solutions form the basis for an operational definition of pD entirely analogous to the definition of pH now adopted internationally.

Fundamentally, this pD scale is based on the concept of a standard deuterium gas electrode for which the standard potential is zero at all temperatures. Consequently, pD values and pH values cannot be strictly compared. The pa_D values for the phosphate buffer in deuterium oxide and the pa_H values for a phosphate buffer of the same molal composition in water are plotted as a function of temperature in Fig. 8–13. It will be noted that the pa_D values lie about 0.6 unit higher than the pa_H values. We shall have more to say concerning the behavior of the glass electrode in deuterium oxide in Chapter 11.

CHAPTER 9

the measurement of
hydrogen ion concentration

Equilibrium Constants

A significant fraction of pH measurements has as its object the estimation of an equilibrium constant for an acid-base process. The thermodynamic equilibrium constant K on the molal scale for the reaction $A \rightleftharpoons B + H^+$ is formulated

$$K = a_{H^+} \frac{m_B \gamma_B}{m_A \gamma_A} \tag{1}$$

where, for generality, the charges are omitted. Its calculation from pH ($-\log a_H$) requires not only a knowledge of the equilibrium concentrations of the acid and its conjugate base but also an evaluation of the activity coefficients of these species. Inasmuch as the concentrations (but not the activities) are usually known, many investigators[1] have chosen to use the "incomplete" ("unvollständige") constant K'' defined by

$$K'' = a_H \frac{m_B}{m_A} = K \frac{\gamma_A}{\gamma_B} \tag{2}$$

This quantity has the inherent disadvantage that it reflects, through the inclusion of γ_H, the influence of small changes in ionic environment, although interionic effects on γ_B and γ_A are presumably excluded. For this reason K'' is sometimes regarded as less useful than K or k, the concentration constant,

$$k = \frac{m_H m_B}{m_A} = K \frac{\gamma_A}{\gamma_H \gamma_B} \tag{3}$$

[1] J. N. Brønsted, Chem. Rev., 5, 293 (1928).

The thermodynamic constant K is dependent on temperature, pressure, the scale of concentration, and the chosen standard state; K'' and k, on the other hand, vary with the ionic strength and composition of the medium. A knowledge of k in a particular medium, however, may often lead readily to the useful quantity m_H. Conversely, an experimental measure of m_H may permit the meaningful concentration constant k to be derived. Depending on the choice of standard state, activities may depart widely from actual concentrations. When the composition of the solvent medium is fixed, therefore, the use of m_H and k has much to recommend it.

Biological Media

For measurements in certain aqueous media, a considerable difference of opinion exists as to the relative usefulness of the hydrogen ion activity and the hydrogen ion concentration.[2] Some investigators engaged in assessing the acid-base balance of the human body, for example, advocate " unlogging " the pH number and the application of various activity coefficient equations to convert the activity so obtained into concentrations expressed in millimoles of hydrogen ion per liter. Physiological processes, it is argued, are governed by the concentrations of electrolytes rather than by their activities. On the other hand, it is often pointed out that the pH value is linearly related to the chemical potential of hydrogen ions and is thus the parameter of importance in fixing the state of any equilibrium process.[3] These matters have been considered in detail in a symposium on concepts and methods of acid-base measurement in biomedicine.[4]

As we shall see in this chapter, the direct measurement of hydrogen ion concentrations with a pH meter of the glass-calomel type is possible when both the standard and unknown solutions are closely similar in their constitution. The solvent must be the same in both, and the nature and concentration of the active (acidic and basic) species must be small relative to a background of neutral electrolytes, the same in each solution, which provide a fixed ionic strength. Isotonic saline, a 0.16 M solution of sodium chloride, and many biological media meet the requirements of a constant ionic medium rather well. The establishment of standards for c_H or m_H in isotonic saline and other neutral salt media thus becomes a matter of considerable importance for biomedical and clinical research.

---

[2] O. Siggaard-Andersen, *The Acid-Base Status of the Blood*, 2nd ed., Munksgaard, Copenhagen, 1964. See also the review by K. Schwabe, *Verh. Exp. Med.*, **1**, 56 (1962).
[3] G. F. Filley, *Acid-Base and Blood Gas Regulation*, Lea and Febiger, Philadelphia, 1971.
[4] *Current Concepts of Acid-Base Measurement*, Ann. N. Y. Acad. Sci., **133**, 1 (1966).

Constant Ionic Media

Coordination chemists, especially, have given impetus to the experimental determination of hydrogen ion concentrations. Many of the most useful chelating agents are polycarboxylic acids; among these may be mentioned nitrilotriacetic acid (NTA) and ethylenediaminetetraacetic acid (EDTA). When the ligand is the conjugate base of a weak acid, the formation of metal-ligand complexes may be accompanied by a release of protons. The competition between metal ions and protons for the coordination sites of the ligand is a function of the dissociation constants of the acidic groups and the stability constant of the metal complex.

In his classic study,[5] Jannik Bjerrum showed that the step-wise equilibrium constants for complexes formed from a metal and a ligand that is the conjugate base of a weak acid can be obtained from the pH of solutions containing known total concentrations of metal and the ligand acid. Thus, for example, the stability of the calcium complex of EDTA (CaY^{2-}) can be studied by means of measurements of the concentration of hydrogen ion in the equilibrium system

$$Ca^{2+} + H_2Y^{2-} \rightleftharpoons CaY^{2-} + 2H^+$$

It is evident that the stability of the complex is a function of pH when a competition between protons and metal ions exists.

In order to facilitate the determination of the relative stabilities of metal-ligand complexes, an experimental procedure for estimating hydrogen ion concentrations was sought. Following the work of Bjerrum[5] and Schwarzenbach,[6] coordination chemists quite generally have evaluated the step-wise equilibrium constants from measurements of hydrogen ion concentration in "constant ionic media."[7] These procedures rely upon the effectiveness of the neutral salt medium in "swamping out" changes in the activity coefficients of the charged species when small changes in concentration occur and in equalizing the activity coefficients of ions of like charge present at small concentrations in the relatively concentrated ionic medium. It is claimed[8] that this device originated with Bodländer but was first described in published work by Grossmann.[9]

[5] J. Bjerrum, *Metal Ammine Formation in Aqueous Solution*, P. Haase and Son, Copenhagen, 1941.

[6] G. Schwarzenbach, *Helv. Chim. Acta*, **33**, 947 (1950).

[7] A discussion of the advantages and limitations of media of constant ionic strength is given in Chapter 2 of F. J. C. Rossotti and H. Rossotti, *The Determination of Stabiilty Constants*, McGraw-Hill Book Co., New York, 1961. See also H. Ohtaki and G. Biedermann, *Bull. Chem. Soc. Japan*, **44**, 1515 (1971), and J. Lagrange, *Ann. Chim. (Paris)*, **6**, 125 (1971).

[8] G. Biedermann and L. G. Sillén, *Ark. Kemi*, **5**, 425 (1953).

[9] H. Grossmann, *Z. Anorg. Chem.*, **43**, 356 (1905).

The Brønsted principle of specific ionic interaction[10] leads one to expect that the activity coefficients of ionic solutes, present in low concentrations in a moderately concentrated neutral salt medium, will remain unchanged as small alterations of composition occur at substantially constant ionic strength. Unfortunately, the "swamping" of activity coefficient variations is not complete. Long-range electrostatic interactions between the ions are doubtless greatly reduced, but specific interactions with the supposedly inert ionic medium are certainly often of importance. For example, the data of Prue and Read[11] demonstrate some variability in the activity coefficient of hydrochloric acid (molality m_1) in concentrated sodium perchlorate solutions (molality m_2), as shown in Table 9–1.

TABLE 9–1. ACTIVITY COEFFICIENTS OF HYDRO-
CHLORIC ACID (MOLALITY m_1) IN CONCENTRATED
SOLUTIONS OF SODIUM PERCHLORATE (MOLALITY
m_2) AT 25 °C

m_1	$\gamma_\pm$ $(m_1 + m_2 = 2.8545)$	$\gamma_\pm$ $(m_2 = 2.8545)$
0.01	1.056	1.057
0.1	1.065	1.087

When constant ionic strength is maintained, the activity coefficient changes by only 1 per cent as m_1 increases from 0.01 to 0.1; when the molality (m_2) of the neutral salt is fixed at 2.8545 mol kg^{-1}, however, the same increase in m_1 produces a change of 3 per cent in the activity coefficient.

These changes in the activity coefficient can be kept small by reducing the concentration of reactive species relative to the concentration of the constant ionic medium. Nevertheless, complete swamping of activity coefficient variations is probably rarely achieved, and it has been claimed[12] that only about half of the equilibrium constants determined in constant ionic media will have lasting significance. Care in the design of the measurement system is essential if precise valid results are to be obtained. Furthermore, it must be recognized that the equilibrium constants and other thermodynamic data apply only in the selected salt medium and are not necessarily indicative of relative stabilities in the usual aqueous standard state. Biedermann and

[10] J. N. Brønsted, *J. Amer. Chem. Soc.*, **44**, 877 (1922).
[11] J. E. Prue and A. J. Read, *J. Chem. Soc. (A)*, 1812, (1966).
[12] T. F. Young and A. C. Jones, *Ann. Rev. Phys. Chem.*, **3**, 287 (1952).

Sillén[8, 13] have evaluated carefully the best experimental designs and hav examined the assumptions used in calculating ion concentrations from e.m. measurements in $NaClO_4$-$HClO_4$ ionic media.

Cells Without Liquid Junction

We have already indicated that the cell without liquid junction

$$Pt; H_2(g), Buffer\ soln., Cl^-, AgCl; Ag$$

is extremely useful in establishing standard reference solutions for pH and i deriving other related quantities, such as activity coefficients and dissociatio constants. Unfortunately, neither m_H nor $m_H \gamma_\pm^2$ in solutions of unknow composition and ionic strength can be obtained readily from the e.m.f. o cells of this type. It is possible, however, to determine values that close] approximate c_H or m_H by e.m.f. measurements of certain other cells. Hitcl cock[14] has discussed a suggestion of Scatchard that the hydrogen ion co centration be determined by measurements of the e.m.f. of a cell reversib not only to hydrogen ions but also to another univalent cation. The determ nation rests on the approximate equivalence of the activity coefficients of th two ions of the same charge in solutions of moderate concentrations.

Redlich and Klinger[15] have studied a cell of this type, namely

$$Pt; H_2(g), Soln.\ X, TlBr(satd.), AgBr; Ag$$

the e.m.f. of which is given by

$$E = E^\circ - \frac{RT}{F} \ln m_H \gamma_H \frac{K_{sp}}{a_{Tl}} \qquad (\text{·}$$

$$= E^\circ - \frac{RT}{F} \ln \frac{K_{sp}}{m_{Tl}} - \frac{RT}{F} \ln m_H \frac{\gamma_H}{\gamma_{Tl}} \qquad (\text{·}$$

When the solubility m_{Tl} and the solubility product constant K_{sp} of thallo\u{} bromide in the solution are known, $m_H \gamma_H / \gamma_{Tl}$ can be determined. Th quantity should be approximately equal to m_H at low and moderate conce trations, for the hydrogen ion and the thallous ion bear the same charge an their concentrations are small compared with the total salt concentration.

Another cell reversible to two univalent cations has been investigated b Maronny and Valensi.[16] The cell utilized potassium amalgam and hydroge

[13] See also Chapter 7 of F. J. C. Rossotti and H. Rossotti, *The Determination of Stabili Constants*, McGraw-Hill Book Co., New York, 1961.

[14] D. I. Hitchcock, *J. Amer. Chem. Soc.*, **58**, 855 (1936).

[15] O. Redlich and H. Klinger, *J. Amer. Chem. Soc.*, **61**, 2983 (1939).

[16] G. Maronny and G. Valensi, *J. Chim. Phys.*, **49**, C91 (1952).

electrodes and was used by these authors as a basis for the establishment of pH standards.

When the constant ionic medium contains an anion such as chloride to which the reference electrode is reversible, it is possible to determine hydrogen ion concentrations, within certain limitations, from the e.m.f. of cells without liquid junction. For example, Hitchcock and Peters[17] have determined the hydrogen ion concentration of acetate and phosphate buffer solutions at 25 and 38 °C from measurements of the cell

$$\text{Glass electrode; Buffer soln.} + \text{NaCl}(I = 0.16), \text{AgCl; Ag} \qquad (6)$$

The concentration of buffer was low, so that the molality of chloride and the ionic strength were both close to 0.16. Under these circumstances, the activity coefficients of hydrogen and chloride ions were regarded as being unchanged, regardless of the nature of the buffer ions. Hence,

$$E = E^{\circ\prime} - \frac{RT}{F} \ln (c_H c_{Cl}) \qquad (7)$$

where c represents concentration in mol dm^{-3}. The standard of reference used to determine $E^{\circ\prime}$ was a mixture of hydrochloric acid (0.01 M) and sodium chloride (0.15 M) in which $c_H c_{Cl}$ is 0.0016. Similar cells with a hydrogen electrode or a glass electrode combined with a silver-silver chloride electrode have been used to determine acidity constants of ligands and the stabilities of metal-ligand complexes in 0.1 M KCl.[18, 19]

Because of the absence of a liquid junction, this procedure seems deserving of wider application than it has received. It should be especially suitable for measurements in solutions of sodium or potassium chlorides of molality 0.5 to 2 mol kg^{-1}. At lower ionic strengths, there is a possibility that variations in activity coefficients are imperfectly swamped out, while at higher chloride concentrations the solubility of silver chloride in the constant ionic medium may be troublesome. Moreover, this method appears to offer a promising means of establishing reference buffer solutions for hydrogen ion concentration with which to standardize the usual glass-calomel cell with liquid junction in constant ionic media of fixed chloride concentration.

Cells without liquid junction can also be used to obtain hydrogen ion concentrations in constant ionic media consisting of perchlorate or nitrate salts in the manner described by McBryde.[20] In order to utilize the silver-silver chloride electrode for measurements of this sort, a constant low

---

[17] D. I. Hitchcock and R. Peters, *J. Amer. Chem. Soc.*, **68**, 1753 (1946).

[18] G. Schwarzenbach, A. Willi, and R. O. Bach, *Helv. Chim. Acta*, **30**, 1303 (1947).

[19] R. F. Lumb and A. E. Martell, *J. Phys. Chem.*, **57**, 690 (1953).

[20] W. A. E. McBryde, *Analyst*, **94**, 337 (1969); **96**, 739 (1971).

concentration (0.001 M) of sodium chloride was added to each of the solu
tions. This amount was so small that it had no perceptible effect on the pH

If the glass electrode responds perfectly to hydrogen ions, changes in the
e.m.f. E of the glass electrode-silver chloride electrode combination will obey
the relationship

$$E = E^{\circ\prime} - \frac{RT \ln 10}{F} \log(m_H \gamma_H \gamma_{Cl}) \tag{8}$$

under these conditions, from which

$$\frac{EF}{RT \ln 10} = C + p(a_H \gamma_{Cl}) \tag{9}$$

The constant C on the right side of equation 9 includes both the standard
e.m.f. of the cell and the molality of chloride. Indeed, when the hydrogen
electrode is substituted for the glass electrode, this constant become
$(E^{\circ}F)/(RT \ln 10) - \log m_{Cl}$, or $(0.2224/0.05916) + 3$ volts when the tem
perature is 25 °C and the molality of the chloride ion is 0.001 mol kg^{-1}
Ordinarily, the standard e.m.f. of a cell with glass electrode is unknown. The
constant C can be calculated readily, however, from a measurement of E in
buffer solutions of known $p(a_H \gamma_{Cl})$. Values of this acidity function have been
tabulated in Table 7 of the Appendix.[21] The feasibility of deriving precise pH
values from titrations performed in cells without liquid junction in solutions of
constant ionic strength and fixed chloride ion concentration was demonstrated
many years ago.[22]

If a constant ionic medium is chosen and it is assumed further that the
activity coefficients of hydrogen and chloride ions have fixed values therein
equation 9 can be rewritten

$$\frac{EF}{RT \ln 10} = C' + pm_H \tag{10}$$

In McBryde's terminology, $m_H = [H]$, and antilog $p(a_H \gamma_{Cl})$ is designated
simply H'. The ratio H'/[H] thus represents $\gamma_H \gamma_{Cl}$ in the neutral salt medium
Values of this ratio in potassium nitrate solutions and sodium perchlorate
solutions at various ionic strengths are summarized in Table 9–2. The
solutions studied were composed of strong acids (nitric or perchloric) or
weak acid buffers, and the pH varied from 1 to 7. The e.m.f. was measured
with three different pH electrometers with glass electrodes from three
different sources—Electronic Instruments Ltd., Radiometer, and Beckman
The assigned uncertainties represent the standard error (s.e.) of the mean of
9 to 11 determinations.

[21] R. G. Bates and R. Gary, J. Res. Nat. Bur. Stand., 65A, 495 (1961).
[22] R. G. Bates, G. L. Siegel, and S. F. Acree, J. Res. Nat. Bur. Stand., 30, 347 (1943).

TABLE 9-2. Values of H'/[H] in Solutions of Potassium Nitrate and Sodium Perchlorate from Measurements of Cells without Liquid Junction

	(Data of McBryde[20])			
	Potassium Nitrate		Sodium Perchlorate	
I	Mean	s.e.	Mean	s.e.
0.05	0.677	0.003	0.677	0.004
0.1	0.621	0.004	0.635	0.004
0.2	0.547	0.003	0.590	0.003
0.5	0.500	0.003	0.576	0.004
1.0	0.505	0.004	0.670	0.008
3.0	—	—	2.024	0.017

Cells with Liquid Junction

For practical reasons, most measurements of hydrogen ion concentration or activity are made with the usual glass electrode pH meter utilizing a reference electrode of the calomel or silver-silver chloride type and a salt bridge which makes a liquid junction with the test solution. When protonation constants are being determined by potentiometric titrations with alkali, it is often a routine practice[23] to titrate acetic acid and to use the known dissociation constant of this acid in a practical standardization of the cell.

When cells with liquid junction are employed, constant ionic media, especially those containing high concentrations of potassium chloride, offer some advantages over dilute solutions. McBryde[20] attributes the enhanced precision he observed with potassium chloride to the effectiveness of this salt in reducing and stabilizing the liquid-junction potential. Sodium perchlorate solutions gave pH readings of excellent stability when the salt bridge of potassium chloride was separated from the solution containing perchlorate by a tube filled with 3.5 M ammonium chloride in 3 per cent agar. The precision of the results, however, was inferior to that found when the constant ionic medium was composed of potassium chloride.

As we have seen in Chapter 2, the operational definition of pH acquires its form from the Nernst equation for the pH cell with hydrogen and calomel electrodes and a liquid junction. The difference of pH, defined formally as

---

[23] G. Schwarzenbach, E. Kampitsch, and R. Steiner, *Helv. Chim. Acta*, **28**, 828, 1133 (1945); **29**, 364 (1946).

$-\log a_H$, between an "unknown" solution X and a reference standard S would be correctly given by the operational formula

$$pH(X) = pH(S) + \frac{(E_X - E_S)F}{RT \ln 10} \qquad (11)$$

if the residual liquid-junction potential $E_j(X) - E_j(S) = \Delta E_j$ were zero. The same equation would hold, of course, if an error-free glass electrode were substituted for the hydrogen electrode. Gross differences in composition or ionic strength between the unknown X and the standard S of conventional hydrogen ion activity (a_H) will enhance the residual liquid-junction potential. It will be remembered that this quantity was expressly omitted in formulating the operational definition of the pH of an unknown, pH(X). Hence it is to be expected that the measured pH will not correctly represent the "true" hydrogen ion activity but will be in error by the amount $[E_j(X) - E_j(S)]/[RT \ln 10)/F]$, which will be designated $\Delta \bar{E}_j$. Thus,

$$pH(X) = pa_H + \Delta \bar{E}_j = pm_H + \Delta \bar{E}_j - \log \gamma_H \qquad (12)$$

or

$$pH(X) = pm_H - A \qquad (13)$$

The quantity A may be regarded as a correction term, by application of which experimental pH values can be converted into hydrogen ion concentrations. From measurements with a variety of acid-base systems in constant ionic media, it has been found [20, 24, 25] that A is a constant characteristic of the salt medium. Evidently this constancy is an indication of the effectiveness of the medium in equalizing the liquid-junction potential $(\bar{E}_j)$ and in reducing variations in the activity coefficients (γ_H). Rossotti and Rossotti[26] have made the useful suggestion that constancy of the e.m.f. as the concentration of the salt bridge is altered may provide a confirmation that liquid-junction errors are not present. Some values of A, taken from the papers of Rosenthal and coworkers[24] and Critchfield and Johnson,[27] are collected in Table 9–3.

[24] D. Rosenthal and J. S. Dwyer, J. Phys. Chem., 66, 2687 (1962); Anal. Chem., 35, 161 (1963). J. S. Dwyer and D. Rosenthal, J. Phys. Chem., 67, 779 (1963). D. Rosenthal, I. T. Oiwa, A. D. Saxton, and L. R. Lieto, ibid., 69, 1588 (1965). W. Proudlock and D. Rosenthal, ibid., 73, 1695 (1969).

[25] H. M. Irving, M. G. Miles, and L. D. Pettit, Anal. Chim. Acta, 38, 475 (1967).

[26] F. J. C. Rossotti and H. Rossotti, The Determination of Stability Constants, p. 147, McGraw-Hill Book Co., New York, 1961.

[27] F. E. Critchfield and J. B. Johnson, Anal. Chem., 31, 570 (1959).

TABLE 9–3. VALUES OF THE CONSTANT A IN SALT SOLUTIONS AT 25 °C; $A = pm_H - pH(X)$.

Salt	$c/\text{mol dm}^{-3}$	A (pH units)
LiCl	1	0.12
	2	0.35
	3	0.64
	4	0.94
	5	1.25
	6	1.61
	7	1.95
	8	2.30
NaCl	4	0.73
NaI	4	0.90
$NaClO_4$	6	1.84
$NaNO_3$	6	0.82
KCl	4	0.40
$CaCl_2$	2.67	1.11
	4	1.97
$MgBr_2$	2.67	1.58

Magnitude of the Liquid-Junction Potential

The influence of the liquid-junction potential on measurements of hydrogen ion concentration is clarified by an important study of Biedermann and Sillén.[28, 29] These authors utilized cells of the type

$$\text{Hg; Hg}_2\text{Cl}_2,\text{3 M NaClO}_4 \quad \left| \quad \begin{array}{ll} \text{Soln. X} & \text{Electrode} \\ \text{HClO}_4\ (c_H) & \text{reversible to} \\ \text{NaClO}_4\ (3-c_H), & \text{H}^+, \text{Me}^{n+}, \text{Cl}^-, \\ & \text{or Br}^- \end{array} \right. \quad (14)$$

For some of the measurements, small amounts of metal or halide ions were present in Solution X; under these circumstances the concentrations of sodium perchlorate and perchloric acid were lowered slightly to maintain the ionic strength constant at 3 mol dm^{-3}. Eight types of electrode, reversible to H^+, Cu^{2+}, Hg_2^{2+}, Ag^+, Cl^-, Br^-, and the redox couple Fe^{3+}, Fe^{2+} were

[28] G. Biedermann and L. G. Sillén, *Ark. Kemi*, **5**, 425 (1953). F. J. C. Rossotti and H. Rossotti, *Acta Chem. Scand.*, **10**, 779, 957 (1956).
[29] See also G. Faraglia, F. J. C. Rossotti, and H. Rossotti, *Inorg. Chem. Acta*, **4**, 488 (1970), and Chapter 7 of the book by Rossotti and Rossotti (reference 13).

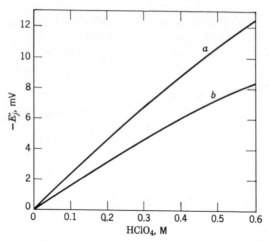

FIG. 9–1. Apparent liquid-junction potentials between 3M NaClO₄ and mixtures of NaClO₄ with HClO₄ (total concentration 3M). Data of Biedermann and Sillén.[28] Curve a, electrodes reversible to Cl^- and Br^-; curve b, electrodes reversible to H^+, Cu^{2+}, Ag^+, Hg_2^{2+}, and Fe^{2+}-Fe^{3+}.

used. The apparent liquid-junction error $(-E_j')$ plotted in Fig. 9–1 is defined, in the case of the hydrogen electrode, for example, by

$$E_j' = E_j + \frac{RT \ln 10}{F} \log y_H \qquad (15)$$

It thus includes both the liquid-junction potential E_j and changes in the activity coefficient y_H (molar concentration scale) produced by the alteration of solution composition at constant ionic strength.

Ignoring the results obtained with the quinhydrone electrode, which are suspect in strong salt solutions containing acid, there is evidence that the activity factors influence the anions (Cl^-, Br^-) differently from the cations. Even though the charge of the latter varied, a single line was obtained for all. Biedermann and Sillén suggest that the activity factors for cations do not change when H^+ is substituted for Na^+ at constant perchlorate concentration, whereas the activity factors for anions are altered by this substitution. This conclusion, it may be observed, is consistent with the postulates on which the theory of specific ionic interaction is based.[10]

It thus follows that E_j' for the cations reflects only the changes in liquid-junction potential, E_j. From the Henderson equation,[30] it may be shown that the potential across the liquid junction

---

[30] P. Henderson, Z. Physik. Chem., **59**, 118 (1907); **63**, 325 (1908).

$$3 \text{ M NaClO}_4 \mid \text{HClO}_4 \, (c_H), \text{NaClO}_4(3 - c_H)$$

should be given by

$$E_j = - \frac{RT \ln 10}{F} \log\left(1 + \frac{dc_H}{3}\right) \tag{16}$$

where d is determined by the molar conductivities of the two electrolytes

$$d = \frac{\Lambda_{\text{HClO}_4} - \Lambda_{\text{NaClO}_4}}{\Lambda_{\text{NaClO}_4}} \tag{17}$$

The experimental data were reproduced very well by $d = 1.95$ over the entire range of c_H from 0 to 0.6 M. This value is considerably lower than 3.3 calculated by equation 17 from the conductivities of 3 M $HClO_4$ and 3 M $NaClO_4$. The difference may signifiy that activity factors, though equally applicable to all cations, are not absent. Alternatively, it may reflect deficiencies in the Henderson equation for the liquid-junction potential.

The results of Biedermann and Sillén[28] and of Rossotti and Rossotti[28, 29] lead to the conclusion that E_j in 3 M sodium perchlorate is given by

$$\frac{E_j}{V} = -0.0165 \, \frac{c_H}{\text{mol dm}^{-3}} \tag{18}$$

and, in 1 M sodium perchlorate,

$$\frac{E_j}{V} = -0.063 \, \frac{c_H}{\text{mol dm}^{-3}} \tag{19}$$

These equations are valid up to $c_H = 0.1$ M. They indicate that the liquid-junction error caused by the presence of 0.01 M perchloric acid in 3 M sodium perchlorate ($pc_H = 2$) is about 0.003 pH unit; in 1 M sodium perchlorate it is about 0.010 pH unit. The high mobility of hydrogen ion is, of course, responsible for this exaggerated effect of the composition of the solution on the liquid-junction potential in this instance. At $pc_H = 3$ and above, both constant ionic media should effectively reduce the liquid-junction potential to negligible values.

Standardization of Titration Curves

It is perhaps worth noting once more that solutions of strong acids, due to their anomalous liquid-junction potentials, make poor standards for pH. Nevertheless, it has long been recognized that this restriction applies with less force to a solution of a strong acid in the presence of neutral salt at a higher concentration. Thus "Veibel's solution" (0.01 M HCl + 0.09M KCl) was known to be superior to 0.1 M HCl as a reference solution for pH measurements. Most constant ionic media can be relied upon to be effective in stabilizing the liquid-junction potential; hence, the determination of the correction factor A as described above seems justified. It is difficult, however, to assess the validity of these correction terms, determined in solutions of strong acids, at pH values above 4.

If both the "unknown" solution X and the standard solution S contain solutes in small concentration in the same constant ionic medium, the activity coefficient of hydrogen ion will be nearly the same in both solutions; the residual liquid-junction potential is also small. Thus it follows that the operational definition of pH (equation 11) can be rewritten

$$pm_H(X) = pm_H(S) + \frac{(E_X - E_S)F}{RT \ln 10} \tag{20}$$

under these conditions.

To measure hydrogen ion concentrations (m_H) in constant ionic media experimentally, therefore, one must have reference points at which the hydrogen ion concentration is accurately known. The composition with respect to the predominant electrolytes must, of course, be the same in the standards as in the "unknown"solutions. Preferably, the reference solutions should cover a wide range of acidity. It is not difficult to set up standards in the range pm_H from 1 to 3; in this range a strong acid added in the appropriate concentration to the constant ionic medium serves very well. In his series of papers under the title "Komplexone," Schwarzenbach[31] has, for example, used hydrochloric acid or perchloric acid to prepare reference standards for the determination of stability constants in 0.1 M potassium chloride and 0.1 M potassium nitrate.

It is often true that the structure of the liquid junction has an influence on the value of A (equation 13) for a particular cell assembly. For this reason, a scale of pc_H or pm_H is most commonly standardized by determining the applicable value of A by titration of a strong acid with a strong base, closely observing the conditions under which the "unknown" titration is to be performed. The following specific instructions, quoted from Irving, Miles, and Pettit,[25] can be recommended. The formulation has been changed slightly to correspond to that of the present discussion:

Adjust the pH using a primary standard buffer solution. Use a standardised solution of a strong base to titrate a solution of a strong acid, of comparable concentration, ionic background, and ionic strength to that of the test solution.

---

[31] See, for example, G. Schwarzenbach et al., Helv. Chim. Acta, 28, 828 (1945); 29, 364 (1946); 30, 1303 (1947); and 31, 456 (1948).

Determine A by subtracting the pH values as read on the meter from the values of pc_H calculated at various points on the titration curve. Add the correction factor (A) to all values of pH obtained with the cell subsequent to standardisation: this will then give $pc_H = -\log_{10} c_H$. It should be noted that this correction factor will be applicable only to the particular cell and the experimental conditions in use.

In a refined procedure, an attempt should be made to determine A values where pc_H of the unknown and standard are nearly the same; then the liquid-junction errors identified by Biedermann and Sillén[28] will disappear, even if the pc_H is 2 or less. At higher pH values, from pH 3 to 11, it seems probable that the constant ionic medium is effective in assuring constancy of E_j, but there remains a need to affirm the response of the glass electrode in this pH region. If the concentration dissociation constant k for a weak acid or base is known in the appropriate medium, a titration of the acid with strong alkali (or base with strong acid) will relate pH(X) to pc_H or pm_H and furnish values of the correction factor A (equation 13).

Powell and Curtis[32] titrated ethylenediamine with perchloric acid in sodium perchlorate solutions of five different ionic strengths, namely 0.35, 0.20, 0.15, 0.10, and 0.04M. The values of k for the two basic groups in sodium chloride solutions at these ionic strengths were taken from the paper of Everett and Pinsent.[33] In spite of the fact that the values of k had been determined in chloride media instead of in perchlorate media, the e.m.f. of the glass-saturated calomel pH assembly was found to vary almost linearly with pc_H in the range from 4 to 10. Furthermore, the data for all five constant ionic media of five different ionic strengths fell on a single curve, suggesting that $E_j + 0.0591 \log y_H$ is independent of ionic strength in this range of acidity. This calibration curve was used to determine pc_H and hence pk for other substituted ethylenediamines in perchlorate solutions.

In later work,[34] the glass-calomel cell was standardized in sodium chloride solutions of ionic strengths 0.04, 0.10, 0.15, and 0.20 by titrating hydrochloric acid, acetic acid, and ethylenediammonium chloride with a standard solution of sodium hydroxide. The values of k for acetic acid in sodium chloride solutions were taken from the work of Harned and coworkers.[35] Again, all of the data for the three acid-base systems at four ionic strengths were colinear, within experimental error, being expressed by

$$\text{pH(X)} = 0.9951 \, pc_H + 0.088 \tag{21}$$

An interesting conclusion of the work was that the difference between the

[32] H. K. J. Powell and N. F. Curtis, *J. Chem. Soc.*, (*B*), 1205 (1966).

[33] D. H. Everett and B. R. W. Pinsent, *Proc. Roy. Soc.* (*London*), **A215**, 416 (1952).

[34] G. R. Hedwig and H. K. J. Powell, *Anal. Chem.*, **43**, 1206 (1971).

[35] H. S. Harned and R. A. Robinson, *J. Amer. Chem. Soc.*, **50**, 3157 (1928); H. S. Harned and F. C. Hickey, *ibid.* **59**, 1284 (1937).

measured (operational) pH value and pc_H could be accounted for only very poorly by log y_H calculated by the Davies[36] equation.

These simple relationships are surprising and suggest that further study of these phenomena would be rewarding. From equation 12

$$pH - pm_H = \Delta\bar{E}_j - \log \gamma_H = \bar{E}_j(X) - \bar{E}_j(S) - \log \gamma_H \qquad (22)$$

it appears that changes of ionic strength, which surely affect log γ_H, must have a compensatory effect on the residual liquid-junction potential, $\Delta\bar{E}_j$. This is not to deny, however, that the constant ionic medium is effective in equalizing the potential $\bar{E}_j(X)$ at liquid junctions of the type

$$\text{Ionic medium, X} \mid \text{Satd. KCl}$$

as the solute X is changed in nature or composition while the ionic medium remains the same. Nevertheless, since $\bar{E}_j(S)$ for the junction

$$\text{pH standard} \mid \text{Satd. KCl}$$

is a constant, the colinearity expressed by equation 21 indicates that $\Delta\bar{E}_j$ is decreasing at the same rate that $-\log \gamma_H$ is increasing when the composition of the ionic medium changes. In sodium chloride solutions of $m = 0.04$ and $m = 0.20$, $-\log \gamma_H$ is estimated to have the values 0.064 and 0.098 respectively, if Kielland's value[37] of 9 Å for the ion-size parameter of hydrogen ion (see page 49) is accepted. The corresponding decrease in liquid-junction potential vs. saturated KCl as the constant ionic medium (NaCl) increases in concentration from 0.04 to 0.2M would thus be a reasonable 2 mV. It is to be expected that log γ_H will pass through a minimum as the concentration of sodium chloride increases from 0.2 to 1M, yet $\bar{E}_j(X)$ should continue to decrease as the concentration of the ionic medium moves toward that of saturated KCl. It is therefore probable that the colinearity demonstrated by Hedwig and Powell is limited to a relatively narrow range of ionic strengths and would not be found at ionic strengths of 1M and above.

Standard Reference Solutions

When values of the hydrogen ion concentration from acid-base titration curves are needed for the purpose of calculating equilibrium constants, standardization of the cell assembly by means of parallel titrations of a strong acid, or weak acids of known k, has much to recommend it. In other applications, however, it would be more convenient to standardize the glass

[36] C. W. Davies, J. Chem. Soc., 2093 (1938); Ion Association, p. 41, Butterworths, London, 1962.

[37] J. Kielland, J. Amer. Chem. Soc., 59, 1675 (1937).

electrode pH meter with two or three reference solutions of known $-\log m_H$ exactly as is done in the measurement of the operational pH. Within certain limits, values of pm_H for reference buffer solutions in constant ionic media can be assigned, although the data needed are at present available for only a few salt solutions. It is perhaps worthwhile to examine the procedures by which pm_H in these media can be derived.

In 1928, Harned and Robinson[38] showed that e.m.f. measurements of cells of the type

$$Pt; H_2(g), HA(m_1), MCl_n(m_2), AgCl; Ag \qquad (23)$$

can be used to determine the concentration constant k for the ionization of the weak acid HA in salt solutions at the limit of $m_1 = 0$. The molality m_1 of the weak acid was varied while m_2, the molality of a completely dissociated chloride salt MCl_n, was kept constant.

The e.m.f. of cell 23 can be expressed by

$$E = E^\circ - \frac{RT \ln 10}{F} \log (nm_2\, m_H\, \gamma_H\, \gamma_{Cl}) \qquad (24)$$

One can also write

$$\gamma_H \gamma_{Cl} \equiv \gamma_\pm^2 = (_s\gamma_\pm^\circ \cdot {}_m\gamma_\pm)^2 \qquad (25)$$

where $\gamma_\pm$, the mean activity coefficient of hydrochloric acid in the mixture of weak acid and chloride salt is expressed in terms of the activity coefficient of HCl at zero molality in the salt solution $(_s\gamma_\pm^\circ)$ and the medium effect caused by the presence of the weak acid $(_m\gamma_\pm)$. It should be noted that $_m\gamma_\pm$ approaches unity as m_1 goes to zero; likewise $_s\gamma_\pm^\circ$ can be obtained by extrapolating to $m_1 = 0$ data for cells of type 23 in which HA is replaced by HCl. Values of $_s\gamma_\pm^\circ$ for hydrochloric acid in a number of salt solutions are available in the literature.[39]

Grouping the known quantities in these equations together, one can now write an expression for the " apparent " hydrogen ion concentration, m_H':

$$-\log m_H' = -\log m_H - 2 \log {}_m\gamma_\pm = \frac{(E - E^\circ)F}{RT \ln 10} + \log nm_2 + 2 \log {}_s\gamma_\pm^\circ \qquad (26)$$

and for the " apparent " concentration dissociation constant, k':

$$k' = \frac{m_H'^2}{m_1 - m_H'} \qquad (27)$$

[38] H. S. Harned and R. A. Robinson, *J. Amer. Chem. Soc.*, **50**, 3157 (1928).
[39] See, for example, H. S. Harned and B. B. Owen, *The Physical Chemistry of Electrolytic Solutions*, 3rd ed., Chapter 14, Reinhold Publishing Corp., New York, 1958.

The true values of m_H and k can, of course, be calculated from m_H' and k' if the medium effect of HA on the mean activity coefficient $\gamma_\pm$ of hydrochloric acid is known. If this is not the case, it is still possible to obtain k in the salt solution by plotting $-\log k'$ as a function of m_1 and extrapolating the curve to $m_1 = 0$. The thermodynamic dissociation constant K is related to k in a salt solution by

$$K = k \cdot \frac{{}_s\gamma_H {}_s\gamma_A}{{}_s\gamma_{HA}} \equiv k\gamma_\pm^2(A) \tag{28}$$

This procedure has been used to determine k and hence the activity coefficient function $\gamma_\pm(A)$ for acetic acid,[35, 38] lactic acid,[35] and a few weak bases[38] solutions of several unassociated chloride salts.

The accurate measurement of hydrogen ion concentrations by the method of equation 26 depends on the availability of values for the medium effect of HA on the activity coefficient of HCl. The nature of the medium effect has been discussed elsewhere.[40] The medium effect ${}_m\gamma_i$ is formally defined in terms of the molar Gibbs energy change $\Delta G_t°(i)$ for the transfer of species i from the standard state in water to the standard state in a second solvent or solvent mixture:

$$\Delta G_t°(i) = RT \ln {}_m\gamma_i \tag{29}$$

This quantity is a function only of standard states. The medium effect for hydrochloric acid, for example, expresses the limit reached at $m = 0$ of the ratio of the activities of hydrochloric acid in a mixed solvent s (composed of water and HA, for example) and in the pure water solvent at the same molality (m) of HCl. It is most readily obtained from the change of $E°$ for the cell

$$\text{Pt}; H_2(g), \text{HCl in solvent } s, \text{AgCl}; \text{Ag}$$

as a nonelectrolyte such as HA is added to the aqueous solvent:

$$\log {}_m\gamma_\pm = \frac{({}_wE° - {}_sE°)F}{2RT \ln 10} \tag{30}$$

For single ionic species, the transfer energy and medium effect cannot be measured; they can only be estimated by various nonthermodynamic procedures.[41]

---

[40] See chapters 7 and 8 and also R. G. Bates in *Hydrogen-Bonded Solvent Systems*, A. K. Covington and P. Jones, eds., p. 49, Taylor and Francis, London, 1968.

[41] These methods have been thoroughly reviewed by O. Popovych, *Crit. Rev. Anal. Chem.*, **1**, 73 (1970).

Some values of the medium effect of nonelectrolytes on $\gamma_\pm$ for hydrochloric acid are collected in Table 9–4.[40] It may be judged that nonelectrolytes at molalities of the order of 0.01 to 0.02 mol kg^{-1} cause only a small medium effect. Actual values may be estimated as follows. If log $_m\gamma_\pm$ varies linearly with the weight percentage of nonelectrolyte, 0.01 mole of nonelectrolyte per kilogram of solvent corresponds to values of log $_m\gamma_\pm$ less than 0.001 for each of the nonelectrolytes for which data are given.

It thus seems evident that the "apparent" hydrogen ion concentration measured by cell 23 for very dilute solutions of a weak acid HA in constant salt media will lie close to the true hydrogen ion concentration. Unfortunately, however, solutions of weak acids make poor reference solutions for pc_H or pm_H; the buffer capacity of the mixture 0.01M with respect to both HA and its salt NaA is, for example, much higher than that of 0.01M HA alone (see page 110). To determine the true m_H of a dilute buffer mixture by equation 26 is a matter of greater difficulty. It requires not only a knowledge of the medium effect but also of the effect of the added NaA on $_s\gamma_\pm{}^\circ$. In other words, one needs to know how much the "trace" activity coefficient of hydrochloric acid in, say, 1 molal sodium chloride is affected by the presence of a small amount, possibly 0.01 mol kg^{-1}, of a second salt. We should like, therefore, to ascertain the error involved in assuming that $_s\gamma_\pm{}^\circ$ in the mixture NaCl (1m), NaX (0.01m) is the same as in the salt solution NaCl (1m) where its value is known. This error arises from two sources: the increase of ionic strength from 1.00 to 1.01, and the substitution of "foreign" anions X$^-$ for an equivalent amount of Cl$^-$.

Sufficient data for mixed electrolytes have been obtained to permit these effects to be estimated. According to "Harned's rule," the logarithms of the activity coefficients of two electrolytes present in a mixture at constant total ionic strength vary linearly with the molalities. Thus, for HCl (component 1) in a solution of MCl (component 2),

$$\log \gamma_1 = \log \gamma_{1(0)} - \alpha_{12} m_2 \tag{31}$$

and

$$\log \gamma_2 = \log \gamma_{2(0)} - \alpha_{21} m_1 \tag{32}$$

where $\gamma_{1(0)}$ designates the activity coefficient of HCl in a solution of HCl alone and $\gamma_{2(0)}$ that of MCl in a solution of MCl alone. For our purpose, equation 31 and the values of α_{12} are sufficient. The pertinent data for mixtures of hydrochloric acid and sodium chloride or potassium chloride are given in Table 9–5. It may be noted that α_{12} varies with the size and degree of hydration of the cation M$^+$; at a molality of 1 mol kg^{-1}, it increases from 0.005 for lithium chloride to 0.100 for cesium chloride. The "trace" activity coefficients of hydrochloric acid in salt solutions are readily calculated by

TABLE 9-4. MEDIUM EFFECTS OF NONELECTROLYTES ON THE MEAN ACTIVITY COEFFICIENTS OF HALOGEN ACIDS IN AQUEOUS SOLUTION AT 25 °C. VALUES OF $\log {}_m\gamma_\pm$.

Nonelectrolyte % (w/w)	Acetic Acid	Ethanol	Ethylene Glycol	Mannitol	Methanol	Dioxane
			$\log {}_m\gamma_\pm$ in the Presence of			
			Hydrochloric Acid			
10	0.101	0.065	0.062	0.050	0.058	
20	0.216	0.125			0.112	0.164
30			0.164			
40	0.509					
45						0.498
50		0.323	0.277		0.269	
60	0.437					
80		0.796			0.619	
82						2.230
100		2.505	1.681		1.965	
			Hydrobromic Acid			
10	0.088		0.030		0.038	
20	0.200				0.064	0.096
30			0.108			
40	0.483					
50			0.197		0.146	
60	0.927					
90					0.746	
100			1.429		1.774	
			Hydriodic Acid			
10	0.072				0.012	
20	0.165				0.033	−0.007
40	0.419					
50					−0.024	
60	0.795					
90					0.447	

TABLE 9–5. VALUES OF α_{12} FOR MIXTURES OF HYDROCHLORIC ACID
AND SODIUM CHLORIDE OR POTASSIUM CHLORIDE AT 25 °C; log γ_1
FOR HYDROCHLORIC ACID IN SALT SOLUTIONS

Total Molality, mol kg^{-1}	α_{12} HCl–NaCl	α_{12} HCl–KCl	log γ_1 (in NaCl)	log γ_1 (in KCl)
0.1	0.043	0.077	−0.103	−0.107
0.5	0.037	0.062	−0.139	−0.152
1.0	0.032	0.056	−0.124^a	−0.148
2.0	0.031	0.057	−0.058	−0.110
3.0	0.031	0.062	0.026	−0.067
4.0	0.030	0.066	0.126	−0.018

^{a} At $m = 1$: −0.097 in LiCl; −0.192 in CsCl.

equation 31 from the activity coefficients $\gamma_{1(0)}$ of the acid in its pure aqueous solution,[42, 43] together with the appropriate values of α_{12}. They are given in the last two columns of Table 9–5.

It may be judged from the data in Table 9–5 that an increase in ionic strength caused by the addition of 0.01 mole kg^{-1} of buffer salt to a constant ionic chloride medium is unlikely to change log $\gamma_\pm{}^\circ$ by more than a few thousandths of a unit at most. Furthermore, it may be noted that the complete substitution of lithium chloride by cesium chloride, an extreme example, alters log $\gamma_\pm{}^\circ$ at $m = 1$ by only 0.1 unit; hence, it seems justifiable to assume that the replacement of a small fraction of a constant chloride medium by a buffer salt such as NaA will be without appreciable effect.

It therefore seems possible to derive values of $-\log m_H$ in dilute buffer media from the e.m.f. of cells of the type

$$\text{Pt}; \text{H}_2(g), \text{HA}(0.01m), \text{NaA}(0.01m), \text{MCl}_n(m_2), \text{AgCl}; \text{Ag} \qquad (33)$$

with an accuracy greater than 0.01 unit. Within these limits of concentration and accuracy, the molality of hydrogen ion or pm_H can also be calculated from the concentration dissociation constant k for the weak acid in the salt solution concerned. Thus,

$$\text{p}m_H \equiv -\log m_H = \text{p}k - \log \frac{m_{HA}}{m_{A^-}} = \text{p}k - \log \frac{0.01 - m_H}{0.01 + m_H} \qquad (34)$$

[42] H. S. Harned and R. W. Ehlers, *J. Amer. Chem. Soc.*, **54**, 1350 (1932); **55**, 2179 (1933).
[43] R. G. Bates and V. E. Bower, *J. Res. Nat. Bur. Stand.*, **53**, 283 (1954).

The data for buffer solutions of acetic acid and sodium acetate in chloride salt solutions, given in Table 9–6, have been derived in this way from the concentration dissociation constants of acetic acid found by Harned and Robinson[35] and Harned and Hickey.[35] Although the latter authors also studied the dissociation of lactic acid in salt solutions, the strength of this acid is comparable with that of acetic acid; hence, lactic acid offers no advantages over acetic acid as a standard for hydrogen ion concentration.

TABLE 9–6. HYDROGEN ION CONCENTRATIONS OF BUFFER SOLUTIONS COMPOSED OF ACETIC ACID ($m = 0.01$) AND SODIUM ACETATE ($m = 0.01$) IN CHLORIDE SALT SOLUTIONS AT 25 °C

Ionic Strength	pm_H in			
	LiCl	NaCl	KCl	BaCl$_2$
0.11	4.56	4.55	4.54	4.54
0.21	4.50	4.51	4.51	4.50
0.51	4.45	4.48	4.48	4.45
1.01	4.44	4.50	4.52	4.42
2.01	4.47	4.61	4.66	4.44
3.01	—	4.74	—	—

Unfortunately, data for the dissociation of other weak acids in salt solutions are few in number. A similar treatment of the values of Everett and Pinsent[33] for ethylenediamine (B), however, can furnish useful standard values of pm_H in the alkaline region. These authors determined the thermodynamic pK from values of pk obtained from e.m.f. measurements in a symmetrical cell, the two halves of which contained solutions of identical ionic strength and closely matching composition:

$$\text{Pt}; \text{H}_2, \text{BH}^+, \text{B}, \text{NaCl} \left| \begin{array}{c} \text{KCl} \\ 3.5\,\text{M} \end{array} \right| \text{HCl}(m_1), \text{NaCl}(m_2), \text{H}_2; \text{Pt} \qquad (35)$$

The ionic strength was kept the same in the right half cell (r) as in the left (l).

Measurements of the e.m.f. of this cell provided values of the second dissociation constant (deprotonation of BH^+), whereas similar cells containing BH_2^{2+} and BH^+ in place of the acid-base pair BH^+, B furnished values for the first dissociation constant (deprotonation of BH_2^{2+}). The e.m.f. E of this cell is a measure of the "apparent" hydrogen ion concentration m_H' in the left-hand compartment:

$$E = \frac{RT \ln 10}{F} \log \frac{m_1 \gamma_H(r)}{m_H \gamma_H(l)} + E_j \tag{36}$$

or

$$\log m_H' \equiv \log m_H \frac{\gamma_H(l)}{\gamma_H(r)} = \log m_1 - \frac{(E - E_j)F}{RT \ln 10} \tag{37}$$

The value of m_H' approaches the true m_H as the liquid-junction potentials between the two half-cell solutions and the concentrated bridge solution approach equality and as the activity coefficient of hydrogen ion in the two half cells becomes equal. Both of these results are achieved by reducing the contribution of the buffer and of the hydrochloric acid to the total ionic strength of the solutions in the two half cells. By extrapolation of pk' (equation 27) to the limit of pure sodium chloride, pk in the salt solution is obtained, and pK is found an extrapolation of pk to $m_2 = 0$.

When the molality m_2 of the "constant ionic medium" greatly exceeds the buffer molality, it may be expected that m_H can be calculated from the known value of pk in that salt medium without serious error. Furthermore, the value of m_H' derived from the e.m.f. by equation 26 can be expected to lie close to the true m_H under these conditions. In this manner, hydrogen ion concentrations have been derived from the data of Everett and Pinsent[33] in 0.10, 0.15, 0.20, and 0.30M solutions of sodium chloride at 20 and 30 °C. They are listed in Table 9–7. Unfortunately, measurements were not made at 25 °C. The buffer solutions were 0.02M with respect to both ethylenediamine

TABLE 9–7. HYDROGEN ION CONCENTRATIONS OF
BUFFER SOLUTIONS COMPOSED OF ETHYLENEDIA-
MINE, B (0.02M), AND ETHYLENEDIAMINE HYDRO-
CHLORIDE, BHCl (0.02M), IN SODIUM CHLORIDE
SOLUTIONS AT 20 AND 30 °C

Ionic Strength	pc_H	
	20 °C	30 °C
0.10	10.10	9.80
0.15	10.11	9.82
0.20	10.13	9.83
0.30	10.15	9.86

(B) and its monoprotonated conjugate (BH^+). The pk values and solute concentrations were based on the concentration (mol dm^{-3}) scale; hence, the values given are properly termed pc_H. Since c/m is 0.9970–0.0183m in sodium chloride solutions at 25 °C,[44] pm_H is smaller than pc_H by 0.002 unit in 0.1 M sodium chloride and by 0.003 unit in 0.2 M sodium chloride at this temperature.

Hydrogen Ion Concentration in Mixed Solvents

To avoid the difficulties imposed by limited solubility in water, coordination chemists often resort to mixed solvent media such as water-dioxane mixtures for the determination of stability constants.[45, 46] A useful empirical procedure for the conversion of measured operational pH values to $-\log c_H$ (pc_H) in these media has been described by Van Uitert and Haas.[47] It consists essentially of calibrating the glass-calomel pH assembly with solutions of a strong acid. If the acid is completely dissociated in the mixed solvent, c_H in these reference solutions is known and the correction U_H can be determined:

$$\log U_H \equiv pc_H - pH \qquad (38)$$

From the discussion of medium effects given above and in earlier chapters, it is evident that U_H is composed of activity coefficients and liquid-junction potential contributions:

$$-\log U_H = \Delta \bar{E}_j - \log {}_s\gamma_H - \log {}_m\gamma_H \qquad (39)$$

or, in terms of the function δ defined on page 244,

$$-\log U_H = \delta - \log {}_s\gamma_H \qquad (40)$$

Although δ has been shown to be, in first approximation, a constant for each solvent composition, U_H includes, in addition, the interionic attraction term ${}_s\gamma_H$, and hence varies with ionic strength. The uses of δ and $\log U_H$ may be compared as follows: subtraction of δ from the experimental operational pH (determined with aqueous pH standards) furnishes values of pa_H* or pH* which are activity units; subtraction of $-\log U_H$ from pH yields pc_H at a particular ionic strength. Both correction terms vary, of course, with the composition of the mixed solvent.

In their careful study, Van Uitert and Haas first tested the hydrogen ion response of their glass electrodes in 75 volume per cent dioxane by plotting

---

[44] H. S. Harned and B. B. Owen, *The Physical Chemistry of Electrolytic Solutions*, 3rd ed., appendix, Table 12-1-1A, Reinhold Publishing Corp., New York, 1958.

[45] M. Calvin and K. W. Wilson, *J. Amer. Chem. Soc.*, **67**, 2003 (1945).

[46] F. J. C. Rossotti and H. Rossotti, *J. Phys. Chem.*, **68**, 3773 (1964).

[47] L. G. Van Uitert and C. G. Haas, *J. Amer. Chem. Soc.*, **75**, 451 (1953).

experimental pH values for solutions of hydrochloric acid in the mixed solvent as a function of the e.m.f. of a cell with hydrogen and silver-silver chloride electrodes and without a liquid junction. A straight line with the proper Nernst slope was found to extend from pH 2 to pH 12. This check on the behavior of the glass electrode, too often overlooked, is recommended whenever this electrode is used for careful measurements in nonaqueous and mixed solvents.

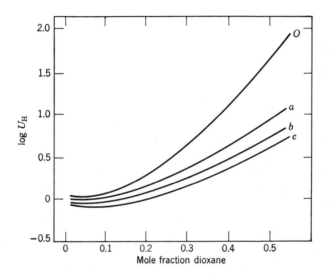

FIG. 9–2. Plot of log U_H as a function of the mole fraction of dioxane in the mixed solvent (after Van Uitert and Haas[47]). Concentration of 1 : 1 electrolyte: curve a, 0.00586M; curve b, 0.0172M; curve c, 0.0312M; curve o, 0 (extrapolated).

The variation of log U_H with solvent composition was demonstrated by measurements in solvent media in which the mole fraction of dioxane was varied; the variation with ionic strength was revealed by measurements made on a series of solutions containing hydrochloric acid and sodium chloride in each solvent mixture studied. The results at three ionic strengths are shown by the curves a, b, and c in Fig. 9–2. When allowance was made for the activity coefficient $_s\gamma_H$, all three curves were brought into coincidence (curve o). These limiting values of log U_H, as equation 40 shows, correspond to $-\delta$ in water-dioxane mixtures.

This approach is of great potential usefulness for the standardization of measurements of pm_H or pc_H in mixed solvent media. An allowance must be made, however, for the effect of ionic strength on $_s\gamma_H$. Alternatively, a constant ionic medium can be used, and the value of $_s\gamma_H$, assumed to be constant, can be included along with $_m\gamma_H$ in the correction term, log U_H.

The values of log U_H, however, are determined in solutions of strong acids; they include therefore a liquid-junction potential term $(\bar{E}_j)$ which may be abnormal relative to the approximately constant value expected at higher pH. Here again, buffer solutions of known hydrogen ion concentration would contribute greatly to the accuracy of measurement. As yet, this area of research remains largely unexplored.

CHAPTER 10

cells, electrodes,
and techniques

The experimental basis for the electrometric measurement of pH will be considered in this chapter. The properties of the most useful electrodes will be discussed and the experimental aspects of the liquid junction examined We shall also be concerned with the techniques of preparing galvanic cells. The accurate measurement of the electromotive force of these cells will be considered in Chapter 12.

A cell with liquid junction, which may be represented by

$$\text{Electrode reversible} \quad \begin{vmatrix} \text{Salt} \end{vmatrix} \quad \begin{vmatrix} \text{Reference} \end{vmatrix}$$
$$\text{to hydrogen ion; Soln. X} \begin{vmatrix} \text{bridge} \end{vmatrix} \text{electrode} \qquad (1)$$

is most commonly used for measurements of pH. The reference electrode is usually a calomel electrode. Of the several electrodes reversible to hydrogen ion, the glass electrode is by far the most convenient and versatile. A separate chapter will be devoted to the properties of this electrode.

The Hydrogen Electrode

The hydrogen electrode is the ultimate standard for the determination of pH values, but owing to the experimental difficulties associated with it other electrodes reversible to hydrogen ion are commonly employed for routine pH measurements. Nevertheless, the performance of these practical secondary electrodes, of which the glass, quinhydrone, and antimony electrodes are the most useful, is always evaluated in terms of the hydrogen electrode. The defects of these electrodes—the salt error of the quinhydrone electrode, the sodium ion error of the glass electrode, the nonlinear response of the antimony electrode, and the like—are revealed by direct comparison of the behavior of these electrodes with that of the hydrogen electrode.

279

The hydrogen electrode is formed by bubbling pure hydrogen gas over a wire or small foil, the surface of which is able to catalyze the reaction

$$H^+ + e \; \rightleftharpoons \; \tfrac{1}{2}H_2(g) \tag{2}$$

and hence to establish equilibrium between hydrogen molecules (or atoms) and the hydrogen ions (or protons) in the solution in which the metal electrode is immersed. In equation 2, e represents the electron. The most useful surface catalyst is platinum black, although palladium black and bright platinum, palladium, or iridium have been employed successfully. The catalytic activity varies with the electrode material and with the condition of its surface. The activity of palladium black and the bright surfaces is lower than that of platinum black. When reduction of the solute is not of concern, it is sometimes advantageous to employ two electrodes of widely different catalytic activities as an index of the attainment of equilibrium, as recommended by Lorch.[1]

The hydrogen electrode cannot be employed in the presence of certain "poisons," materials that inhibit the reversibility of the electrode process (equation 2). Some of these are cyanide ion, hydrogen sulfide, arsenic compounds, and the cations of metals more noble than hydrogen—for example, silver and mercury. Some anions interfere; nitrate may be reduced to ammonium in strongly acid solutions, although it causes no difficulty in dilute solutions of weak acids.[2] Nitrophenols, benzoic acid, and other aromatic compounds may be reduced by hydrogen in the presence of finely divided platinum. Reduction is usually accelerated by elevated temperatures and can sometimes be retarded if thinly plated electrodes are used.[3] In some cases the lower catalytic activity of metals other than platinum permits useful results to be obtained. Hamer and Acree[4] found that electrodes coated with finely divided palladium furnish reproducible and constant

---

[1] E. Lorch, *Ind. Eng. Chem., Anal. Ed.*, **6**, 164 (1934). S. N. Das and D. J. G. Ives, *J. Chem. Soc.*, 1619 (1962), have used a pair of hydrogen electrodes, one of which is half platinized and half bright, to confirm the attainment of equilibrium and to detect the onset of poisoning.

[2] R. G. Bates, G. L. Siegel, and S. F. Acree, *J. Res. Nat. Bur. Stand.*, **30**, 347 (1943); R. G. Bates, P. T. Diamond, M. Eden, and S. F. Acree, *ibid.*, **37**, 251 (1946).

[3] D. A. MacInnes, D. Belcher, and T. Shedlovsky, *J. Amer. Chem. Soc.*, **60**, 1094 (1938); I. M. Kolthoff and H. A. Laitinen, *pH and Electro Titrations*, 2nd ed., Chapter 6, John Wiley and Sons, New York, 1941.

[4] W. J. Hamer and S. F. Acree, *J. Res. Nat. Bur. Stand.*, **33**, 87 (1944). A palladium hydride electrode has been described by P. Nylén, *Svensk Kem. Tidskr.*, **48**, 76 (1936); *Z. Elektrochem.*, **43**, 915, 921 (1937). The electrode is formed by depositing palladium black on small platinum foils which are then immersed for 2 minutes in 10 per cent formic acid. It has been used successfully for electrometric titrations of acids and bases in oxygen-free solutions. No supply of hydrogen gas is required.

potentials in solutions of potassium hydrogen phthalate, where reduction of the phthalate often makes the platinum black electrode unsuitable.

OPERATION

If the partial pressure (in atmospheres) of hydrogen gas at the electrode is p_{H_2}, the half-cell potential E_H of reaction 2 is written

$$E_H = E_H{}^\circ + \frac{RT}{F} \ln \frac{a_H}{\sqrt{p_{H_2}}} \qquad (3)$$

Inasmuch as the standard potential of the hydrogen electrode $E_H{}^\circ$ is, by convention, assigned a value of zero at all temperatures, equation 3 becomes

$$E_H = \frac{RT}{F} \ln a_H \qquad (3a)$$

when $p_{H_2} = 1$, the usual standard state.

The partial pressure of hydrogen in the typical experiment is not exactly 1 atm; hence the observed e.m.f. must be corrected to the reference pressure of 760 mm of dry hydrogen. The potential of the electrode becomes more negative as the pressure of gas increases, as equation 3 shows. If the partial pressure of hydrogen gas in equilibrium with the electrode is less than 760 mm and the electrode is at the left of the cell scheme, the correction must be added, because the e.m.f. increases with the gas pressure. The calculation is made as follows: Two hydrogen electrodes, at partial pressures of p_{H_2} mm and 760 mm, respectively, are imagined to be immersed in the same solution. The electrode reaction at each is given by equation 2, and the thermodynamic process that takes place when current is drawn is

$$H_2(g, 760 \text{ mm}) = H_2(g, p_{H_2} \text{ mm}) \qquad (4)$$

The potential difference between the two electrodes is the correction ΔE to be added to the observed e.m.f.:

$$\Delta E = \frac{RT}{2F} \ln \frac{760}{p_{H_2}} \qquad (5)$$

Hainsworth, Rowley, and MacInnes[5] have shown that the change of the potential of the electrode with pressure obeys this equation closely to pressures as high as 100 atm of hydrogen.

In aqueous solutions containing no volatile solutes such as ammonia or carbon dioxide, the partial pressure of hydrogen gas at the electrode surface is obtained by subtracting the tension of water vapor p_{H_2O} from the

[5] W. R. Hainsworth, H. J. Rowley, and D. A. MacInnes, *J. Am. Chem. Soc.*, **46**, 1437 (1924).

corrected barometric pressure P and adding what is known as the excess pressure or *depth effect*. Hills and Ives[6] found that the effective partial pressure of gas at an electrode mounted at or beneath the surface of a solution is not determined so much by the depth of immersion of the electrode as by the depth of the jet through which the gas enters. The concentration of dissolved gas is apparently dependent on the pressure of gas at the point where the gas goes into solution. Mixing is thorough and the solution at the surface is slow to release its excess dissolved gas.

According to Hills and Ives, the excess pressure is four-tenths of the hydrostatic pressure of the column of solution above the jet. If h is the depth of the jet in millimeters, the excess pressure becomes $0.4h/13.6$ mm Hg. Hence equation 5 shows that a jet depth of 5 cm causes an error of about 0.03 mV in the e.m.f. at room temperature and nearly 0.2 mV at 95 °C. However, if the depth of the jets in the cells used for the determination of the standard potential was also about 5 cm, the greater part of the error cancels out.

The pressure of water vapor over aqueous solutions of low and moderate concentrations can be assumed, for all practical purposes, to be the same as the vapor pressure of pure water at the appropriate temperature (see Appendix, Table 3). The complete expression for p_{H_2} (in mm Hg) is therefore

$$p_{H_2} = P - p_{H_2O} + \frac{0.4h}{13.6} \tag{6}$$

Table 10-1 lists the hydrogen electrode corrections ΔE in millivolts for barometric pressures from 720 to 775 mm Hg and temperatures from 0 to 95 °C. These values, which must be added to the observed e.m.f. to correct it to the standard reference partial pressure of 760 mm, were computed by equation 5. The partial pressure of hydrogen was obtained from equation 6 without the last term, which embodies the depth effect.

THEORY

As equation 3a indicates, the potential of the hydrogen electrode is a formal measure of the proton activity of the solution. The electrode gives reproducible potentials in some anhydrous media, for example mixtures of glacial acetic acid and acetic anhydride,[7] as well as in many aqueous and partially aqueous solutions. The solvent does not appear to participate in the electrode reaction, and the hydrogen electrode is presumed to be reversible to protons, even though it is unlikely that free protons exist to any appreciable extent in most solvent media. Nevertheless, the stability and reproducibility of the

---

[6] G. J. Hills and D. J. G. Ives, *Nature*, **163**, 997 (1949).

[7] J. Russell and A. E. Cameron, *J. Amer. Chem. Soc.*, **60**, 1345 (1938); O. Tomicek and A. Heyrovsky, *Collection Czech. Chem. Commun.*, **15**, 984 (1950). See also Chapter 7.

TABLE 10–1. Pressure Corrections for the Hydrogen Electrode in Millivolts

Barometric Pressure mm Hg	Temperature, °C																			
	0	5	10	15	20	25	30	35	40	45	50	55	60	65	70	75	80	85	90	95
720	0.71	0.76	0.82	0.89	0.99	1.13	1.30	1.52	1.81	2.18	2.67	3.30	4.12	5.18	6.60	8.51	11.17	15.06	21.35	34.56
725	0.63	0.67	0.73	0.81	0.91	1.03	1.20	1.42	1.71	2.08	2.56	3.18	3.99	5.05	6.45	8.34	10.96	14.79	20.95	33.66
730	0.55	0.59	0.65	0.72	0.82	0.94	1.11	1.32	1.61	1.97	2.45	3.06	3.87	4.91	6.30	8.17	10.76	14.53	20.57	32.81
735	0.47	0.51	0.56	0.63	0.73	0.85	1.02	1.23	1.51	1.87	2.34	2.95	3.74	4.78	6.15	8.00	10.56	14.27	20.19	32.01
740	0.39	0.43	0.48	0.55	0.64	0.76	0.92	1.13	1.41	1.77	2.23	2.83	3.62	4.65	6.01	7.83	10.36	14.01	19.82	31.24
745	0.31	0.34	0.39	0.46	0.55	0.67	0.83	1.04	1.31	1.66	2.12	2.72	3.50	4.52	5.86	7.67	10.16	13.77	19.46	30.51
6	0.29	0.33	0.38	0.45	0.53	0.65	0.81	1.02	1.29	1.64	2.10	2.70	3.48	4.49	5.83	7.63	10.12	13.72	19.39	30.37
7	0.28	0.31	0.36	0.43	0.52	0.63	0.79	1.00	1.27	1.62	2.08	2.68	3.45	4.46	5.80	7.60	10.08	13.67	19.31	30.23
8	0.26	0.30	0.34	0.41	0.50	0.62	0.78	0.98	1.25	1.60	2.06	2.65	3.43	4.44	5.77	7.57	10.04	13.62	19.24	30.09
9	0.24	0.28	0.33	0.40	0.48	0.60	0.76	0.96	1.23	1.58	2.04	2.63	3.40	4.41	5.74	7.54	10.00	13.57	19.17	29.95
750	0.23	0.26	0.31	0.38	0.47	0.58	0.74	0.94	1.21	1.56	2.02	2.61	3.38	4.39	5.72	7.50	9.96	13.52	19.10	29.81
1	0.21	0.25	0.30	0.36	0.45	0.57	0.72	0.92	1.19	1.54	2.00	2.58	3.35	4.36	5.69	7.47	9.93	13.47	19.03	29.68
2	0.20	0.23	0.28	0.34	0.43	0.55	0.70	0.90	1.17	1.52	1.97	2.55	3.33	4.33	5.66	7.44	9.89	13.42	18.96	29.54
3	0.18	0.22	0.26	0.33	0.42	0.53	0.68	0.89	1.15	1.50	1.95	2.53	3.30	4.31	5.63	7.40	9.85	13.37	18.90	29.41
4	0.16	0.20	0.24	0.31	0.40	0.51	0.67	0.87	1.13	1.48	1.93	2.51	3.28	4.28	5.60	7.37	9.81	13.33	18.83	29.28
755	0.15	0.18	0.23	0.29	0.38	0.49	0.64	0.85	1.12	1.46	1.91	2.49	3.25	4.26	5.57	7.34	9.77	13.28	18.76	29.14
6	0.13	0.17	0.21	0.28	0.36	0.48	0.63	0.83	1.10	1.44	1.89	2.47	3.23	4.23	5.55	7.31	9.74	13.23	18.69	29.01
7	0.12	0.15	0.20	0.26	0.35	0.46	0.61	0.81	1.08	1.42	1.87	2.45	3.20	4.20	5.52	7.28	9.70	13.18	18.62	28.88
8	0.11	0.14	0.18	0.24	0.33	0.44	0.59	0.79	1.06	1.40	1.85	2.42	3.18	4.18	5.49	7.24	9.66	13.13	18.56	28.76
9	0.09	0.12	0.16	0.23	0.31	0.42	0.58	0.78	1.04	1.38	1.83	2.40	3.17	4.16	5.46	7.21	9.62	13.09	18.49	28.63

TABLE 10-1. *Pressure Corrections (continued)*

	0	5	10	15	20	25	30	35	40	45	50	55	60	65	70	75	80	85	90	95
760	0.07	0.10	0.15	0.21	0.30	0.41	0.56	0.76	1.02	1.36	1.79	2.39	3.14	4.13	5.43	7.18	9.58	13.04	18.41	28.50
1	0.06	0.09	0.13	0.20	0.28	0.39	0.54	0.74	1.00	1.34	1.77	2.36	3.12	4.10	5.40	7.15	9.55	12.99	18.35	28.38
2	0.04	0.07	0.12	0.18	0.26	0.37	0.52	0.72	0.98	1.32	1.76	2.34	3.09	4.08	5.38	7.12	9.51	12.95	18.29	28.25
3	0.03	0.06	0.10	0.16	0.24	0.36	0.50	0.70	0.96	1.30	1.74	2.32	3.07	4.05	5.35	7.08	9.47	12.90	18.22	28.13
4	0.01	0.04	0.08	0.15	0.23	0.34	0.49	0.69	0.94	1.28	1.72	2.30	3.05	4.03	5.32	7.05	9.44	12.85	18.16	28.00
765	−0.01	0.02	0.07	0.13	0.21	0.32	0.47	0.67	0.92	1.26	1.70	2.28	3.02	4.00	5.29	7.02	9.40	12.80	18.09	27.89
6	−0.02	0.01	0.05	0.11	0.19	0.31	0.45	0.65	0.91	1.24	1.68	2.26	3.00	3.98	5.26	6.99	9.36	12.76	18.03	27.76
7	−0.04	−0.01	0.04	0.10	0.18	0.29	0.43	0.63	0.89	1.22	1.66	2.23	2.98	3.95	5.24	6.96	9.32	12.71	17.96	27.64
8	−0.05	−0.02	0.02	0.08	0.16	0.27	0.42	0.61	0.87	1.20	1.64	2.21	2.95	3.93	5.21	6.93	9.29	12.66	17.90	27.53
9	−0.07	−0.04	0.00	0.06	0.14	0.26	0.40	0.59	0.85	1.18	1.62	2.19	2.93	3.90	5.18	6.90	9.25	12.62	17.83	27.41
770	−0.08	−0.05	−0.01	0.05	0.13	0.24	0.38	0.57	0.83	1.16	1.60	2.17	2.91	3.88	5.16	6.87	9.21	12.57	17.77	27.29
1	−0.10	−0.07	−0.03	0.03	0.11	0.22	0.36	0.56	0.81	1.14	1.58	2.14	2.88	3.85	5.13	6.84	9.18	12.53	17.70	27.17
2	−0.11	−0.08	−0.04	0.01	0.09	0.20	0.34	0.54	0.79	1.12	1.56	2.12	2.86	3.83	5.10	6.80	9.14	12.48	17.64	27.06
3	−0.13	−0.10	−0.06	0.00	0.08	0.19	0.32	0.52	0.77	1.10	1.54	2.10	2.84	3.80	5.07	6.77	9.10	12.44	17.58	26.95
4	−0.14	−0.11	−0.08	−0.02	0.06	0.17	0.31	0.50	0.75	1.09	1.52	2.08	2.81	3.78	5.04	6.74	9.07	12.39	17.51	26.83
775	−0.16	−0.13	−0.09	−0.04	0.04	0.15	0.29	0.48	0.74	1.07	1.50	2.06	2.78	3.75	5.02	6.71	9.03	12.35	17.45	26.72

hydrogen electrode potentials show that the proton activity a_H has a definite value in these media.[8] It is not necessary to know whether the protons participating in the equilibrium at the electrode surface are free or exist in combination with solute and solvent species from which they are readily detached.

A picture of the mechanism of the reversible hydrogen electrode has been provided by Hickling and Salt.[9] Any electrode at which hydrogen gas is present may be regarded as an atomic hydrogen electrode. However, there are two energy barriers encountered in the conversion of hydrogen molecules to protons.[10] The first, which is dependent on the type of cathode metal, accompanies the dissociation of diatomic hydrogen molecules into hydrogen atoms. The second depends on the reducing power of the atoms formed, that is, the ease with which electrons can be removed from them. The catalytic surface of the electrode presumably eliminates both energy barriers, raises the pressure of atomic hydrogen to the equilibrium pressure for the process $H_2 \rightleftharpoons 2H$, and makes the hydrogen electrode reversible.

If the electrode surface is a poor catalyst for the combination of hydrogen atoms, the equilibrium pressure of atomic hydrogen will not be maintained. This reaction, rather than the equilibrium $H \rightleftharpoons H^+$, is regarded as the rate-determining step responsible for hydrogen overvoltage. Similarly, electrode poisons hinder the attainment of equilibrium by decreasing the activity of the catalytic surface. The harmful effect of oxygen on the reversible hydrogen electrode may result from removal of atomic hydrogen by oxidation more rapidly than it can be replaced.

DEVELOPMENT OF THE HYDROGEN ELECTRODE

The first report of the use of a hydrogen electrode to follow changes of acidity appears to be a paper published by Böttger in 1897.[11] He obtained better results with palladinized gold electrodes than with electrodes of platinized platinum. G. N. Lewis used electrodes of iridium on platinum or gold bases as early as 1905.[12] Shortly thereafter the hydrogen electrode was used

[8] G. Kortüm, *Z. Elektrochem.*, **48**, 145 (1942).

[9] A. Hickling and F. W. Salt, *Trans. Faraday Soc.*, **38**, 474 (1942). Later illuminating discussions of the mechanism of the hydrogen electrode have been given by G. J. Hills and D. J. G. Ives in Chapter 2, *Reference Electrodes*, D. J. G. Ives and G. J. Janz, eds., Academic Press, New York, 1961, and by A. M. Feltham and M. Spiro, *Chem. Rev.*, **71**, 177 (1971). See also J. Horiuti, *Transactions of the Symposium on Electrode Processes*, edited by E. Yeager, p. 17, John Wiley and Sons, New York, 1961.

[10] W. D. Bancroft and J. E. Magoffin, *J. Franklin Inst.*, **224**, 475 (1937); G. Okamoto, *J. Fac. Sci., Hokkaido Univ.*, Ser. *III*, **2**, 115 (1937).

[11] W. Böttger, *Z. Physik. Chem.*, **24**, 253 (1897).

[12] Private communication, 1945.

to measure hydrogen ion concentrations of borate mixtures.[13] In 1911 and 1913, Acree and his coworkers[14] reported detailed studies of the properties of the platinum-hydrogen electrode and the application of this electrode to the study of the hydrolysis of aniline hydrochloride. Hildebrand's paper in 1913[15] brought the hydrogen electrode to the attention of scientists outside the field of physical chemistry by extending its application to many different types of reactions.

In spite of the wide use of the glass electrode in practical pH measurements, the hydrogen electrode is in no sense outmoded. Perley[16] has urged that all chemical laboratories be equipped to make measurements with the hydrogen electrode, in order that the accuracy and stability of other pH electrodes and reference buffer solutions can readily be checked from time to time and unrecognized errors of the secondary pH methods revealed.

TYPES OF HYDROGEN CELLS

When the solutions contain no oxidizing substances or specific poisons, the hydrogen electrode is very useful in practical pH measurements. The chief remaining disadvantages are the slow attainment of equilibrium and the relatively large volume of solution usually required.

Clark's rocking cell[17] requires only a small volume of hydrogen. It is charged with hydrogen and rocked to saturate the solution with the gas. Traces of oxygen are reduced catalytically at the platinum surface and hence have little effect on the measurement when equilibrium has been attained.

With the Hitchcock and Taylor cell,[18] shown in Fig.10–1, constancy of the e.m.f. to 0.1 mV is attained in 15 to 30 minutes. The electrode consists of a spiral of platinum wire coated with platinum black. The cell has a water jacket to control the temperature of the solution. A glass bead in the top of the bulb where the liquid junction is formed minimizes disturbance of the boundary by the bubbling hydrogen. The micro-electrode described by Frediani[19] is fashioned from a capillary stopcock and is said to furnish

---

[13] C. L. A. Schmidt and C. P. Finger, *J. Phys. Chem.*, **12**, 406 (1908); see also T. B. Robertson, *ibid.*, **11**, 442 (1907). The hydrogen electrode was also employed in Europe during this period, notably by Bjerrum and Sørensen.

[14] N. E. Loomis and S. F. Acree, *Am. Chem. J.*, **46**, 585, 621 (1911); C. N. Myers and S. F. Acree, *ibid.*, **50** 396 (1913).

[15] J. H. Hildebrand, *J. Amer. Chem. Soc.*, **35**, 847 (1913).

[16] G. A. Perley, *Trans. Electrochem. Soc.*, **92**, 485 (1947).

[17] W. M. Clark, *The Determination of Hydrogen Ions*, 3rd ed., Chapter 14, The Williams and Wilkins Co., Baltimore, Md., 1928; see also D. A. MacInnes, D. Belcher, and T. Shedlovsky, *J. Amer. Chem. Soc.*, **60**, 1094 (1938).

[18] D. I. Hitchcock and A. C. Taylor, *J. Amer. Chem. Soc.*, **59**, 1812 (1937).

[19] H. A. Frediani, *Ind. Eng. Chem., Anal. Ed.*, **11**, 53 (1939).

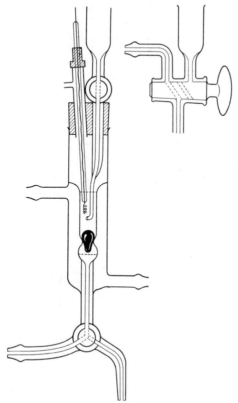

FIG. 10–1. Hydrogen electrode of Hitchcock and Taylor.

accurate pH measurements with 5 to 60 mm^3 of solution. The salt bridge is a cotton thread saturated with potassium chloride solution.

The electrode designed by Hildebrand,[20] shown in Fig. 10–2, has been widely used. This immersion-type electrode has the distinct advantage of furnishing reasonably accurate potentials in systems that are not closed; the pH of the sample can be measured in an open beaker. The electrode is so designed that the platinum foil or wire is partly covered by the solution. Relatively large volumes of hydrogen are required. Hovorka and Evans[21] prepared satisfactory hydrogen electrodes from porous graphite containing

[20] J. H. Hildebrand, *J. Amer. Chem. Soc.*, **35**, 847 (1913).
[21] F. Hovorka and R. D. Evans, *Trans. Electrochem. Soc.*, **80**, 193 (1941).

about 1.5 per cent by weight of finely divided platinum. In the operation of this electrode hydrogen is forced through the porous structure.

If the system is closed, a foil electrode immersed in the solution over a jet of slowly bubbling hydrogen is suitable. With a calomel electrode of the immersion type, the cell may be of a very simple design. Perley[22] has designed

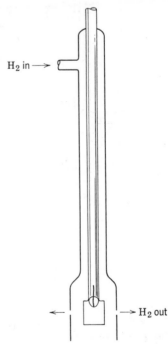

H₂ in →

→ H₂ out

FIG. 10–2. The Hildebrand hydrogen electrode.

a cell in which equilibrium with small samples is rapidly attained. His cell requires only 2 ml of solution, and a measurement accurate to 0.02 pH unit can be made 5 minutes after the flow of hydrogen is begun. As may be seen in Fig. 10–3, the sample is contained in a glass tube which is immersed in a saturated solution of potassium chloride. The level of the latter is slightly higher than that of the sample; hence the chloride solution streams slowly through the small aperture in the bottom of the sample tube, forming the liquid junction.

---

[22] G. A. Perley, *Trans. Electrochem. Soc.*, **92**, 485 (1947).

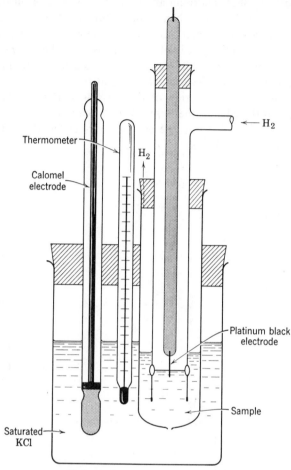

FIG. 10–3. Perley's hydrogen electrode cell.

The hydrogen electrode has also been employed in reference half-cells. A cell consisting of two hydrogen electrodes, one immersed in a standard reference buffer solution, was designed by Bates, Pinching, and Smith.[23] This cell is shown in Fig. 3–2 (page 56).

PREPARATION OF HYDROGEN ELECTRODES

It is perhaps worth while to describe in some detail the preparation of the platinum bases for hydrogen electrodes and the methods for providing a

[3] R. G. Bates, G. D. Pinching, and E. R. Smith, *J. Res. Nat. Bur. Stand.*, **45**, 418 (1950).

catalytically active surface. Sheet platinum about 0.125 mm thick is cut into pieces of about 1 cm square. A piece of No. 26 platinum wire 2 cm in length is spot-welded to the foil near the center of one edge. The welding is accomplished by placing the foil, with the wire in place, on a piece of asbestos board and heating the spot to be welded with a small gas-oxygen flame. A sharp blow with a hammer joins the two pieces of white-hot metal.

Figure 10–4 illustrates the construction of the glass seal. Flint glass is melted over the wire and the edge of the foil to form a bead about 4 mm in diameter. This bead is then sealed into the end of an 8-cm length of flint glass tubing of 5 mm outside diameter.

FIG. 10–4. Base for hydrogen electrode.

New electrodes are cleaned before use by brief immersion in a cleaning mixture prepared by combining three volumes of 12 M hydrochloric acid with one volume of 16 M nitric acid and four volumes of water. This mixture, which can also be employed repeatedly for stripping the platinum black from used electrodes, is sometimes called " 50 per cent aqua regia."

It is often stated that the base metal should, for best results, be very smooth. For this reason, the foil is sometimes polished with emery or laid upon a flat surface and "ironed" smooth with a glass rod. The platinum bases are also sometimes plated lightly with gold to prevent diffusion of hydrogen into the base metal.[24] The author has never observed any significant differences in the behavior of freshly platinized electrodes with etched, smooth, or gold-plated bases. There remains a possibility however, that differences in the duration of activity of the surface do exist.

Although platinization is the best means of activating the surface of the electrode, it is sometimes desirable (in particular when reducible substances are present in the solution) to employ bright platinum electrodes which have been activated mechanically.[25] The foil is supported on a glass plate and "ironed" with the rounded end of a glass rod. After slight etching in warm 50 per cent aqua regia and washing in concentrated nitric acid and in water, the electrode is ready for use; it will retain its activity for several hours.

PLATINIZATION

Feltham and Spiro[26] have made a careful examination of the properties of platinized platinum electrodes and have recommended the following conditions for platinization of the bases for hydrogen electrodes. The plating solution should consist of a 3.5 per cent (w/v) solution of chloroplatinic acid to which has been added lead acetate trihydrate in the amount of 0.005 per cent (w/v). A deposition time of 5 minutes should be adequate at a current density of 30 mA cm^{-2}. Feltham and Spiro find that good stirring is essential and that no gas should be evolved at the cathode. Chlorine gas evolved at the platinum anode should not be allowed to interact with the cathode; plating cells of H-type or U-type construction are effective in preventing the chlorine from reaching the cathode. After plating, the surface coat should appear uniform and very black; streaking should be absent.

The presence of lead has a marked effect on the character of the electro-deposited platinum, possibly through an increase of the grain size.[27] Platinum black deposited from plating solutions that contain no lead is rarely smooth and adherent. Nevertheless, the presence of traces of lead on the surface of the electrode may often be undesirable, and excellent results have been obtained with electrodes activated by plating from solutions containing no lead. Hills and Ives[28] found that electrodes which had been electrolyzed for 10 to 20 minutes at a current density of 10 to 20 mA cm^{-2} in a solution of

[24] S. Popoff, A. H. Kunz, and R. D. Snow, *J. Phys. Chem.*, **32**, 1056 (1928).

[25] D. Feakins and C. M. French, *Chem. Ind. (London)*, 1107 (1954); *J. Chem. Soc.*, 3168 (1956); 2284 (1957).

[26] A. M. Feltham and M. Spiro, *Chem. Rev.*, **71**, 177 (1971).

[27] H. T. S. Britton, *Hydrogen Ions*, 4th ed., Vol. I, Chapter 3, D. Van Nostrand Co., Princeton, N.J., 1956.

[28] G. J. Hills and D. J. G. Ives, *J. Chem. Soc.*, 305 (1951); Chapter 2 in *Reference Electrodes*, D. J. G. Ives, and G. J. Janz, eds., Academic Press, New York, 1961.

2 per cent platinic chloride in 2 M hydrochloric acid were satisfactory in every way. Although very active, these electrodes were not at all black but were covered instead with a light gray or golden film. When the presence of traces of lead is likely to produce undesirable results, Feltham and Spiro[2] recommend removal of the surface lead by soaking the electrode for 24 hours in aerated 1 M perchloric acid.

The finished electrodes should be stored in water. Cells with hydrogen electrodes are sometimes evacuated to remove adsorbed gases from the electrodes before the cell is filled.[29] The electrodes become dry and remain so for a few minutes without impairment of their usefulness. Nevertheless, if dry electrodes are exposed to air for any considerable period, their catalytic activity is lost, and they must be replatinized before use.[30]

PLATINUM SOLUTION

The following procedure for the preparation of the platinizing solution has been found satisfactory: Scrap platinum, 1.6 g, is cleaned in hot concentrated nitric acid, rinsed with water, and ignited to red heat in a Bunsen flame. The metal is cut into small fragments to facilitate solution and is digested in warm aqua regia (mixture of three volumes of concentrated hydrochloric acid with one volume of concentrated nitric acid) until completely dissolved. The acid platinum mixture is evaporated to dryness on a steam bath and the residue taken up in about 20 ml of concentrated hydrochloric acid. The evaporation and addition of hydrochloric acid are repeated twice. The final (fourth) evaporation should be stopped before the crystals are completely dry. The residue of chloroplatinic acid hexahydrate, $H_2PtCl_6 \cdot 6H_2O$, remaining after the final evaporation is dissolved in 100 ml of distilled water, and 5 mg of lead acetate trihydrate is added, if desired. The solution should be perfectly clear.

PALLADIUM ELECTRODES

The very interesting behavior of palladium hydrogen electrodes has been clarified by the work of Hoare and Schuldiner.[31] When platinum electrodes coated with palladium are immersed in aqueous solutions containing dissolved hydrogen, they achieve the same potentials as platinized electrodes. Pure palladium electrodes, however, take up hydrogen spontaneously only in the amount required to saturate the α-phase of the metal, in which the

---

[29] This vacuum technique was used by Harned and his coworkers; see also W. J. Hamer and S. F. Acree, *J. Res. Nat. Bur. Stand.*, **23**, 647 (1939); R. G. Bates and S. F. Acree, *ibid.*, **30**, 129 (1943).

[30] C. N. Myers and S. F. Acree, *Am. Chem. J.*, **50**, 396 (1913).

[31] J. P. Hoare and S. Schuldiner, *J. Electrochem. Soc.*, **102**, 485 (1955); **103**, 237 (1956); **104**, 564 (1957); S. Schuldiner and J. P. Hoare, *ibid.*, **103**, 178 (1956); **105**, 278 (1958).

original lattice structure persists. In this equilibrium state, the electrode has a potential about 50 mV positive to that of the usual hydrogen electrode. Once formed, the α-palladium hydrogen electrode can be used in solutions free of hydrogen gas[32] and should prove useful for electrometric measurements in closed systems and in solutions susceptible to reduction by hydrogen. Further uptake of hydrogen does not occur spontaneously in aqueous solutions saturated with the gas but can be accomplished by electrolytic charging. Under these conditions, the β-palladium hydrogen electrode, identical in potential with the platinum electrode, is formed.

PALLADIUM SOLUTION

According to Hamer and Acree,[33] a solution suitable for the deposition of palladium black is prepared by the addition of 5 ml of a 1 M solution of hydrochloric acid and 40 ml of water to 50 ml of a solution of chloropalladous acid containing 10 per cent of palladium (as Pd). However, the composition of the palladium solution can be varied within fairly wide limits. Britton[34] recommends a 1 to 2 per cent solution of palladous chloride which is about 1 M with respect to hydrochloric acid.

The author has prepared suitable plating solutions by electrolytic corrosion of palladium foil in 12 M hydrochloric acid. Two strips of the metal are dipped into about 50 ml of the acid and a direct current of about 500 mA passed between them with frequent reversal of polarity. When about 0.5 g of the metal has passed into solution, the electrodes are removed, the volume of the solution reduced to 2 to 5 ml on a steam bath, and 50 ml of water and 40 mg of lead acetate trihydrate added. To avoid contamination with platinum and other metals, only palladium anodes should be used in the plating solution.

Palladium black is a rather poor substitute for platinum black. Whenever the absence of reducible substances permits them to be used, platinized electrodes are to be preferred on account of their superior reproducibility and rapid attainment of equilibrium. Each type loses its activity slowly when kept in solutions containing dissolved air. The coating should be replaced when the electrode is slow to reach equilibrium.

HYDROGEN

The active surface of the electrode should preferably be completely covered by the solution when in use. It should be supplied with pure hydrogen gas at the rate of one to two bubbles per second from a jet about 1 mm in

32 S. Schuldiner and J. P. Hoare, *Can. J. Chem.*, **37**, 228 (1959).
33 W. J. Hamer and S. F. Acree, *J. Res. Nat. Bur. Stand.*, **33**, 87 (1944).
34 H. T. S. Britton, *Hydrogen Ions*, 4th ed., Vol. I, Chapter 3, D. Van Nostrand Co., Princeton, N.J., 1956.

diameter. To avoid changes of concentration the gas can be passed through a saturator containing water or cell solution before it enters the cell.

The electrolytic hydrogen available commercially in cylinders is a convenient source of the gas. In the analysis of samples from six different cylinders of electrolytic hydrogen, the National Bureau of Standards found oxygen in amounts ranging from 0.15 to 0.20 per cent and nitrogen in amounts of 0.02 to 0.05 per cent. The gas from one cylinder contained 0.01 per cent of carbon dioxide, but none was detected in samples from the other five. The potential of the hydrogen electrode is quite sensitive to traces of oxygen. The oxygen and carbon dioxide should be removed if the most accurate results are required.

A tube of soda-lime or Ascarite, guarded with plugs of fine glass wool to prevent dust from entering the line, is suitable for the absorption of carbon dioxide. Traces of oxygen are reduced by the hydrogen in contact with certain catalytic surfaces. Three types of catalyst have been used extensively: finely divided copper heated to 450 °C, palladium on an asbestos base heated to 200 °C, and a platinum catalyst that functions at room temperature. The latter is available commercially in the form of a cartridge with brass fittings at both ends.

These catalysts convert any oxygen present to water vapor, leaving the effluent hydrogen slightly moist. The absorbent for carbon dioxide, if used, should be located between the cylinder and the catalytic purifier; otherwise the absorption of water by the soda-lime will rapidly lower the efficiency of the absorbent or even block the passage of gas through the line.

The Quinhydrone Electrode

The quinhydrone electrode is an important secondary hydrogen ion electrode. It was apparently first applied to pH measurements by Biilmann[35] and has been developed largely through his efforts and those of his co-workers.[36]

Quinhydrone itself is an equimolecular compound of benzoquinone, OC_6H_4O, and hydroquinone, HOC_6H_4OH, slightly soluble in water. These two substances, which will be referred to as Q and H_2Q, respectively, form with hydrogen ion a reversible oxidation-reduction system.

$$Q + 2H^+ + 2e \rightleftharpoons H_2Q \tag{7}$$

Consequently, an inert electrode (usually gold or platinum) dipping in a

[35] E. Biilmann, *Ann. Chim.*, [9] **15**, 109 (1921); *Bull. Soc. Chim.*, **41**, 213 (1927).

[36] E. Biilmann and H. Lund, *Ann. Chim.*, [9] **16**, 321, 339 (1921); E. Biilmann and A. L. Jensen, *Bull. Soc. Chim.*, **41**, 151 (1927).

solution containing dissolved quinone and hydroquinone and hydrogen ion assumes a potential E_Q dependent on the activities of these three substances. The electrode potential is given in terms of the IUPAC convention by

$$E_Q = E_Q{}^\circ + \frac{RT}{2F} \ln \frac{a_Q a_H{}^2}{a_{H_2Q}} \tag{8}$$

or

$$E_Q = E_Q{}^\circ + \frac{RT}{F} \ln a_H + \frac{RF}{2F} \ln \frac{m_Q}{m_{H_2Q}} \frac{\gamma_Q}{\gamma_{H_2Q}} \tag{8a}$$

Berthoud and Kunz[37] have determined the solubility S of quinhydrone in water at 15, 25, and 35 °C and also the equilibrium constant K_Q for the dissociation of dissolved quinhydrone into hydroquinone and quinone:

$$K_Q = \frac{a_{H_2Q} a_Q}{a_{Q \cdot H_2Q}} \approx \frac{\alpha^2 S}{1 - \alpha} \tag{9}$$

where α is the fraction dissociated. Their results are given in Table 10–2.

TABLE 10–2. SOLUBILITY AND DISSOCIATION CONSTANT OF QUINHYDRONE

t °C	Solubility mol dm^{-3}	K_Q
15	0.01224	0.223
25	0.01870	0.259
35	0.02788	0.291

It is evident that about 4 g of solid quinhydrone dissolves in a liter of pure water at 25 °C and that about 93 per cent of the dissolved compound is dissociated into its two components. Measurements of light absorption show no evidence of semiquinone molecules, HOC_6H_4O.[38] The rather involved kinetics of the electrochemical oxidation of hydroquinone and the reduction of quinone have been studied by Rosenthal, Lorch, and Hammett.[39]

In acid solutions the concentrations of hydroquinone and quinone are nearly equal. As long as the activity coefficients of these two substances also

---

[37] A. Berthoud and S. Kunz, *Helv. Chim. Acta*, **21**, 17 (1938); see also S. P. L. Sørensen, M. Sørensen, and K. Linderstrøm-Lang, *Ann. Chim.*, **16**, 283 (1921).

[38] C. Wagner and K. Grünewald, *Z. Elektrochem.*, **46**, 265 (1940).

[39] R. Rosenthal, A. E. Lorch, and L. P. Hammett, *J. Amer. Chem. Soc.*, **59**, 1795 (1937).

remain equal, as they presumably do at low ionic strengths, the last term of equation 8a is zero, and the quinhydrone electrode behaves as a true hydrogen ion electrode (compare equation 3a):

$$E_Q = E_Q^\circ + \frac{RT}{F} \ln a_H \qquad (10)$$

The quinhydrone electrode is simple to construct, comes to equilibrium more rapidly than the hydrogen electrode, is not as readily disturbed by poisons or oxidizing agents, and can be employed in the presence of substances reduced by hydrogen. The pH of solutions containing dissolved gases can be measured with the quinhydrone electrode, and the electrode can be used in the presence of air, although best results are obtained under oxygen-free conditions. It is likewise applicable in many nonaqueous and partly aqueous media, including alcohol-water mixtures, phenols, acetone, and formic acid.[40] The chief disadvantage of the quinhydrone electrode is its limitation to solutions of pH less than 8. It also gives erroneous results in the presence of proteins, certain oxidizing agents, and high concentrations of salts, and it is not stable for long periods of time, particularly at temperatures above 30 °C. A useful summary of the properties, theory, and behavior of quinhydrone electrodes has been given by Janz and Ives.[40]

ERRORS OF QUINHYDRONE ELECTRODE

Any influence that tends to make the last term of equation 8a depart from zero will cause an error in the hydrogen ion response of the quinhydrone electrode as expressed by equation 10. Two defects of the electrode, namely the alkaline error and the salt error, can be explained in terms of changes in the ratios of the concentrations or the activity coefficients that appear in this term. The departures in alkaline solutions, which usually become apparent at pH values above 7 to 8, can be attributed to oxidation of the hydroquinone and to the acidic dissociation of hydroquinone. Both of these reactions cause the ratio m_Q/m_{H_2Q} to increase; as a consequence, the electrode potential becomes too positive and the apparent pH will be too low.

The salt error, usually of secondary importance in pH measurements, results from unequal changes in the activity coefficients of hydroquinone and quinone when a neutral salt or other substance is added to the solution saturated with quinhydrone. The ratio of concentrations in the last term of equation 8a remains unaltered on addition of salt. The potential of the

---

[40]References to the original literature are given by G. J. Janz and D. J. G. Ives in Chapter 6, *Reference Electrodes*, D. J. G. Ives and G. J. Janz, eds., Academic Press, New York, 1961.

electrode depends on a ratio of activities, however, and this ratio changes with the ionic strength.

The logarithm of the activity coefficient of a nonelectrolyte usually varies linearly with the concentration (or ionic strength) of added salt.[41] Hence $\log(\gamma_Q/\gamma_{H_2Q})$ and the observed salt error are also linear functions of the salt concentration. We are concerned here with a ratio of activity coefficients of two neutral species, a quantity that is little affected by addition of salt. Stonehill[42] found, for example, that the salt error in solutions of nitric acid of concentrations c up to 1 M is $0.00529c$ V at 25 °C. Thus, the error in 1 M nitric acid amounts to less than 0.09 pH unit.

The salt error of the quinhydrone electrode has been studied extensively through solubility measurements. However, Hovorka and Dearing[43] and Gabbard[44] have compared the quinhydrone electrode directly with the hydrogen electrode in salt solutions by means of the cell

$$Pt; H_2, H^+, \text{Salt Soln.}, Q \cdot H_2Q; Au \tag{11}$$

Gabbard studied the dependence of the salt error on pH by measuring the e.m.f. of cells of type 11 which contained 0.1 to 2.0 M solutions of potassium, magnesium, and barium chlorides and magnesium sulfate, slightly buffered at pH 3.3 to 8.1. The error appeared to be independent of pH in solutions of pH less than 4.5 but increased rapidly, for a given concentration of salt, as the pH rose above 5.5. These results have been explained by the salt effect on the ionization of hydroquinone. In dilute solutions, ionization is appreciable only above pH 7. However, high salt concentrations promote the ionization of hydroquinone, like that of other uncharged acids, through a lowering of the activities of the ions which are in equilibrium with the undissociated molecules.

Hovorka and Dearing found that ΔpH, the salt error in pH units, is nearly proportional to c_s, the normal concentration of salt:

$$\Delta pH = kc_s \tag{12}$$

The values of k for a number of substances, based on the work of these investigators and of Stonehill,[45] are summarized in Table 10–3. Negative values of k and ΔpH signify that the pH indicated by the potential of the quinhydrone electrode is too high. In mixtures of salts the errors are approximately additive.

---

[41] M. Randall and C. F. Failey, *Chem. Rev.*, **4**, 291 (1927).

[42] H. I. Stonehill, *Trans. Faraday Soc.*, **39**, 67 (1943).

[43] F. Hovorka and W. C. Dearing, *J. Amer. Chem. Soc.*, **57**, 446 (1935).

[44] J. L. Gabbard, *J. Amer. Chem. Soc.*, **69**, 533 (1947).

[45] H. I. Stonehill, *Trans. Faraday Soc.*, **39**, 67 (1943).

TABLE 10-3. SALT ERRORS OF THE QUINHYDRONE ELECTRODE AT
25 °C. VALUES OF THE CONSTANT k (EQUATION 12)

(Data of Hovorka and Dearing and of Stonehill)			
Substance Added	k	Substance Added	k
HCl	-0.0616	$SrCl_2$	-0.0379
HNO_3	-0.0895	$BaCl_2$	-0.0438
LiCl	-0.0353	H_2SO_4	-0.0314
NaCl	-0.0413	Li_2SO_4	0.0269
KCl	-0.0372	Na_2SO_4	0.0227
KNO_3	-0.0645	K_2SO_4	0.0238
$MgCl_2$	-0.0346	$MgSO_4$	0.0206
$CaCl_2$	-0.0367	Mannitol	0.0237

As we have stated, the basis for the quinhydrone salt error is a change in the ratio of the activities of quinone and hydroquinone in the solution saturated with the compound, quinhydrone. The product of the activities of these two substances in the saturated solution must, however, remain constant at a given temperature, as equation 9 shows. If the solution were saturated not only with quinhydrone but also with hydroquinone or with quinone, no salt error would exist.[46] The two electrodes formed in this manner are called the hydroquinhydrone and quinoquinhydrone electrodes. They have not been widely used in the measurement of pH.

PREPARATION OF QUINHYDRONE ELECTRODES

Considerable attention has been given to devising the best methods of preparing quinhydrone, cleaning and drying the gold or platinum electrodes, washing the quinhydrone with cell electrolyte, and other techniques.[47] Quinhydrone is available commercially. On account of the slight volatility of quinone, however, commercial quinhydrone should be recrystallized from water at 70 °C to assure that the two components are present in equimolecular proportions. The product is dried at room temperature.

---

[46] E. Biilmann and H. Lund, Ann. Chim., [9] 16, 321, 339 (1921); S. P. L. Sørensen, M. Sørensen, and K. Linderstrøm-Lang, ibid., [9] 16, 283 (1921).
[47] E. Biilmann and A. L. Jensen, Bull. Soc. Chim., 41, 151 (1927); J. L. R. Morgan, O. M. Lammert, and M. A. Campbell, J. Amer. Chem. Soc., 53, 454 (1931), Trans. Electrochem. Soc., 61, 409 (1932); O. M. Lammert, J. L. R. Morgan, and M. A. Campbell, J. Amer. Chem. Soc., 53, 597 (1931); H. S. Harned and D. D. Wright, ibid. 55, 4849 (1933); W. J. Clayton and W. C. Vosburgh, ibid., 59, 2414 (1937). See also reference 40.

Both bright platinum and gold electrodes are satisfactory. The half-cell should be so designed that the electrode is in contact with some of the excess solid, thus assuring saturation of the solution in the vicinity of the electrode. Electrodes should be cleaned with hot chromic acid and washed with water and absolute alcohol.[48] Clayton and Vosburgh[49] dried the electrodes by placing them, still wet with alcohol, into the half-cell, which was then evacuated with an oil pump. The vacuum was broken with dry nitrogen. Air-free electrolyte, containing quinhydrone in suspension, was immediately forced in and stirred within the cell by pure nitrogen. Cells prepared in this way came to agreement within 0.005 mV within 1 minute after introduction of the solution, provided that no difference of temperature existed. If the electrodes were dried in air, about an hour was required for agreement to 0.01 mV.

A reference half-cell consisting of a quinhydrone electrode in a solution of potassium tetroxalate has been described by Schomaker and Brown.[50] A reference quinhydrone electrode in a mixture 0.01 M with respect to hydrochloric acid and 0.09 M with respect to potassium chloride was recommended earlier by Veibel.[51] Cells for the measurement of pH with a quinhydrone electrode and a saturated calomel reference electrode have been described by Perley[52] and Sanders.[53] The quinhydrone electrode is well suited for measurements with small samples.[54]

POTENTIAL OF THE QUINHYDRONE ELECTRODE

Harned and Wright[55] measured the e.m.f. of the cell

$$\text{Pt};\text{Q}\cdot\text{H}_2\text{Q},\text{HCl}(0.01\text{ M}),\text{AgCl};\text{Ag} \qquad (13)$$

at intervals of 5 degrees from 0 to 40 °C. By combining their data with the e.m.f. of the hydrogen-silver chloride cell containing 0.01 M hydrochloric acid, one can calculate the standard potential of the quinhydrone electrode, that is, the e.m.f. of the cell

$$\text{Pt};\text{H}_2(1\text{ atm.}),\text{H}^+(a=1),\text{Q}\cdot\text{H}_2\text{Q};\text{Pt} \qquad (14)$$

The value obtained at 25 °C (0.69992 V) differs by 0.31 mV from that found

[48] J. L. R. Morgan, O. M. Lammert, and M. A. Campbell, *Trans. Electrochem. Soc.*, **61**, 409 (1932).

[49] W. J. Clayton and W. C. Vosburgh, *J. Amer. Chem. Soc.*, **59**, 2414 (1937).

[50] V. Schomaker and D. J. Brown, *Ind. Eng. Chem., Anal. Ed.*, **9**, 34 (1937).

[51] S. Veibel, *J. Chem. Soc.*, **123**, 2203 (1923).

[52] G. A. Perley, *Proc. Am. Assoc. Textile Chem. Colorists*, 832 (1937).

[53] G. P. Sanders, *Ind. Eng. Chem., Anal. Ed.*, **10**, 274 (1938).

[54] J. A. Pierce, *J. Biol. Chem.*, **117**, 651 (1937); A. Itano, *Compt. Rend. Trav. Lab. Carlsberg, Sér. Chim.*, **22**, 235 (1938).

[55] H. S. Harned and D. D. Wright, *J. Amer. Chem. Soc.*, **55**, 4849 (1933).

by Hovorka and Dearing (0.69961 V).[56] The value at 25 °C listed in Table 10–4 is the mean of these two. The temperature coefficient of E^0, as found by Harned and Wright, was used to compute the standard potential at the other temperatures.

Stonehill[57] has determined the standard potentials of the hydroquin-hydrone and quinoquinhydrone electrodes at 25 °C. He found 0.6141 V for the former and 0.7459 V for the latter.

TABLE 10–4. STANDARD POTENTIALS
OF THE QUINHYDRONE ELECTRODE

t °C	E^0 (cell 14) V
0	0.71798
5	0.71437
10	0.71073
15	0.70709
20	0.70343
25	0.69976
30	0.69607
35	0.69237
40	0.68865

The Antimony Electrode

It has long been known that the potential of an electrode consisting of a stick of antimony cast in air is a function of the pH of the solution in which the stick is immersed.[58] The potential is probably developed by an oxidation-reduction reaction involving antimony and the thin layer of antimonous oxide (Sb_2O_3) that is usually present on the surface of the metal. The electrochemical process can be formulated as follows:

$$Sb_2O_3(s) + 6H^+ + 6e \; \xrightleftharpoons \; 2Sb(s) + 3H_2O(l) \qquad (15)$$

[56] F. Hovorka and W. C. Dearing, J. Amer. Chem. Soc., **57**, 446 (1935). The unit of e.m.f used in this book is that formerly called the absolute volt. The international volt no longer exists, and the units in which E is expressed need cause concern only in the case of data recorded prior to 1948. These older results can be corrected to the modern scale in the following way: $E(abs\ V) = 1.00033E(int\ V)$.

[57] H. I. Stonehill, Trans. Faraday Soc., **39**, 67 (1943).

[58] I. M. Kolthoff and B. D. Hartong, Rec. Trav. Chim., **44**, 113 (1925); E. J. Roberts and F. Fenwick, J. Amer. Chem. Soc., **50**, 2125 (1928); H. T. S. Britton and R. A. Robinson, J. Chem. Soc., **458**, (1931).

Antimony and antimony oxide, both solids, are regarded as being in their standard states of unit activity. As long as the activity of water remains approximately unity, as it does in dilute aqueous solutions, the potential of the electrode, E_{Sb}, should be a function of the activity of hydrogen ions:

$$E_{Sb} = E_{Sb}{}^\circ + \frac{RT}{F} \ln a_H \qquad (16)$$

In actual practice, the antimony electrode is found not to furnish results of high accuracy. The condition of the metal, whether cast or electrodeposited, polished or etched, and that of the oxide, have some influence on the behavior of the electrode. The usefulness of the antimony electrode in pH measurements can be considerably enhanced by careful calibration of its potential with a series of standard buffer solutions.[59] Even so, the electrode is not completely reversible, and its potential is found to vary not only with the the hydrogen ion activity but with the concentration of dissolved oxygen,[60] with the composition of the buffer solution,[61] and with stirring of the electrolyte.[62]

Britton and Robinson[63] studied several forms of electrode and chose cast antimony bars as the most satisfactory. These were polished with emery paper immediately before use. Pure precipitated antimonous oxide was suspended in the solution in which these electrodes were immersed. Hovorka and Chapman[64] prepared satisfactory electrodes by deposition of pure antimony from its solution in hydrofluoric acid. Engel[65] has described an electrode composed of powdered antimony held together with a binder. Cast antimony electrodes, mounted in a plastic sleeve with only the tip protruding, are available commercially. Levin[66] has described a convenient semimicro electrode which has been found to vary with pH in a satisfactorily linear fashion between pH 2 and 11.[67] Only the tip of a bar of metal cast within a

---

[59] H. T. S. Britton and R. A. Robinson, *J. Chem. Soc.*, 458 (1931); G. A. Perley, *Ind. Eng. Chem., Anal. Ed.*, **11**, 319 (1939).

[60] Y. Kauko and L. Knappsberg, *Suomen Kemistilehti*, **12B**, 17 (1939); *Z. Elektrochem.*, **45**, 760 (1939); see also the paper by Perley.

[61] H. Yoshimura, *Japan. J. Med. Sci., III Biophys.*, **4**, 131 (1936); M. Tamai, *J. Biochem. (Tokyo)*, **29**, 307 (1939).

[62] K. Fischbeck and F. Eimer, *Z. Elektrochem.*, **44**, 845 (1938); J. T. Stock, W. C. Purdy, and L. M. Garcia, *Chem. Rev.*, **58**, 611 (1958).

[63] H. T. S. Britton and R. A. Robinson, *J. Chem. Soc.*, 458 (1931).

[64] F. Hovorka and G. H. Chapman, *J. Amer. Chem. Soc.*, **63**, 955, 2024 (1941).

[65] F. Engel, German Patent 727,846 (Oct. 8, 1942).

[66] I. Levin, *Chemist-Analyst*, **41**, 89 (1952).

[67] L. M. Garcia, W. C. Purdy, and J. T. Stock, *Chemist-Analyst*, **47**, 36 (1958). A cell with two electrodes of the Levin type, useful for both direct potentiometric titrations and derivative titrimetry, has been described by J. T. Stock and W. C. Purdy, *ibid.*, **47**, 37 (1958).

narrow glass tube is exposed to the solution, and the surface can be renewed when necessary by polishing with an abrasive cloth which removes a small amount of the antimony and the surrounding glass wall as well. A pH meter or other electronic amplifier may have to be used to measure the e.m.f. of cells comprising the relatively highly resistant micro forms of the antimony electrode.

CHARACTERISTICS OF THE ANTIMONY ELECTRODE

The advantages and disadvantages of the antimony electrode in pH measurements, electrometric titrations, and industrial pH control have been summarized elsewhere.[68] The rapid response and the simplicity and rugged construction of this electrode make it useful for continuous industrial recording when great accuracy is not required. It can be used at elevated temperatures and in alkaline solutions, and its low resistance recommends it for installations where the humidity of the atmosphere is high and the electronic amplifiers necessary for measurements of glass electrode potentials may be subject to large leakage losses. It is very well suited for use as an end point indicator in titrations and can be employed in cyanide and sulfite solutions where the hydrogen and quinhydrone electrodes are inapplicable. Under controlled conditions, it will furnish useful results in the presence of reducing sugars,[69] alkaloids,[70] gelatin, and 3 per cent agar.[71] It has also been used successfully for titrations in aqueous ethanol.[72] Inasmuch as water participates in the electrode reaction (equation 15), the titration curve presumably suffers some displacement as the activity of water is changed. For this reason, the composition of the solvent should be kept constant throughout the course of the titration.

There is no general agreement as to the extent of the linear portion of the curve of potential vs. pH. According to Roberts and Fenwick,[73] the theoretical slope is maintained from pH 1 to 10 within ± 0.01 unit when equilibrium is approached from the alkaline side and no unstable solid phase is present. More commonly, the e.m.f.-pH slope is found to depart from the theoretical value. Hovorka and Chapman[74] give pH 2.2 and 8.0 as its limits. Ball[75]

[68] G. A. Perley, *Ind. Eng. Chem., Anal. Ed.*, **11**, 316, 319 (1939); J. T. Stock, W. C. Purdy, and L. M. Garcia, *Chem. Rev.*, **58**, 611 (1958); D. J. G. Ives in Chapter 7, *Reference Electrodes*, D. J. G. Ives and G. J. Janz, eds., Academic Press, New York, 1961.

[69] U. Beretta, *Rend. Accad. Sci. (Napoli)*, [4] **8**, (1938).

[70] R. Carreras, *Afinidad*, **19**, 341 (1942).

[71] R. Spychalski, *Roczniki Chem.*, **19**, 441 (1939).

[72] N. A. Izmailov and V. P. Pivnenko, *Referat. Zh. Khim.*, **4**, 65 (1941).

[73] E. J. Roberts and F. Fenwick, *J. Amer. Chem. Soc.*, **50**, 2125 (1928).

[74] F. Hovorka and G. H. Chapman, *J. Amer. Chem. Soc.*, **63**, 955, 2024 (1941).

[75] T. R. Ball, *Trans. Electrochem. Soc.*, **72**, 139 (1937).

found the change of potential to be linear between pH 2 and 7 within a tolerance of ± 0.01 pH unit. The linear region could be extended to pH 10 by replacement of the antimonous oxide with antimonous sulfide. Izmailov and Pivnenko[76] found two linear branches in the e.m.f.-pH curve, the first extending from pH 2.5 to 9.0 and the second from pH 10 to 13. A break in the curve occurred between pH 8.5 and 10.0. This behavior is explained by the detailed study of the solubilities and amphoteric properties of highly purified antimonous and antimonic oxides conducted by Tourky and El Wakkad.[77] These investigators found that antimonous oxide has an isoelectric point at pH 8.6. The oxide behaves as an ortho base below this point and as a meta acid above. They discovered that the antimony electrode, under strictly air-free conditions, yields a potential-pH curve that actually consists of four parallel branches, separated by slight breaks at pH values of about 2, 5, and 8. The potential-pH plot is thus actually a titration curve, first of the tri-acid base and then of the monobasic acid. They regard their treatment of the electrodes with hydrogen at high temperatures as having removed the adsorbed oxygen which usually masks the amphoteric properties of the hydrous oxide and leads to a plot of potential vs. pH that is roughly linear. This evidence suggests that the antimony electrode should be viewed as a metal-metal oxide-oxygen electrode.

In common with other oxidation-reduction electrodes, the antimony electrode is disturbed by oxidizing and reducing agents.[78] Oxidizing agents shift the potential to the positive side and reducing agents to the negative side. There is also a marked sensitivity to anions of the hydroxy acids (tartrates, citrates) that form complexes with antimony,[79] to metaphosphate and oxalate,[80] and to traces of certain cations. Perley[81] has found, for example, that 0.5 ppm of cupric ion can cause an error of 0.2 pH unit. Complex-formers may impair the performance of the electrode by removing the oxide film from the surface.

The antimony electrode is a simple immersion electrode and can be used with cells of the simplest design. Electrodes of very small size behave well. In one interesting application, antimony electrodes were used to study the changes of pH in the stomach during digestion.[82] The pieces of cast metal

[76] N. A. Izmailov and V. P. Pivnenko, *Referat. Zh. Khim.*, **4**, 65 (1941).
[77] A. R. Tourky and S. E. S. El Wakkad, *J. Chem. Soc.*, 752, 756, 759 (1948).
[78] K. Fischbeck and F. Eimer, *Z. Elektrochem.*, **44**, 845 (1938).
[79] H. T. S. Britton, *Hydrogen Ions*, 4th ed., Vol. I, Chapter 6, D. Van Nostrand Co., Princeton, N.J., 1956; W. Blum, A. O. Beckman, and W. R. Meyer, *Trans. Electrochem. Soc.*, **80**, 249 (1941).
[80] M. Tamai, *J. Biochem.* (*Tokyo*), **29**, 307 (1939).
[81] G. A. Perley, *Ind. Eng. Chem., Anal. Ed.*, **11**, 319 (1939).
[82] R. C. Erb and K. L. Senior, *J. Am. Osteopath. Assoc.*, **38**, 96 (1938); H. W. Haggard and L. A. Greenberg, *Science*, **93**, 479 (1941).

were 5 mm in length and 1 mm in diameter, with leads consisting of three strands of No. 43 copper wire, and could be swallowed without difficulty. The circuit was completed to a calomel reference electrode through saline in a basin in which the subject's foot was placed.

STANDARD POTENTIAL OF THE ANTIMONY ELECTRODE

The standard potential $E_{Sb}°$ and temperature coefficient of the antimony electrode have been considered by Fischbeck and Eimer, by Perley, by Hovorka and Chapman,[83] and in two later reviews.[84] However, these potentials are of little practical value in the measurement of pH, in view of the numerous factors influencing the potential and the consequent difficulty of preparing any one "standard" electrode. The most useful procedure is to construct a calibration curve of pH vs. e.m.f. for the particular antimony cell with which measurements are to be made. The standard buffers chosen should not contain oxalates, tartrates, or citrates. The slope of the pH-e.m.f. curve for the antimony electrode is often only approximately the value demanded by equation 16, that is, 0.05916V per pH unit at 25 °C.

The temperature coefficient of the cell composed of antimony and saturated calomel electrodes is rather large, amounting to about 1 mV per degree at pH 3 and to over 3 mV per degree at pH 12.[85] The temperature coefficient in unbuffered solutions appears to be somewhat different from that in well-buffered media.

Some of the properties of the hydrogen, quinhydrone, antimony, and glass electrodes are compared in Table 10–5.[86]

Other Hydrogen Ion Electrodes

A few other hydrogen ion electrodes deserve brief mention, although their accuracy is in general low. The potentials of tungsten and tellurium rods change with pH, and these electrodes can be employed successfully as end point indicators in acid-base titrations. It is presumed that the mode of operation is to be ascribed, like that of the antimony electrode, to the presence of a surface coating of oxide. Britton and Dodd[87] found the e.m.f.-pH slope

[83] K. Fischbeck and F. Eimer, *Z. Elektrochem.*, **44**, 845 (1938); G. A. Perley, *Ind. Eng. Chem., Anal. Ed.*, **11**, 319 (1939); F. Hovorka and G. H. Chapman, *J. Amer. Chem. Soc.*, **63**, 955 (1941).

[84] J. T. Stock, W. C. Purdy, and L. M. Garcia, *Chem. Rev.*, **58**, 611 (1958); D. J. G. Ives in Chapter 7, *Reference Electrodes*, D. J. G. Ives and G. J. Janz, eds., Academic Press, New York, 1961.

[85] G. A. Perley, *Ind. Eng. Chem., Anal. Ed.*, **11**, 319 (1939).

[86] R. G. Bates, *J. Electroanal. Chem.*, **2**, 93 (1961).

[87] H. T. S. Britton and E. N. Dodd, *J. Chem. Soc.*, 829 (1931).

TABLE 10-5. PROPERTIES OF pH-RESPONSIVE ELECTRODES

Property	Hydrogen Electrode	Quinhydrone Electrode	Antimony Electrode	Glass Electrode
pH range	Unlimited	0–8	0–11	0–13 (with corrections)
pH response	Theoretical	Theoretical	Variable	Nearly theoretical (pH 0–11)
Precision (pH)	±0.001	±0.002	±0.1 (when properly calibrated)	±0.002
Convenience of measurement	Low	Medium	High	High
Time required for measurement (minutes)	30–60	5	3	<1
Versatility	Low	Medium	Medium	High
Cost of equipment	Medium	Low	Low	High
Electrical resistance	Low	Low	Low	High
Disadvantages	Strong reducing action, air must be excluded	Limited pH range, salt error	Defective response, not completely reversible	Variable asymmetry potential, high resistance, alkaline error
Interferences	Oxidizing agents, reducible organic substances, noble metal ions, SO_2 CN, unbuffered solutions	Proteins, some amines	Some oxidizing agents, Cu ion, anions of hydroxy acids	Dehydrating solutions, some colloids, fluorides, surface deposits on the electrode

for tungsten electrodes to vary considerably over the pH range 2 to 12. Several forms of tungsten electrodes have been patented.[88]

The pH response of bismuth electrodes has been examined with some care.[89] Schwabe found, for example, that the bismuth electrode is well suited to the approximate measurement of pH in the alkaline region. The slope of a plot of e.m.f. at 20 °C vs. pH from 3 to 14 was 0.051 instead of the expected

---

[88] W. Kordatzki and P. Wulff, German Patent 626,151 (Feb. 21, 1936); M. Schlötter, German Patent 701,788 (Dec. 24, 1940); F. Engel, German Patent 727,846 (Oct. 8, 1942). Sodium tungsten bronzes function as indicating electrodes for hydrogen ion and several metal cations; see M. A. Wechter, H. R. Shanks, G. Carter, G. M. Ebert, R. Guglielmino, and A. F. Voight, *Anal. Chem.*, **44**, 850 (1972).

[89] D. N. Mehta and S. K. K. Jatkar, *J. Indian Inst. Sci.*, **18A**, 109 (1935); K. Schwabe, *Z. Elektrochem*, **53**, 125 (1949); K. Schwabe and B. Philipp, *ibid.*, **55**, 411 (1951); H. M. Sammour and A. A. Moussa, *J. Chem. Soc.*, 1762 (1958).

0.058 volts per pH unit. The variations among the potentials of individual electrodes of three types were about 5 mV; deviations were greatest in poorly buffered solutions near the neutral pH.

Perley and Godshalk[90] have described a modified oxygen electrode suitable for pH measurements in solutions free of molecular hydrogen and variable oxidation-reduction systems. The metallic electrode is fashioned of nonporous iridium, rhenium, osmium, or ruthenium. It must not be contaminated with occluded hydrogen. The potential as measured against a calomel electrode is said to change linearly from pH 0 to 14 in solutions containing molecular oxygen and is reproducible at temperatures from 0 to 100 °C.

The hydrogen ion function of electrodes of manganese dioxide has been studied by Cahoon and by Johnson and Vosburgh.[91] Cahoon found that the potential of an electrode of African pyrolusite is a linear function of pH in the range 1 to 11.4. The electrolytes studied contained zinc chloride, ammonium chloride, and ammonium hydroxide. The slope of the line is within 2 per cent of the theoretical value. Johnson and Vosburgh found that the potential of the manganese dioxide electrode in solutions of ammonium chloride depends not only on the pH but on the ammonia concentration as well. They showed that the manganese dioxide must be brought into exchange equilibrium with the electrolyte if constant and reproducible electrodes are to result.

Hall and Conant[92] obtained reproducible potentials with the chloranil electrode in glacial acetic acid. The electrode was formed by immersing a piece of platinum in the solution under study which had been saturated both with tetrachlorohydroquinone and tetrachlorobenzoquinone (chloranil). Their study is described on pages 200–202.

Calomel Reference Electrodes and Liquid Junctions

The calomel electrode is the reference electrode most commonly used to measure the changes of potential of hydrogen ion electrodes. Basically, the calomel element consists of mercury, mercurous chloride (calomel), and chloride ion:

---

[90] G. A. Perley and J. B. Godshalk, British Patent 567,722 (Feb. 28, 1945); U.S. Patent 2,416,949 (March 4, 1947). An electrode of graphite in a silicone-rubber matrix has been found to have a pH response; E. Szepesváry and E. Pungor, *Anal. Chim. Acta*, **54**, 199 (1971).

[91] N. C. Cahoon, *Trans. Electrochem. Soc.*, **68**, 177 (1935); R. S. Johnson and W. C. Vosburgh, *J. Electrochem. Soc.*, **99**, 317 (1952).

[92] N. F. Hall and J. B. Conant, *J. Amer. Chem. Soc.*, **49**, 3047 (1927); J. B. Conant and N. F. Hall, *ibid.*, **49**, 3062 (1927).

$$Cl^-,Hg_2Cl_2(s);Hg(l) \qquad (17)$$

The chloride solution is saturated with calomel at the surface of the mercury, and an excess of solid calomel assures that it will remain so in spite of temperature changes. The three well-known types of calomel electrode are identified by the concentration of the potassium chloride solution used. They are the "0.1 M calomel electrode," "the 3.5 M calomel electrode," and "the saturated calomel electrode." Electrolytic connection between the reference electrode and the solution of unknown pH (compare cell 1) is established through a salt bridge. In the United States, the bridge solution is usually a saturated solution of potassium chloride (4.16 M or 4.80 molal at 25 °C) and in Europe often the 3.5 M solution. Inasmuch as the calomel electrode and the salt bridge are frequently a single unit, it will be convenient to consider the experimental aspects of both in this section.

The ideal reference electrode has a constant and reproducible potential that is not markedly affected by temperature changes and shows little hysteresis. The ideal salt bridge would always generate the same diffusion potential, or, better still, no difference of potential, across the liquid junction,

$$\text{Bridge} \mid \text{Soln. X} \qquad (18)$$

no matter what the composition or pH of solution X might be. As we have seen, no perfect salt bridge is known. A concentrated solution of potassium chloride appears to function satisfactorily at intermediate pH values but leaves much to be desired at low and high pH.

In most respects, the calomel electrode meets these requirements satisfactorily. The 0.1 M calomel electrode is reproducible, and its potential (with respect to the standard hydrogen electrode) changes but little with changes of temperature. On the other hand, the temperature coefficient of the e.m.f. of the corresponding saturated calomel cell is much larger. The potentials of saturated electrodes are not as reproducible as those of the unsaturated types and are subject to larger hysteresis errors. There is a small shift in the potential of calomel electrodes during the first days or weeks after preparation. This change can probably be attributed to a slow disproportionation:

$$Hg_2Cl_2 = Hg + HgCl_2 \qquad (19)$$

A comparison of the properties of the 0.1 M and saturated calomel electrodes makes it seem surprising that the saturated electrode has largely replaced the 0.1 M electrode. The explanation is probably to be found in the necessity for using a very concentrated potassium chloride solution in the salt bridge. If the reference electrode itself is of the saturated type, the inconvenience of a second liquid junction can be avoided. The saturated calomel electrode is used in practical pH measurements at high temperatures as well as low, but its useful life is known to be short at temperatures above 70 °C.

PREPARATION OF CALOMEL ELECTRODES

Hills and Ives[93] have studied the conditions for the preparation of reversible calomel electrodes in unsaturated chloride solutions. Their study of the best means of preparing this important reference electrode is the most careful that has been undertaken. Five factors with a bearing on the reproducibility and constancy of the electrode were identified. These are[94]

1. Dissolved oxygen. The transitory effect is reversible. There is also a slower reaction (see below) which lowers the chloride concentration at the electrode.
2. Amount of solid phase. A very thin layer is sufficient, but there should be enough calomel to assure that the surface of the mercury is covered completely at the highest temperature at which the electrode will be used.
3. Penetration of solution between the mercury and the wall of the vessel. This is most undesirable and can be prevented by rendering the interior walls of the vessel hydrophobic. For this purpose a 1 per cent solution of silicone fluid in carbon tetrachloride is suitable.
4. State of the calomel. Electrolytic calomel[95] may be used, but it is slightly less satisfactory than very finely divided chemically precipitated calomel, intimately mixed with a dispersion of mercury globules.
5. Reaction of calomel with mercury in the dry state. This discovery is perhaps the most interesting result of the study made by Hills and Ives. When the calomel and mercury were allowed to interact in the dry state before solution was added, greatly improved electrodes resulted. The "pearly skin" formed on the surface of the mercury persists after the solution is added and appears to be essential to full reversibility.

Hills and Ives confirmed the observation of Randall and Young and of Müller and Reuther[96] that there is a marked effect of dissolved oxygen on the potentials of calomel electrodes in acid solutions. However, it is believed that the calomel electrode, like the silver-silver chloride electrode,[97] is rela-

---

[93] G. J. Hills and D. J. G. Ives, *Nature*, **165**, 530 (1950); *J. Chem. Soc.*, 311 (1951).

[94] G. J. Hills and D. J. G. Ives in Chapter 3, *Reference Electrodes*, D. J. G. Ives and G. J. Janz, eds., Academic Press, New York, 1961. See also S. N. Das and D. J. G. Ives, *J. Chem. Soc.*, 1619 (1962).

[95] J. H. Ellis, *J. Amer. Chem. Soc.*, **38**, 740 (1916). Directions for the preparation of electrolytic calomel have been given by G. Mattock, *pH Measurement and Titration*, Chapter 7, The Macmillan Co., New York, 1961.

[96] M. Randall and L. E. Young, *J. Amer. Chem. Soc.*, **50**, 989 (1928); F. Müller and H. Reuther, *Z. Elektrochem.*, **48**, 220 (1942).

[97] E. R. Smith and J. K. Taylor, *J. Res. Nat. Bur. Stand.*, **20**, 837 (1938).

tively insensitive to oxygen when the solution is a neutral chloride. The greater part of the oxygen effect in acid solutions disappears when the dissolved gas is removed. A slow residual change, irreversible in character, is apparently due to the oxidation of mercury,

$$2Hg + 2HCl + \tfrac{1}{2}O_2 = Hg_2Cl_2 + H_2O \tag{20}$$

The procedure of Hills and Ives is easily modified for the preparation of saturated calomel electrodes.[98] A little saturated potassium chloride solution is added to the dry mixture of calomel and mercury, along with a few crystals of potassium chloride. A portion of the thin paste, prepared by stirring the mixture with a glass rod, is transferred to the surface of the mercury in the vessel. Coarse crystals of potassium chloride are placed upon the layer of calomel, and the vessel is filled with a solution of potassium chloride from which dissolved air has been removed.

PURIFICATION OF SODIUM AND POTASSIUM CHLORIDES

The large effect of traces of bromide on the potentials of both calomel and silver-silver chloride electrodes makes the removal of this impurity from chlorides used in electrochemical studies a matter of some importance. Pinching and Bates[99] added 0.1 mole per cent of potassium bromide to reagent-grade sodium chloride and potassium chloride and studied the effectiveness of the usual purification procedures in removing this impurity. After three crystallizations from water, 0.003 to 0.005 per cent of bromide remained. Three precipitations of the chlorides with alcohol lowered the bromide to 0.002 to 0.003 per cent, but only 0.001 per cent remained after three precipitations with hydrogen chloride gas. Only the latter method gave a satisfactory yield of product. When chlorine gas was bubbled through the saturated chloride solution for about 10 minutes, the solution boiled to expel halogens, and the chloride precipitated once with gaseous hydrogen chloride, a 60 to 85 per cent yield of salt containing only 0.001 mole per cent of bromide was obtained.

The presence of bromide in chloride salts can be detected by the uranine color test.[100] For accurate work, bromide should be removed if present in excess of 0.005 mole per cent. The recommended procedure for the purification of sodium and potassium chlorides calls for treatment of the saturated solution with chlorine, followed by precipitation of the salts with hydrogen

---

[98] A saturated calomel electrode prepared with sodium chloride has been described by A. D. E. Lauchlan and J. E. Page, *Nature*, **151**, 84 (1943).

[99] G. D. Pinching and R. G. Bates, *J. Res. Nat. Bur. Stand.*, **37**, 311 (1946).

[100] F. L. Hahn, *Compt. Rend.*, **197**, 245 (1933); R. G. Aickin, *Australian Chem. Inst. J. Proc.*, **4**, 267 (1937); G. D. Pinching and R. G. Bates, *J. Res. Nat. Bur. Stand.*, **37**, 311 (1946). See also F. L. Hahn, *J. Amer. Chem. Soc.*, **57**, 2537 (1935).

chloride gas. The gas can be generated conveniently by dropping concentrated hydrochloric acid into concentrated sulfuric acid. It should be passed through a plug of glass wool to remove spray. The salt is collected on a filter of sintered glass and washed with small portions of water. After a preliminary drying at 110 °C, the salt is crushed and dried at 180 °C.

The chloride salts prepared in this way, although substantially bromide-free, contain occluded acid that is completely removed only by fusion. Unfortunately, fusion in moist air may leave the product slightly alkaline. This hydrolysis is usually more extensive with sodium chloride than with potassium chloride. The total alkalinity after fusion may amount to 0.001 to 0.01 per cent.

Figure 10–5 is a diagram of the apparatus constructed by Pinching and Bates[99] for the fusion of alkali chlorides in an atmosphere of dry nitrogen. It consists essentially of a platinum crucible b and a platinum trough c that fit loosely within a clear quartz tube a. The latter is provided with a Pyrex glass cap d through which the dry gas is admitted. The crucible is charged with salt, the cover wired in place, a slow stream of dry nitrogen passed for 5 minutes to displace the moisture-laden air, and the quartz tube finally supported in the top of a crucible furnace heated to 1000 °C. When all the salt has fused, the tube is withdrawn from the furnace, tipped to allow the molten salt to flow into the cooler trough, and allowed to cool with the nitrogen flowing. Salt fused in this way usually contains no more than 0.001 per cent of free alkali.

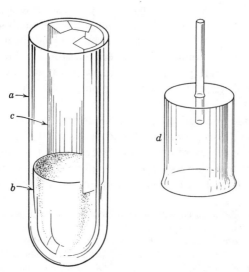

FIG. 10–5. Apparatus for fusing sodium and potassium chlorides.

THE BRIDGE SOLUTION

Since the days of the first measurements of cells with transference, the problem of evaluating or eliminating the liquid-junction potential has occupied the attention of physical chemists. The early work of Tower,[101] who attempted to reduce the interfacial potential by interposition of solutions of potassium chloride, potassium nitrate, and other salts, was doubtless stimulated by Planck's integration of the differential equation for the liquid-junction potential in 1890.[102] The Planck equation and the Henderson equation presented seventeen years later[103] showed that the magnitude of the liquid-junction or diffusion potential is a function of the mobilities, concentrations, and valences of the ions on both sides of the boundary. The interfacial potential should be small if a concentrated bridge solution of an equitransferent salt (that is, a solution in which the cationic and anionic transference numbers are both 0.5) were interposed. The accurate computation of the liquid-junction potential is impossible, and its estimation is difficult and unsatisfactory. Consequently, one is forced to seek experimental means of reducing the magnitude of this potential and, particularly, of making its value reproducible from one measurement to the next.

Bjerrum[104] conducted extensive studies of the elimination of diffusion potentials, choosing bridge solutions of potassium chloride because of the approximate equality of the transference numbers of potassium and chloride ions in concentrated solutions of this salt. The observed decrease in the diffusion potential with increasing concentration of the bridge solution led him to suggest an extrapolation method for the substantial elimination of the potential. Measurements of the e.m.f. of the cell with two different bridge solutions, 1.75 M and 3.5 M (called "half-saturated" and "saturated"), are required. The observed difference is applied to the measurement with the 3.5 M solution to obtain the e.m.f. for an infinitely concentrated bridge solution, assuming a linear variation of e.m.f. with the reciprocal of the bridge concentration. The Bjerrum extrapolation is illustrated on page 19.

Other investigators have tried to find highly concentrated salt mixtures that would minimize the liquid-junction potential more successfully than does potassium chloride. Acree and his coworkers[105] studied many concentrated

---

[101] O. F. Tower, *Z. Physik. Chem.*, **18**, 17 (1895).

[102] M. Planck, *Ann. Physik*, [3] **39**, 161 (1890); [3] **40**, 561 (1890).

[103] P. Henderson, *Z. Physik. Chem.*, **59**, 118 (1907); **63**, 325 (1908).

[104] N. Bjerrum, *Z. Physik. Chem.*, **53**, 428 (1905); **59**, 336, 581 (1907); *Z. Elektrochem.*, **17**, 58, 389 (1911); see also L. Michaelis and A. Fujita, *Biochem. Z.*, **142**, 398 (1923); A. H. W. Aten and J. van Dalfsen, *Rec. Trav. Chim.*, **45**, 177 (1926).

[105] N. E. Loomis and S. F. Acree, *Amer. Chem. J.*, **46**, 585, 621 (1911); L. J. Desha and S. F. Acree, *ibid.*, **46**, 638 (1911); G. M. Kline, M. R. Meacham, and S. F. Acree, *Bur. Stand. J. Res.*, **8**, 101 (1932).

salt solutions but found only one, a mixture of potassium chloride (3 M) and potassium nitrate (1 M), that was superior to saturated potassium chloride for measurements between pH 3 and 10. Likewise, Grove-Rasmussen[106] considered an equimolar mixture of potassium chloride and potassium nitrate to be more nearly equitransferent than potassium chloride alone. He has suggested the use of a bridge solution 1.8 M with respect to each of these salts.

Work in the author's laboratory has failed to reveal any differences in the pH(X) − pH(S) (measurement of which yields the operational pH value) attributable to the concentration of the bridge solution of potassium chloride, when the latter is 3.0 M, 3.5 M, 4.0 M, or saturated and the temperature is between 10 and 40 °C. The unsaturated solutions are a considerable convenience, as crystallization of salt is avoided. The accumulation of solid salt at the bridge-solution interface impedes the establishment of reproducible liquid-junction potentials, increases the electrical resistance of the cell, and aggravates hysteresis errors on change of temperature. For general use, the author recommends the 3.5 M electrode together with a bridge of the same concentration.

Semple[107] has likewise found that the pH at 38 °C for many buffer solutions is independent of the concentration of the bridge solution within rather wide limits. However, this is not true for blood samples. Presumably a concentrated solution of potassium chloride, contaminating the sample of blood, brings about precipitation of the plasma proteins and crenation of the blood cells. For this reason. Semple[107] and Maas[108] have studied the use of a bridge of 0.152 M or 0.160 M sodium chloride (isotonic saline) for blood measurements.

It can be shown by the Henderson equation (Chapter 3) that liquid-junction potentials at a dilute bridge solution are much less uniform than at the boundary with a concentrated bridge solution. Hence the isotonic saline bridge cannot be recommended for use with a variety of test solutions of widely ranging compositions and pH. The experiments of Sibbald and Leonard[109] support this view. In measurements of the pH of numerous blood samples, these authors obtained very consistent results with a flowing junction formed at a porous ceramic plug. In a comparison of four bridge solutions, a 10 per cent solution of potassium nitrate was found to be nearly

[106] K. V. Grove-Rasmussen, *Acta Chem. Scand.*, **2**, 937 (1948); **3**, 445 (1949).

[107] S. J. G. Semple, *J. Appl. Physiol.*, **16**, 576 (1961).

[108] A. H. J. Maas, *Clin. Chim. Acta*, **28**, 373 (1970); *J. Appl. Physiol.*, **30**, 248 (1971). See also R. N. Khuri and C. R. Merril, *Phys. Med. Biol.*, **9**, 541 (1964).

[109] P. G. Sibbald and J. E. Leonard, *Proc. San Diego Biomed. Eng. Symp.*, 1961. See also O. Siggaard-Andersen, *J. Clin. Lab. Invest.*, **13**, 205 (1961), and R. J. Linden and J. R. Ledsome, *Phys. Med. Biol.*, **8**, 333 (1963).

as satisfactory as a saturated bridge solution of potassium chloride for measurements of the pH of blood.

Maas and others[108] have shown that the pH of blood measured with an isotonic saline (0.160 M NaCl) bridge solution is lower by more than 0.1 pH unit than that found with the use of the saturated potassium chloride bridge. At 38 °C, the differences ΔpH, that is [pH (satd. KCl) − pH (0.160 M NaCl)], were as follows:

Solution	ΔpH
Blood phosphate buffer	+0.005
0.025 M NaHCO$_3$ + 0.135 M NaCl	+0.153
Cerebrospinal fluid	+0.163
Plasma	+0.125
Blood	+0.106

As Maas found, blood pH data reproducible to ±0.002 pH unit can nonetheless be obtained with the bridge of 0.160 M sodium chloride if the pH cell is properly standardized to compensate for the rather large residual liquid-junction potential. In another approach, Khuri and Merril have avoided the liquid junction completely in blood pH measurements by using a sodium ion-selective electrode as a reference for determining the glass electrode potential.

STRUCTURE OF THE LIQUID-LIQUID BOUNDARY

Accurate pH measurements are possible only when the liquid-junction potential between the bridge and the test solution matches that between the bridge and the standard solution. The physical structure of the junction between the concentrated salt bridge and solution X is of primary importance in determining the reproducibility, constancy, and to some extent the magnitude of the liquid-junction potential. Diffusion of the ions in both directions across the well-designed and properly formed junction rapidly sets up a steady state characterized by a reproducible, constant difference of potential.

The potential at the boundary between two solutions of the same salt at different concentrations has been shown, both by theory and by experiment, to be independent of the manner in which the junction is formed. However, when different ions are present in the two bounding solutions, the potential depends on the composition of the transition layers between the two end solutions. The concentrated bridge solution nullifies to some extent, although not entirely, the variations resulting from indefinite boundary structures.

Nevertheless, if the most reproducible junction potential is sought, it must be possible to duplicate accurately the physical structure of the boundary.

Cumming and Gilchrist[110] classified liquid junctions as " sharp," " mixed," or " diffused." Guggenheim[111] identified four classes of junctions and studied them from both the experimental and the theoretical standpoints. The following paragraphs describe these boundaries and summarize Guggenheim's findings. The structures of the four types are shown schematically in Fig. 10–6. In each diagram, B represents the heavier solution.

1. The continuous-mixture junction can be formed by mixing the two end solutions in continuously varying proportions, so that the composition at any point in the transition layer is a linear combination of these two solutions. The potential of this type of junction is given by the Henderson formula. Diffusion produces no true steady state, yet instability can be minimized by making the transition layer sufficiently long.
2. The free-diffusion junction is very simply made by bringing the two solutions together to form an initially sharp boundary within a vertical tube or at a stopcock of wide bore. If cylindrical symmetry is achieved, the transition layer increases in length with time, but the potential remains constant. Unfortunately, there is no explicit formula for the potential of this type of junction. It is believed that many freshly formed junctions of this type correspond closely to the continuous-mixture model and hence can be treated by the Henderson equation. Free-diffusion junctions formed with a congealed bridge of potassium chloride and agar-agar are also reproducible and stable. The e.m.f. is slightly different from that found without agar-agar.
3. The constrained-diffusion junction, the potential of which is given by the Planck formula, is difficult to realize experimentally. This type of junction is established at a porous plug or membrane kept washed on the two sides with the two end solutions. Diffusion of the end solutions within the membrane sets up a steady state with a transition layer of unchanging length.
4. Sharp junctions of indefinite type are formed by dipping the side tubes of half-cells into a U-tube containing the bridge solution. Cylindrical symmetry, an important requisite for reproducibility, is usually absent, and definite conditions for a steady state are lacking. Hence these junctions are often subject to irregular fluctuations amounting to several millivolts. Maclagan[112] concluded that the most stable junctions are formed within a tube rather than at its tip.

[110] A. C. Cumming and E. Gilchrist, *Trans. Faraday Soc.*, **9**, 174 (1913).

[111] E. A. Guggenheim, *J. Amer. Chem. Soc.*, **52**, 1315 (1930).

[112] N. F. Maclagan, *Biochem. J.*, **23**, 309 (1929).

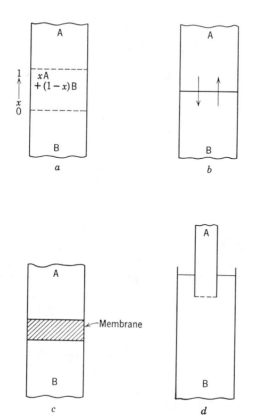

FIG. 10–6. Structures of liquid-liquid boundaries. (*a*) Continuous-mixture junction, (*b*) free-diffusion junction, (*c*) constrained-diffusion junction, (*d*) junction of indefinite structure.

Another type of junction, important for its reproducibility, is difficult to place in any one of the foregoing classes. It is the flowing junction,[113] one form of which is shown in Fig. 10–7. The two solutions flow in *a* and *b* toward one another, meet at *c* to form the boundary, and flow away in parallel streams. A steady state is established which depends but little on the rate of flow, provided that the flow is not so great as to produce turbulence.

[113] A. B. Lamb and A. T. Larson, *J. Amer. Chem. Soc.*, **42**, 229 (1920); see also D. A. MacInnes, *The Principles of Electrochemistry*, Chapter 13, Reinhold Publishing Corp., New York, 1939. J. V. Lakhani, *J. Chem. Soc.*, 179 (1932), found that constant and reproducible junctions are also formed by the meeting of fine jets of the two liquids.

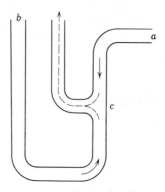

FIG. 10–7. Flowing junction.

It is likely that the flowing junction is best regarded as a special case of the continuous-mixture boundary.

Experimental studies of heteroionic junctions have dealt principally with the boundary between solutions of hydrochloric acid and potassium chloride, with and without interposed bridge solutions. The potential at the junction

$$HCl(0.1\ M) \mid KCl(0.1\ M) \tag{21}$$

is about 27 mV at 25 °C,[114] and variations of potential with changes in the structure of the boundary are easily observed.

The most important reproducible junctions are of the continuous-mixture, free-diffusion, and flowing types. The data of Guggenheim[115] show that the potential of junction 21 when set up as a continuous-mixture boundary differs by only 0.4 mV from the potential of the corresponding free-diffusion boundary. The potential of the flowing junction appears to differ from these by about 1 mV.[116] The effect of interposing bridge solutions of potassium chloride on the potentials of the first two types of junctions is shown in Fig. 10–8, which is a plot of data obtained by Guggenheim.

An important practical conclusion can be drawn from Guggenheim's study, namely, that all cylindrically symmetrical steady-state junctions have potentials that are nearly the same, when a concentrated solution of potassium

---

[114] D. A. MacInnes, *The Principles of Electrochemistry*, Chapter 13, Reinhold Publishing Corp., New York, 1939; Dover, New York, 1961.

[115] E. A. Guggenheim, *J. Amer. Chem. Soc.*, **52**, 1315 (1930).

[116] D. A. MacInnes, *The Principles of Electrochemistry*, Chapter 13, Reinhold Publishing Corp., New York, 1939; Dover, New York, 1961.

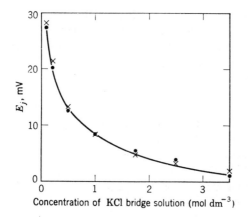

FIG. 10–8. Effect of concentration of the bridge solution on the potential E_j of the junction element HCl(0.1M) | Bridge | KCl(0.1M) (data of Guggenheim). Crosses, continuous-mixture junctions; dots, free-diffusion junctions.

chloride is used as a bridge. This result is in essential agreement with the findings of Ferguson, Van Lente, and Hitchens.[117] These investigators, furthermore, have shown that flowing junctions between hydrochloric acid (0.1 M or 0.01 M) and saturated potassium chloride have somewhat different potentials from static junctions. Scatchard[118] found that the potential across a flowing junction between saturated potassium chloride and solutions of hydrochloric acid did not vary with acid concentration when the molality of the latter was less than 0.1.

For the highest stability and reproducibility, the junction between solution X (cell 1) and a concentrated solution of potassium chloride should probably be formed within a cylindrical tube. Junctions formed within a vertical length of 1-mm capillary seem to be particularly stable.[119] Unfortunately, junctions of this type, because of their inconvenience, are not widely used in practical pH measurements. The liquid junction formed by the streamlined flow of the concentrated bridge solution into the solution under test, as provided in several commercially available reference electrodes, is probably somewhat less reproducible than the confined junction.

[117] A. L. Ferguson, K. Van Lente, and R. Hitchens, *J. Amer. Chem. Soc.*, **54**, 1279, 1285 (1932).

[118] G. Scatchard, *J. Amer. Chem. Soc.*, **47**, 696 (1925).

[119] D. J. Alner and J. J. Greczek, *Lab. Practice*, **14**, 721 (1965); D. J. Alner, J. J. Greczek, and A. G. Smeeth, *J. Chem. Soc. (A)*, 1205 (1967).

Considerable effort has been applied to the design of bridge tubes for practical reference electrodes, with the aim of establishing the liquid junction conveniently and rapidly. Perley[120] has listed eleven different devices that have been used in the construction of salt bridges:

1. An inverted U-tube of small diameter.
2. Glass tubes filled with agar-agar saturated with potassium chloride.
3. A siphon fitted with a glass stopper.
4. A plug ground into the bottom of a glass tube.
5. A diffusion opening of small diameter.
6. A rubber tube with external clamp.
7. Asbestos fiber sealed in a glass tube.
8. Small cones of cotton fitted within a siphon.
9. Layers of sand in a glass tube.
10. Wicks of various sorts.
11. Porous plugs.

Of the many forms of junction, the continuous impeded-flow type and the static junctions with partial cylindrical symmetry are probably the most reproducible. The chloride solution of the flow-type bridge should stream into the test solution at a constant rate, and the lines of flow should be free of turbulence. Streaming should be confined to a single leakage path; the existence of several leakage paths through which the bridge solution may flow at varying rates is not conducive to the establishment of a reproducible steady state. Relatively large fluctuations in potential have been observed when these junctions were inadequately flushed between measurements.

Several forms of calomel half-cells are discussed by Clark in his monograph.[121] There has been a later trend toward the convenient dip-type or immersion-type reference electrodes in commercial equipment and, still more recently, toward junctions formed at ceramic plugs or sintered-glass disks. The utility of porous glass for this purpose has been demonstrated.[122]

In the glass-calomel assembly designed by Cannon, an annular space around the glass electrode is filled with the bridge solution, which flows slowly through a ground-glass collar on the stem of the glass electrode, forming the liquid junction. The solution of potassium chloride constitutes an electrostatic shield for the glass electrode. Similar combination electrodes are

---

[120] G. A. Perley, *Trans. Electrochem. Soc.*, **92**, 497 (1947).

[121] W. M. Clark, *The Determination of Hydrogen Ions*, 3rd ed., Chapter 15, The Williams and Wilkins Co., Baltimore, Md., 1928.

[122] W. N. Carson, Jr., C. E. Michelson, and K. Koyama, *Anal. Chem.*, **27**, 472 (1955).

available commercially, as are compact, miniature electrodes. Other reference electrodes of small size have been described.[123]

A cylindrically symmetrical reproducible liquid junction, formed in a tube above a stopcock, is provided in the glass-calomel assembly of Coates.[124] This cell and that of Smith[125] can be refilled and the junction formed without removal of the apparatus from the constant-temperature aath in which it is immersed.

ZoBell and Rittenberg[126] designed a calomel cell and junction tube which provided a continuous impeded flow of the bridge solution through a small orifice partially closed by an asbestos fiber. In the reference cell described by Manassi and Co.,[127] the junction is formed at a porous plate. Inasmuch as the plate is located at the tip of a vertical tube with the heavier solution above, convection causes streaming through the pores. Hence the junction is probably to be regarded as a modified flowing type rather than a constrained-diffusion type. Perley[120] has described a salt-bridge tube with a single leakage path formed by a crack of controlled dimensions in the base of the tube. In the design of flow-type bridges, care must be exercised to make the aperture large enough for adequate flow, yet small enough to avoid contamination of the test solution with potassium chloride.

Four commercial reference electrodes are shown in Fig. 10–9. The salt-bridge tube of the Leeds and Northrup electrode screws into the cap of the electrode and can be readily replaced, if desired, by another tube for measurements at another temperature. If measurements are made regularly at a series of temperatures, extra salt-bridge tubes can be maintained at these temperatures ready for use. The Beckman fiber electrode is convenient for measurements in solutions that do not tend to clog the small orifice and thus interrupt the flow of bridge solution. The sleeve-type junction is preferred for emulsions, for solutions of proteins and soaps, and for nonaqueous solutions of high resistance. Beckman Instruments also supplies a reference electrode in which the junction is formed at a palladium annulus. The Philips electrode utilizes a piece of porous stone to separate the inner calomel element from the salt bridge and another to establish the liquid junction. The junction is flushed by turning the bridge tube with respect to the element. The electrode can be removed entirely from the glass envelope to replenish the bridge solution. A Coleman reference electrode is provided with a

---

[123] E. H. Shaw, Jr., *J. Chem. Educ.*, **18**, 330 (1941); J. E. Land, *Chemist-Analyst*, **32**, 87 (1943).

[124] G. E. Coates, *J. Chem. Soc.*, 489 (1945).

[125] G. S. Smith, *Trans. Faraday Soc.*, **45**, 752 (1949).

[126] C. E. ZoBell and S. C. Rittenberg, *Science*, **86**, 502 (1937).

[127] V. Manassi and Co., Italian Patent 434,455 (April 27, 1948).

[128] G. A. Perley, *Trans. Electrochem. Soc.*, **92**, 497 (1947).

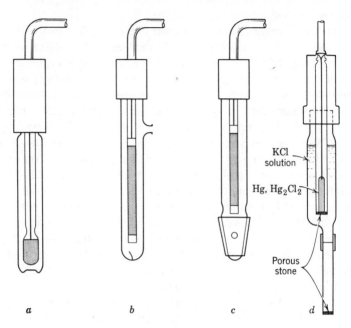

FIG. 10–9. Four forms of calomel reference electrodes. (*a*) Leeds and Northrup, (*b*) Beckman fiber type, (*c*) Beckman sleeve type, (*d*) Philips.

precision-ground plunger, a slight movement of which is sufficient to renew the liquid junction.

Silver-silver chloride reference electrodes, containing no mercury, are also available commercially. These were originally designed particularly for use in the photographic industry. In certain applications, these electrodes can also prove useful in minimizing temperature effects in pH measurements.

ERRORS OF CALOMEL ELECTRODES AND LIQUID JUNCTIONS

We have already mentioned the marked influence of bromide upon the potentials of calomel electrodes. The presence of bromide and other contaminants usually influences the potential of the calomel electrode without affecting its constancy. This and similar factors are therefore of little concern when the pH assembly is standardized with a buffer solution of known pH, but they must be considered when it is desired to prepare a standard reference electrode and to compute pH values with the aid of standard potentials. This was once the accepted procedure. However, it is much easier to prepare a standard buffer solution than to prepare a reproducible standard reference electrode. The availability of buffer standards of known pH makes it no longer necessary to secure reference electrodes which meet rigid specifications of design, purity of the components, and the like.

Hysteresis. The hysteresis errors of the saturated calomel electrode have been studied by Wingfield and Acree,[129] who determined the rate of change of the potentials of saturated calomel half-cells while the temperatures of these cells were being increased at the rate of 4 to 5 °C per hour. During the change interval, differences of 0.5 to 0.9 mV (0.01 to 0.02 pH unit) were observed between the measured potentials and the equilibrium values corresponding to the temperature of the potassium chloride solution within the half-cell. Two or three hours after a temperature change of 8 °C, the potentials of the half-cells were still 0.2 to 0.3 mV from their equilibrium values. Slightly smaller errors were observed following a decrease of temperature.

In his excellent discussion of temperature effects in pH cells, Mattock[130] has observed that a lack of true thermal equilibrium is often responsible for a part of the lag attributed to hysteresis. It seems likely that the rest is caused by the slow reversal of reaction 19 and by the time required for solubility equilibrium to be reestablished.

Instability of Calomel. The detailed composition of a saturated solution of calomel is the resultant of no less than nine separate temperature-sensitive equilibria.[131] One of the most important of these is the disproportionation of mercurous ion to form mercuric ion and mercury (equation 19). This equilibrium is established relatively slowly at room temperature. Up to 24 hours may be required for initial equilibrium in dilute solutions of hydrochloric acid. Das and Ives[132] have shown that the time required may be shortened by increasing the ratio of electrode surface area to volume of half-cell solution.

It is possible that spontaneous disproportionation proceeds unchecked at elevated temperatures, leading to instability of the calomel electrode at these temperatures. Leonard[133] has found that the average life of commercial calomel electrodes above 100 °C is only 9 hours. Nevertheless, calomel electrodes have been used for short periods in solutions of hydrochloric acid at

---

[129] B. Wingfield and S. F. Acree, *J. Am. Leather Chemists' Assoc.*, **31**, 403 (1936); *J. Res. Nat. Bur. Stand.*, **19**, 163 (1937). See also F. Strafelda and B. Polej, *Chem. Prumysl*, **7**, 240 (1957).

[130] G. Mattock, *pH Measurement and Titration*, Chapters 7 and 9, The Macmillan Co., New York, 1961.

[131] G. J. Hills and D. J. G. Ives, Chapter 3 in *Reference Electrodes*, D. J. G. Ives and G. J. Janz, eds., Academic Press, New York, 1961.

[132] S. N. Das and D. J. G. Ives, *J. Chem. Soc.*, 1619 (1962).

[133] J. E. Leonard, *Symposium on pH Measurement*, p. 16, ASTM Tech. Publ. 190, Philadelphia, Pa., 1957.

temperatures as high as 250 °C, even though instability of the electrode in 1 M solutions of potassium chloride was observed to commence at about 70 °C.[134] At the other extreme, calomel reference electrodes have been found to behave satisfactorily at temperatures as low as -30 °C when polyalcohol such as glycerol were added to the saturated aqueous solution of potassium chloride.[135]

Suspension Effect. It has often been observed that different results are obtained for the pH of soils, depending on whether the sample is an aqueous extract of the soil or a thick slurry. If the glass electrode and salt bridge are immersed in the sediment of an acid soil, the measured pH is considerably lower than that found by immersion of the electrodes in the filtrate o supernatant liquid.[136] This phenomenon, known as the "suspension effect" or "Pallmann effect," may be caused by abnormal junction potentials in colloidal systems.[137] The colloidal particles are regarded as affecting markedly the relative rates of diffusion of potassium and chloride ions. This view has been supported by measurements of the transference numbers of these ions in mixtures of potassium chloride and a cation exchanger material by a modified Hittorf method.

Nevertheless, the origin of the suspension effect may lie elsewhere. In his review of an earlier edition of this book, Longsworth[138] reminds the reader of Overbeek's suggestion that a Donnan effect may be responsible. Colloidal particles with ion-exchange properties may be immobilized by gravity as effectively as by a semipermeable membrane, giving rise to a Donnan potential difference across the interface. In Longsworth's view, if identical calomel electrodes are used to make contact with the suspension and the supernatant liquid and the cell potential is more than a few millivolts, the bulk of the potential is probably to be assigned to the interface rather than to the liquid junction between the slurry and the potassium chloride of the bridge solution.

[134] M. H. Lietzke and J. V. Vaughen, *J. Amer. Chem. Soc.*, **77**, 876 (1955). On the other hand, in a comparison of the "skin calomel electrode" of Hills and Ives[93] with the silver silver chloride electrode, J. V. Dobson, R. E. Firman, and H. R. Thirsk, *Electrochim. Acta* **16**, 793 (1971), found the e.m.f. to be stable at temperatures as high as 200 °C. Although the solubility of calomel was high, no extensive disproportionation of calomel was detected
[135] L. Van den Berg, *Anal. Chem.*, **32**, 628 (1960).
[136] H. Pallmann, *Koll.-Chem. Beih.*, **30**, 334 (1930); C. du Rietz, thesis, Tekn. Högskolan Stockholm, 1938; L. E. Davis, thesis, University of California, 1941; R. Loosjes, *Chem. Weekblad*, **46**, 902 (1950); H. R. Kruyt, editor, *Colloid Science*, Vol. I, p. 184, Elsevier Publishing Co., New York, 1952.
[137] H. Jenny, T. R. Nielsen, N. T. Coleman, and D. E. Williams, *Science*, **112**, 164 (1950); J. Overbeek, *J. Colloid Sci.*, **8**, 593 (1953). On the other hand, E. Eriksson, *Science*, **113**, 418 (1951), does not believe that abnormal liquid-junction potentials exist in these systems
[138] L. G. Longsworth, *J. Amer. Chem. Soc.*, **86**, 3912 (1964).

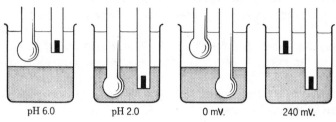

pH 6.0   pH 2.0   0 mV.   240 mV.

FIG. 10–10. The suspension effect.

In general, the pH of pastes and suspensions appears to be lower than the presumed pa_H of the mixture. The glass electrode will yield the same potential whether it is immersed in the sediment or in the liquid of a two-phase system, but the position of the reference electrode is critical. An extreme example of this phenomenon is shown in Fig. 10–10, taken from the paper of Jenny, Nielsen, Coleman, and Williams.

Bower[139] was able to show that the magnitude of the suspension effect decreases as the concentration of the bridge solution is lowered. A suspension effect was also observed when the pH-sensitive glass electrode was replaced by a sodium ion-sensitive glass electrode. In order to avoid the liquid junction and its attendant uncertainties in colloidal systems of constant sodium ion concentration, the use of a sodium ion–sensitive glass electrode as a reference has been suggested by Merril and by Khuri.[139]

Unfortunately, it seems doubtful that either of the two experimental pH values, determined with the reference electrode in the supernatant liquid or in the slurry, is very useful. As LaMer has pointed out,[140] " The pH that is effective is the value corresponding to that for the localized part of the solution between the charged particles of the slurry which often differs by as much as several pH units from that of the supernatant liquid in equilibrium with the slurry."

Miscellaneous Structural Features. It is advisable to examine capillary junctions occasionally for evidence of interrupted or retarded flow, for errors from this source may be quite large. Another error, usually of little concern in practical pH measurements, is caused by the changes in heat content that accompanying mixing by diffusion at the liquid junction. Hamer[141] found

---

[139] C. A. Bower, *Soil Sci. Soc. Proc.*, **25**, 18 (1961); C. R. Merril, *Nature*, **192**, 1087 (1961); R. N. Khuri and C. R. Merril, *Phys. Med. Biol.*, **9**, 541 (1964).

[140] V. K. LaMer, *J. Phys. Chem.*, **66**, 973 (1962).

[141] W. J. Hamer, *J. Amer. Chem. Soc.*, **57**, 662 (1935). With solutions of hydrochloric acid, the error, although measurable, is much smaller; see G. Scatchard and T. F. Buehrer, *ibid.*, **53**, 574 (1931).

that careful control of the temperature of junctions between solutions of sulfuric acid was necessary to prevent rather large disturbances from heat effects.

The Residual Liquid-Junction Potential. Unfortunately, pH measurements made with the best reference electrodes and the most reproducible liquid junctions are subject to the residual liquid-junction potential error discussed in Chapter 3. This error results from a difference in the liquid-junction potentials at the boundaries

$$\text{Bridge,KCl} \mid \text{Standard buffer (S)} \qquad (22)$$

and

$$\text{Bridge,KCl} \mid \text{Test soln. (X)} \qquad (22a)$$

When this difference exists, the junction potential E_j is incompletely balanced out, and the residual $E_j(S) - E_j(X)$ produces an error in the measured pH.

Test solutions have a great variety of compositions, and it is remarkable that the pH of so many of them can be given a useful physical significance. That the interpretation of pH values in aqueous solutions is indeed successful in so many instances is due in part to the efficacy of the concentrated bridge solution in nullifying the potential changes that would otherwise result from the differences of ionic concentration and mobility. Ionic mobilities, with the exception of those of hydrogen and hydroxide ions, are rather uniform.[142] It is to be expected, therefore, that the residual liquid-junction potential will be small when the ionic strengths of solutions S and X are not greatly different and when the hydrogen and hydroxyl ions account for only a small fraction of the ionic strength of the test solution. In general, this latter restriction means that pH(X) lies between 2.5 and 11.5.

The largest junction errors are probably incurred when the medium departs substantially from a simple dilute solution of ions in water. These departures may be caused by differences in the composition of the solvent due to the presence, for example, of alcohol or acetone in the test solution, or by the presence of other phases, as in an emulsion or suspension. In these cases, no simple interpretation of measured pH in terms of hydrogen ion concentration or activity is possible (see Chapter 8).

---

[142] The limiting equivalent conductances of 48 cations and anions are listed by H. S. Harned and B. B. Owen in their monograph (*The Physical Chemistry of Electrolytic Solutions,* 3rd ed., Chapter 6, Reinhold Publishing Corp., New York, 1958). All but ferrocyanide (111), ferricyanide (100), tetrapropylammonium (23.5), tetrabutylammonium (19.2), oxalate (24), hydrogen (349.8), and hydroxide (197.6) fall in the range 30 to 80.

POTENTIALS OF CELLS WITH CALOMEL REFERENCE ELECTRODES

The standard electrode potential $E°$ of the mercury-mercurous chloride electrode, that is, the e.m.f. of the cell

$$H_2 ; Pt, H^+ (a = 1), Cl^- (a = 1), Hg_2Cl_2 ; Hg \qquad (23)$$

is 0.26804 V at 25 °C, according to data summarized by Hills and Ives.[143] For pH measurements, however, $E°$ is much less useful than $E°' + E_j$ for the 0.1 M, 3.5 M, and saturated calomel electrodes, which find common use as reference electrodes. The quantity $E°' + E_j$ is the e.m.f. that the cell

$$Pt ; H_2, H^+ (a = 1) \mid KCl \text{ soln., calomel electrode} \qquad (24)$$

would have if the liquid-junction potential were the same as in the corresponding cell containing a buffer solution of moderate acidity. The reaction taking place in cell 24 can be written

$$\tfrac{1}{2}H_2 + \tfrac{1}{2}Hg_2Cl_2 = H^+ (a = 1)$$
$$+ \; Cl^- \text{ (in KCl soln.)} + Hg \pm \text{ ion transfer} \qquad (25)$$

The evaluation of $E°' + E_j$ has been discussed in Chapter 3.

Let us consider the computation of $pH(pa_H)$ from measurements of the e.m.f. E of cell 1, in which the reference is a calomel electrode and the hydrogen ion electrode may be the hydrogen gas electrode, glass, quinhydrone, or antimony. From the two half-cell reactions we may write, by rearrangement,

$$pa_H = \frac{E + E_h° - (E°' + E_j)}{(RT \ln 10)/F} = \frac{E - E_{hcal}°}{(RT \ln 10)/F} \qquad (26)$$

where $E_h°$ is the true standard potential (on the normal hydrogen scale) of the electrode reversible to hydrogen ion, and $E_{hcal}°$ is $E°' + E_j - E_h°$. For the gaseous hydrogen electrode, $E_h°$ is, of course, zero at all temperatures.

It must be remembered that the sign of E, the measured e.m.f., is determined by the direction in which current flows spontaneously through the cell *as written*. If oxidation takes place at the left electrode of cell 1, that electrode will be negative and the right electrode will be positive. The e.m.f. will therefore be regarded as positive according to the IUPAC convention outlined in Chapter 1. If the activities of the components of the cell are such

---

[143] This is the mean of the six determinations reported by G. J. Hills and D. J. G. Ives, Chapter 3 in *Reference Electrodes*, D. J. G. Ives and G. J. Janz, eds., Academic Press, New York, 1961. The standard deviation computed from the six values is 0.1 mV. More recently, a value of 0.26816 V at 25 °C has been found by S. R. Gupta, G. J. Hills, and D. J. G. Ives, *Trans. Faraday Soc.*, **59**, 1874 (1963).

that current flows spontaneously in the opposite direction, oxidation occurring at the right electrode (which is thus negative), the e.m.f. in equation 26 will have to be given a negative sign. The sign of E in cell 1 is likely to cause confusion only when the quinhydrone electrode is used. The e.m.f. of the quinhydrone-calomel cell is negative at low pH values and positive at high, passing through zero at about pH 6.2 (0.1 M calomel reference) or pH 7.7 (saturated calomel reference).

The standard potentials of glass electrodes and antimony electrodes depend on the design and past history of the individual electrode. Hence the constant E°_{hcal} must be calculated from equation 26 by the measurement of E with a standard solution of known pa_H in the cell. As already stated, this determination of E°_{hcal} for the particular cell system at hand has much to recommend it and should be employed for all but the most reproducible assemblies.

The standard potentials of hydrogen-calomel and quinhydrone-calomel cells from 10 to 40 °C are given in Table 10–6. The values at 25 °C for the hydrogen-0.1 M calomel cell and for the hydrogen-saturated calomel cell are best estimates consistent with the difference of 0.0912 V between the potentials of the two reference electrodes.[144] The values for the hydrogen-0.1 M calomel cell at other temperatures were based on the temperature coefficient found by Chateau. The e.m.f. given for cells with a 0.1 M reference electrode corresponds to an assembly in which a bridge of saturated potassium chloride separates the reference electrode and the test solution.

The data for the cells in which a hydrogen electrode is combined with a 3.0 M, 3.5 M, or 4.0 M calomel reference electrode (and for the temperature coefficient of the cell with a saturated reference electrode) were calculated by equation 26 from the e.m.f. of cells containing the standard equimolal phosphate buffer (a mixture of potassium dihydrogen phosphate and disodium hydrogen phosphate, each at a molality of 0.025).[145] The pH(S) (that is, pa_H) of this solution at several temperatures is given in Table 4–6. In each case the bridge was a solution of potassium chloride of the same concentration as that in the calomel electrode chamber. The standard potentials of the cells with quinhydrone electrodes were obtained by combining the e.m.f. of the corresponding hydrogen cell with data for the hydrogen-quinhydrone cell.[146] For the temperature coefficient of the hydrogen-quinhydrone cell, the results of Harned and Wright were used.

---

[144] The data were taken from D. I. Hitchcock and A. C. Taylor, *J. Amer. Chem. Soc.*, **59**, 1812 (1937); from D. A. MacInnes, D. Belcher, and T. Shedlovsky, *ibid.*, **60**, 1094 (1938); from R. G. Bates, G. D. Pinching, and E. R. Smith, *J. Res. Nat. Bur. Stand.*, **45**, 418 (1950); and from H. Chateau, *J. Chim. Phys.*, **51**, 590 (1954).

[145] M. Paabo and R. G. Bates, unpublished measurements.

[146] H. S. Harned and D. D. Wright, *J. Amer. Chem. Soc.*, **55**, 4849 (1933); F. Hovorka and W. C. Dearing, *ibid.*, **57**, 446 (1935).

TABLE 10-6. Standard Potentials (E°_{heal} in Volts) of Some Cells Used for pH Measurements

Tempera-ture °C	H₂, 0.1 M Calomel	H₂, 3.0 M Calomel	(Liquid-junction potential included) H₂, 3.5 M Calomel	H₂, 4.0 M Calomel	H₂, Satd. Calomel	Q·H₂Q, 0.1 M Calomel	Q·H₂Q, Satd. Calomel
10	0.3362	0.2602	0.2556	—	0.2543	—	−0.4564
12	0.3362	—	—	—	—	−0.3731	—
15	0.3362	—	0.2520	—	0.2511	−0.3710	−0.4560
20	0.3359	0.2569	0.2501	0.2459	0.2479	−0.3676	−0.4555
25	0.3356	0.2549	0.2481	0.2438	0.2444	−0.3642	−0.4554
30	0.3351	0.2530	—	—	0.2411	−0.3607	−0.4550
35	0.3344	—	0.2448	0.2402	0.2376	−0.3573	−0.4548
38	0.3338	0.2496	0.2439	0.2393	0.2355	−0.3552	−0.4547
40	0.3336	0.2487	—	—	0.2340	—	−0.4547

The temperature coefficients of E°_{hcal} for all four cells consisting of a hydrogen electrode and a concentrated calomel electrode were satisfactorily reproducible, but a slow drop of E°_{hcal} with age of the electrode was observed.[14] For example, the cell with a 3.0 M reference electrode decreased in e.m.f about 0.8 mV in one month, but the 3.5 M electrode appeared to drop only 0. to 0.3 mV in the same time, and no change in the potential of the 4.0 M electrode was observed in two weeks. The values listed in Table 10-6 corre spond to the e.m.f. of the cell about one week after the preparation of a new reference electrode. The data for the hydrogen-saturated calomel cell agree well with those of Baxter.[147] The average difference at eight temperature is 0.2 mV.

Inasmuch as the values in Table 10-6 include the liquid-junction potential there is some variation at low and high pH. In general, each value of E°_{hca} given in the table may be expected to increase by about 1 mV below pH and to decrease by about the same amount above pH 12.[148]

Silver-Silver Chloride Electrodes

The silver-silver chloride electrode is very reproducible and fairly easy to prepare. It is often used as an inner reference in glass electrodes and occasion ally also as an outer reference electrode in cells of type 1. It has found exten sive application in cells without liquid junction, where its relative freedom from disturbing side reactions has made it a valuable tool not only for study ing the thermodynamic constants of solutions of chlorides but also for the accurate determination of the dissociation constants of weak acids, bases and ampholytes.[149] The purity of the electrode reaction,

$$AgCl(s) + e \rightleftharpoons Ag(s) + Cl^- \qquad (27$$

has been affirmed in solutions with pH ranging from 0 to more than 13 The properties of this useful electrode have been summarized in admirably thorough fashion by Janz.[150]

[147] W. P. Baxter, quoted by M. Dole, *The Glass Electrode*, pp. 156, 166, John Wiley and Sons, New York, 1941.

[148] See Chapter 3 and the following papers: D. I. Hitchcock and A. C. Taylor, *J. Amer Chem. Soc.*, **59**, 1812 (1937); W. J. Hamer, *Trans. Electrochem. Soc.*, **72**, 45 (1937); F. Müller *Korrosion u. Metallschutz*, **18**, 253 (1942); G. G. Manov, N. J. DeLollis, and S. F. Acree *J. Res. Nat. Bur. Stand.*, **34**, 115 (1945).

[149] H. S. Harned and B. B. Owen, *The Physical Chemistry of Electrolytic Solutions*, 3rd ed. Chapter 15, Reinhold Publishing Corp., New York, 1958.

[150] G. J. Janz, Chapter 4 in *Reference Electrodes*, D. J. G. Ives and G. J. Janz, eds., Aca demic Press, New York, 1961. See also G. J. Janz and H. Taniguchi, *Chem. Rev.*, **53**, 397 (1953).

The potential of the silver-silver chloride electrode is very sensitive to traces of bromide in the solutions,[151] is altered by dissolved oxygen in acid solutions,[152] and is subject to an aging effect during the first 20 to 30 hours after the electrode is prepared.[153] Pinching and Bates found that 0.01 mole per cent of bromide, present as an impurity in potassium chloride, is sufficient to alter the potentials of silver-silver chloride electrodes immersed in the chloride solution by 0.1 to 0.2 mV. The electrodes in the solution with bromide are negative with respect to those immersed in a pure chloride solution of the same concentration. The potentials are not greatly affected by traces of iodide or cyanide. Light of ordinary intensities does not have a marked effect on the potentials of silver-silver chloride electrodes, although exposure to direct sunlight should probably be avoided. There is evidence, however, that the saturated Ag; AgCl reference electrode is considerably more sensitive to light than are silver chloride electrodes immersed in dilute chloride solutions (see page 390).

The effect of air can doubtless be ascribed to the reaction

$$2Ag + 2HCl + \tfrac{1}{2}O_2 = 2AgCl + H_2O \tag{28}$$

which causes a slight decrease in the concentration of chloride ion within the interstices of the electrode and makes the potential slightly more positive than that of an air-free electrode (compare equation 20). The slow approach to initial electrochemical equilibrium can be explained by concentration polarization.[154] The duration of change is dependent on the porosity of the electrode and the stirring of the solution. Freshly prepared electrodes are positive to aged electrodes. Figure 10–11, taken from the paper of Smith and Taylor, is a typical aging curve for silver-silver chloride electrodes of the thermal-electrolytic type in a 0.05 M oxygen-free solution of potassium chloride.

Silver chloride is appreciably soluble in concentrated solutions of chlorides. Nevertheless, a silver chloride electrode can be employed successfully as a

[151] E. Güntelberg, *Studier over Elektrolyt-Aktiviteter*, G. E. C. Gads Forlag, Copenhagen (dissertation, 1938); G. D. Pinching and R. G. Bates, *J. Res. Nat. Bur. Stand.*, **37**, 311 (1946).

[152] E. Güntelberg, *Z. Physik. Chem.*, **123**, 199 (1926); J. K. Taylor and E. R. Smith, *J. Res. Nat. Bur. Stand.*, **22**, 307 (1939).

[153] D. A. MacInnes and K. Parker, *J. Amer. Chem. Soc.*, **37**, 1445 (1915); E. R. Smith and J. K. Taylor, *J. Res. Nat. Bur. Stand.*, **20**, 837 (1938); *Roczniki Chem.*, **18**, 762 (1938).

[154] E. R. Smith and J. K. Taylor, *J. Res. Nat. Bur. Stand.*, **20**, 837 (1938); *J. Amer. Chem. Soc.*, **64**, 3053 (1942). The attainment of equilibrium has been found by J. H. Ashby, J. E. Crook, and S. P. Datta, *Biochem. J.*, **56**, 190 (1954), to be accelerated by heating the electrodes in water at 50 °C for 2 hours.

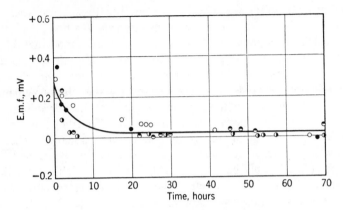

FIG. 10–11. Aging of silver-silver chloride electrodes.

reference electrode in saturated potassium chloride if the solution in the electrode chamber is also saturated with precipitated silver chloride. Saturated silver chloride reference electrodes are available commercially.

PREPARATION OF SILVER-SILVER CHLORIDE ELECTRODES

Three common types of electrodes are recognized: (1) the electrolytic type, (2) the thermal-electrolytic type, and (3) the thermal type. The electrolytic type is formed by the electrodeposition of silver on a platinum wire, foil, or gauze from a cyanide plating solution. The surface of the carefully washed deposit is then converted to silver chloride by electrolysis in a chloride solution. The thermal-electrolytic type[155] is prepared by forming a mass of porous silver on a helix of platinum wire by the thermal decomposition of a paste of silver oxide (or silver oxalate[156]) and water. A part of the silver formed in this way is then converted to silver chloride by electrolysis. The thermal type of electrode[157] consists of an intimate mixture of silver and silver chloride formed by the thermal decomposition of a paste of silver oxide, silver chlorate, and water. These three types of electrode have been shown to have identical potentials at equilibrium.[158] Silver-silver chloride electrodes have also been prepared by the thermal decomposition of mixtures

[155] G. N. Lewis, *J. Amer. Chem. Soc.*, **28**, 158 (1906); A. A. Noyes and J. H. Ellis, *ibid.*, **39**, 2532 (1917); H. S. Harned, *ibid.*, **51**, 416 (1929).

[156] D. T. Ferrell, Jr., I. Blackburn, and W. C. Vosburgh, *J. Amer. Chem. Soc.*, **70**, 3812 (1948).

[157] C. K. Rule and V. K. LaMer, *J. Amer. Chem. Soc.*, **58**, 2339 (1936).

[158] E. R. Smith and J. K. Taylor, *J. Res. Nat. Bur. Stand.*, **20**, 837 (1938).

of silver oxide and precipitated silver chloride and mixtures of silver oxide and silver perchlorate.[159]

According to Keyes and Hara,[160] the partial pressure p of oxygen in equilibrium with a mixture of silver and silver oxide is given by

$$\log p = 6.2853 - \frac{2859}{T} \tag{29}$$

with p expressed in atmospheres and T in kelvins. The decomposition pressure of silver oxide is therefore equal to 0.21 atm (the approximate partial pressure of oxygen in air) at a temperature of about 138 °C, and silver oxide should begin to decompose in air at this temperature. Until a temperature of 300 °C is exceeded, however, its rate of decomposition is slow, and for complete and rapid decomposition a temperature of at least 400 °C is recommended. Silver chloride is formed from silver chlorate at temperatures above 270 °C. Electrodes of the thermal type should not be heated above 455 °C, the melting point of silver chloride.

The thermal-electrolytic type of electrode has probably been used more extensively than any other form. The author has found the following to be a convenient method for preparing electrodes of this type.

The base for each electrode is a helix of No. 26 platinum wire about 7 mm in length and about 2 mm in diameter, sealed in a tube of flint glass (see Fig. 10–12). The bases are cleaned in warm 6 M nitric acid and a thick paste of well-washed silver oxide (see below) and water is applied to each helix. The electrodes are suspended in a crucible furnace heated to about 500 °C and allowed to remain there for 10 minutes or until they are completely white. A second layer of silver is formed in a similar manner with a slightly thinner paste to make the surface smooth. The silver on each electrode should weigh about 150 to 200 mg.

Each silver electrode is mounted in a cell of modified U-tube design and electrolyzed in a 1 M solution of twice-distilled hydrochloric acid for 45 minutes at a current of 10 mA. Silver is the positive electrode, and a platinum wire serves as the negative electrode. If the current efficiency is 100 per cent, 15 to 20 per cent of the silver will be converted to silver chloride. Thick coats of silver chloride tend to make the electrodes sluggish and should be avoided. The completed electrodes are placed in a 0.05 M solution of hydrochloric

---

[159] F. Ishikawa and S. Matsuo, *Sci. Rpt. Tohoku Univ., First Ser.*, **33**, 228 (1949). Excellent results, particularly in mixed solvents, have been obtained by E. L. Purlee and E. Grunwald, *J. Phys. Chem.*, **59**, 1112 (1955); *J. Chem. Phys.*, **27**, 990 (1957), with electrodes formed by anodizing silver mirror surfaces in a dilute solution of hydrochloric acid.

[160] F. G. Keyes and H. Hara, *J. Amer. Chem. Soc.*, **44**, 479 (1922).

FIG. 10–12. Base for silver-silver chloride electrode.

acid overnight. Their potentials are then intercompared and individual electrodes that differ from the average of the group by more than 0.1 mV are
rejected. The electrodes are immersed in distilled water for use within a few
days. Semimicroelectrodes of this type, formed from 16 mg of silver, have
also been found to give excellent results.[161]

If highly reproducible electrodes are to be obtained, two points must
receive special attention. The first is the purity of the hydrochloric acid,
and the second is the washing of the silver oxide. Reagent-grade acid should
be distilled twice in an all-glass still, retaining each time only the middle
fraction of the distillate. The best silver-silver chloride electrodes are light
gray to white. Dark gray electrodes are sometimes obtained immediately
after fresh acid has been added to the electrolytic cells. The cause of this
behavior has not been determined.

---

[161] R. G. Bates, G. D. Siegel, and S. F. Acree, *J. Res. Nat. Bur. Stand.*, **30**, 347 (1943).

PREPARATION OF SILVER OXIDE

Dissolve 338 g (2 moles) of silver nitrate in 3 liters of water. Dissolve slightly less than 2 moles (80 g) of sodium hydroxide in 400 ml of water, and add the solution drop by drop to the vigorously stirred solution of silver nitrate. Silver should be present in slight excess at the end of the precipitation. The product should be washed 30 to 40 times with distilled water. For this step, a tall glass-stoppered cylinder, in which the oxide can be shaken vigorously with water, is convenient. The necessity for extensive washing is seen in Fig. 10–13, where the conductivity of the wash water, as found by the author and V. E. Bower in a typical preparation of the oxide, is plotted as a function of the number of washings. The conductivity becomes substantially constant, indicating that the soluble electrolytes have been largely removed, after the oxide has been washed about 25 times.

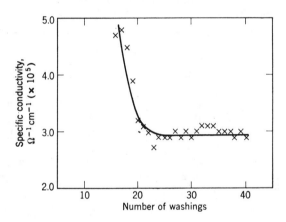

FIG. 10–13. Washing of silver oxide.

Ferrell, Blackburn, and Vosburgh[162] preferred to form the silver electrodes by thermal decomposition of silver oxalate instead of silver oxide. The oxalate was purified by dissolving it in ammonia and reprecipitating it with oxalic acid, and a substantial part of the laborious washing procedure was thereby avoided. Silver oxalate spatters considerably during decomposition. If the temperature of decomposition is too high, it yields a fluffy, poorly adherent mass of silver that does not cover the wire support uniformly. The

---

[162] D. T. Ferrell, Jr., I. Blackburn, and W. C. Vosburgh, *J. Amer. Chem. Soc.*, **70**, 3812 (1948).

spattered silver should be removed from the glass tubes before the electrodes are chloridized.

POTENTIALS OF SILVER-SILVER CHLORIDE ELECTRODES

The standard potential of the silver-silver chloride electrode in volts from 0 to 95 °C, that is, the e.m.f. of the cell

$$Pt; H_2, H^+(a = 1), Cl^-(a = 1), AgCl; Ag \qquad (30)$$

is listed in Table 10–7. These values are based on the e.m.f. measurements of solutions of hydrochloric acid made by Bates and Bower.[163] The temperature range has been extended to 275 °C by Greeley, Smith, Stoughton, and Lietzke.[164] Their results at 25, 60, and 90 °C are in good agreement with the values given in Table 10–7.

TABLE 10–7. STANDARD POTENTIAL E° OF THE SILVER-SILVER CHLORIDE ELECTRODE FROM 0° TO 95 °C, IN VOLTS

t °C	E°	t °C	E°
0	0.23655	40	0.21208
5	0.23413	45	0.20835
10	0.23142	50	0.20449
15	0.22857	55	0.20056
20	0.22557	60	0.19649
25	0.22234	70	0.18782
30	0.21904	80	0.1787
35	0.21565	90	0.1695
		95	0.1651

Differences of 0.1 to 0.2 mV among independent determinations of E° suggest that the electrode potential is a function of the method of preparation and prior treatment of the electrode. This conclusion is supported by the observation that individual determinations of the activity coefficient of hydrochloric acid display a uniformity of a considerably higher order than do the values of E°. A practical standardization of this electrode in 0.01 m hydrochloric acid, based on accepted values of γ_{HCl}, has therefore been

---

[163] R. G. Bates and V. E. Bower, *J. Res. Nat. Bur. Stand.*, **53**, 283 (1954). See also H. S. Harned and R. W. Ehlers, *J. Amer. Chem. Soc.*, **54**, 1350 (1932); **55**, 2179 (1933).
[164] R. S. Greeley, W. T. Smith, Jr., R. W. Stoughton, and M. H. Lietzke, *J. Phys. Chem.*, **64**, 652 (1960).

recommended.[165] The mean activity coefficient of hydrochloric acid should be taken as 0.904 at 25 °C and 0.908 at 0 °C.

Manov, DeLollis, and Acree[166] used the silver-silver chloride electrode in a saturated solution of potassium chloride as a reference in their study of comparative liquid-junction potentials. With the aid of pa_H values obtained from cells without liquid junction, they concluded that $E^{\circ\prime} + E_j$ for the saturated silver chloride reference electrode has the value 0.1992 V in 0.1 M hydrochloric acid, about 0.1981 V in buffer solutions, and 0.1969 V in a mixture of calcium hydroxide and sodium chloride. The values at 10 to 40 °C given in Table 10–8 were obtained by combining $E^{\circ\prime} + E_j$ for the 3.5 M and saturated calomel electrodes (Table 10–6) with the e.m.f. of the cell: $Ag; AgCl, Cl^-, Hg_2Cl_2; Hg$ as given by Pouradier and Chateau.[167]

TABLE 10–8. STANDARD POTENTIALS ($E^{\circ\prime} + E_j$) OF THE 3.5 M AND SATURATED SILVER-SILVER CHLORIDE ELECTRODES, IN VOLTS

t, °C	3.5 M KCl	Satd. KCl
10	0.2152	0.2138
15	0.2117	0.2089
20	0.2082	0.2040
25	0.2046	0.1989
30	0.2009	0.1939
35	0.1971	0.1887
40	0.1933	0.1835

Thallium Amalgam-Thallous Chloride Electrode

Fricke[168] has shown that the reference electrode

$$KCl \text{ (satd.)}, TlCl(s); Tl \text{ (40 per cent amalgam)}$$

termed the "Thalamid" electrode, is superior to either the saturated calomel electrode or the saturated silver chloride reference electrode for many purposes, especially when measurements are made over a range of temperatures.

[165] R. G. Bates, E. A. Guggenheim, H. S. Harned, D. J. G. Ives, G. J. Janz, C. B. Monk, J. E. Prue, R. A. Robinson, R. H. Stokes, and W. F. K. Wynne-Jones, *J. Chem. Phys.*, **25**, 361 (1956); **26**, 222 (1957).

[166] G. G. Manov, N. J. DeLollis, and S. F. Acree, *J. Res. Nat. Bur. Stand.*, **34**, 115 (1945).

[167] J. Pouradier and H. Chateau, *Compt. Rend.*, **237**, 711 (1953).

[168] H. K. Fricke in *Beiträge zur Angewandten Glasforschung*, E. Schott, ed., p. 175 Wissenschaftliche Verlagsgesellschaft, Stuttgart, 1960; *Zucker*, **14**, No. 7 (1961).

Although the calomel and silver-silver chloride electrodes both showed appreciable hysteresis after heating to 100 °C, the Thalamid electrode resumed its equilibrium potential with scarcely any time lag. Thalamid reference electrodes, manufactured by Jenaer Glaswerk Schott und Gen. (Mainz), are suitable for use at temperatures as high as 135 °C. They are unsatisfactory for measurements below 0 °C, the temperature at which the amalgam solidifies completely.

The standard potential of the saturated Thalamid electrode is the e.m.f. of the cell

$$Pt;H_2,H^+(a = 1) \mid KCl \text{ (satd. soln.)}, TlCl(s);Tl \text{ (40 per cent amalgam)} \quad (31)$$

including the liquid-junction potential. Baucke[169] has determined this potential, $(E^{o\prime} + E_j)$, from 5 to 90 °C by measuring the e.m.f. of cells of type 31 with solutions of assigned pa_H (compare equation 26). Solutions used were 0.1 M hydrochloric acid, the NBS standard phosphate buffer (molality of each phosphate = 0.025), and the NBS standard borax buffer (molality 0.01). In addition, some measurements were made with the NBS secondary standard, potassium tetroxalate (molality 0.05). The values obtained in the phosphate buffer solution are summarized in Table 10–9.

TABLE   10–9.   STANDARD   POTENTIALS
$(E^{o\prime} + E_k)$ OF THE SATURATED THALAMID
REFERENCE ELECTRODE

	(After Baucke)
t, °C	$E^{o\prime} + E_j$, V
5	−0.5624
10	−0.5652
15	−0.5687
20	−0.5727
25	−0.5767
30	−0.5806
35	−0.5846
38	−0.5872
40	−0.5889
50	−0.5971
60	−0.6057
70	−0.6144
80	−0.6229
90	−0.6309

[169] F. G. K. Baucke, *J. Electroanal. Chem.*, **33**, 135 (1971).

It is not surprising that the values of $E^{\circ\prime} + E_j$ reported by Baucke for the acidic solutions differ by 1 to 3 mV (0.02 to 0.05 pH unit) from those obtained with the neutral phosphate solution; these values are -0.5733 V for the solution of hydrochloric acid and -0.5754 V for the tetroxalate solution at 25 °C. The borax solution gave a value of -0.5794 V, however; this difference of 2.7 mV (0.045 pH unit) seems large in the light of the residual liquid-junction potentials apparent on page 86. The results reported in Chapter 4 were obtained with capillary junctions of cylindrical symmetry, while Baucke formed the liquid junction at a ceramic diaphragm. It seems likely, therefore, that the differences in $E^{\circ} + E_j$ reflect the variations inherent in junctions of markedly different structure.

Temperature Coefficients of Cells and Electrodes

The changes of the e.m.f. of certain cells with changes of temperature have been listed in the tables of the foregoing sections. These e.m.f. values are standard potentials; that is, the hydrogen ion is presumed to be present at the hypothetical unit activity. Furthermore, these potentials apply only when there is temperature equilibrium throughout the cell. Two questions of practical importance for pH measurements suggest themselves: (1) How does the temperature coefficient of the cell e.m.f. change with the pH of the solution? and (2) What is the error which may result from changing the temperature of one electrode and not that of the other?

The overall temperature coefficient of glass-calomel pH cells is dependent not only on the properties of the electrode and solution within the glass bulb[170] and the calomel electrode outside, but also on the pH of the test solution. The manufacturers of pH equipment have designed glass electrodes that compensate to a large extent the temperature coefficient of the saturated calomel reference electrode for certain conditions deemed to be average—for example, at pH 7. However, it is difficult to nullify the temperature effect over the entire pH range, or for all types of test samples, without an infinite number of inner solutions. It must also be remembered that the pH of the test samples themselves changes with the temperature. This matter will be considered further in the next chapter.

The approximate effect of pH changes on the temperature coefficient of the cell with hydrogen gas electrode and the saturated calomel reference electrode from 10 to 40 °C has been reported by Perley.[171] His values are summarized in Table 10–10. It should be emphasized that these figures represent the temperature coefficient of the electrode system itself and have

---

[170] When the inner electrode is reversible to hydrogen ion (for example, a quinhydrone electrode), the e.m.f. and temperature coefficient are practically independent of the composition of the inner solution.

[171] G. A. Perley, *Trans. Electrochem. Soc.*, **92**, 485 (1947).

TABLE 10–10. dE/dT FOR THE CELL
PT;H$_2$, SOLN. X | KCl (SATD.), CALOMEL
AT FIVE VALUES OF pH

(Data of Perley)	
pH	dE/dT, K
2.0	-0.00036
4.0	$+0.00003$
7.0	0.00062
9.0	0.00102
11.0	0.00142

nothing to do with the effect of temperature on the pH of the solution in the cell. They can be obtained through differentiation of equation 26 with respect to temperature at constant pa_H. The term 0.0592(dpa_H/dT) must be added, however, to obtain the temperature coefficient of the e.m.f. of the complete cell at 25 °C.

There is no way by which the true temperature coefficient of the potential of a glass, hydrogen, or calomel electrode can be determined. Fortunately, this information is not needed in order to obtain useful estimates of the errors that may result from lack of temperature uniformity throughout the pH cell. Wingfield and Acree,[172] for example, determined the equilibrium potentials of saturated calomel electrodes at temperatures between 22.5 and 31.3 °C with respect to similar electrodes maintained at 25 °C. In this way the error in potential caused by inadequate control of the temperature of the saturated reference electrode was found to be 0.25 millivolts per kelvin. The electrode at the higher temperature is positive.[173] Thus the apparent temperature coefficient of the potential of the saturated calomel electrode is $+0.25$ mV K^{-1}. The measurements of Bjerrum and Unmack gave $+0.22$ mV K^{-1}, and those of Ewing and of Fales and Mudge[174] led to a value of $+0.20$ mV K^{-1}.

The effect of temperature changes on the potentials of electrodes reversible to hydrogen ion can now be estimated. The data of Table 10–6 give about -0.67 mV K^{-1} for the temperature coefficient of the e.m.f. of the cell

$$Pt; H_2, H^+(a = 1) \mid KCl \text{ (satd.)}, Hg_2Cl_2; Hg \qquad (32)$$

[172] B. Wingfield and S. F. Acree, *J. Res. Nat. Bur. Stand.*, **19**, 163 (1937).
[173] N. Bjerrum and A. Unmack, *Kgl. Danske Videnskab. Selskab, Mat.-Fys. Medd.*, **9**, No. 1 (1929).
[174] W. W. Ewing, *J. Amer. Chem. Soc.*, **47**, 301 (1925); H. A. Fales and W. A. Mudge, *ibid.*, **42**, 2434 (1920).

Hydrogen is the negative electrode and calomel the positive electrode, so that

$$\frac{dE}{dT} = -\frac{dE_H}{dT} + \frac{dE_{cal}}{dT} \tag{33}$$

Hence the potential of the standard hydrogen electrode also appears to become more positive with increase of temperature; the apparent temperature coefficient is $+0.92$ mV K^{-1}. Similarly, the data of Table 10–10 suggest that the apparent temperature coefficient of the hydrogen electrode is about -0.37 mV K^{-1} at pH 7. In the same way, the temperature coefficient of the 0.1 M calomel electrode is found to be $+0.8$ mV K^{-1}, or about three times that of the saturated electrode. The temperature coefficient of the quin-hydrone electrode appears to be nearly zero at pH 1 and about -1.0 mV K^{-1} at pH 7, whereas that of the silver-silver chloride electrode in 0.1 M hydrochloric acid is about $+0.55$ mV K^{-1}.

CHAPTER 11

glass electrodes

The development of the convenient and versatile glass electrode has been responsible, more than any other single factor, for the widespread application of pH measurements in the control of industrial and commercial processes as well as in research. Although the pH function of the glass electrode was discovered in the first decade of this century,[1] the intensive study of the electrode and its many possible applications dates from about 1922, when Hughes published the results of his comparison of the glass and hydrogen electrodes.[2] Further impetus came with the completion a few years later of an important investigation by MacInnes and Dole.[3]

Dole, Kratz, and Eisenman have written excellent comprehensive monographs on the properties and uses of glass electrodes.[4] The book of Kratz surveys the developments to the year 1948, and later, less comprehensive, reviews discuss the most important contributions made to the subject prior to 1964.[5] The book edited by Eisenman is the best available source of information on the theory of the glass electrode and on the relationship between glass composition and specificity. In this chapter an attempt is made to describe briefly the properties that bear most directly on the usefulness of

---

[1] M. Cremer, *Z. Biol.*, **47**, 562 (1906); F. Haber and Z. Klemensiewicz, *Z. Physik. Chem.*, **67**, 385 (1909); G. Borelius, *Ann. Physik*, **45**, 929 (1914); **50**, 447 (1916).

[2] W. S. Hughes, *J. Amer. Chem. Soc.*, **44**, 2860 (1922); *J. Chem. Soc.*, 491 (1928).

[3] D. A. MacInnes and M. Dole, *Ind. Eng. Chem., Anal. Ed.*, **1**, 57 (1929); *J. Amer. Chem. Soc.*, **52**, 29 (1930).

[4] M. Dole, *The Glass Electrode*, John Wiley and Sons, New York, 1941; L. Kratz, *Die Glaselektrode und Ihre Anwendungen*, Verlag von Dr. D. Steinkopff, Frankfurt am Main, 1950; G. Eisenman, ed., *Glass Electrodes for Hydrogen and other Cations*, Marcel Dekker, New York, 1967.

[5] R. G. Bates, Chapter 5 in *Reference Electrodes*, D. J. G. Ives and G. J. Janz, eds., Academic Press, New York, 1961; G. Mattock, *pH Measurement and Titration*, Chapter 6, The Macmillan Co., New York, 1961; K. Schwabe and H. D. Suschke, *Angew. Chem.*, **76**, 39 (1964).

the glass electrode in pH determinations and to review the studies that shed light on the mechanism of the response to changes of pH.

The pH Response

In one of its most common forms, the glass electrode consists of a thin glass bulb inside which is mounted a reference electrode immersed in a solution of constant pH containing the ion to which the inner reference electrode is reversible. The MacInnes and Dole electrode,[3] shown in Fig. 11–1, is formed quite simply, by fusing a membrane of pH-responsive glass across the end of a glass tube. The inner cell is often composed of a silver chloride or calomel electrode in hydrochloric acid or a buffered chloride solution. The potential of the glass electrode relative to an external reference and the change of potential with temperature are determined by the types of inner electrode and inner solution.

The manner in which the potential of the glass electrode responds to changes of pH can be ascertained by direct comparison with the hydrogen electrode or by measurement of the e.m.f. of a cell consisting of a glass electrode and an outer reference calomel electrode with salt bridge. This cell

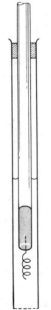

FIG. 11–1. Glass electrode of MacInnes and Dole.

is similar to that often used for the determination of pH. It may be represented by the following scheme:

$$\text{Ag; AgCl, HCl(0.1 M)} \,||\, \text{Glass} \,||\, \text{Test soln.} \,|\, \text{Satd. KCl, Calomel} \qquad (1)$$

where the liquid junction is indicated by a single vertical line and the boundaries of the glass membrane are indicated by double lines. The pH response of the glass electrode is expressed in volts (or millivolts) per pH unit. If E_1 and E_2 are the values of the electromotive force of cell 1 in test solutions of pH equal to pH_1 and pH_2, respectively, the pH response R_{pH} is given by

$$R_{pH} \equiv \frac{E_2 - E_1}{pH_2 - pH_1} \qquad (2)$$

The ideal pH response is shown by a glass electrode behaving in exactly the same manner as the standard hydrogen gas electrode towards changes of hydrogen ion activity. If the platinum–hydrogen electrode were substituted for the glass electrode in cell 1 and the potential across the liquid junction remained constant,

$$E_2 - E_1 = \frac{RT \ln 10}{F} (pH_2 - pH_1) \qquad (3)$$

As equation 3 shows, the ideal pH response is thus $(RT \ln 10)/F$ or $2.3036 RT/F$ volts per pH unit, that is, 0.05420 V at 0 °C, 0.05916 V at 25 °C, and 0.07305 V at 95 °C (see Appendix, Table 1). Unfortunately, no glass electrode yet constructed has the theoretical response in all types of test solutions and over the entire practical pH range.

For many years the best pH-responsive glass available was Corning 015 glass. This glass has the desirable properties of low melting point, high hygroscopicity, and relatively high electrical conductivity. The potential E_g of the ideal glass electrode is given by

$$E_g = E_g^{\circ} + \frac{RT}{F} \ln a_H \qquad (4)$$

The potential of the 015 electrode follows equation 4 very well between pH 1 and 9.

It is convenient to characterize the response of glass electrodes by means of the "electromotive efficiency" β_e. This quantity is a fraction less than 1, defined by[6]

$$\beta_e = \frac{E_X - E_S}{E_X' - E_S'} \qquad (5)$$

---

[6] *Glass Electrodes*, British Standard Specification 2586 (1955), British Standards Institution, 2 Park St., London, W.1.

where E_X, E_S are the e.m.f. values of the glass–calomel assembly containing two solutions X and S, respectively, and E_X', E_S' are the e.m.f. values observed for the same solutions when the glass electrode is replaced by a hydrogen electrode. In the intermediate range of pH, the potential of the glass electrode may be, and usually is, a linear function of pH, but the response R_{pH} differs somewhat from the theoretical value. Specifications for the electromotive efficiency of glass electrodes of three different types, taken from the British standard, are summarized in Table 11–1.

TABLE 11–1. Electromotive Efficiency of Glass Electrodes

Type of Electrode	pH Range	Electromotive Efficiency, β_e
General purpose	1–10	>0.995
High pH	9–11	>0.98
	11–12	>0.95
	12–13	>0.90
Wide pH range	1–10	>0.995
	10–11	>0.98
	11–12	>0.95
	12–13	>0.90

If the electromotive efficiency is $1 - y$, the error in the measured pH will be y unit for each unit difference between the pH of solutions X and S. It is not uncommon for individual glass electrodes to have electromotive efficiencies as low as 0.995 ($R_{pH} = 0.05886$ V/pH unit at 25 °C); hence an electrode standardized in the standard phthalate buffer at pH 4.01 may read low by more than 0.02 unit in the standard borax buffer, pH(S) = 9.18.

Impaired response is especially serious in alkaline solutions and in strongly acidic solutions. Under these conditions the potential of the electrode is not a linear function of pH and also varies with time. The error is positive in alkaline solutions containing sodium and other cations and negative in concentrated solutions of acids.[7] The departures in alkaline solutions are attributable to the development of a partial response to cations; in strongly acidic solutions there is some evidence of anion penetration of the pH-sensitive surface layer of the glass. These errors will be discussed further in later sections of this chapter.

---

[7] The terms "negative error" and "positive error" indicate that the glass electrode is too negative or too positive in the electrochemical sense.

Departures from the ideal hydrogen ion function are often expressed as ΔE, called the "voltage departure," under any given set of conditions. MacInnes and Belcher[8] compared the response of glass electrodes directly with that of the hydrogen electrode in a cell such as

$$\text{Pt}; \text{H}_2, \text{Test soln.} \,||\, \text{Glass} \,||\, \text{HCl}(0.1 \text{ M}), \text{AgCl}; \text{Ag} \tag{6}$$

Many different solutions and several different electrodes may be used on the reference side of the glass membrane, that is, on the right of cell 6. The electromotive force $E_6{}^\circ$ for a test solution in which the glass electrode is known to be error-free is accordingly the reference potential for the determination of the error in other test solutions:

$$\Delta E = E_6 - E_6{}^\circ \tag{7}$$

This determination of ΔE will include *changes* in the potential of the glass electrode due to asymmetry of the two surfaces of the glass. Dole and his associates[9] preferred to measure the glass electrode and hydrogen electrode potentials separately with reference to a calomel electrode, in order to discover any changes of electromotive force with time.

The criteria of Hughes[10] for the selection of glass electrodes were low electrical resistance, low electrode errors, good stability of electromotive force, and asymmetry potentials that are both small and constant. The composition of the glass, its hygroscopicity, chemical durability, and thickness are all related in some way to the pH response. However, the role of these properties in the mechanism of operation of the glass electrode has not been fully elucidated.

Structure of Electrode Glasses

Studies of glasses by means of x rays reveal a network of oxygen atoms or ions held together in irregular chains by silicon atoms.[11] Each silicon is presumably associated with four oxygens, and each oxygen is shared by two SiO_4 groups to form a three-dimensional network. The oxygen atoms or ions are relatively large (about 1.4 Å in diameter as compared with 0.4 Å for silicon), and hence make up the bulk of the network. The holes in the three-dimensional pattern are occupied by cations, held there more or less strongly by the electrostatic fields of the neighboring oxygen ions. However, as a

---

[8] D. A. MacInnes and D. Belcher, *J. Amer. Chem. Soc.*, **53**, 3315 (1931).

[9] M. Dole and B. Z. Wiener, *Trans. Electrochem. Soc.*, **72**, 107 (1937); see also Dole's monograph (reference 4).

[10] W. S. Hughes, *J. Amer. Chem. Soc.*, **44**, 2860 (1922); *J. Chem. Soc.*, 491 (1928).

[11] W. H. Zachariasen, *J. Amer. Chem. Soc.*, **54**, 3841 (1932); B. F. Warren, *Chem. Rev.*, **26**, 237 (1940).

result of the irregularity of the silicon–oxygen lattice, the ions occupying the many holes in the lattice possess many different levels of energy. In other words, the work required to move an ion or to detach it from the lattice may be different for each individual ion. The " anionic field strength " of the glass sites will be a determining factor. Figure 11–2 is a representation of the structure of silica glass.[12] The relative sizes of the ions are indicated by circles of different diameters.

Perley[12] has considered the relation between the composition of the glass and the pH function of electrodes blown from the glass. He compares the glass structure shown in Fig. 11–2 with the structure of zeolites, from which water can be removed and in which the interstitial ions can be reversibly replaced without damage to the lattice structure. The exchange of alkali metal ions within the silicon–oxygen network for ions in the solution in which the glass electrode is immersed plays a large part in the development of the pH response and the error of the electrode in certain alkaline solutions. If a network structure could be produced in which the exchange of hydrogen ions and cations of the alkali and alkaline earth series could be controlled, the ideal glass electrode might become a reality.

Hughes made a systematic study of the electrical properties of glasses as a function of the composition. On the basis of the four criteria already mentioned, he selected a glass containing roughly 72 per cent by weight SiO_2, 8 per cent CaO, and 20 per cent Na_2O as most suitable for use in the fabrication of electrodes. In a somewhat more extensive investigation, MacInnes and Dole[13] came to the conclusion that the glass having the lowest possible melting point in the ternary system, SiO_2-CaO-Na_2O, was the best for pH determinations. The eutectic composition is 72.2 mole per cent SiO_2, 6.4 per cent CaO, and 21.4 per cent Na_2O. This glass has a low electrical resistance and a low asymmetry potential. The Corning Glass Works manufactures this glass under the designation Corning 015, and the Jenaer Glaswerk Schott und Gen. produces it under the designation Schott 4073[III]. Other pH glasses containing lithium, cesium, barium, and lanthanum oxides have been developed.[14] Electrodes with greatly reduced alkaline errors can be made from these and similar glasses.

Cationic exchange, and hence alkaline errors, depend in part on the size of the ion, the strength of the anionic electrostatic forces in the glass, and the solvating power of the cations that penetrate the glass surface. Most anions

[12] G. A. Perley, *Anal. Chem.*, **21**, 391, 394 (1949).

[13] D. A. MacInnes and M. Dole, *J. Amer. Chem. Soc.*, **52**, 29 (1930).

[14] S. I. Ssokolof and A. H. Passynsky, *Z. Physik. Chem.*, **A160**, 366 (1932); G. A. Perley, British Patent 574,029 (Dec. 18, 1945) and U.S. Patent 2,497,235 (Feb. 14, 1950); H. H. Cary and W. P. Baxter, U.S. Patent 2,462,843 (March 1, 1949).

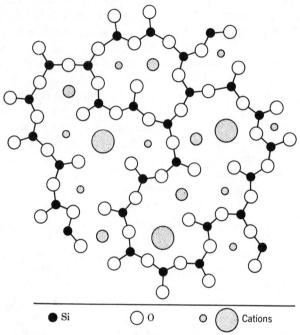

FIG. 11–2. Structure of silica glass (after Perley).

are larger than the simple alkali and alkaline earth cations and can accordingly penetrate the vitreous network with less ease. Furthermore, the repulsions of the oxygen ions surrounding the interstices of the lattice make the introduction of anions statistically improbable. These conclusions are amply supported by observation. It has long been believed that negative ions are without influence on the behavior of the glass electrode in alkaline solutions, and a careful study by Dole, Roberts, and Holley of glass electrode errors in the presence of fluoride, chloride, borate, and hydroxide ions has affirmed this belief.[15] Quite small concentrations of molecular hydrofluoric acid, however, were found to modify the electrode surface in such a way that the alkaline errors due to cations were greatly increased.

---

[15] M. Dole, R. M. Roberts, and C. E. Holley, Jr., *J. Amer. Chem. Soc.*, **63**, 725 (1941). However, anions may play a part in producing the negative errors observed in strongly acidic media.

Hygroscopicity and pH Response

It is well known that the water content of the glass membrane has a marked effect on the pH function of the electrode. Haber and Klemensiewicz[16] obtained poor results with electrodes that had been allowed to become dry. Some electrodes that were practically devoid of hydrogen electrode function developed a pH response after treatment with superheated water under pressure. MacInnes and Belcher[17] found that the electrical resistance of glass electrodes at 25 °C increased 230 per cent, on the average, when the electrodes were dried for 10 days over phosphorus pentoxide. The resistance returned slowly to its original value when the electrodes were immersed in water. Perley[18] has found that electrodes of the newer lithia–silica glasses are influenced less by drying agents than those of 015 glass. The lithia glasses are known to absorb only about one-ninth as much water as do the soda and potash glasses.[19]

Hughes[20] noted that repeated or prolonged heating of a glass electrode at high temperatures sometimes results in complete loss of the hydrogen electrode function. The effect of heat treatment has been clarified somewhat by the work of Hubbard and Rynders,[21] who compared the pH response of new unleached electrodes of 015 glass before and after annealing, finding no appreciable voltage departures from pH 1.9 to 9.2. The electrodes were annealed for 10 minutes between the critical temperature (500 °C) and the deformation temperature (550 °C). A striking difference was observed, however, in the behavior of electrodes that had been leached in a 0.1 M solution of hydrochloric acid for 6 hours at 80 °C. Leached electrodes, which showed a normal pH response after leaching, were annealed in the same manner as the unleached electrodes but were found thereafter to display rather large permanent voltage departures. The pH response of a typical electrode dropped from 59 mV per pH unit to 22 mV per pH unit. If these "dead" electrodes were immersed in a solution of hydrofluoric acid (1 : 1) for a few seconds, the normal electrode function was promptly restored.

These experiments emphasize the importance of the structure of the surface of the glass membrane to the pH response. The loss of pH function on heat treatment appears to be due to the formation of a thin nonhygroscopic silica-rich layer, and the imperfect pH response of the treated electrode

[16] F. Haber and Z. Klemensiewicz, *Z. Physik. Chem.*, **67**, 385 (1909).
[17] D. A. MacInnes and D. Belcher, *J. Amer. Chem. Soc.*, **53**, 3315 (1931).
[18] G. A. Perley, *Anal. Chem.*, **21**, 559 (1949).
[19] D. Hubbard, *J. Res. Nat. Bur. Stand.*, **36**, 365 (1946).
[20] W. S. Hughes, *J. Chem. Soc.*, 491 (1928).
[21] D. Hubbard and G. F. Rynders, *J. Res. Nat. Bur. Stand.*, **40**, 105 (1948); **41**, 163 (1948).

resembles that of electrodes fabricated from glasses of low hygroscopicity. From measurements with an interferometer, Hubbard and Rynders estimated the thickness of the inhibiting layer to be less than 5.8×10^{-6} cm. Films of conductors such as metallic silver and of nonconductors such as vaseline were also effective in inhibiting the pH function. Indeed, it is possible to make pH measurements with a cell consisting of two glass electrodes, one of which has been partially coated with lacquer or a similar substance. The success of this procedure depends on the reduction in the pH response of one of the electrodes produced by the surface coating.[22]

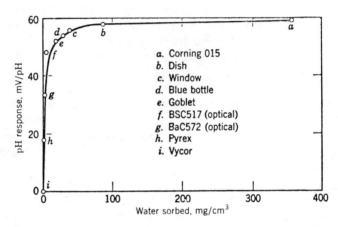

FIG. 11–3. Hygroscopicity and pH response of nine glasses (after Hubbard).

The hygroscopicity of glasses of equal surface area can be compared successfully by exposing weighed samples of the powdered glass to humid atmospheres for a suitable time and determining the increase of weight.[23] The correlation between the water sorption of a glass and the pH response of electrodes made from the glass is a very direct one, as Fig. 11–3 shows. Satisfactory pH electrodes cannot be made from optical glass or from chemically resistant glasses suitable for the manufacture of laboratory glassware. On the other hand, the highly hygroscopic glasses like Corning 015 display a superior pH response. The response of lithia–silica glasses to deuterium ion has also been correlated with the sorption of heavy water.[24] The hygroscopicity of many glasses of different compositions has been

[22] D. W. Lübbers, *Naturwissenschaften*, **49**, 493 (1962).
[23] See, for example, D. Hubbard, *J. Res. Nat. Bur. Stand.*, **36**, 511 (1946).
[24] D. Hubbard and G. W. Cleek, *J. Res. Nat. Bur. Stand.*, **49**, 267 (1952).

measured by Hubbard and his coworkers and compared with electrode function.[25] In all cases glasses of low hygroscopicity were found to yield electrodes with pH responses appreciably below the theoretical 59 mV per pH unit at 25 °C.

The ability of a glass electrode to function satisfactorily as an indicator of pH changes is intimately associated with the water content of the glass, but the role of water in the mechanism of the electrode is still a matter for speculation. Water may facilitate the movement of ions in the glass, lowering the electrical resistance to practical levels, or it may lower the energy barrier for the transfer of protons from the solution to the gel layer. The proton is probably associated with at least one molecule of water within the glass network as well as in the solution. Nevertheless, a hydronium ion does not migrate, intact, through the surface. The process appears to be a proton transfer, and electron transfer between solution and glass is normally unimportant. The glass electrode is thus to be regarded as a "protode."[26]

Chemical Durability and Voltage Departures

The ideal glass electrode must be sufficiently durable to give many hours of service in corrosive media at high temperatures as well as low. Its rate of attack should be so low that accurate measurements of the pH of water and slightly buffered solutions can be made. Unfortunately, some glasses of adequate hygroscopicity and satisfactory pH response are so soluble in aqueous solutions that they are useless for practical pH work. The electrical resistance of very thin glass membranes is sometimes found to drop after prolonged immersion of the membrane in water. This decrease is presumably the result both of penetration of water into the lattice and thinning of the membrane by solution of a part of the glass.

As the electrical resistance of electrode glasses increases so does the chemical resistance, yet ideally the glass electrode should have both a low electrical resistance and a high chemical durability. In the fabrication of glass electrodes, the electrical resistance of the membrane is lowered to practical levels at some sacrifice of corrosion resistance. The processes of corrosion and leaching may be expected to bring a glass electrode to the end of its useful life in 9 months to 2 years.

The deterioration of the pH-sensitive glass membrane probably proceeds somewhat as follows. Water sorbed by the conditioned membrane of the electrode penetrates ever more deeply, extracting the alkaline constituents

---

[25] See *J. Res. Nat. Bur. Stand.*, **37**, 223 (1946); **41**, 273 (1948); **44**, 247 (1950); **45**, 430 (1950); and **46**, 168 (1951).

[26] T. Shedlovsky, *Science*, **113**, 561 (1951); W. H. Beck, K. Grove-Rasmussen, and W. F. K. Wynne-Jones, *J. Physiol.*, **121**, 1 (1953). See also W. H. Beck, *Nature*, **190**, 712 (1961).

from the interior of the glass. The strong alkali thus produced contributes to the destruction of the silicon–oxygen network in the pH-sensitive surface, and the decomposition proceeds inward from both surfaces until breakdown occurs. Repeated leaching and drying hasten the process. According to Kratz,[27] the useful life of a glass electrode can be prolonged as much as 30 per cent by keeping the electrode immersed in water or a buffer solution when it is not in use.

Lengyel and Blum[28] determined the solubility of a number of electrode glasses by immersion of bulbs of known area in water at 80 °C. After 3 hours, the alkalinity of the water was determined by titration with a 0.0005 M solution of sulfuric acid to the end point of the indicator iodoeosin. A soda glass of a composition very close to that of Corning 015 glass was found to yield 5.6 mg of Na_2O per square decimeter, as compared with 0.2 mg for ordinary borosilicate glass. Kratz[29] has measured the susceptibility of electrode glasses to chemical attack by observing the rate of increase in the conductance of a sample of water in which the electrode was immersed.

Above pH 9, the silicate structure of the glass may be attacked by the alkali. This may be regarded as a surface reaction between hydrous silica, or metasilicic acid, and sodium hydroxide. As Britton has pointed out,[30] the first step in the neutralization of metasilicic acid,

$$H_2SiO_3 + NaOH = NaHSiO_3 + H_2O \tag{8}$$

should occur at pH values between 9 and 10.4, and attack should proceed rapidly at higher alkalinities.

By the application of an interferometric method designed to indicate the relative solubilities of optical glass specimens, Hubbard, Hamilton, and Finn[31] found that the attack of 015 glass does indeed begin at about pH 8.5 to 9. Attack was expressed as a number of fringes. The results shown in Fig. 11–4 (lower curve) were obtained in the Britton–Robinson buffer solutions of pH 2 to 12. The upper curve shows that the voltage departures or pH errors parallel the attack curve. It was concluded that a glass should have a uniform durability over an extended pH range if electrodes formed from it are to display the correct pH response.

---

[27] L. Kratz, *Glastech. Ber.*, **20**, 305 (1942).

[28] B. Lengyel and E. Blum, *Trans. Faraday Soc.*, **30**, 461 (1934).

[29] L. Kratz, *Glastech. Ber.*, **20**, 305 (1942). See also the monograph of Kratz (reference 4).

[30] H. T. S. Britton, *Hydrogen Ions*, 4th ed., Vol. I, Chapter 7, D. Van Nostrand Co., Princeton, N.J., 1956.

[31] D. Hubbard, E. H. Hamilton, and A. N. Finn, *J. Res. Nat. Bur. Stand.*, **22**, 339 (1939). See also D. Hubbard and E. H. Hamilton, *ibid.*, **27**, 143 (1941); D. Hubbard and G. F. Rynders, *ibid.*, **39**, 561 (1947); and D. Hubbard, M. H. Black, S. F. Holley, and G. F. Rynders, *ibid.*, **46**, 168 (1951).

The glass was found to swell slightly when immersed in buffer solutions of pH less than 7. This phenomenon is indicated as negative attack in Fig. 11–4. Durability measurements were also made in strong solutions of hydrochloric acid, sulfuric acid, and phosphoric acid. The rate of swelling was found to be repressed in solutions of pH less than 2, and no swelling at all was observed in 10 M sulfuric acid. The fact that swelling was also repressed at pH 3 to 4 by high concentrations of salt suggests that it is associated with penetration of water into the silicate network.

These findings established a relationship between the voltage departures, or electrode errors, of the 015 glass electrode and changes in the outer structure of the glass membrane. Other glasses that exhibited no swelling in the "super-acid" region exhibited no voltage departures above a pH of −1. Furthermore, the durability of the glass was found to be greater in a solution of ammonia of pH 13.3 than in a more dilute ammonia solution of pH 12.5. This change in durability with concentration is opposite that found with strong alkalies, and the corresponding voltage departures were also found to be opposite in sign to those normally encountered in alkaline solutions.

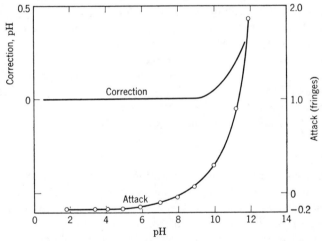

FIG. 11–4. Chemical durability and pH errors of the 015 glass electrode.

Electrical Resistance

There is apparently no connection between the resistance of the membrane and the pH response.[32] Pyrex glass, for example, does not develop a satisfactory pH function even when the resistance of the membrane is low, and

---

[32] B. Lengyel and E. Blum, *Trans. Faraday Soc.*, **30**, 461 (1934).

electrodes of 015 glass fail to develop the theoretical response if the membrane is too thick. There appears to be a critical thickness for each glass, above which voltage departures occur. This threshold thickness, which is a function of hygroscopicity, ranges from 54 to 130 μm.[33]

The high electrical resistance of glass electrodes and the large change of resistance with temperature are responsible for many of the experimental difficulties encountered in the measurement of the potentials of glass electrodes. This resistance is commonly from 10 to 500 megohms at room temperature and may be considerably greater at lower temperatures. It often changes markedly when the electrode is subjected to chemical attack.[34]

Some of the results of Eckfeldt and Perley[35] for the change of d-c resistance with temperature for electrodes of Corning 015 glass and Leeds and Northrup glasses 379 and 399 are plotted in Fig. 11–5. It is evident that the resistance rises sharply as the temperature is lowered. The value at 8 °C may be 300 times that at 75 °C. The compositions of the three glasses in mole per cent are given below the figure.

THE *IR* DROP

Let us consider first the general problem of measuring the electromotive force E_g of a cell containing glass and calomel electrodes immersed in an aqueous solution. The resistance of the calomel electrode and of most common aqueous solutions amounts to no more than a few thousand ohms. Hence the internal resistance R_g of the cell is substantially the resistance of the glass electrode itself. In the measurement of the electromotive force, one must inescapably draw a certain amount of current from the cell. The passage of a direct current I_g lowers the measured electromotive force, E_{meas}, below the true E_g by the amount of the "*IR* drop":

$$E_{meas} = E_g - I_g R_g \qquad (9)$$

If the e.m.f. is measured by the compensation method, that is, with a potentiometer and galvanometer, the cell is opposed by a potential exactly equal to its own *at the point of balance*. Thus I_g is zero and the true "open-circuit" value of E_g is measured. However, in order to fix the point of balance sufficient current must be drawn from the cell to give observable deflections of the galvanometer. If an accuracy of ± 2 mV in the e.m.f. of a cell having a resistance of 100 megohms is to be achieved, the galvanometer must be sensitive to a current of only 2×10^{-11} A. The most sensitive galvanometers

[33] R. G. Goldman and D. Hubbard, *J. Res. Nat. Bur. Stand.*, **48**, 370 (1952); J. J. Diamond and D. Hubbard, *ibid.*, **47**, 443 (1951).

[34] H. Richter and H. G. Rosenthal, *Z. Elektrochem.*, **45**, 79 (1939).

[35] E. L. Eckfeldt and G. A. Perley, *J. Electrochem. Soc.*, **98**, 37 (1951).

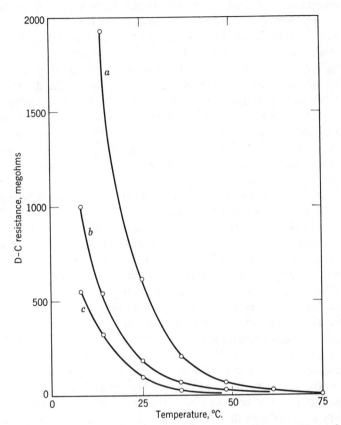

FIG. 11-5. Resistance of glass electrodes as a function of temperature (data of Eckfeldt and Perley). Curve *a*, L & N 379 glass. Composition in mole per cent: Li_2O, 25; Cs_2O, 2; CaO, 2; BaO, 5; La_2O_3, 3; SiO_2, 63. Curve *b*, Corning 015 glass. Composition in mole per cent: Na_2O, 21.4; CaO, 6.4; SiO_2, 72.2. Curve *c*, L & N 399 glass. Composition in mole per cent: Li_2O, 28; Cs_2O, 2; BaO, 5; La_2O_3, 2; SiO_2, 63.

available commercially have a sensitivity of about 5×10^{-10} A per scale division.

When vacuum-tube electrometers became available, it was possible to achieve an accuracy of ± 1 mV through considerably higher resistances. The inconvenient and fragile electrodes of relatively low resistance were therefore gradually replaced by more durable types whose resistances ranged commonly from 50 to 500 megohms. However, the electrometer tube of the pH meter draws a small grid current even at the point of balance. In this respect

the electronic amplifier differs from the galvanometer used as a null indicat
in low resistance potentiometer circuits. Hence the term $I_g R_g$ in equation 9
not zero, and there is a real difference between the true e.m.f. and the mea
ured value if either the resistance or the grid current is unusually large. A
Fig. 11–5 shows, an electrode intended for use at high temperatures may ha
much too high a resistance at low temperatures to permit accurate measur
ments of the e.m.f. to be made. At high temperatures, the durability an
alkaline errors of the glass electrode are of most concern; at low temperatur
the problem is chiefly one of high resistance. Some of the pH meters utilizin
"chopper" circuits or vibrating condensers operate with almost no curre
drain (see Chapter 12). Nevertheless, electrodes of excessively high resistanc
display a slow and erratic response and are unsuitable for use even with thes
highly sensitive measuring instruments.

An electrode that has become dehydrated may have such a high resistanc
that the IR drop is abnormally large. The resistance of the test solutior
usually negligible compared with that of the glass membrane, becomes im
portant in solvents of low dielectric constant where little ionization occurs
A large grid current may precede failure of the electrometer tube.

The leads to electrodes of high resistance must be carefully guarded fror
electrical leakage, as must the stem of the electrode itself. Furthermore, hig
humidities may cause erratic behavior not observed with systems of lo
resistance. The stems of electrodes can be made water-repellent and thu
protected from losses resulting from moisture adsorption by treatment wit
certain silicone preparations[36, 37] such as General Electric Dri-Film, Beck
man Desicote, or Hopkin and Williams Repelcote. These materials have bee
applied to the pH-sensitive tip of the electrode as well, in order to facilitat
cleaning. Little is known concerning the effect of this treatment on th
asymmetry potential and the voltage-departure pattern of the electrode. Th
pH response and electrical resistance are only slightly affected.[36] The pro
tective coating is unaltered by prolonged immersion of the electrode in aci
solutions (except hydrofluoric acid) but is slowly destroyed by contact wit
alkaline solutions.

RESISTANCE AND COMPOSITION

Although conduction through the glass is very slight, it appears likely tha
sodium or lithium ions carry most of the current. Haugaard found that the
mobility of hydrogen ions in the glass phase is much lower than that o

---

[36] P. T. Gilbert, Jr., *Science*, **114**, 637 (1951).
[37] E. L. Eckfeldt and G. A. Perley, *J. Electrochem. Soc.*, **98**, 37 (1951).

sodium ions.[38] Presumably, the hydrogen ions that penetrate into the glass become bonded more tightly to the silicon–oxygen network than the relatively labile sodium ions. This conclusion has been confirmed by Schwabe and Dahms[39] who used tritium labeling to show that hydrogen ions make almost no contribution to conduction through the glass, even at elevated temperatures.

The d-c resistances of glass electrodes are frequently as much as thirty times the values obtained with alternating current, as shown by MacInnes and Belcher.[40] Eckfeldt and Perley[37] examined both the a-c and d-c methods for measuring the electrical resistance of glass electrodes and concluded that the d-c resistance should be regarded as the true ohmic resistance of the glass.[41] Glass is primarily a dielectric and exhibits only slight electrical conductivity. Hence it is not surprising that the resistance measurement is markedly affected by such factors as dielectric absorption and dielectric loss, that is, the loss of electric energy as heat due to a changing electric field. In the a-c measurement there is a loss of energy within the glass which adds a conductivity component not present in the d-c measurement. This leads not only to a lower apparent resistance in the a-c measurement but also to a change of the a-c resistance with frequency. MacInnes and Belcher found the a-c resistance at 3380 Hz to be about one-half that at 1020 Hz.

Many (but not all) lithia glasses have higher d-c resistances than do soda glasses. Kratz has made the interesting observation that glasses containing both sodium and lithium, or sodium, lithium, and potassium, have about the same a-c resistances as soda glasses, although their d-c resistances are much greater. Dole[42] explains this fact as follows: The presence of small amounts of lithium ion causes breaks in the chains of sodium ions that carry current through the glass. If a lithium ion of large activation energy and high stability occupies a hole in the silicon–oxygen network normally occupied by a sodium ion and into which a sodium ion must migrate in passing through the glass, the d-c electrical resistance will be increased. The a-c resistance might not be greatly affected, because the ions trapped in the vitreous network would merely oscillate under the applied potential instead of migrating out of their individual locations.

---

[38] G. Haugaard, *Nature*, **140**, 66 (1937); *Compt. Rend. Trav. Lab. Carlsberg*, **22**, 199 (1938); *J. Phys. Chem.*, **45**, 148 (1941). See also F. Quittner, *Ann. Physik*, [4] **85**, 745 (1928).

[39] K. Schwabe and H. Dahms, *Monatsber. Deut. Akad. Wiss.*, **1**, 279 (1959).

[40] D. A. MacInnes and D. Belcher, *J. Amer. Chem. Soc.*, **53**, 3315 (1931).

[41] The d-c measurement was also preferred by MacInnes and Belcher and by L. Kratz, *Z. Elektrochem.*, **49**, 474 (1943).

[42] M. Dole, *Symposium on pH Measurement*, p. 41, ASTM Tech. Publ. 73, Philadelphia, Pa., 1947.

Relatively small changes in the composition of the glass may have a marked effect on the resistances of glass electrodes, possibly through a change in the hygroscopicity. As shown in Fig. 11–5, electrodes of L & N 379 glass have resistances about six times those of electrodes of L & N 399 glass, yet the compositions differ only in the replacement of 3 mole per cent of lithium oxide by 2 mole per cent of calcium oxide and 1 mole per cent of lanthanum oxide. It must be remembered, however, that the resistance of a glass electrode is a function of the thickness of the membrane and the size of the glass bulb. Hence the comparative resistance of two glass electrodes is not always an accurate index of the comparative resistivity of glasses of the two different compositions. Although the electrodes of 399 glass contain lithium, they have resistances lower than electrodes of Corning 015 glass and are well suited to measurements at low temperatures.

MEASUREMENT OF D–C RESISTANCE

At room temperature or above, the d-c resistance of most glass electrodes can be measured readily without interference from dielectric absorption. At lower temperatures, however, the resistance is often very difficult to measure, because the conduction current may not reach a constant value for hours. Under these conditions, an adjustable potential difference in series with the cell has been used advantageously to hasten the establishment of the equilibrium current.[37] In view of the fact that the properties of glass are altered to a large extent by heat treatment, it is not surprising that highly reproducible values for the resistance of different electrodes made from glass of the same composition are not usually obtainable. By annealing glass electrodes, their resistances may be increased by a factor of 4 or 5.

Figure 11–6 is a schematic diagram of the circuit used by Eckfeldt and Perley for the measurement of d-c resistance. The potential of the glass cell is represented by E_g, and R_g is the cell resistance which is to be measured. The resistance R_c is a standard resistor of known value, and V is an adjustable known source of potential difference for which a low-resistance potentiometer can be used. An electronic amplifier and a potentiometer are first employed to measure E_g in the usual way before the cell is made a part of the circuit shown in the figure. Then the cell is connected to R_c and V and the amplifier-potentiometer combination used once more to measure E_r. Inasmuch as

$$R_g = R_c \frac{V + E_g - E_r}{E_r} \tag{10}$$

it is possible to determine R_g in this way. The conditions are most favorable when the calibrated resistor approximately equals the resistance of the glass, or when $V + E_g$ is approximately $2E_r$. If resistance measurements are to be

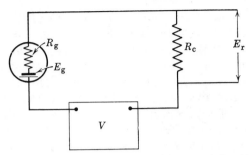

FIG. 11–6. Circuit for the measurement of the d-c resistance of glass electrodes (Eckfeldt and Perley).

made at a number of temperatures, therefore, several calibrated resistors should be used to cover the range most suitably.

The results of Eckfeldt and Perley fit the Rasch and Hinrichsen equation[43] for the change of resistance with temperature:

$$\ln R_g = A + \frac{B}{T} \tag{11}$$

Plots of $\ln R_g$ as a function of the reciprocal of the thermodynamic temperature $1/T$ were found to be straight lines, the slopes of which are represented by B in equation 11. This empirical relationship suggests that the differential change of $\ln R_g$ with T can be represented by an expression of the familiar form

$$\frac{d \ln R_g}{dT} = \frac{-Q}{RT^2} \tag{12}$$

where Q (with the dimensions of energy) replaces RB; R is the gas constant. Lengyel and Blum[44] regard Q as an "activation energy" for the conducting ion in the glass, hence the energy required for the ion to surmount the potential barrier between locations of equilibrium within the glass structure. These Q values, however, do not show any direct correlation with the composition of the glass.

---

[43] E. Rasch and F. W. Hinrichsen, *Z. Elektrochem.*, **14**, 41 (1908).

[44] B. Lengyel and E. Blum, *Trans. Faraday Soc.*, **30**, 461 (1934). See also the later study of the electrical conductivity of glass as influenced by changes in composition; B. Lengyel, M. Somogyi, and Z. Boksay, *Z. Physik. Chem. (Leipzig)*, **209**, 15 (1958).

Composition of Electrode Glasses

Within the regions of adequate hygroscopicity and chemical durability, the performance of glass electrodes can be altered and improved by changes in the composition of the glass. This approach has led to the development of glasses which are considerably superior to Corning 015 glass in pH response at high alkalinities and to the fabrication of electrode glasses with excellent response to alkali cations.[45] The stability, electrical conductivity, and sodium errors of electrodes are believed to depend on the ionic properties of the modifier elements (for example, the alkali and alkaline earth cations) in the holes of the glass network, as well as upon the anionic field strength prevailing there.[12] These cations are held strongly in the network of some glasses and weakly in others. Changes in the composition of the glass affect the ease with which ionic transfer between glass and solution can occur. If the surface of a glass electrode were impermeable to all ions except hydrogen ion, the alkaline error would presumably be of little consequence. However, such a glass, if it were made, might have other serious disadvantages, such as an excessive electrical resistance or low durability. The study of the characteristics of electrodes as a function of the composition of the glass is a logical means of achieving a realistic balance among the several properties of pH electrodes.

Electrode glasses usually have at least three constituents, namely SiO_2, R_2O, and MO (or M_2O_3), where R is an alkali metal and M is a bivalent or trivalent metal, the latter preferably one of the rare earth group. Oxides of both the MO and M_2O_3 types may be present, and R_2O may represent a mixture of two alkali metal oxides, such as lithium oxide and cesium oxide. Similarly, the MO constituent may be a mixture of more than one oxide. The usual composition limits are as follows: SiO_2, 60 to 75 mole per cent; R_2O, 17 to 32 mole per cent; and MO (or M_2O_3), 3 to 16 mole per cent. Although barium and strontium oxides are more effective than the smaller MO constituents (calcium, magnesium, and beryllium) in reducing the alkaline error, the addition of these substances to the glass raises the electrical resistance considerably.[46, 47]

It appears that some of the heavy metal oxides behave as "network modifiers," exerting their effect through a change in the length of the silicon–

---

[45] J. O. Isard, Chapter 3 in *Glass Electrodes for Hydrogen and Other Cations*, G. Eisenman, ed., Marcel Dekker, New York, 1967; G. Eisenman, *ibid.*, Chapter 7.

[46] In a study of sodium barium silicate glasses, G. T. Petrovskii and F. Cuta, *Collection Czech. Chem. Commun.*, **26**, 2289 (1961), report a linear pH response up to pH 13.5 for electrodes containing 7.5 to 10.0 mole per cent BaO and 14 to 17.5 mole per cent Na_2O.

[47] The properties of glass electrodes containing BeO have been studied by H. J. C. Tendeloo, A. E. Mans, I. Kateman, and F. H. van der Voort, *Rec. Trav. Chim.*, **81**, 505 (1962).

oxygen chains rather than by determining the size of the interstitial sites available for cation exchange in the conditioned electrode. For example, Schwabe[48] has found that an excellent electrode glass can be made by the substitution of uranium oxide for 5 to 6 per cent of the SiO_2 in the MacInnes–Dole formulation (Corning 015). This glass has ten times the conductivity of 015 glass at 25 °C and has a linear E/pH plot from pH 1 to 10 or even 12. Useful high-temperature electrodes can also be made by substituting titanium dioxide for a part of the SiO_2 in the 015 formula. Simon and Wegmann[49] replaced 1 to 4 mole per cent of the SiO_2 by germanium oxide and found an improvement in the workability of the glass. The electrochemical properties of the glass membrane were almost unaffected.

In 1932, Ssokolof and Passynsky[50] reported the results of a study of the errors of glass electrodes of different compositions, noting that the alkaline errors were large when the solution contained the same cation as did the glass, or a cation of the same group but of smaller atomic number. Furthermore, they discovered that electrodes prepared from glass composed of 72.2 mole per cent SiO_2, 18.2 mole per cent Li_2O, and 9.6 mole per cent CaO displayed nearly the ideal pH response when immersed in solutions of alkali hydroxides other than lithium hydroxide. Lithium ion is not a common constituent of test solutions, and consequently a lithium ion error is of less concern, from the practical point of view, than a sodium ion error. Indeed, electrodes of lithia glass have been found to display a linear response up to pH 12.5 in solutions of sodium hydroxide and to respond rapidly to wide sudden changes of pH in buffered solutions.[51]

By modification of the original formula of Ssokolof and Passynsky, several other electrode glasses suitable for the measurement of pH in highly alkaline media have been developed. For the construction of electrodes with low sodium errors, Cary and Baxter recommend a glass of the following composition: SiO_2, 68 mole per cent; Li_2O, 25 mole per cent; CaO, 7 mole per cent.

Perley[52] found that cesium or rubidium oxide, present with the lithium oxide, is effective in further reducing the sodium ion error. For the MO constituent, the oxides of calcium, strontium, or barium can be used. Electrodes blown from these glasses are said to be stable in aqueous solutions up to 90 °C. Further studies showed that an alkaline earth oxide is not essential if lanthanum oxide is present. A pH-responsive glass composed of 65 mole

[48] K. Schwabe, *Chem. Tech.* (*Berlin*), **6**, 301 (1954).

[49] W. Simon and D. Wegmann, *Helv. Chim. Acta*, **41**, 2099 (1958).

[50] S. I. Ssokolof and A. H. Passynsky, *Z. Physik. Chem.*, **A160**, 366 (1932).

[51] H. H. Cary and W. P. Baxter, U.S. Patent 2,462,843 (March 1, 1949); G. A. Perley, *Anal. Chem.*, **21**, 559 (1949).

[52] G. A. Perley, British Patent 574,029 (Dec. 18, 1945); U.S. Patent 2,444,845 (July 6, 1948).

per cent SiO_2, 28 mole per cent Li_2O, 3 mole per cent Cs_2O, and 4 mole per cent La_2O_3 has a relatively low sodium error.[53]

The Asymmetry Potential

Cremer[54] first recognized that both sides of a glass membrane do not behave exactly alike. If identical solutions are placed inside and outside the bulb of the electrode, there is usually a small potential difference between suitable identical electrodes placed in these two solutions. Inasmuch as the potentials at the two electrodes are of opposite sign and equal magnitude, this potential difference, usually a few millivolts, must arise from asymmetry of the glass membrane.

Beck and Wynne-Jones[55] expressed the view that the proton response of the glass electrode is established instantaneously, and they ascribed all changes of the potential with time to a changing asymmetry potential. Further experiments have shown[56] that this view may need modification under certain conditions.

CAUSES OF ASYMMETRY

Any influence capable of altering the composition and ion-exchange properties of one surface of the membrane preferentially is able to affect the asymmetry potential. Kratz[57] has summarized some of the most important causes:

1. Alteration of the water-sorptive capacity. Loss of alkali in the flame while the glass bulb is being fabricated.

---

[53] G. A. Perley, *Anal. Chem.*, 21, 394 (1949); U.S. Patent 2,497,235 (Feb. 14, 1950).

[54] M. Cremer, *Z. Biol.*, 47, 562 (1906).

[55] W. H. Beck and W. F. K. Wynne-Jones, *J. Chim. Phys.*, 49, C97 (1952). It is certainly true that the response of a glass electrode in well-buffered solutions within its optimum response region may be exceedingly rapid. A. Distèche and M. Dubuisson, *Rev. Sci. Inst.*, 25, 869 (1954), found the response so fast that they had difficulty in producing, mechanically, a pH change quickly enough to approximate a "rectangular step." See also A. Distèche, *Mém. Acad. Royale Belgique*, 32, No. 1 (1960). The response is much slower in poorly buffered media than in buffer solutions. Also, in a study of flowing solutions, A. L. Giusti, Jr., M.S. thesis, St. Louis Univ., 1960, found that less time is required when the solution changes from dilute acid (pH 4.7 to 5.3) to water than from water to dilute acid. As J. Meier and G. Schwarzenbach, *Helv. Chim. Acta*, 40, 907 (1957), showed, accurate pH values for very rapidly flowing solutions can only be obtained with the glass electrode when the conductivity is not too low. An ionic strength of 0.1 was found to be satisfactory.

[56] W. H. Beck, J. Caudle, A. K. Covington, and W. F. K. Wynne-Jones, *Proc. Chem. Soc.*, 110 (1963).

[57] L. Kratz, *Die Glaselektrode und Ihre Anwendungen*, Verlag von Dr. D. Steinkopff, Frankfurt am Main, 1950.

2. Dehydration of the swollen surface layer. Drying, or prolonged immersion in dehydrating solutions.
3. Destruction of the swollen surface layer. Mechanical damage (grinding or polishing); chemical attack (etching by alkalies or hydrofluoric acid).
4. Disturbance of the exchange capacity for hydrogen ions (adsorption of foreign ions; grease films, proteins and surface-active agents).

Unequal strains in the two surfaces of a glass bulb contribute to asymmetry.[58] If the holes in the silicon–oxygen network at one surface become markedly distorted and different in size and shape from those at the other surface, the equilibria for the transfer of ions between glass and solution are affected and electrical asymmetry results. Annealing reduces but does not eliminate the asymmetry potential.[59] If both surfaces are polished, the asymmetry potential is low, but polishing of only one surface produces a relatively large asymmetry potential.[60]

Chemical attack of the glass also brings about differences between the two surfaces. When an etched electrode is allowed to dry, the asymmetry potential is sometimes found to increase. Kratz explains this action by removal of the soluble alkaline constituents of the glass surface, leaving a layer of swollen hydrous silica. On drying, this silica-rich layer shrinks, placing the membrane under mechanical tension.[61] These results emphasize once again the importance of the surface condition in determining the electrical properties of the electrode. Hamilton and Hubbard[62] found that differences in the ability of the inner and outer surfaces of an electrode to absorb the dye Victoria Blue B could be correlated with differences in surface condition produced by chemical attack of the glass. It was suggested that the asymmetry potential of the glass electrode results from differences in the ability of the two surfaces to absorb ions.

MEASUREMENT OF ASYMMETRY POTENTIAL

Although the asymmetry potential of a glass electrode may drift slightly from day to day, it is not ordinarily subject to large and sudden fluctuations. It may be regarded, for a short series of measurements, as a constant of the assembly, as is the potential at the junction of the salt bridge with the pH standard or with the test solution. In most pH meters, these constant potentials are compensated by adjustment of the standardizing knob or zero

[58] H. Yoshimura, *Bull. Chem. Soc. Japan*, **12**, 443 (1937).
[59] L. Kratz, *Glastech. Ber.*, **20**, 15 (1942).
[60] W. Bräuer, *Z. Elektrochem.*, **47**, 638 (1941).
[61] L. Kratz, *Z. Elektrochem.*, **46**, 253 (1940).
[62] E. H. Hamilton and D. Hubbard, *J. Res. Nat. Bur. Stand.*, **27**, 27 (1941).

adjustor when the instrument is made to read the correct pH for the standard solution.

The asymmetry potential can be measured by a cell like the following:

$$\text{Ag}; \text{AgCl}, 0.1 \text{ M HCl} \,\|\, \text{Glass} \,\|\, 0.1 \text{ M HCl}, \text{AgCl}; \text{Ag} \qquad (13)$$

If the two silver–silver chloride electrodes have first been immersed in the same solution of hydrochloric acid and shown to agree in potential, the e.m.f. of cell 13 is the asymmetry potential of the glass electrode.

Commercial glass electrodes are usually sealed, and hence their asymmetry potentials are not readily measured. Changes of asymmetry potential from time to time can, however, be detected by measurement of the potential of the electrode with respect to a reproducible calomel electrode or a silver–silver chloride electrode in a solution of hydrochloric acid:

$$\text{Glass electrode} \,\|\, 0.1 \text{ M HCl}, \text{AgCl}; \text{Ag} \qquad (14)$$

ELIMINATION OF THE ASYMMETRY POTENTIAL

In the most accurate applications of the glass electrode to electrochemical measurements the effect of a changing asymmetry potential with time has been eliminated by an extrapolation.[63] In the study of the cell

$$\text{Glass} \,\|\, \text{HCl}(m), \text{AgCl}; \text{Ag}$$

for example, the e.m.f. was measured first as a function of time in a reference solution of the acid. When the time variation of the e.m.f. (usually linear) had been established, the glass electrode was transferred rapidly into another cell containing a second solution of the acid after a quick washing with a portion of that solution. The e.m.f. of the cell before and after transfer was plotted as a function of time, and the two plots were extrapolated to the time of transfer. At this moment alone the difference between the e.m.f. of the two cells is independent of the asymmetry potential, as the latter is identical in both.

This technique presupposes that no "jump" of asymmetry potential occurs when the external environment of the glass membrane is suddenly altered. However, the excellent agreement between the activity coefficients derived from the glass electrode cell and those obtained with the hydrogen electrode is adequate proof of the validity of the procedure.

---

[63] A. K. Covington and J. E. Prue, *J. Chem. Soc.*, 3696, 3701 (1955); 1567 (1957); A. J. Zielen, *J. Phys. Chem.*, **67**, 1474 (1963). Studies of solutions of hydrochloric acid have been made at 0°C by A. K. Covington *J. Chem. Soc.*, 4441 (1960), by the same techniques, and at 65°C by H. L. Clever and R. M. Reeves, *J. Phys. Chem.*, **66**, 2268 (1962), who used a modified procedure. The dissociation constants of weak acids have been determined in this way by E. J. King and J. E. Prue, *J. Chem. Soc.*, 275 (1961).

Development of the Surface Potential

When a freshly blown glass bulb is first immersed in a solution, hydrogen ions from the solution exchange with alkali metal ions from the glass membrane, finding points of high stability in the surface. The conditioned electrode behaves thereafter as a " protode," or site for proton transfer between the solution and the reservoir of protons in the glass surface; the potential of the surface changes as protons are acquired or lost. The glass electrode process is not one of electron exchange, that is, of oxidation-reduction. Hence the glass electrode is the only hydrogen ion electrode not disturbed by oxidizing and reducing agents.

Haugaard[67] has presented the following picture of the interaction of the 015 glass membrane with neutral, acidic, or weakly alkaline solutions: A freshly blown, dry glass surface first takes up water, hydrolyzing some of the sodium silicate of which the glass is composed. A skeleton of silicic acid is formed in the surface, affording an easy passage of hydrogen ions to the glass surface as sodium ions pass into the solution. In the body of the glass there remains a layer of intact sodium salt. This layer moves slightly toward one surface or the other with the passage of feeble currents, the direction of movement depending upon the direction of the current.

In Haugaard's experiments, a 0.02 M solution of hydrochloric acid was placed both inside and outside a thin bulb of 015 glass, and a small known current was passed for a time sufficiently long to bring about measurable changes in the concentrations of the two solutions. The amount of sodium chloride formed was determined by evaporation of the solution, and the decrease of acidity was found by titration. In this manner it was shown that hydrogen ion was carried into the glass on one side and sodium migrated out of the glass on the other, both in accordance with Faraday's laws, that is, in the amount of one equivalent per faraday of electricity. Haugaard also showed that the potential of a freshly blown glass electrode varies linearly with the logarithm of the amount of hydrogen ion taken up per unit surface area as the fresh glass surface comes to equilibrium with the solution. The slope of the line was approximately 0.059 V/pH, as demanded by theory.

The ions that determine the potential are those which cross the boundary or double layer. MacInnes and Belcher[65] expressed the belief that the behavior of the glass electrode can be satisfactorily explained by the ability of the hydrogen ion or proton to pass the electrolyte–glass boundary more

---

[64] G. Haugaard, *Nature*, **140**, 66 (1937); *J. Phys. Chem.*, **45**, 148 (1941). See also K. Horovitz, *Z. Physik*, **15**, 369 (1923).

[65] D. A. MacInnes and D. Belcher, *J. Amer. Chem. Soc.*, **53**, 3315 (1931).

easily than any other positive ion. Dole[66] showed that the glass does not behave as a semipermeable membrane but that a potential difference is developed independently on each surface. Attempts have also been made to treat the potential between glass and solution as a liquid-junction or diffusion potential, but this approach was not successful.

The Alkaline Error

The pH response of most glass electrodes is imperfect at both ends of the pH scale. The error ΔE of the 015 electrode is positive in strong solutions of the alkali hydroxides and in alkaline solutions containing high concentrations of the cations of the alkali and alkaline earth series. It is negative at very low pH. The error at both ends of the pH scale is time-dependent. A positive ΔE means that the response of the glass electrode to changes of pH is falling below the ideal hydrogen electrode response as a partial response to alkali metal ions is developed. Hence the pH value of highly alkaline mixtures, as measured with the glass electrode, will be too low. The alkaline error of the 015 electrode increases rapidly as the temperature rises above 30 °C but is only slightly smaller at 10 °C than at 25 °C.

The pH response of the 015 glass electrode in aqueous solutions at temperatures from 0 to 30 °C is unimpaired between pH 1 and 9.5. At 50 °C, departures appear at pH 8 and, when the temperature is 60 °C, at pH 5 to 6.[67] The error in a solution 1 M with respect to sodium ion between pH 12 and 13 amounts to 70 to 90 mV (1.1 to 1.5 pH unit) at 25 °C. Somewhat smaller errors are found in solutions 1 M with respect to lithium ion, and the errors caused by potassium ions are of still less concern.

The departures caused by cations not present in the glass increase with a decrease in the radius of the ion. This observation is consistent with the view that the smallest ions experience the least difficulty in penetrating the silicon–oxygen network. Thus the smallest errors will be observed when the glass contains alkali metal cations of small radius and when the solution contains cations of large radius.[68] Magnesium ions are said not to cause voltages departures,[69] and the errors caused by ammonium and barium ions are small.[70] The lithium ion error and the potassium ion error of the 015 glass

[66] M. Dole, *J. Amer. Chem. Soc.*, **53**, 4260 (1931); see also D. Hubbard and R. G. Goldman, *J. Res. Nat. Bur. Stand.*, **48**, 428 (1952).

[67] M. Dole and B. Z. Wiener, *Trans. Electrochem. Soc.*, **72**, 107 (1937); W. C. Gardiner and H. L. Sanders, *Ind. Eng. Chem., Anal. Ed.*, **9**, 274 (1937); V. A. Pchelin, *Zh. Fiz. Khim.*, **13**, 490 (1939).

[68] S. I. Ssokolof and A. H. Passynsky, *Z. Physik. Chem.*, **A160**, 366 (1932).

[69] E. S. Amis and J. L. Gabbard, *J. Amer. Chem. Soc.*, **59**, 557 (1937).

[70] D. Hubbard, E. H. Hamilton, and A. N. Finn, *J. Res. Nat. Bur. Stand.*, **22**, 339 (1939); M. Dole, *J. Phys. Chem.*, **36**, 1570 (1932).

electrode are often about one-half and one-fifth, respectively, of the corresponding sodium ion error.

Figure 11–7 is a plot of the corrections at 25 °C (in pH units to be added) for the 015 electrode in solutions of hydrochloric acid and sulfuric acid as a function of pH, and for the 015 electrode and three commercial electrodes in alkaline solutions containing sodium ion at a concentration of 1 M. The data are from the papers of Dole[71] and MacInnes and Belcher,[65] and from the bulletins of the manufacturers.

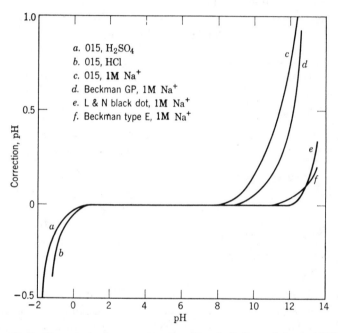

a. 015, H_2SO_4
b. 015, HCl
c. 015, 1M Na^+
d. Beckman GP, 1M Na^+
e. L & N black dot, 1M Na^+
f. Beckman type E, 1M Na^+

FIG. 11–7. Glass electrode corrections in acid and alkaline solutions at 25 °C.

The Acid Error

In strong aqueous solutions of salts and strong acids at pH less than 1, and in certain nonaqueous solutions, the pH response of the glass electrode displays a departure opposite in sign to that found in alkaline solutions. As a result of this so-called negative error, the pH measured in this region is somewhat greater than the true pH. Unlike the alkaline error, the error in acid solutions changes but little with the temperature.

---

[71] M. Dole, *J. Amer. Chem. Soc.*, **54**, 2120, 3095 (1932).

The magnitude of the negative error in solutions of strong acids is indicated by the results of Goldman and Hubbard[72] in Table 11-2. The errors are given in mV (ΔE) as well as in pH units (ΔpH). As the acid error is time-dependent, the values given are only an approximation. The onset of the acid error was found to be accompanied by some reduction in the thickness of the swollen layer of hydrous silica in the surface of the glass membrane.[73]

TABLE 11-2. ERRORS OF THE 015 GLASS ELECTRODE IN STRONGLY ACID SOLUTIONS

Solution		(Goldman and Hubbard) ΔE, mV	ΔpH
H_2SO_4	3.0 M	10	0.17
	12.7 M	100	1.7
	18.0 M	525	8.9
H_3PO_4	12.1 M	10	0.17
	14.7 M	20	0.34
HCOOH	19.2 M	10	0.17
	23.5 M	32	0.54
CH_3COOH	13.4 M	12	0.20
	17.4 M	220	3.7

Dole found negative errors as large as 0.7 pH unit in 98 per cent ethanol and made the suggestion that these departures were caused by a lowering of the water activity.[71] By keeping the pH substantially constant and lowering the activity of water by the addition of either salt or alcohol, Dole observed errors of the 015 glass electrode at moderate pH values that were similar in direction to the errors found in strongly acid solutions. He attempted to explain the observed departures by assuming that the proton, in migrating from the solution into the surface of the glass, carries one molecule of water with it. By contrast, water does not participate in the hydrogen electrode reaction. This view is no longer generally accepted.

---

[72] R. G. Goldman and D. Hubbard, *J. Res. Nat. Bur. Stand.*, **48**, 370 (1952).
[73] D. Hubbard, E. H. Hamilton, and A. N. Finn, *J. Res. Nat. Bur. Stand.*, **22**, 339 (1939). It is perhaps significant that glass fibers have been found to have an "isoelectric point" (near pH 2.9). See M. J. O'Leary and D. Hubbard, *J. Res. Nat. Bur. Stand.*, **55**, 1 (1955).

From the earliest days, characterization of the acid error has been plagued by conflicting observations. Thus Hubbard, Hamilton, and Finn[73] were able to make 015 glass electrodes that showed no departures in sulfuric acid solutions up to 5 M. In solutions of hydrochloric acid, departures set in at concentrations as low as 1 M. Moreover, electrodes of a Jena glass have been reported to be error-free at pH values as low as -1.3.[74] In contrast with Dole's observations, Beck and Wynne-Jones[75] found errors of less than 0.05 pH unit five minutes after glass electrodes were immersed in a solution of 0.1 M hydrochloric acid in 95 per cent ethanol. Indeed, Izmailov and his coworkers[76] failed to detect an error attributable to a lowering of the water activity in aqueous methanol, ethanol, or acetone.

Beck, Caudle, Covington, and Wynne-Jones[77] called attention to a difference between the behavior of glass electrodes in dilute acid solutions and in concentrated acid solutions. In dilute solutions the potentials are steady and reproducible (apart from the slow change attributable to the changing asymmetry potential), but in concentrated solutions there is a lack of reproducibility and a considerable drift toward more negative values. Occasionally there is a reversal in the direction of drift, and a steady value is rarely reached even after many hours. There is a critical concentration above which the second type of behavior is regularly observed; this molality is dependent on the acid anion and is 1 mol kg^{-1} for HCl; 5 for HBr; 7 for H_2SO_4; and 10 for H_3PO_4, CH_3COOH, and $HClO_4$.

The origin of the error in acidic solutions is not yet clear. The negative error was thought by Beck and Wynne-Jones[75] to be a manifestation of the changing asymmetry potential in strongly acid media. It has also been attributed to the existence (or formation) of proton acceptor sites of abnormally low energy in the gel surface; at high acidities proton exchange between the solution and these "reserve" sites may occur.[78]

Early in the development of the glass electrode, MacInnes and Belcher[65] speculated that penetration of the glass by small anions might explain the negative errors. Schwabe and his coworkers have confirmed the fact that absorption of acid from strongly acid solutions does actually take place.[79] With the aid of radioactive tracers, they have shown that the acid error is

[74] B. Lengyel and J. Vincze, *Glastech. Ber.*, **19**, 359 (1941).

[75] W. H. Beck and W. F. K. Wynne-Jones, *J. Chim. Phys.*, **49**, C97 (1952).

[76] N. A. Izmailov and M. A. Bel'gova, *Zh. Obshch. Khim.*, **8**, 1873 (1938); N. A. Izmailov and T. F. Frantsevich-Zabludovskaya, *ibid.*, **15**, 283 (1945).

[77] W. H. Beck, J. Caudle, A. K. Covington, and W. F. K. Wynne-Jones, *Proc. Chem. Soc.*, 110 (1963).

[78] Z. Boksay, B. Csákvári, and B. Lengyel, *Z. Physik. Chem. (Leipzig)*, **207**, 223 (1957).

[79] K. Schwabe and G. Glöckner, *Z. Elektrochem.*, **59**, 504 (1955); K. Schwabe, H. Dahms, Q. Nguyen, and G. Hoffmann, *ibid.*, **66**, 304 (1962).

proportional to the uptake of acid from solutions of the halogen acids,[80] which presumably increases the activity of hydrogen ions in the glass phase. On the contrary, no absorption of sulfuric acid or phosphoric acid was observed, and the error in strong solutions of these acids was attributed to lowered water activity in the manner of Dole. However, it now seems unlikely that, under less extreme conditions, a transfer of water through the membrane plays any part in the electrochemical process responsible for the development of the pH response.

Theory of the Glass Electrode

A quantitative explanation of the dependence of the response of the glass electrode on pH, concentration of foreign cations, and temperature is an exacting test of any theory. Some time ago, Dole[81] and Nicolsky[82] derived equations that met these requirements reasonably well. The first was based on statistical reasoning and the second on thermodynamics. These equations accounted successfully for the behavior of the glass electrode both in its substantially error-free region and in the region where an appreciable sodium ion function is developed. They were less successful in describing the actual behavior in the transition region where the electrode responds to both hydrogen ion and alkali cations. More recently, the theory has been refined and extended and the factors determining selectivity clarified, largely through the work of Eisenman,[83] Nicolsky, Shultz, and their coworkers.[84] As a result of these crucial studies, it is now possible to account well for the response of glass electrodes and to understand the relationship between glass composition and specificity for various cations. Much of this information is also pertinent to the behavior of ion-selective electrodes in general.

In the region where the pH response is error-free, the only electrochemical process of concern is the exchange of protons between the solution and the gel layer in the hydrated glass surface. The potential E_g then follows the

---

[80] Similar conclusions were reached by N. A. Izmailov and A. G. Vasil'ev, *Zh. Fiz. Khim.*, **29**, 1866, 2145 (1955), who also used tracer methods. It was suggested by Boksay, Csákvári, and Lengyel (reference 78) that the acid error might be explained by penetration of the glass surface by undissociated molecules.

[81] M. Dole, *J. Chem. Phys.*, **2**, 862 (1934). The derivation of Dole's equation has been refined by N. Landqvist, *Acta Chem. Scand.*, **9**, 595 (1955).

[82] B. P. Nicolsky, *Zh. Fiz. Khim.*, **10**, 495 (1937); *Acta Physicochim. URSS*, **7**, 597 (1937); *Zh. Fiz. Khim.*, **27**, 724 (1953).

[83] See G. Eisenman, Chapters 5, 7, and 9 in *Glass Electrodes for Hydrogen and Other Cations*, G. Eisenman, ed., Marcel Dekker, New York, 1967.

[84] For a summary and references to the original literature, see B. P. Nicolsky, M. M. Shultz, A. A. Belijustin, and A. A. Lev, Chapter 6 in the volume edited by G. Eisenman (reference 83).

Nernst formula for the hydrogen electrode, although the standard potential is, of course, not usually zero. When a mixed electrode function develops, there is a corresponding alteration of the exchange equilibrium at the glass surface, with cations M^+ replacing some of the hydrogen ions occupying stable sites in the glass phase:

$$H^+(gl) + M^+(soln.) \rightleftharpoons M^+(gl) + H^+(soln.) \tag{15}$$

or, in the general case,

$$i(gl) + j(soln.) \rightleftharpoons j(gl) + i(soln.) \tag{16}$$

With certain assumptions having to do with the number of exchange sites available to the two cations in the glass surface and with the activity coefficients of ions in the glass, Nicolsky[82] was able to derive the equation

$$E_g = \text{const.} + \frac{RT}{F} \ln (a_i + K' a_j) \tag{17}$$

where the constant K' was identified with the equilibrium constant K_{ij} for the exchange process, equation 16.

Although Nicolsky's equation was moderately successful in describing the change in potential of the glass electrode in mixtures of cations i and j, Eisenman, Rudin, and Casby[85] found it possible to represent a wider range of experimental data by an equation of the form

$$E_g = \text{const.} + \frac{nRT}{F} \ln [a_i^{1/n} + (K' a_j)^{1/n}] \tag{18}$$

In equation 18, K' and n are empirical parameters characteristic of the pair of ions and the composition of the glass.

Theoretical studies now support the form of equation 18. It has been demonstrated that the voltage response of the glass electrode is to be regarded as the sum of two terms:[86]

$$E_g = E_{ex} + E_d \tag{19}$$

namely the ion-exchange potential E_{ex} and a diffusion potential E_d. Derivations of the ion-exchange potential (phase-boundary potential),[87] on the assumption that glass is a perfect cation exchanger with n-type nonideal characteristics ($d \ln a/d \ln c = n$), led to an expression for E_{ex} of the same

---

[85] G. Eisenman, D. O. Rudin, and J. U. Casby, *Science*, **126**, 831 (1957).

[86] K. H. Meyer and J. R. Sievers, *Helv. Chim. Acta*, **19**, 649 (1936); T. Teorell in *Progress in Biophysics and Biophysical Chemistry*, J. A. V. Butler and J. T. Randall, eds., Vol. 3, Pergamon Press, New York, 1953.

[87] G. Eisenman, *Biophys. J.*, **2**, Pt. 2, 259 (1962).

form as equation 18, in which K' is identified with K_{ij}. Furthermore, it was shown[88] that the diffusion potential E_d can be expressed by

$$E_d = -\frac{nRT}{F} \ln \frac{c_i(2) + c_j(2)(u_j/u_i)}{c_i(1) + c_j(1)(u_j/u_i)} \tag{20}$$

where $c(2)$ and $c(1)$ are the ionic concentrations in the two surfaces of the glass membrane and u_j/u_i is the ratio of the mobilities of the two cations in the glass, considered to be a constant. With the aid of the exchange constant K_{ij}, the ion concentrations in the glass could be related to ion activities in the bounding solutions. Addition of E_{ex} and E_d (equation 19) does not alter the form of equation 18 but demonstrates that the constant K', obtained empirically, is actually a combination of K_{ij}, the exchange equilibrium constant for equation 16, and the mobility ratio u_j/u_i:

$$K' = K_{ij}\left(\frac{u_j}{u_i}\right)^n \tag{21}$$

For many pairs of univalent cations and many glass compositions, n is found to be 1. For two potassium-selective glasses of approximate compositions Na_2O (27 mole per cent), Al_2O_3 (4 mole per cent), SiO_2 (69 mole per cent), Eisenman found $n = 1$ for Na^+-K^+, Na^+-H^+, and K^+-H^+ mixtures. This was true even though the value of K' was 10.3 for one glass and 8.5 for the other. This relationship of equation 18 was valid over the pH range 6 to 11 and for cation concentrations from 0.01 to 0.1 M. For these potassium-selective glasses, potassium ion has a greater tendency to penetrate the glass surface than does sodium ion and is more strongly bound in the hydrated glass surface. It is, however, less mobile than sodium ion in the glass. Thus, the ion-exchange contribution E_{ex} is in the same direction as the total potential E_g, but the diffusion-potential contribution E_d is in the opposite direction. The ion-exchange potential matches the form of the total potential reasonably well, and for this reason equation 17 is fairly successful in accounting for the errors of pH electrodes in solutions of alkali metal cations at high pH. It is interesting to note that the diffusion potential is constant and equal to $(RT/F) \ln (u_i/u_j)$ when the glass membrane separates pure solutions of j and i, regardless of their concentrations.

Conti and Eisenman[88] have shown that (u_j/u_i), K_{ij}, and n can be evaluated from the analysis of current-potential relationships for glass electrodes. Furthermore, Eisenman[89] has found that the "anionic field strength" of the glass is a principal determinant of specificity. In his words,

---

[88] G. Karreman and G. Eisenman, *Bull. Math. Biophys.*, **24**, 413 (1962); F. Conti and G. Eisenman, *Biophys. J.*, **5**, 247 (1965); R. H. Doremus, Chapter 4 in *Glass Electrodes for Hydrogen and Other Cations*, G. Eisenman, ed., Marcel Dekker, New York, 1967.
[89] G. Eisenman in *Advances in Analytical Chemistry and Instrumentation*, C. N. Reilley, ed., Vol. 4, p. 213, Wiley-Interscience, New York, 1965.

The essential requirement for producing cation selectivity is the introduction of a structural element into the lattice in a coordination state higher than its oxidation state. Such an introduction produces a fixed anionic site of a field strength appropriate to cation selectivity whenever the disproportion between oxidation state and coordination leads to a residual bonding strength of the oxygen of 1/2 or less.

Hydration of the glass is a factor fixing the magnitude of specificity without altering the sequence of relative ionic effects. If the anionic groups in the glass have a high field strength, the smaller cations are preferred by the glass over the larger cations.[90] As the field strength is lowered, the cationic preference shifts regularly until the larger cations are preferred at very low field strengths. To some extent, glass compositions can be tailored to provide the desired cationic selectivity.[91] Considerable progress has been made towards an understanding of the role of ion exchange in determining ionic specificity. On the contrary, little is yet known about the effect of relative ionic mobilities on the selectivity of glasses.

Glasses containing aluminum oxide or boric oxide are particularly responsive to metallic ions.[91] Most alkali cation-sensitive glass electrodes available commercially are fabricated from glasses containing Al_2O_3 as one constituent. In general, these electrodes are highly responsive to silver ion and have, as well, a residual hydrogen ion function. To assure a suitable response to alkali cations, therefore, it is often necessary to buffer the solution at a pH of 7 or higher. For details on the characteristics of cation-sensitive glass electrodes, the reader is referred to the monograph edited by Eisenman.

These studies make it evident that the pH electrode and the alkali ion-selective glass electrode are extreme members of a continuous series of cation-responsive electrodes. Anionic sites of high field strength prefer hydrogen ions over the alkali metal cations, while low field strength sites have the opposite preference. A good glass for pH electrodes should have a composition such that all of the alkali metal ions within the surface are easily exchanged for hydrogen ions; this occurs when the glass has a strong preference for hydrogen ions.[90] In other words, the product $(1/K_{HM})(u_H/u_{M^+})a_H$ should be considerably larger than the activity of any of the other univalent ions M^+ present in the solution. This means that K_{HM} (compare equation 15) should be very small, in view of the fact that the mobility of hydrogen ions in the glass is generally lower at ordinary temperatures than that of alkali cations. Under these conditions, the diffusion potential contribution to the

---

[90] R. H. Doremus, Chapter 4 in *Glass Electrodes for Hydrogen and Other Cations*, G. Eisenman, ed., Marcel Dekker, New York, 1967.

[91] J. O. Isard, Chapter 3 in the volume edited by G. Eisenman (reference 90). See also reference 84 and the extensive series of investigations by Nicolsky, Shultz, and their co-workers to which reference is made.

glass electrode measurement will be absent and the electrode error-free. The alkaline error at high pH sets in when $(1/K_{HM})(u_H/u_{M^+})a_H$ becomes of the same order of magnitude as a_{M^+}. When this occurs, both the ion-exchange potential and the diffusion potential depend upon a_{M^+}, and the electrode assumes a mixed response.

Using radioactive tracers and electrodes activated with thermal neutrons in a reactor, Schwabe and Dahms[92] have studied the kinetics of ion exchange between solution phase and glass phase in the alkaline region for glasses of different compositions. Their results support the general concept of an ion exchange mechanism which accounts not only for the pH response but also for the alkali error. They conclude that the activity of hydrogen ions is unity and constant in the gel layer when the electrode displays a pure hydrogen function. The same is true of sodium ions in an electrode that has a pure sodium ion function, but in the transition region the activity coefficients of both ions change in a regular manner.

Schwabe and Dahms developed the following equation for ΔpH, the alkali error in pH units:

$$\Delta pH + n \log (1 - 10^{-\Delta pH}) = pH + \log a_{Na^+} + \log K \qquad (22)$$

In addition to the ion-exchange constant K for equilibrium 15, this equation contains only the constant n, which is said to denote the ratio of the activities of hydrogen and sodium ions in the glass phase. The constant n is obtained from a single measurement of ΔpH in the transition region. Equation 22 was shown to give an excellent representation of the alkali errors of four glasses, including Corning 015 and Schwabe's titanium glass, up to values of ΔpH as high as 4.0. For 015 glass, $\log K$ is -11.6 and n is 2.6.

The theory of Oláh[93] also accounts for the alkaline errors in terms of a dual response of the electrode to protons and to other cations. In Oláh's equation, the voltage departure is expressed in terms of the dissociation constants for silicic acid and alkali silicates in the gel layer.

The Glass Electrode in Nonaqueous and Partially Aqueous Media

It appears that a certain amount of water is necessary for proper functioning of the 015 glass electrode. The performance of the electrode usually becomes less satisfactory as the membrane becomes partially dehydrated. It is not yet clear to what extent the role of water in the glass surface can be assumed by other amphiprotic solvents.

---

[92] K. Schwabe and H. Dahms, *Z. Elektrochem.*, **65**, 518 (1961) A method for determining concentration profiles of ions in the surface layers of glass electrodes has been described by H. Bach and F. G. K. Baucke, *Electrochim. Acta*, **16**, 1311 (1971).

[93] C. Oláh, *Periodica Polytech.*, **4**, 141 (1960). See also B. P. Nicolsky and M. M. Shultz, *Zh. Fiz. Khim.*, **36**, 1327 (1962).

The usefulness of the glass electrode in nonaqueous solutions may be impaired by a defective response and sometimes by the high resistance of the medium. In spite of these difficulties, glass electrodes have been reported to function satisfactorily in organic solvents of dielectric constant as low as 2.3.[94] Lykken[95] has used glass electrodes successfully in benzene–isopropyl alcohol solvents after saturation of the glass surface with water. After the electrode has been immersed in nonaqueous media, its pH response can usually be fully restored by soaking in water.

Little difficulty is encountered in the use of the glass electrode in ethanol–water mixtures containing less than 90 weight per cent of ethanol.[96] At higher concentrations of alcohol, and in 50 to 90 per cent acetone, some contraction of the rectilinear portion of the e.m.f.-pH curve has been found, as well as a time variation of the potential. Nevertheless, successful results in pure acetone have been reported.[97] In 40 per cent ethanol, the theoretical slope is maintained from pH 3 to 9.5, but in 50 and 70 per cent ethanol deviations appear at pH 8 and at 7, respectively.[98] The glass electrode is stable in methanol.[99]

Moreover, the glass electrode has been found to respond satisfactorily to hydrogen ion in hydrogen peroxide.[100] It appears to be capable of furnishing reliable data in formic acid,[101] although a gradual loss of sensitivity has been observed. The glass electrode appears to be capable of a reproducibility of ± 5 mV or better in acetic acid.[102] Nevertheless, "acid errors" as large as 70 mV were sometimes observed when chloride ion was present.[103] These errors have been ascribed to penetration of chloride ion into the surface

---

[94] A. Gemant, *J. Chem. Phys.*, **12**, 79 (1944).

[95] L. Lykken, *Symposium on pH Measurement*, p. 71, ASTM Tech. Publ. 73, Philadelphia, Pa., 1947; see also L. Lykken, P. Porter, H. D. Ruliffson, and F. D. Tuemmler, *Ind. Eng. Chem., Anal. Ed.*, **16**, 219 (1944).

[96] See, for example, W. H. Beck and W. F. K. Wynne-Jones, *J. Chim. Phys.*, **49**, C97 (1952); J. P. Morel, P. Séguela, and J. C. Pariaud, *Compt. Rend.*, **253**, 1326 (1961); J. P. Morel, thesis, Clermont-Ferrand, 1969.

[97] F. Aufauvre and A. Comte, *Compt. Rend.*, **263C**, 618 (1966); F. Aufauvre, thesis, Clermont-Ferrand, 1969.

[98] N. A. Izmailov and M. A. Bel'gova, *Zh. Obshch. Khim.*, **8**, 1873 (1938); N. A. Izmailov and T. F. Frantsevich-Zabludovskaya, *ibid.* **15**, 283 (1945).

[99] A. L. Bacarella, E. Grunwald, H. P. Marshall, and E. L. Purlee, *J. Org. Chem.*, **20**, 747 (1955); C. D. Ritchie and P. D. Heffley, *J. Amer. Chem. Soc.*, **87**, 5402 (1965); C. D. Ritchie and G. H. Megerle, *ibid.*, **89**, 1452 (1967).

[100] A. G. Mitchell and W. F. K. Wynne-Jones, *Trans. Faraday Soc.*, **51**, 1690 (1955); J. R. Kolczynski, E. M. Roth, and E. S. Shanley, *J. Amer. Chem. Soc.*, **79**, 531 (1957).

[101] A. M. Shkodin, N. A. Izmailov, and N. P. Dzuba, *Zh. Obshch. Khim.*, **20**, 1999 (1950); **23**, 27 (1953).

[102] N. A. Izmailov and A. M. Aleksandrova, *Zh. Obshch. Khim.*, **20**, 2127 (1950); D. Wegmann, J. P. Escarfail, and W. Simon, *Helv. Chim. Acta*, **45**, 826 (1962).

[103] A. T. Cheng, R. A. Howald, and D. L. Miller, *J. Phys. Chem.*, **67**, 1601 (1963).

layers of the glass. Useful results have been obtained with glass electrodes in acetonitrile,[104] in quinoline and pyridine,[105] in dimethylformamide,[106] and in *m*-cresol.[107] A glass electrode for measurements in acetonitrile has been described.[108] As inner solution, a buffered acetonitrile solution is used.

Coetzee and Bertozzi[109] found that the glass electrode behaves well in sulfolane (tetramethylenesulfone), provided contamination of the sulfolane with traces of water is carefully avoided. In dimethylsulfoxide, a slow response at high pH values has been observed, but the response time can be shortened by replacement of the internal solution with metallic mercury or with a silver electrode dipping in a 0.05 M solution of silver perchlorate in dimethylsulfoxide.[110] The electrode is then capable of responding to changes in hydrogen ion activity extending over 25 powers of 10. To obtain the best results with the glass electrode in dimethylformamide, Ritchie and Megerle[106] used as an inner electrode a silver wire dipping in a solution consisting of 0.001 M silver perchlorate in dimethylformamide with added picric acid and tetraethylammonium perchlorate. The external surface of the glass electrode was etched for 3 minutes in a 20 per cent aqueous solution of ammonium bifluoride. An internal mercury electrode in contact with 0.01 M mercurous ion and 0.1 M perchloric acid in dimethylformamide has also been used.[111] The apparent pK values of an extensive series of organic acids and bases in the mixed solvent system water-glycol monomethyl ether (methylcellosolve) have also been determined with the glass electrode.[112]

Information on the alkaline error of the glass electrode in nonaqueous media is scanty. Harlow[113] has observed an irregularity in titration curves obtained with the glass electrode in pyridine, which appears to originate in "titration" of the pH-sensitive glass with the traces of potassium ion that

---

[104] E. Römberg and K. Cruse, *Z. Elektrochem.*, **63**, 404 (1959); J. F. Coetzee and G. R. Padmanabhan, *J. Phys. Chem.*, **66**, 1708 (1962); I. M. Kolthoff and M. K. Chantooni, Jr., *J. Amer. Chem. Soc.*, **87**, 4428 (1965).

[105] F. N. Kozlenko, *Zh. Fiz. Khim.*, **33**, 1866 (1959); M. Bos and E. A. M. F. Dahmen, *Anal. Chim. Acta*, **53**, 39 (1971); **55**, 285 (1971).

[106] M. Tézé and R. Schaal, *Bull. Soc. Chim. France*, 1372 (1962); C. D. Ritchie and G. H. Megerle, *J. Amer. Chem. Soc.*, **89**, 1447 (1967); J. Juillard, thesis, Clermont-Ferrand, 1968.

[107] M. Bos and E. A. M. F. Dahmen, *Anal. Chim. Acta*, **57**, 361 (1971).

[108] J. Badoz-Lambling, J. Desbarres, and J. Tacussel, *Bull. Soc. Chim. France*, 53, (1962); J. Desbarres, *ibid.*, 2103 (1962).

[109] J. F. Coetzee and R. J. Bertozzi, *Anal. Chem.* **43**, 961 (1971).

[110] C. D. Ritchie and R. E. Uschold, *J. Amer. Chem. Soc.*, **89**, 1721 (1967); I. M. Kolthoff and T. B. Reddy, *Inorg. Chem.*, **1**, 189 (1962).

[111] G. Demange-Guérin and J. Badoz-Lambling, *Bull. Soc. Chim. France*, 3277 (1964).

[112] W. Simon, G. H. Lyssy, A. Mörikofer, and E. Heilbronner, *Zusammenstellung scheinbarer Dissoziationskonstanten im Lösungsmittelsystem Methylcellosolve/Wasser*, Part 10, Juris-Verlag, Zurich, 1959; P. F. Sommer and W. Simon, Part 11, Juris-Verlag, Zurich, 1961.

[113] G. A. Harlow, *Anal. Chem.*, **34**, 148 (1962).

contaminate the titrant. The alkaline error in acetic acid was studied by Wegmann, Escarfail, and Simon.[102] An error in dimethylformamide has been noted by Juillard.[106] There is some evidence that lithium ion interferes with the performance of the glass electrode in acetonitrile, as pointed out by Badoz-Lambling, Desbarres, and Tacussel.[108] Satisfactory behavior has been found in "acidic" water–hydrazine mixtures (*i.e.*, those containing $N_2H_5^+$ ions), but there is an error when the medium is alkaline.[114] Because of the existence of a sodium ion error in ethylenediamine, Deal and Wyld[115] have suggested that potassium or quaternary ammonium salts be used instead of sodium salts in this medium.

In the opinion of Morel, Séguela, and Pariaud,[96] the glass electrode is perfectly reversible to hydrogen ions in mixed solvents containing at least 20 per cent (w/w) of water. In their experience, the standard potential of the electrode is independent of the nature and composition of the solvent mixture under these restricted circumstances.

The Glass Electrode in Deuterium Oxide

There is considerable evidence[116-119] that the glass electrode responds in a theoretical manner to deuterium ion in deuterium oxide (heavy water). An empirical approach to the standardization of a pD scale for solutions in deuterium oxide (D_2O) was proposed by Glasoe and Long,[116] and this scale has found extensive use. When the pH of solutions of DCl in D_2O, determined by means of the usual pH cell with glass and aqueous calomel electrodes, together with aqueous pH standards, was compared with that of HCl solutions of identical concentrations in H_2O, a constant difference of 0.4 unit in the pH was found. The same figure was obtained from measurements with NaOD and NaOH in D_2O and H_2O, respectively. The scale of pD at 25 °C was thus expressed by

$$pD = pH \text{ meter reading} + 0.40 \tag{23}$$

The essential correctness of the constant 0.40 in equation has been confirmed a number of times.[120] A careful comparison of glass electrodes with

[114] D. Bauer, *Bull. Soc. Chim. France*, 3302 (1965); thesis, Paris, 1967.

[115] V. Z. Deal and G. E. A. Wyld, *Anal. Chem.*, **27**, 47 (1955).

[116] P. K. Glasoe and F. A. Long, *J. Phys. Chem.*, **64**, 188 (1960).

[117] P. R. Hammond, *Chem. Ind. (London)*, **7**, 311 (1962).

[118] V. Gold and B. M. Lowe, *Proc. Chem. Soc.*, 140, (1963).

[119] A. K. Covington, M. Paabo, R. A. Robinson, and R. G. Bates, *Anal. Chem.* **40**, 700 (1968).

[120] See, for example, E. Mikkelsen and S. O. Nielsen, *J. Phys. Chem.*, **64**, 632 (1960); P. Salomaa, L. Schaleger, and F. A. Long, *J. Amer. Chem. Soc.*, **86**, 1 (1964), and reference 119.

the deuterium gas electrode in a variety of buffer solutions and solutions of a strong acid (DCl) or a strong base (NaOD), was made by Covington, Paabo, Robinson, and Bates.[119] The true value of pD was assumed to be the conventional pa_D determined from cells without liquid junction which contained deuterium electrodes and silver-silver chloride electrodes (see Chapter 8). Four commercial glass electrodes from three different manufacturers showed no significant differences in response. The average of 46 determinations of the difference $pa_D - pH$ in 12 different solutions in D_2O was 0.45 (molal scale) and 0.41 (molar concentration scale), with a standard deviation of 0.031. The constant found for DCl solutions was about 0.03 unit smaller than the average of all of the results.

Internal Reference Cells

INNER ELECTRODES AND SOLUTIONS

In order to complete the electrical circuit for the measurement of the changes in potential at the outer surface of the glass membrane, a solution of constant hydrogen ion concentration is usually placed within the glass bulb and an appropriate reference electrode immersed in it. The inner solution should be one that will not attack the glass, and both the solution and the electrode must be stable over extended periods of time and wide ranges of temperature. Together they should provide a constant reference potential that displays little hysteresis and the temperature coefficient of which opposes the effect of temperature changes on the other elements of the pH cell.

The inner cell is most commonly a silver-silver chloride or calomel electrode immersed in dilute hydrochloric acid or a buffered chloride solution. A platinum wire coated with mercury and immersed in a solution of perchloric acid and mercurous perchlorate is also used. The buffer capacity must be fairly high, as the inner solution will neutralize alkali leached from the glass. The inner electrode may be quite small, inasmuch as the currents that flow through the glass membrane are quite insufficient to cause polarization. The inner electrode and reference solution chosen should have a favorable effect on the temperature coefficient of the e.m.f. of the pH cell. The composition of the inner reference solution is selected with a consideration of the desired electrode potential. Commercial electrodes are often designed so that the glass-calomel assembly will have an e.m.f. of zero in test solutions of a particular pH, perhaps 4 or 7.

It is evident that the e.m.f. of the internal reference cell cannot reasonably be expected to be entirely free of fluctuations and slow irreversible changes. These unavoidable variations of the potential are compensated, along with the asymmetry potential, by adjusting the standardization control or asymmetry potential knob of the pH meter. If the drifts are too large, they may exceed the correcting range of the instrument, and the electrode must be replaced. It is therefore important that the inner solution be stable and that the potential of the inner electrode remain constant for long periods.

The inner solution can be dispensed with entirely.[121] Electrical contact is made with the inner surface of the glass membrane by filling the bulb with mercury or by coating the inside with a thin layer of silver. Patents covering the use of metals and alloys for this purpose have been issued.[122] In one arrangement the outer surface of the glass was coated with the metal, and the test solution and salt bridge were placed within. The metal thus provided the necessary electrical contact with the glass and served as an electrical shield at the same time. Metal-connected electrodes have not been found as satisfactory as those with inner reference electrodes and have never been adopted widely.

TEMPERATURE EFFECTS

We consider next the effect of temperature changes on the e.m.f. of the glass-calomel cell. At a given temperature the e.m.f. of the entire cell

$$\text{Inner reference} \,\|\, \text{Glass} \,\|\, \text{Test soln.} \,|\, \text{Satd. KCl;Calomel} \qquad (24)$$

depends on the type of inner electrode and the composition of the internal solution, as well as on the pH of the test solution. In the region where both surfaces of the glass membrane are behaving as hydrogen electrodes (that is, are error-free), cell 24 can be considered as a combination of the cell

$$H_2(1 \text{ atm}); \text{Test soln.} \,|\, \text{Satd. KCl;Calomel} \qquad (25)$$

with another cell, the components of which are completely within the glass bulb. The latter may be one of the following:

$$Ag; AgCl, HCl(0.1 \text{ M}); H_2(1 \text{ atm}) \qquad (26)$$

$$Ag; AgCl, \text{Buffer}, Cl^- ; H_2(1 \text{ atm}) \qquad (27)$$

or

$$Hg; Hg_2Cl_2, HCl(0.1 \text{ M}); H_2(1 \text{ atm}) \qquad (28)$$

The effect of changes of temperature on the e.m.f. of the pH cell assembly (cell 24), at temperature equilibrium in both the initial and final states, is the sum of the temperature effects on cell 25 and the appropriate inner cell. It is manifestly impossible to design a cell system whose e.m.f. is uniformly insensitive to temperature changes, for the temperature coefficient of cell 25

[121] M.R. Thompson, *Bur. Stand. J. Res.*, **9**, 833 (1932).

[122] P. A. Kryukov and A. A. Kryukov, Russian Patent 51,509 (July 31, 1937); H. Bender and D. J. Pye, U.S. Patent 2,117,596 (May 17, 1938).

varies with the pH and composition of the test solution. This temperature coefficient is, for example, more than twice as large for a test solution of borax as for a test solution of potassium hydrogen phthalate, as the following data for cell 25 illustrate:

Test Solution	dE/dt mV K^{-1}
0.05 M KH phthalate (pH 4.01)	+0.33
0.025 M KH$_2$PO$_4$, 0.025 M Na$_2$HPO$_4$ (pH 6.86)	+0.54
0.01 M Na$_2$B$_4$O$_7$ (pH 9.18)	+0.76

Nevertheless, the temperature effect can be compensated fairly successfully under certain "average" conditions, perhaps at pH 7, by selection of a suitable inner cell. Cell 26 is not a good choice, as its temperature coefficient of e.m.f., like that of 25, is positive ($dE/dt = +0.18$ mV K^{-1} near 25°C). Cell 28 ($dE/dt = -0.2$ mV K^{-1}) is more suitable. The temperature coefficient of cell 27 is usually negative also and can be varied over a rather wide range by selection of different buffer–chloride mixtures, as shown by the following data:

Solution	dE/dt mV K^{-1}
0.05 M K tetroxalate, 0.01 M KCl (pH 1.7)	−0.16
0.01 M KH tartrate, 0.01 M NaCl (pH 3.6)	−0.41
0.05 M KH phthalate, 0.005 M KCl (pH 4.0)	−0.71
0.031 M KH$_2$PO$_4$, 0.031 M Na$_2$HPO$_4$, 0.004 M NaCl (pH 6.8)	−1.06

Citrate-buffered chloride solutions (pH 4.0) constitute the inner solution of some commercial glass electrodes.

Fluctuations of temperature often affect the electrodes of the pH cell unequally, and a well-designed assembly with a small overall temperature coefficient may still be subject to relatively large errors under conditions of unequal heating or cooling. Smirnov[123] and Kratz[124] have recommended

---

[123] A. A. Smirnov, *Bull. Acad. Sci. URSS Sér. Biol.*, 172 (1944).
[124] L. Kratz, *Kolloid Z.*, **86**, 51 (1939).

that a calomel electrode, similar to the external reference electrode, be used within the glass bulb, because of the relatively low temperature errors of such a symmetrical arrangement. The "Thalamid" electrode, consisting of thallium amalgam (40 per cent) and a saturated solution of thallous chloride (see page 335), has been used in a symmetrical cell also.

The standard potential of the pH cell changes with temperature, and for this reason it is usually necessary that the temperature of standardization be the same as the temperature of measurement. However, symmetrical cells in which both inner and outer reference electrodes are identical and in contact with solutions of nearly the same chloride ion activity have certain special advantages. A cell of this type would be particularly advantageous if, in addition, the pH of the inner chloride solution were unaffected by changes of temperature. If both the outer (o) and inner (i) surfaces of the glass membrane behave as hydrogen electrodes, the e.m.f. E_X of the cell is

$$E_X = \left[E_j - k \log \frac{(a_{Cl})_o}{(a_{Cl})_i} - k\text{pH}_i \right] + k\text{pH}_X \qquad (29a)$$

$$= E_c{}^\circ + k\text{pH}_X \qquad (29b)$$

where $E_c{}^\circ$ is written for the expression in square brackets, k is $(RT \ln 10)/F$, and pH_X is the pH of the outer (unknown) solution.

When both the inner and outer reference electrodes are in contact with a saturated solution of potassium chloride, the ratio of the chloride ion activities probably changes but little with temperature, even though the inner solution also contains buffer substances. Moreover, Fricke[125] has found that the pH of a buffer solution 1 M with respect to both acetic acid and sodium acetate and saturated with potassium chloride is nearly constant in the temperature range 20 to 100 °C. The pH of dilute solutions of strong acids is also nearly independent of temperature.

It is evident that a symmetrical chloride cell with this acetate-buffered chloride inner solution should have an e.m.f. of zero at a fixed value of pH_X, regardless of the temperature, as illustrated in Fig. 11–8. The e.m.f./pH lines (which have slopes k) intersect at the point $E = 0$, pH = 4.75; this pH is near that of the inner solution. This point of intersection is known as the "Isothermenschnittpunkt" or "isopotential pH." Although there is a variation of the standard potential as the temperature is changed (altering the value of k), this potential is nearly proportional to k, provided E_j is negligible and the ratio of chloride ion activities is independent of the temperature, conditions which seem to have been fulfilled in Fricke's measurements. Thus,

$$E_X = k[\text{pH}_X - C'] \qquad (30)$$

---

[125] H. K. Fricke in *Beiträge zur Angewandten Glasforschung*, E. Schott, ed., p. 175, Wissenschaftliche Verlagsgesellschaft, Stuttgart, 1960; *Zucker*, **14**, No. 7 (1961).

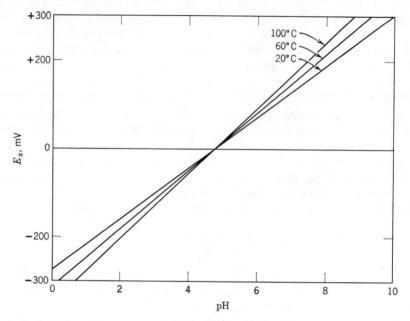

FIG. 11–8. Isotherms for a symmetrical pH cell when the pH of the solution within the glass bulb is nearly unaffected by changes of temperature (after Fricke).

where C' is a constant independent of temperature; its value can be determined by standardization at a single temperature. However, the possibility of changes in the asymmetry potential and in the liquid-junction potential should not be overlooked. It should be noted here that an isopotential pH may exist for unsymmetrical cells as well; in this case, however, it does not usually lie on the $E = 0$ axis (see page 414).

It would be most advantageous if the pH cell could be designed in such a manner that E_c° (equation 29b) were truly independent of temperature changes. This possibility has been examined by Bates.[126] If the liquid-junction potential E_j and the ratio of activities of chloride ion in contact with the inner electrode and the outer reference electrode were largely unaffected by changes of temperature, it would be sufficient to select an inner solution for which $d(k\text{pH}_i)/dT = 0$. When the inner solution is a buffer consisting of equal concentrations of a weak acid (dissociation constant K) and its salt, $\text{pH}_i \approx pK$ and $d(\text{pH}_i)/dT \approx dpK/dT$. Inasmuch as

[126] R. G. Bates, Chapter 1 in *Glass Microelectrodes*, M. Lavallée, O. F. Schanne, and N. C. Hébert, eds., John Wiley and Sons, New York, 1969.

$$dk/dT = (RT \ln 10)/F = k/T$$

compensation of temperature effects on E_c° could be achieved by choosing an inner buffer solution that meets the condition

$$\frac{\mathrm{d}pK}{\mathrm{d}T} = -\frac{pK}{T} \tag{31}$$

By introduction of the van't Hoff equation, $\mathrm{d}pK/\mathrm{d}T$ can be expressed in terms of ΔH°, the enthalpy of ionization of the weak acid (see page 123). At 25 °C, therefore, the condition can be stated as follows:

$$\frac{\Delta H^\circ}{pK} = 1.36 \text{ kcal mol}^{-1} \tag{32}$$

It was shown that only cation acids such as protonated amines give promise of fulfilling this requirement at room temperature. Most suitable choices would appear to be base-salt buffers in the ammonia, n-butylamine, and tris(hydroxymethyl)aminomethane ("tris") systems, for which $\Delta H^\circ/pK$ has the values 1.349, 1.305, and 1.409 kcal mol^{-1}, respectively.

Another approach to the selection of the most suitable inner solution was also suggested.[126] It is based on a direct comparison of the temperature coefficients of those terms in the expression for the e.m.f. of the pH cell that are temperature-sensitive but which are independent of the value of pH_X. These are the e.m.f. E_i of the inner cell (cells 26 to 28) and $E^{\circ\prime} + E_j$ for cell 25. Together these terms make up the standard potential E_X° of the glass-calomel assembly:

$$E_X = (E_i + E^{\circ\prime} + E_j) + k pH_X = E_X^\circ + k pH_X \tag{33}$$

To nullify temperature effects on the standard potential, one must place within the glass electrode an electrode-solution element so chosen that temperature changes on E_i are of equal magnitude and opposite sign to those on $E^{\circ\prime} + E_j$. In other words,

$$\frac{\mathrm{d}E_i}{\mathrm{d}T} = -\frac{\mathrm{d}(E^{\circ\prime} + E_j)}{\mathrm{d}T} \tag{34}$$

From Table 10–6, $\mathrm{d}(E^{\circ\prime} + E_j)/\mathrm{d}T$ is found to be -0.67 mV K^{-1} for the saturated calomel electrode and -0.39 mV K^{-1} for the 3.5 M calomel electrode at 25 °C. To compensate for the effect of temperature changes on the reference electrode, therefore, one should choose an inner electrode system for which $\mathrm{d}E_i/\mathrm{d}T$ is as close as possible to $+0.67$ mV K^{-1} (saturated calomel) and $+0.39$ mV K^{-1} (3.5 M calomel). Table 11–3 lists some suitable

TABLE 11–3. TEMPERATURE COEFFICIENT AT 25 °C FOR
INNER CELLS OF THE TYPE Ag;AgX, SOLUTION i, H_2;Pt

Solution i (molality)	dE_i/dT, mV K^{-1}
Ag;AgCl Electrode	
HCl (0.2)	0.28
HCl (0.5)	0.41
HCl (1.0)	0.51
HCl (3.0)	0.65
HCl (4.0)	0.68
NH_4Cl (0.03017), NH_3 (0.03017)	0.28
NH_4Cl (0.1077), NH_3 (0.1077)	0.36
Tris·HCl (0.06046), Tris (0.06130)[a]	0.38
Tris·HCl (0.1015), Tris (0.1029)	0.42
Ag;AgBr Electrode	
HBr (1.0)	0.39
HBr (2.0)	0.52

[a] Tris = tris(hydroxymethyl)aminomethane.

combinations of inner electrode and inner solution.[126] Again, only strong acids (HCl in cell 26) and buffers of the type A^+B° (in cell 27) were found suitable. Some data for the standard potential E_X° of the glass-saturated calomel cell and the glass–3.5 M calomel cell with various inner solutions, normalized to 25 °C, are plotted as a function of temperature in Fig. 11–9. In each case, the inner element comprised a silver-silver chloride electrode.

A solution of either 0.5 M hydrochloric acid or a buffer mixture of tris(hydroxymethyl)aminomethane, "tris", (0.06 M) and its hydrochloride (0.06 M) is successful in making the standard potential of the cell assembly with 3.5 M calomel reference electrode nearly constant over a range of 30 °C. Furthermore, an inner solution of 3.0 M hydrochloric acid provides adequate compensation for temperature effects on the standard potential of the external saturated calomel reference electrode. As equation 33 shows, a pH assembly

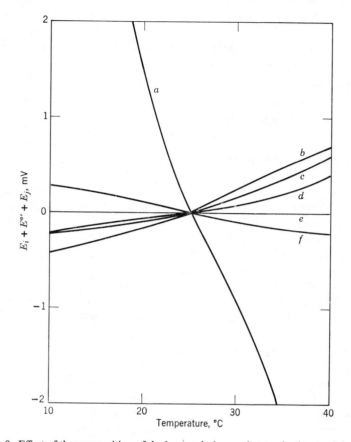

FIG. 11–9. Effect of the composition of the inner solution on the standard potential of glass-calomel pH cells.

Satd. calomel	3.5M calomel
a, 0.1M tris buffer	*b*, 0.1M tris buffer
d, 4M HCl	*c*, 0.5M HCl
e, 3M HCl	*f*, 0.06M tris buffer

so constituted will have an isopotential point at $pH_x = 0$. A cell with these characteristics can be standardized at any convenient temperature and used for measurements at another temperature without appreciable loss of accuracy, provided that appropriate adjustments of the Nernst e.m.f./pH slope are made. This adjustment is provided by the manual "Temperature Compensator" of most commercial pH meters, but is sometimes accomplished automatically.

By an empirical approach, Wegmann and Simon[127] have selected buffer solutions which provide an isopotential pH of 7.0 when used in conjunction with an inner silver-silver chloride electrode and a saturated external calomel reference electrode. Two suitable solutions have the following compositions:

1. 0.50 M phenylphosphonic acid; 0.82 M morpholine; 0.1 M hydrochloric acid; in 84 per cent aqueous glycerine.
2. 0.50 M phosphoric acid; 0.623 M diethanolamine; 0.188 M triethanolamine; 0.06 M hydrochloric acid; in 84 per cent aqueous glycerine.

Commercial Glass Electrodes

Commercial glass electrodes are made in many forms for special applications. Electrodes are designed to operate successfully with samples as small as one drop; others require at least 5 ml of solution. There is an optimum sample volume for each size of electrode, where maximum accuracy is achieved without undue waste of the sample. If the volume of test solution is too small relative to the surface area of the bulb or pH-sensitive membrane, changes of pH through adsorption or solution of the glass may become serious. The number of washings necessary is larger with small sample volumes than with large, and agitation or stirring may be somewhat difficult.

Some commercial forms of glass electrode are shown in Fig. 11–10. The MacInnes-Belcher electrode,[128] which is used extensively in biochemical studies, consists of a tube of pH-sensitive glass surrounded by the "inner" solution. The standard or test solution is poured into the funnel-like top and is withdrawn through the stopcock at the bottom, which also establishes the liquid junction. A water jacket for temperature control can be provided. Special micro cells for the pH measurement of blood are supplied by several manufacturers.

Glass electrodes and their leads are usually shielded unless the electrodes are intended for use within a shielded compartment, which may be a part of the instrument itself. The electrode shield is grounded to the case of the meter. The inner cell is tightly sealed from the atmosphere by means of a wax or plastic dielectric and a metal or plastic cap.

Corning 015 glass, formerly the best pH-responsive glass, has been largely replaced, especially for use at high temperatures and high alkalinities, by other glasses with superior characteristics. New stable pH meters of very low

[127] D. Wegmann and W. Simon, *Helv. Chim. Acta*, **47**, 1181 (1964).

[128] D. A. MacInnes and D. Belcher, *Ind. Eng. Chem., Anal. Ed.*, **5**, 199 (1933). The behavior of glass electrodes in biological media has been reviewed by R. G. Bates and A. K. Covington, *Ann. N. Y. Acad.*, **148**, 67 (1968).

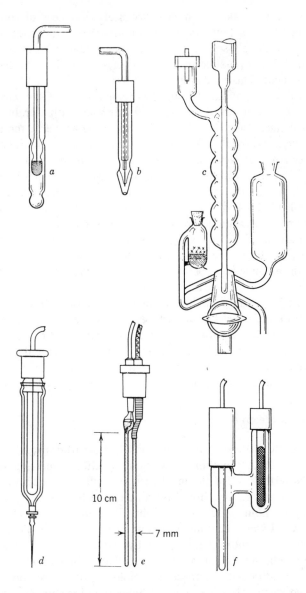

FIG. 11–10. Some commercial types of glass electrode. *a* and *b*, immersion electrodes for general laboratory use; *c*, MacInnes and Belcher electrode; *d*, hypodermic type (Electronic Instruments Ltd.); *e*, L & N miniature electrodes (glass and calomel); *f*, Coleman micro glass-calomel cell.

grid current drain make the higher electrical resistances of some of these newer lithia glasses at low temperatures a matter of little concern.

Other electrode glasses of lower resistance than 015 glass have been developed, and it is possible to use bulbs of such thick walls that breakage is greatly reduced. These durable electrodes resist abrasion and can withstand pressures of several hundred pounds per square inch. Glass electrodes are continually being improved, but an electrode sufficiently durable and error-free for prolonged use at 100 °C in solutions of pH above 12 and at the same time of sufficiently low resistance to provide adequate sensitivity at 0 °C has not yet been made. Nevertheless, a practical solution to the problem of the sodium ion error within specific ranges of temperature has been achieved.

Some of the characteristics of commercial electrode glasses are summarized in Table 11-4.[129] For further information, the reader should consult the literature of the manufacturers. The alkaline errors of several commercial glass electrodes at 25 °C and 50 °C are summarized in Tables 11-5 and 11-6. The corrections listed are to be added to the scale readings. For the 015 electrode, the alkaline error below room temperature is known to be of about the same magnitude as that at 25 °C; hence the corrections given for these electrodes at 25 °C can probably be applied in the range 0 to 25 °C.

The condition of the electrode surface causes some variation in the alkaline error; moreover, the potential varies with time when the error is appreciable. For these reasons the corrections given in the tables should be considered only approximate. When the error of the electrode exceeds 0.5 unit, an accurate pH value cannot be obtained through the application of corrections.

Care of Glass Electrodes

There is some evidence that many dry glass electrodes reach nearly their equilibrium potentials within a few minutes after immersion in a buffer solution. Nevertheless, for the most precise results, new electrodes or those that have been stored dry should be conditioned before use by soaking the bulb for 1 to 2 hours or even overnight. Electrodes intended for measurements below pH 9 may be immersed in water or in phosphate buffer solution of pH near 7, and those designed for use exclusively in alkaline solutions should be conditioned in borax or other alkaline buffer solution. If an inadequately conditioned electrode is employed, the assembly should be standardized frequently with standard buffer solution, for the electrode will not maintain a stable potential during the conditioning period.

Care should be exercised that the pH-sensitive tip of the electrode does not become scratched or cracked through contact with the sample cup or

---

[129] The data were taken in large part from W. Simon and D. Wegmann, *Helv. Chim. Acta*, **41**, 2308 (1958), and from D. Wegmann, G. H. Lyssy, and W. Simon, *ibid.*, **44**, 25 (1960).

TABLE 11-4. CHARACTERISTICS OF SOME COMMERCIAL ELECTRODE GLASSES AT 25 °C

Designation of Glass or Electrode	Composition of Glass	Resistance (megohms)	Correction (pH units to be added) in 0.1 M NaOH	1 M NaOH[a]
Beckman E2	Li_2O, BaO, SiO_2	375	0.03	0.17
Beckman General Purpose	Li_2O, BaO, SiO_2	150	0.43	1.4
Beckman Amber	Li_2O, BaO, SiO_2	550	0.02	0.17
Cambridge Standard	Na_2O, CaO, SiO_2	87	0.7	2.1
Cambridge Alki	Li_2O, BaO, SiO_2	560	0.05	0.25
Corning 015	Na_2O, CaO, SiO_2	90	1.0	2.5
Doran Alkacid	Li_2O, BaO, SiO_2	200	0.07	0.3
Electronic Instruments GHS	Li_2O, Cs_2O, SiO_2	200	0.03	0.16
Ingold U	—	250	0.7	2.3
Ingold T	—	140	0.8	2.2
Ingold UN	Li_2O, SiO_2[b]	30	0.7	1.9
Jena H	—	105	0.8	2.2
Jena U	—	30	0.31	0.7
Jena HT	—	800	0.9	2.2
Jena HA	—	290	0.08	0.25
L & N Blue Dot	Na_2O, CaO, SiO_2	50	—	—
L & N Black Dot	Li_2O, La_2O_3, SiO_2	70	0.02	0.25
L & N White Dot	Li_2O, La_2O_3, SiO_2	250[c]	0.01[c]	0.4[c]
Lengyel 115	Li_2O, BaO, UO_3, SiO_2[d]	15	0.6	1.7
Metrohm H	Li_2O, BaO, SiO_2	1400	0.08	0.15
Metrohm X	Li_2O, CaO, SiO_2	100	0.9	2.2
Metrohm U	Li_2O, BaO, SiO_2	500	0.08	0.25

[a] Time-dependent correction. The value given was recorded 3 to 7 minutes after immersion in 1 M NaOH.
[b] Glass described by K. Schwabe, *Chem. Ing. Tech.*, **29**, 656 (1957).
[c] At 50 °C.
[d] Glass described by B. Lengyel and F. Till, *Egypt J. Chem.*, **1**, 99 (1958).

solids in the test solution. However, the tip can be dried without damage by gentle rubbing with absorbent tissue. Some of the more rugged electrodes can be immersed in suspensions of soils and in abrasive slurries without risk of damage. Standardization of the pH cell with two buffer solutions of different pH will usually reveal the imperfect pH response of a cracked electrode.

Glass electrodes should not be placed in chromic acid solution or in other dehydrating agents.[130] The time of exposure to anhydrous solvents should be short and should be followed by conditioning in water. A translucent film

[130] D. A. MacInnes and L. G. Longsworth, *Trans. Electrochem. Soc.*, **71**, 73 (1937); D. A. MacInnes, *The Principles of Electrochemistry*, Chapter 15, Reinhold Publishing Corp., New York, 1939; Dover Publications, New York, 1961.

TABLE 11–5. SODIUM ION CORRECTIONS FOR SOME COMMERCIAL GLASS ELECTRODES AT 25 °C

			(Correction in pH units to be added to scale reading)				
Scale Reading pH	Beckman General Purpose	Beckman E-2 and Amber	L & N Black Dot	Cambridge Alki[a]	Radio-meter Type A[b]	Radio-meter Type B[b]	Radio-meter Type C[b]
			Na$^+$, 0.1 M				
10.5	0.03	—	—	—	0.15	—	0.02
11.0	0.05	—	—	—	0.28	—	0.04
11.5	0.08	—	—	—	0.50	—	0.06
12.0	0.15	—	—	—	—	—	0.10
12.5	0.25	0.02	—	—	—	0.02	0.16
13.0	0.48	0.03	0.02	—	—	0.03	0.26
13.5	—	0.05	0.11	—	—	0.05	0.40
			Na$^+$, 1 M				
10.0	0.05	—	—	—	0.20	—	0.05
10.5	0.09	—	—	—	0.40	—	0.08
11.0	0.15	—	—	—	—	—	0.14
11.5	0.25	0.02	—	—	—	0.02	0.22
12.0	0.48	0.03	—	—	—	0.03	0.35
12.5	—	0.05	0.02	0.02	—	0.05	0.50
13.0	—	0.10	0.13	0.08	—	0.08	—
13.5	—	0.18	0.34	0.18	—	0.12	—
			Na$^+$, 2 M				
10.0	0.07	—	—	—	0.30	—	0.08
10.5	0.12	—	—	—	0.50	—	0.13
11.0	0.20	—	—	—	—	—	0.20
11.5	0.37	0.02	—	—	—	0.03	0.31
12.0	—	0.04	—	0.02	—	0.04	0.50
12.5	—	0.08	0.05	0.06	—	0.06	—
13.0	—	0.15	0.18	0.12	—	0.10	—
13.5	—	0.27	0.45	0.18	—	0.18	—

[a] Corrections indicated for temperatures from 10 to 50 °C.
[b] At 20 °C.

or deposit sometimes appears on the pH-sensitive membrane after the electrode has been in use for some time. It can often be removed by washing the bulb of the electrode in 6 M hydrochloric acid and rinsing thoroughly with distilled water. The same treatment is advisable after prolonged immersion in alkaline solutions of high sodium ion concentration. A water-repellent silicone coating is often applied to the outer surface of the glass electrode to retard fouling in viscous mixtures.[131]

[131] P. T. Gilbert, Jr., Science, 114, 637 (1951).

TABLE 11–6. SODIUM ION CORRECTIONS FOR SOME COMMERCIAL GLASS ELEC-
TRODES AT 50 °C

	(Correction in pH units to be added to scale reading)			
Scale Reading pH	Beckman General Purpose	Beckman E-2 and Amber	L & N Black Dot	L & N White Dot
		Na^+, 0.1 M		
10.5	0.05	—	—	—
11.0	0.10	—	—	—
11.5	0.25	—	—	—
12.0	0.7	—	0.03	—
12.5	—	0.02	0.07	—
13.0	—	0.06	0.12	0.01
		Na^+, 1 M		
10.0	0.08	—	—	—
10.5	0.20	—	—	—
11.0	0.50	—	—	—
11.5	—	0.02	0.04	—
12.0	—	0.05	0.08	0.02
12.5	—	0.13	0.15	0.09
13.0	—	0.37	0.23	0.18
		Na^+, 2 M		
10.0	0.13	—	—	—
10.5	0.33	—	—	—
11.0	0.8	—	0.01	—
11.5	—	0.03	0.05	—
12.0	—	0.08	0.10	0.04
12.5	—	0.23	0.17	0.13
13.0	—	—	0.26	0.28

Extraction of the soluble components of the glass from the surface of the
membrane may cause a type of deterioration marked by sluggish behavior
and some impairment of the electrode function, as evidenced by failure of
the electrode to respond satisfactorily with two buffers of different pH. If the
treatment with hydrochloric acid fails to correct the difficulty, it is often
possible to rejuvenate the electrode by immersing it for 1 minute in a 20 per
cent solution of ammonium bifluoride at room temperature. A waxed beaker
or waxed paper cup will serve as a container for the fluoride solution. This

treatment dissolves a small amount of the glass surface and should be employed only when other remedial measures fail.

Light Effects

An effect of light on the potentials of commercial glass electrodes[132] and pH cells utilizing a saturated silver-silver chloride reference electrode[133,134] has been reported. In the majority of instances, if not all, the light effect can probably be attributed either to a temperature change or to light sensitivity of the Ag; AgCl internal reference. Furthermore, the results suggest that the silver-silver chloride electrode immersed in a concentrated solution of KCl saturated with silver chloride is markedly more sensitive to illumination than is the same electrode in a dilute chloride solution.

---

[132] A. F. Milward, *Analyst*, **94**, 154 (1969).
[133] G. J. Moody, R. B. Oke, and J. D. R. Thomas, *Analyst*, **94**, 803 (1969).
[134] R. A. McAllister and R. Campbell, *Anal. Biochem.*, **33**, 200 (1970).

CHAPTER 12

measurement of electromotive

force. The pH meter

The type of equipment chosen for the measurement of the electromotive force of pH cells is determined to a large extent by the internal resistance of the cell. In view of the high resistance of the glass electrode, special vacuum-tube electrometers are usually needed to measure glass electrode potentials. The other common electrodes make up cell systems of relatively low resistance, provided that they are not too small in size and that the solution in which they are immersed contains an appreciable concentration of ions. The first part of this chapter deals with the measurement of the e.m.f. of cells of relatively low resistance; the remainder is devoted to a discussion of glass electrode pH meters.

Potentiometers

The most accurate means of measuring the e.m.f. of cells with internal resistances of 10^5 ohms or less is the compensation method, with a galvanometer or electronic null instrument as an indicator of balance. In this procedure, the unknown e.m.f. is compared as directly as possible with the e.m.f. of a standard cell. The instrument by means of which this comparison is made is called a *potentiometer*. Its function is entirely analogous to that of an analytical balance by which unknown masses are compared with a standard of known mass. Lever arms are sometimes used to multiply forces in comparing masses. In the potentiometer or " voltage balance," resistance ratios fulfill a similar function, and a galvanometer indicates that the condition of balance has been achieved. When the two electromotive forces are in balance, no current flows and the " open-circuit voltage" of the cell is obtained. The resistances of cell and leads influence the point of balance only in so far as they affect, through the magnitude of the galvanometer deflections, the ease and accuracy with which this point can be established.

The operation of a simple two-dial potentiometer with a range of 0 to 0.1 V is illustrated by the circuit shown in Fig. 12–1.[1] With the switch S in position 1, the current I flowing from the working cell BA through resistance A is adjusted by means of the variable resistance R until the voltage drop (IA) across A just equals the e.m.f. of the standard cell SC and no current flows through the galvanometer G. The switch is then moved to position 2, and dials P_1 and P_2 are adjusted until balance, as indicated by zero deflection of the galvanometer, is again attained. If M is the resistance between the contact points on P_1 and P_2, one obtains from Ohm's law,

$$\frac{M}{A} = \frac{E}{E_{SC}} \tag{1}$$

because the same current flows through both A and M. Hence, P_1 and P_2 can be calibrated to read in volts. The Leeds and Northrup type K2 potentiometer, the circuit of which is shown in Fig. 12–2, is one of several commercial instruments designed for precise measurements of e.m.f.

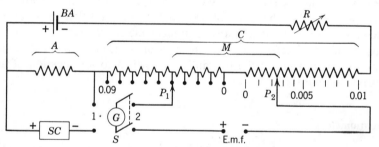

FIG. 12–1. Circuit of a simple potentiometer (after Andress).

Galvanometers

The galvanometer in the potentiometric circuit functions as a null indicator. Its purpose is to indicate the cessation of current flow. The moving-coil type of galvanometer commonly employed in potentiometry consists of a coil of wire suspended in the field of a strong permanent magnet. A flow of current through this coil generates a magnetic field which, by interaction with the

[1] P. M. Andress, *Semi-Precision Potentiometers*, Rubicon Notes No. 1, Rubicon Company, Philadelphia, Pa.

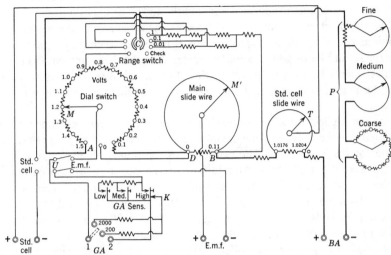

FIG. 12–2. Leeds and Northrup type K2 potentiometer (courtesy of Leeds and Northrup Co.).

field of the permanent magnet, causes the coil to turn on its suspensions. Motion of the coil is indicated by a needle attached to it or by the motion of a beam of light reflected from a small mirror moving with the coil. Electronic null instruments which amplify and detect the null current are also suitable and are often more convenient than galvanometers.

Three major characteristics should be considered in selecting the most suitable galvanometer.[2] These are the sensitivity, the period, and the critical damping resistance. The *current sensitivity* is the current required to cause a standard deflection, for example 1 mm on the scale. The *voltage sensitivity* is the product of the current sensitivity and the total circuit resistance. It represents the potential difference that must be impressed on the circuit of the critically damped galvanometer to produce the standard deflection. The *period* is the time elapsing between two successive passages of the needle or light beam through the rest point.

The *critical damping resistance* is the maximum effective resistance that will eliminate excessive oscillation of the galvanometer coil and produce the critically damped condition. If greatly underdamped, the system undergoes oscillations of considerable magnitude across the rest position, and time is lost waiting for them to cease. If overdamped, the needle or light beam

---

[2] F. K. Harris, *Electrical Measurements*, Chapter 3, John Wiley and Sons, New York, 1952.

travels very slowly, and it is difficult to tell when balance has been achieved. It is convenient to work with a galvanometer that is slightly underdamped and which overshoots the rest point slightly on the first oscillation.

The choice of galvanometer is dictated by the resistance of the cell and the required accuracy of the measurement. It is unwise to employ a galvanometer the sensitivity of which is much greater than the measurement demands. The zero shift is likely to be larger than that of a less sensitive instrument, and the period will be longer than necessary, entailing a needless waste of time.

For the highest possible voltage sensitivity, the external resistance at the terminals of the galvanometer should equal the critical damping resistance rating for the instrument. The resistance of the potentiometer (R_p) may be regarded as about 50 ohms. An additional resistance must usually be placed in the circuit to obtain critical damping. Thus if a galvanometer with a critical damping resistance rating R_d of 1500 ohms is to be used to measure the e.m.f. of cells with internal resistances R_c of 1000 ohms, an additional series resistance of about 450 ohms will be required. If, however, the cells have resistances greater than 1450 ohms, for example, 2000 ohms, it will be necessary to place the added resistance R_x across the terminals of the galvanometer to lower the resistance of the external circuit to the desired 1500 ohms. The necessary value of R_x is calculated by the relation

$$\frac{1}{R_x} = \frac{1}{R_d} - \frac{1}{R_c + R_p} \tag{2}$$

For example, if the resistance of the cell (R_c) is 2000 ohms, a shunt resistance of 5500 ohms will be required to achieve critical damping of this galvanometer.

Another characteristic, the resistance of the moving coil, must be considered when the galvanometer is damped with a shunt resistor. If this resistance across the galvanometer terminals is of the same order of magnitude as the coil resistance, the current passing through the galvanometer will be seriously reduced and the sensitivity greatly impaired. Hence, the galvanometer selected should permit critical damping to be attained with a shunt the resistance of which considerably exceeds that of the moving coil, if series damping is impractical.

Standard Cells

The Weston normal cell is one of the most reproducible and constant galvanic cells known. Consequently, it serves very satisfactorily as a standard of electromotive force. This cell is represented by the following scheme:

Cd(10 per cent amalgam); $CdSO_4 \cdot \frac{8}{3}H_2O(s)$,

$$CdSO_4(\text{satd. soln.}), Hg_2SO_4(s); Hg(l) \quad (3)$$

Its construction is illustrated in Fig. 12–3.

The faults of the Weston standard cell are few;[3] the most serious is the appreciable change of e.m.f. with temperature. Hysteresis, sometimes of concern in older cells, appears to be associated with the dynamics of the crystallization of hydrated cadmium sulfate from the solution. Slow drifts of e.m.f. may be caused by hydrolysis of the mercurous sulfate, accelerated by contact with the glass wall of the vessel. Small amounts of sulfuric acid are sometimes added to the cell electrolyte to prevent hydrolysis of the mercurous salt. Addition of acid, however, gives a lower e.m.f. dependent on the acid concentration.

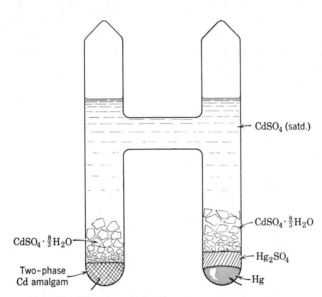

FIG. 12–3. The Weston normal cell.

In terms of the phase rule, the Weston normal cell may be regarded as a system of four components, seven phases, and two surfaces where electric potentials exist. The system is univariant, and the e.m.f. is a function only of temperature. The saturated cell can be used from the freezing point (about $-20\ °C$) to the transition point of hydrated cadmium sulfate (43.6 °C). The

---

[3] G. W. Vinal, *Primary Batteries*, Chapter 6, John Wiley and Sons, New York, 1950.

temperature coefficient of e.m.f. is positive at the lower temperatures and negative at the higher, as shown in Fig. 12–4. This figure is taken from the book by Vinal and the scale of ordinates changed to read (absolute) volts.

The international formula for the e.m.f. (expressed in volts) of the saturated cell at Celsius temperatures t is

$$E_t = 1.01864 - 4.06 \times 10^{-5}(t - 20) - 9.5 \times 10^{-7}(t - 20)^2$$
$$+ 1 \times 10^{-8}(t - 20)^3 \quad (4)$$

Hence the temperature coefficient of e.m.f. in mV K^{-1} is

$$\frac{dE_t}{dt} = -0.0406 - 0.0019(t - 20) + 0.00003(t - 20)^2 \quad (5)$$

If the derivative is set equal to zero and equation 5 solved for t, the maximum e.m.f. is found to occur at a temperature near 3 °C. Between 25 and 35 °C, the e.m.f. decreases by 0.05 and 0.06 mV per °C increase in temperature. Hence control of the temperature of the Weston cell is necessary in accurate work. A thermostated air bath consisting of inner and outer aluminum boxes thermally insulated from each other and from the surroundings has

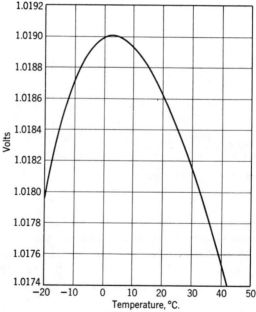

FIG. 12–4. Electromotive force of the Weston normal cell as a function of temperature.

been described by Mueller and Stimson.[4] Details of this air bath are shown in Fig. 12-5. The inner box accommodates four saturated cells and is maintained at a temperature higher than the ambient room temperature at any time of the year. Both boxes are provided with tight-fitting aluminum covers. This device can be equipped with a resistance thermometer, but the author has found a calibrated mercury-in-glass thermometer, inserted in a well in the outer block, to be suitable.

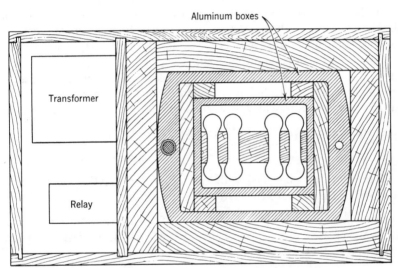

FIG. 12-5. Air thermostat for standard cells (Mueller and Stimson).

THE UNSATURATED WESTON CELL

Weston discovered that the temperature coefficient of the cadmium-mercurous sulfate cell could be lowered by use of a cadmium sulfate solution saturated at 4 °C but unsaturated at higher temperatures. This unsaturated cadmium cell is the form of commercial standard cell most common in the United States today. It is used for all but the most precise electromotive force measurements and is frequently a part of pH meters of the potentiometric type.

The temperature coefficient of the unsaturated cell is negligible for all practical purposes. From the point of view of constancy, however, this cell is much less satisfactory than its saturated counterpart. Vinal has summarized

[4] E. F. Mueller and H. F. Stimson, *J. Res. Nat. Bur. Stand.* **37**, 311 (1946).

the results of 400 observations on the change of e.m.f. of 200 unsaturated cells with time.[5] The e.m.f. of the average unsaturated cell decreased 0.07 mV per year, and the average decrease of 4 per cent of the cells was 0.38 mV per year.

SATURATED CELLS OF LOW TEMPERATURE COEFFICIENT

Vosburgh and his coworkers[6] have studied various modifications of the saturated Weston cell in an effort to lower the temperature coefficient of this cell without sacrificing any of its constancy. It was found that modification of the cell by (1) addition of bismuth to the cadmium amalgam and (2) saturation of the electrolyte with the hydrated double salt of sodium and cadmium sulfates, as well as with the customary hydrated cadmium sulfate, combine to reduce the temperature coefficient by 70 to 80 per cent. Hysteresis is about the same as in the Weston normal cell.

The concentration of bismuth could be varied rather widely. An amalgam containing 11 per cent cadmium by weight, 15 per cent bismuth, and 74 per cent mercury was satisfactory between 10 and 40 °C. The cell is represented

$$Cd(Bi,Hg); CdSO_4 \cdot \tfrac{8}{3}H_2O(s), CdSO_4 \cdot Na_2SO_4 \cdot 2H_2O(s),$$

$$\text{Satd. soln.,} Hg_2SO_4; Hg \quad (6)$$

The electrolyte was acidified with sulfuric acid or acetic acid to reduce hydrolysis of the mercurous sulfate.

The e.m.f. of cell 6 is about 1.0188 V at 25 °C. Near room temperature, the e.m.f. increases by a little more than 0.01 mV per °C increase of temperature, as shown by the expression[6, 7]

$$E_t = 1.0188 + 0.000013(t - 25) \quad (7)$$

The e.m.f. of five of these cells was found to change, on the average, less than 0.03 mV during 26 years;[8] two decreased slightly in e.m.f., two increased slightly, and one showed no change. In view of their stability and low temperature coefficient, these cells show considerable promise as secondary standards of e.m.f.

All standard cells, both primary and secondary, are delicate and should be handled carefully and protected from sudden changes of temperature, from extremes of temperature, and from current drain. The mercurous

---

[5] G. W. Vinal, *Primary Batteries*, Chapter 6, John Wiley and Sons, New York, 1950.

[6] W. C. Vosburgh, M. Guagenty, and W. J. Clayton, *J. Amer. Chem. Soc.*, **59**, 1256 (1937); W. C. Vosburgh and H. C. Parks, *ibid.*, **61**, 652 (1939); W. C. Vosburgh, P. F. Derr, G. R. Cooper, and B. Pettengill, *ibid.*, **61**, 2687 (1939).

[7] This equation gives the temperature coefficient of e.m.f. between 10 and 35 °C with an accuracy of ± 0.02 mV.

[8] W. C. Vosburgh and R. G. Bates, *J. Electrochem. Soc.*, **111**, 997 (1964).

sulfate is slightly light-sensitive and should not be exposed to strong light for any considerable time.

Types of pH Meter

When the internal resistance of the source of unknown e.m.f. is very high, as is that of the glass electrode cell, the current that flows is too small, for small increments of potential, to cause significant deflections of most galvanometers of high sensitivity. These instruments commonly require a current of at least 5×10^{-10} A for a deflection of 1 mm. Although deflection galvanometers capable of detecting about 0.1 mV through a resistance of 100 megohms have been constructed, they have long periods and are troublesome to use. The quadrant electrometer is more sensitive to small differences of potential than the best galvanometers and has proved useful for measuring the potentials of glass electrodes.[9] Nevertheless, as a consequence of the development of the convenient vacuum-tube and solid state electrometers, the quadrant electrometer is no longer used for this purpose.

In general, commercial pH meters can be classified broadly into two main categories, the *potentiometric* or *null-detector type* and the *direct-reading type*. Instruments of the first type are essentially potentiometers which employ an electronic amplifier as a null detector. The second type is basically an electronic voltmeter of the deflection type. In addition, meters are often designated " battery-operated " or " line-operated," or the nature of the amplifier or circuit is specified, as, for example, "chopper-type feedback-stabilized meter."

The potentiometric type of instrument, utilizing the amplifier only for detection of the point of balance, is capable of somewhat greater accuracy than the direct-reading type, for the deflection meter of the latter type may have a calibration error at full scale of 1 per cent or more. On the other hand, for the operation of the potentiometric instrument a constant d-c current must flow through the slide wire. This type of meter is therefore not as readily adapted to the convenience of a-c operation as is the direct-reading type. Furthermore, the direct-reading meter is particularly useful in following the changes of pH that occur during the course of a reaction or an electrometric titration. It is also readily adaptable to automatic recording systems.

The criteria which have influenced the design of pH meters have been discussed by Clark and Perley.[10] They may be summarized in a general way as follows:

[9] D. A. MacInnes and M. Dole, *J. Amer. Chem. Soc.*, **52**, 29 (1930).
[10] W. R. Clark and G. A. Perley, *Symposium on pH Measurement*, p. 34, ASTM Tech. Pub. 190, Philadelphia, Pa., 1957; see also R. G. Bates, *Chimia*, **14**, 111 (1960).

1. The measurement should be unaffected by the high resistance of the pH cell or by large changes in the magnitude of this resistance.
2. The meter should have a provision for temperature compensation, preferably automatic.
3. For many purposes, the pH meter should be operated by alternating current, and its operation should be independent of normal changes in the voltage of the power supply.
4. The circuit constants should not be altered when the vacuum tubes or transistors are changed. In addition, it is often desirable that a recorder be connected to the measuring system and that the precision of the measurement not be affected thereby.

Great progress has been made in electronic instrumentation in recent years. As a result, many excellent pH meters capable of furnishing reproducible pH numbers with a precision of a few thousandths of a unit have been developed. It must be remembered, however, that the fundamental meaning of these measured pH values is considerably less certain than the precision with which the numbers can be obtained.

Amplifiers and Their Characteristics

The basic objective of pH meter design is the accurate measurement of the e.m.f. of a cell of extraordinarily high resistance without drawing an appreciable current. The problem is complicated by the fact that the resistance of the cell is markedly temperature-sensitive. Provision must therefore be made for "swamping out" large unpredictable changes in the cell resistance. The effective resistance must be sufficiently high that changes in the temperature cannot significantly alter the total circuit resistance. This high resistance load (of the order of 10^{11} ohms) limits the choice of amplifier essentially to the voltage-operated types.[11] The direct-coupled d-c amplifier was formerly the one most commonly used, but there is a noticeable trend toward the remarkably stable amplifiers of the frequency-conversion types. The latter utilize contact modulators (choppers) or dynamic condensers, with subsequent a-c amplification of the modulated signal.

D-C AMPLIFIERS

In the simplest application of a vacuum-tube amplifier to pH measurements, the e.m.f. of a pH cell is impressed directly on the grid of a suitable electrometer tube, and the measured output (plate) current is compared with a calibration curve constructed in advance. This system constitutes the

---

[11] G. Hitchcox, *J. Brit. Inst. Radio Eng.*, **13**, 401 (1953).

prototype of the direct-reading pH meter. In a better arrangement, the amplifier is used only as a null detector, that is, as an indicator of the balance point of a potentiometric circuit.

The output of an amplifier is related to the input by

$$\text{Output} = Z + G \text{ (Input)} \tag{8}$$

where Z is the zero point and G is the gain; variations in Z and G cause *zero drift* errors and *proportional* errors, respectively. The performance of d-c amplifiers is governed to a considerable extent by the quality of the electrometer tube, which determines input resistance, grid current, and, to a large extent, zero stability. Changes in the battery voltages or fluctuations in the power supply also contribute to zero drift. Proportional errors result not only from variable amplifier gain but also from electrometer tubes of inadequate input resistance.

Improvements in zero stability have come through the development and selection of better electrometer tubes and improved power sources as well as through more sophisticated circuitry. Mercury batteries and electronically regulated d-c power supplies enhance the stability of the amplifier. Moreover, transistor converters operated by batteries have been used to provide stability in the plate supply voltage and the filament current. Frequent automatic correction for zero drift is a feature of some commercial pH meters. Differential or balanced amplifiers have also been successful in reducing zero drift. Two strictly identical amplifiers are so connected that their responses to external signals are additive while those to internal noise or drift are subtractive.[12]

NEGATIVE FEEDBACK

Negative feedback offers the most useful means of reducing proportional errors. By an application of the feedback principle, calibrated direct-deflection meters that are independent of tube linearity characteristics can be constructed. If the voltage applied at the input of a d-c amplifier is designated E_{in} and the voltage amplification is μ', the signal voltage at the output (E_0) will be $\mu'E_{in}$. If a fraction β of this output voltage is fed back in such a way as to oppose a portion of the input voltage (*negative* or *inverse feedback*), the

---

[12] The novel features of the circuits of commercial pH meters are described in detail by G. R. Taylor, Chapter 10 in G. Mattock, *pH Measurement and Titration*, The Macmillan Co., New York, 1961. See also H. J. Wolf, *Z. Instrumentenk.*, **67**, 147 (1959). The basic principles of high input impedance amplifiers have been described with admirable clarity by S. Z. Lewin, *Anal. Chem.*, **33**, 23A (1961). The essential characteristics and special features of the pH meters from 23 different manufacturers have been summarized by A. Wilson, *pH Meters*, pp. 85–119, Kogan Page, London, and Barnes and Noble, New York, 1970.

new input voltage will be $E_{in} - \beta E_0$ and the new output voltage will be $\mu'(E_{in} - \beta E_0)$. Hence

$$\frac{E_0}{E_{in}} = \frac{\mu'}{1 + \mu'\beta} \tag{9}$$

When β is not too small and the voltage gain is large, it is evident that

$$E_0 = \frac{1}{\beta} E_{in} \tag{10}$$

Thus the output voltage is a linear function of the input.[13] Negative feedback also provides increased stability at the expense of sensitivity.[14]

The negative-feedback principle is utilized in the design of many direct-reading pH meters. In a measuring circuit of the potentiometric type, such as that shown in Fig. 12–1, a constant current I flows through a resistance wire, and the balancing potential is the IR drop across a variable fraction of this resistance. In the negative-feedback type of circuit, the balancing potential is the IR drop across a fixed resistance R through which a variable current flows. This IR drop is opposed in polarity to the e.m.f. of the glass-calomel cell, and the *difference*, or voltage unbalance, is fed into the amplifier. A milliammeter with scales of millivolts and pH registers the output current flowing through R. In this fashion the current in R is automatically maintained at a value that produces an IR drop nearly, but not exactly, equal to the e.m.f. of the glass–calomel cell. This current is accurately linear with respect to the input voltage, regardless of the linearity characteristics of the tube.[15] The standard component in this type of circuit is a fixed resistance; hence a standard cell is not required.

FREQUENCY-CONVERSION AMPLIFIERS

The outstanding recent development in pH instrumentation is undoubtedly the introduction of frequency-conversion amplifiers. Instruments of these types are able to provide excellent long-term stability as well as excellent discrimination. The potential difference at the electrodes, or the difference voltage from a potentiometer, regulates the amplitude modulation of the output from a generator operating at a high and fixed frequency. This is accomplished by means of either a chopper or a vibrating condenser. These devices

---

[13] See, for example, R. H. Müller, R. L. Garman, and M. E. Droz, *Experimental Electronics*, Chapter 7, Prentice-Hall, Inc., New York, 1942.

[14] F. L. Smith, ed., *Radiotron Designer's Handbook*, 3rd ed., Chapter 6, distributed by Radio Corporation of America, Harrison, N.J., 1945.

[15] A. O. Beckman, *The Development of pH Instrumentation*, Reprint No. R-36, Beckman Instruments, Inc., Fullerton, Calif., 1950.

produce a pulsating signal the a-c component of which is extracted by a suitable filter network and fed to an a-c amplifier.

The band width of the amplifier need be no more than 1 Hz on either side of the fixed frequency. The carrier frequency may be selected, if desired, to avoid multiples of the line frequency, thus reducing d-c interference to a minimum. Substantially noise-free amplification can therefore be obtained, and, on reconversion to direct current, the amplitude may be great enough to swamp any subsequent low-frequency noise. Choppers have been designed with noise levels as low as 0.002 microvolt.[11] Negative feedback is often used with frequency-conversion amplifiers to stabilize the gain of the system as a whole.

Chopper circuitry has been adapted for use in pH meters of both the direct-reading and potentiometric types. The circuit of one of the former, the Leeds and Northrup Model 7401 meter, is outlined in Fig. 12–6. The d-c voltage of the electrode system is opposed by a feedback voltage and the resultant converted to alternating current by a mechanical vibrating chopper. The 60 Hz voltage is amplified and then rectified by an electronic phase-sensitive converter. The resulting d-c voltage supplies the stabilizing feedback potential and operates a d-c meter with a linear scale. Provision is made for both automatic and manual temperature compensation, and a jack is provided for direct connection to a standard recorder. Other similar meters supplied by a number of manufacturers feature scale expansion, permitting a range of one or two pH units to be expanded to full scale on the indicating meter.

In the Vibron Electrometer, manufactured by Electronic Instruments Ltd., a vibrating condenser is used to achieve sensitivity to 0.1 mV and high stability in the measurement of e.m.f. in circuits of very high resistances. The vibrating capacitor used in the Philips Universal pH meter, shown schematically in Fig. 12–7, will serve to illustrate the principle of operation of these devices. The coil S is mounted in the field of the permanent magnet P. The diaphragms M confine the movement of the spindle A to the longitudinal direction. One end of the spindle is attached to a plate, C_1, of a condenser. The other condenser plate, C_2, is insulated and fixed rigidly in position.

If an alternating voltage of a certain frequency is applied to the coil S by means of flexible leads, C_1 will move back and forth horizontally with this same frequency, altering the capacitance, C, of the condenser. The alternating potential difference V between the two plates follows the law

$$V = \frac{Q}{C} \tag{11}$$

where Q, the charge on the plates, is directly proportional to the d-c potential between the glass and calomel electrodes, if these are connected to the

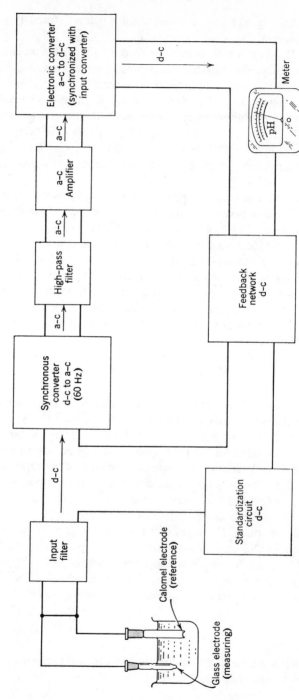

FIG. 12–6. Simplified circuit of Leeds and Northrup Stabilized pH Indicator, No.7401 (courtesy of Leeds and Northrup Co.).

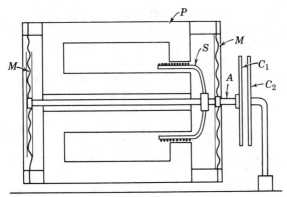

FIG. 12–7. Dynamic condenser (courtesy of North American Philips Co., Inc.).

condenser. Thus, the e.m.f. of the glass-calomel cell regulates Q, which in turn regulates the amplitude of the alternating voltage between the plates. The only current drawn from the cell, in addition to leakage flow, is the extremely small current required to charge the capacitor. The amplifier is of the a-c type, which is usually less complicated and more stable than the d-c type, in which variations in the power supply, emission of the electrometer tube, and the like, appear as part of the output signal. Furthermore, it can be made selective for the single frequency of the vibrating coil, to avoid the possibility of amplifying undesired voltages of other frequencies.

A similar instrument, the vibrating reed electrometer described by LeCaine and Waghorne[16] and others,[17] has been applied successfully to the measurement of glass electrode potentials with an accuracy of 0.01 mV.[18] It is said to be capable of measuring voltages in circuits of resistances as high as 10^{15} ohms. This instrument combines high sensitivity with rugged construction and freedom from zero drift. It is well adapted for use with automatic recorders. The Cary Model 31 vibrating reed electrometer can be readily applied to precise measurements of pH.[19]

[16] H. LeCaine and J. G. Waghorne, *Can. J. Res.*, **19**, 21 (1941).

[17] S. A. Scherbatskoy et al., U.S. Patent 2,349,225 (May 6, 1944); U.S. Patent 2,361,389 (Oct. 31, 1944); H. Palevsky, R. K. Swank, and R. Grenchik, *Rev. Sci. Instr.*, **18**, 298 (1947).

[18] K. A. Kraus, R. W. Holmberg, and C. J. Borkowski, *Report ORNL* 383, Oak Ridge National Laboratory, Oak Ridge, Tenn., 1949.

[19] *Precision Electrochemical Measurements Utilizing the Cary Model 31 Vibrating Reed Electrometer*, Report 31-1, Applied Physics Corp., Monrovia, Calif., 1962.

Commercial pH Meters

Many excellent pH meters are supplied by manufacturers in North America, Great Britain, Europe, and Japan. The choice of one of these instruments for purchase will be dictated largely by convenience and simplicity of operation; the nature of the measurement problem; precision and stability required; the need for portability; cost; and availability of suitable accessories, special electrodes, and the like.

A direct-reading meter is more convenient for titrations than a meter of the potentiometric type, yet the latter may be preferred in measurements of blood pH, for example, where precise discrimination of small changes in pH is necessary. Many direct-reading meters provide scale expansion for added precision and convenience. If stability of the zero point over long periods of time is essential, amplifiers of the frequency-conversion types will often be preferred to unstabilized d-c amplifiers. An instrument that will be used frequently for measurements in nonaqueous media or at low temperatures should draw a very low grid current, especially if standardization at the exact temperature of the unknown solutions is not always practicable. Many portable battery-operated meters are inexpensive and extremely satisfactory when a precision of ± 0.05 to ± 0.1 unit is sufficient. However, the sensitivity and stability required for reproducible measurements to ± 0.01 unit or better are usually obtainable only in the more costly instruments.

A summary of the basic features of more than seventy commercial meters has been given by Taylor in Mattock's monograph.[20] A critical survey of the pH meters supplied by British manufacturers in 1950 has also been published.[21] Likewise, Clark and Perley[22] have discussed in some detail the essential features of a number of meters manufactured in the United States. The characteristics of commercial pH meters have also been tabulated by Wilson.[23]

Two typical pH meters, the Beckman Research pH Meter and the Leeds and Northrup Stabilized pH Indicator (Model 7401), are shown in Fig. 12–8. Both are line-operated meters that make use of contact-modulated (chopper) circuits. The Beckman instrument utilizes a potentiometric circuit, whereas the Leeds and Northrup meter is of the direct-reading type.

---

[20] G. R. Taylor, Chapter 10 in G. Mattock, *pH Measurement and Titration*, The Macmillan Co., New York, 1961.

[21] G. I. Hitchcox, *Mfg. Chemist*, **22**, 93 (1951).

[22] W. R. Clark and G. A. Perley, *Symposium on pH Measurement*, p. 34, ASTM Tech. Publ. 190, Philadelphia, Pa., 1957.

[23] A. Wilson, *pH Meters*, pp. 85–119, Kogan Page, London, and Barnes and Noble, New York, 1970.

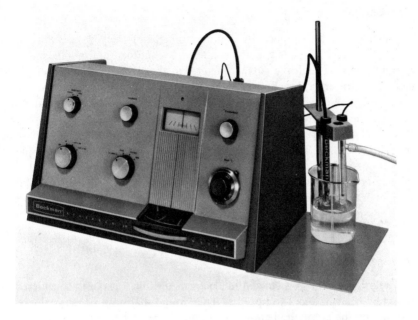

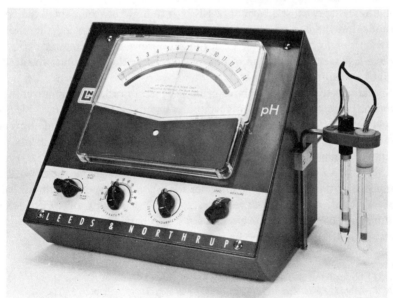

FIG. 12–8. Beckman Research pH Meter, Model 1019 (top) and Leeds and Northrup Stabilized pH Indicator, Model 7401 (bottom) (courtesy of Beckman Instruments, Inc., and Leeds and Northrup Co.).

Grid Current

When the e.m.f. supplied by a potentiometer of the low-resistance type (Figs. 12–1 and 12–2) is exactly balanced by another source of e.m.f., no current flows through the galvanometer and there is no IR drop within the cell. If, however, a direct-coupled amplifier is used in the measurement of the e.m.f. or in detecting the balance point, a small current flows through the cell and the grid circuit of the electrometer tube even at balance. Moreover, this small grid current does not cease when the grid is sufficiently negative to repel all electrons. The electrometer tube which is the heart of the pH meter makes possible the accurate measurement of the e.m.f. of glass-calomel cells with internal resistances that may exceed 1000 megohms. Yet this resistance is so high that extremely small currents, of the order of 10^{-12} A, for example, may cause an IR drop sufficiently large to introduce a significant error into the pH measurement.

SOURCE OF THE GRID CURRENT

The origin of the grid current in electrometer tubes has been considered in detail by Metcalf and Thompson.[24] They found that positive ions, which are emitted by the hot cathode in large numbers, are drawn to the negative control grid and may cause a current of 10^{-12} A to flow between a small tungsten cathode and the control grid. Other sources of grid current are thermionic emission due to heating of the grid, ionization of gas molecules in the tube, surface leakage, and emission of photoelectrons by the grid under the action of light and soft x rays.[24,25]

Special electrometer tubes have been designed in order to minimize, in so far as possible, the stray currents from these sources. In the General Electric FP-54 tetrode (the Pliotron),[26] for example, a space-charge grid prevents positive ions from reaching the control grid. The thoriated filaments are operated at relatively low temperatures, and all the potentials are kept low (less than 8 V) to minimize the ionization of residual gas in the tube. It is claimed that this tube is capable, under ideal conditions, of detecting a current of 10^{-17} A when a galvanometer with a sensitivity of 10^{-10} A per millimeter is used in the plate circuit.[24, 26] The characteristics of the Westinghouse RH-507 electrometer tube have been described by Cherry.[27]

---

[24] G. F. Metcalf and B. J. Thompson, *Phys. Rev.*, **36**, 1489 (1930).

[25] A. L. Albert, *Fundamental Electronics and Vacuum Tubes*, Chapter 5, The Macmillan Co., New York, 1938.

[26] *The Pliotron*, General Electric Bulletin ET1-160A, General Electric Co., Schenectady, N.Y., 1948.

[27] R. H. Cherry, *Trans. Electrochem. Soc.*, **72**, 33 (1937).

Electronic amplifiers with a grid current drain less than 10^{-12} A need not necessarily utilize the relatively expensive electrometer tubes. A number of commercial pH meters use radio receiving tubes and solid state devices exclusively. Several excellent amplifiers of both the d-c and a-c types have been designed to the specifications of radio tubes.[28,29] In general, the output current from the first tube is amplified in one or more stages before reaching the milliammeter or galvanometer. Most of the tubes chosen have at least four elements, including a positively charged screen grid to limit the positive ion current and often a suppressor grid which may be connected directly to the cathode. The filaments, anode, and screen grid are usually operated considerably below their rated voltages, in order to minimize thermionic and photoelectric emission from the control grid caused by heat and soft x rays.

If the input resistance is very high, the supply voltages must be extremely constant. The internal resistance of a failing battery increases rapidly and is a common cause of erratic performance. Nelson[30] examined in some detail the causes, apart from changes in the grid potential, of fluctuations in the filament–anode current. As a result of his study, he was able to reduce instability to a minimum.

MEASUREMENT OF THE GRID CURRENT

The grid current of a pH meter can be determined approximately by measuring the change of apparent e.m.f. of a known source, as indicated by the voltage scale of the meter, with increasing input resistance. A simple circuit is given in Fig. 12–9. Several standard resistors R_S can conveniently be mounted within a shielded box and the desired one chosen by means of a selector switch. Inasmuch as the currents drawn are too small to cause appreciable polarization of the source of constant potential, a standard cell *SC* will serve as a source. The grid lead is shielded, and the box is grounded to the case of the meter. In order to obtain satisfactory results, careful shielding and high insulation resistance are essential.

Let E_S be the e.m.f. of the standard cell, I_{gr} the current flowing in the grid circuit, and e the apparent e.m.f. of the cell as read from the pH meter with R_S in the circuit. The potentials E_S and e are equal for zero series resistance and differ for large values of R_S by the IR drop across this resistance. Hence

$$I_{gr} = \frac{E_S - e}{R_S} \tag{12}$$

[28] M. Dole, *The Glass Electrode*, Chapter 3, John Wiley and Sons, New York, 1941.

[29] S. B. Ellis and S. J. Kiehl, *Rev. Sci. Instr.*, **4**, 131 (1933); F. Müller and W. Dürichen, *Z. Elektrochem.*, **42**, 31 (1936); D. I. Hitchcock and A. Mauro, *Yale J. Biol. Med.*, **22**, 309 (1950); S. Natelson, *Anal. Chem.*, **27**, 1004 (1955).

[30] H. Nelson, *Rev. Sci. Instr.*, **1**, 281 (1930).

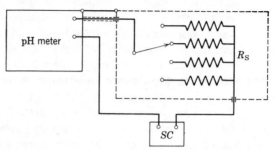

FIG. 12–9. Arrangement for approximate determination of grid current.

Standard resistors rated from 10^9 to 10^{12} ohms usually are sufficient to determine approximately the grid current of commercial pH meters with d-c amplification.

Strictly speaking, frequency-conversion amplifiers draw no " grid current." The charging current drawn from the pH cell by the chopper or vibrating condenser is exceedingly small. The effective resistance of the device is determined largely by the quality of the insulation. Under the most favorable conditions it may be as high as 10^{16} ohms.[31]

Effect of Humidity

A pH meter operates most satisfactorily at relative humidities between 20 and 60 per cent at ordinary temperatures. At very low humidities, static charges may cause unsteady readings, whereas adsorption of moisture at high humidities seriously increases electrical leakage. Most pH meters are completely useless at relative humidities exceeding 90 per cent at 25 °C. The most sensitive pH electrometers are usually the most seriously affected by moisture.

Reduction of humidity errors is largely a matter of proper insulation of the electrode leads, both inside and outside the case of the meter. Glass, quartz, amber, glazed porcelain, low-loss Bakelite, methyl methacrylate, polystyrene, polyethylene, and Teflon are good insulators. Rubber wire covering is best when thickest. Insulators may prove more effective when made water-repellent by treatment with the silicone materials now available.

External surface leakage can be controlled only by providing good insulation and by maintaining the outer surface of the tube clean and dry. Internal

---

[31] G. I. Hitchcox, *J. Brit. Inst. Radio Eng.*, **13**, 401 (1953).

surface leakage is determined largely by the type of mount construction and by the internal insulation.

The practice of locating the electrometer tube within a dry box is effective in reducing leakage from the control grid to ground. A coat of wax on the external surface of the tube envelope is also advantageous. The exposed surface of the tube must be kept free of traces of foreign matter which might form a conducting path.

Temperature Effects

The glass–saturated calomel pH cell with inner electrode reversible to chloride ion may be represented by

$$M; MCl,Buffer,Cl^- \parallel Soln. X \mid KCl \text{ (satd.)},Hg_2Cl_2; Hg \qquad (13)$$

where the double vertical line marks a glass membrane. If both surfaces of the membrane are assumed to behave as hydrogen electrodes, the e.m.f. of this cell is given by (see Chapter 11)

$$E_X = (E_i + E^{o\prime} + E_j) + k\text{pH}_X = E_X{}^\circ + k\text{pH}_X \qquad (14)$$

in which E_i represents the e.m.f. of the inner cell (the part to the left of the glass membrane) and k is written for $(RT \ln 10)/F$, that is, $2.3026RT/F$. The quantity enclosed in parentheses in equation 14 may be regarded as the "standard potential" of cell 13.

TEMPERATURE COMPENSATION

The pH meter is essentially a voltmeter equipped in such a way that values of E_X are properly converted into pH units on the scale of the instrument. Furthermore, the e.m.f. characteristics of the pH cell are temperature-dependent. A change in the temperature of the cell has two effects:

1. The *scale length* is altered. When k varies because of a change in T, the scale length (in volts) corresponding to unit pH change is also altered.
2. The *scale position* is altered. A change in T affects the standard potential through changes in $E^{o\prime} + E_j$ and E_i as well as in k.

The objective of temperature compensation in a pH meter is to nullify changes in e.m.f. from any source except a change in the true pH of the unknown solution X.

Some pH meters of early design were provided with multiple scales to permit pH values to be read directly at several different temperatures. Most pH meters of the present day, however, are equipped with a "temperature compensator" which is adjusted either manually or automatically. The range of adjustment may extend from 0 to 100 °C. In the potentiometric type of meter, the temperature compensator alters the resistance in the slide-wire circuit in such a manner that a change in the e.m.f. of the cell is correctly converted to a difference of pH at the temperature of the measurement, whatever it may be. This variable resistor is necessary in order that the meter can be calibrated, with a single scale, to read directly in pH units at more than one temperature.

A difference of 1 pH unit corresponds to a difference of about 59 mV in the e.m.f. at 25 °C, 66 mV per pH unit at 60 °C, etc. This ratio, millivolts per pH unit, is the value of $(RT \ln 10)/F$ and is often called the "slope factor." Its value from 0 to 100 °C is given in Table 1 of the Appendix.

Automatic temperature compensation is accomplished through the medium of a resistance thermometer with the proper temperature coefficient of resistance. This thermometer is mounted in the electrode holder and is immersed in the test solutions along with the electrodes. As the temperature of the solution changes, the circuit constants are altered accordingly.

A fully adequate compensation of changes in the standard potential (shift of scale position) is less easy to accomplish. Hence it is usually necessary to standardize the pH cell at the temperature at which the unknowns are to be measured; under these conditions the standard potential cancels out. This procedure should always be followed in the most accurate work.

The nature of the correction for scale position or "zero shift" has already been considered in Chapter 11. This problem can be understood most easily in terms of equation 14. Let us consider, for example, cell 13 with an inner element consisting of a silver-silver chloride electrode dipping in a solution of hydrochloric acid, molality 0.1 mol kg^{-1}. The standard potential $E_x°$ at several different temperatures can be calculated (compare equation 14) from $E°' + E_j$ given in Table 10–6,[32] and the e.m.f. of the inner cell (E_i), which has been tabulated elsewhere.[33] By this means, the following values for $E_x°$ are obtained:

t, °C	$-E_x°$, V	t, °C	$-E_x°$, V
0	0.0949	30	0.1102
10	0.1001	40	0.1149
20	0.1053	90	0.1337
25	0.1079		

---

[32] The values at 0 and 90 °C were taken from G. J. Hills and D. J. G. Ives, Chapter 3 in *Reference Electrodes*, D. J. G. Ives and G. J. Janz, eds., Academic Press, New York, 1961.
[33] R. G. Bates and V. E. Bower, *J. Res. Nat. Bur. Stand.*, **53**, 283 (1954).

The value of k increases linearly with rise of the temperature, according to the equation

$$k = 0.05419 + 0.0001984t \qquad (15)$$

whereas E_X° is a quadratic function of t. Hence, the curves representing the change of the total e.m.f. E_X of the cell as a function of pH at different temperatures do not quite meet at a single point (see equation 14 and Fig. 12–10). The situation would be considerably improved if the standard potential were a linear function of t, namely $E_X^\circ = a + bt$. Then the total e.m.f. of the cell would have the same value at all temperatures when the pH of solution X was $-b/0.0001984$. The several curves of Fig. 12–10 would then intersect at a single "Isothermenschnittpunkt" or "isopotential pH"[34] as shown in Fig. 11–8, page 380.

The selection of the most suitable inner solution has already been discussed in Chapter 11. Because the e.m.f. E_i of most inner cells is a quadratic function of temperature, the pH cells in common use often do not have sharply defined isopotential pH values over wide ranges of temperature. Nevertheless, the variation is so small over a range of 20 to 40 °C (as in Fig. 12–10) that temperature compensation for scale position has considerable utility.[35] Under these limited conditions,

$$E = C + k(\text{pH}_X - \text{pH}_{iso}) \qquad (16)$$

where C is a constant independent of the temperature and pH_{iso} is the isopotential pH. The standard potential $C - k\text{pH}_{iso}$ is evidently not independent of temperature changes. Nevertheless, after slope calibration (scale-length adjustment) has been made, scale-position adjustment can be accomplished by making a voltage of $E - C$ correspond to a pH_X equal to pH_{iso} on the scale at each temperature.

Automatic compensation for zero shift has been incorporated into a number of commercial pH meters, and this feature is of considerable utility in industrial measurements where control of the temperature of the electrodes is often impossible. Compensation is achieved in several different ways. In one method, for example, the current flowing through a sensing resistor, dipped in the solution along with the electrodes, automatically controls the magnitude of a bias potential connected in series with the cell. The circuit constants are so chosen that the effective potential introduced matches closely the change of the standard potential. The bias needed, in volts, is evidently $k\text{pH}_{iso}$.

---

[34] This term was suggested by J. Jackson, *Chem. Ind. (London)*, **7**, (1948).

[35] The application of the isopotential concept to temperature compensation has been discussed by G. Mattock, *pH Measurement and Titration*, Chapter 9, The Macmillan Co., New York, 1961. See also Chapter 10 of the same monograph.

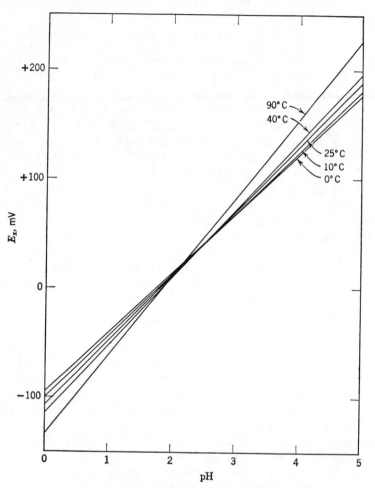

FIG. 12-10. Plot of the e.m.f. of the cell: Ag; AgCl, 0.1MHCl‖ soln. X|KCl (satd.), Hg$_2$Cl$_2$; Hg at five temperatures as a function of the pH of solution X.

IR DROP AND TEMPERATURE

When the grid current exceeds 10^{-12} A or when the resistance of the glass-calomel cell is unusually high, differences between the temperature of the standard and that of the unknown may cause another kind of error. This error, illustrated by the data in Table 12-1, is associated with the IR drop through the cell. The grid current of most modern pH meters is so small that this error is now rarely of concern.

TABLE 12–1. Change of *IR* Drop in a Typical Glass-Calomel Cell with Temperature

t, °C	*R*, megohms	IR Drop at Grid Current of		
		1×10^{-12} A	1×10^{-11} A	5×10^{-11} A
10	1000	1.0 mV	10.0 mV	50.0 mV
		0.02 pH	0.18 pH	0.89 pH
25	200	0.2 mV	2.0 mV	10.0 mV
		0.00 pH	0.03 pH	0.17 pH
40	60	0.1 mV	0.6 mV	3.0 mV
		0.00 pH	0.01 pH	0.05 pH

Let us consider a meter with a relatively large grid current, for example 5×10^{-11} A. The resistance of an electrode of Corning 015 glass may be about 1000 megohms at 10 °C, 200 megohms at 25 °C, and 60 megohms at 40 °C. The *IR* drop may therefore amount to 50 mV (0.89 pH unit) at 10 °C. Nevertheless, it is possible to balance out even this large drop in potential, for the resistance of the cell remains about the same, *at constant temperature*, in the standardization process and the pH measurement. The change in the resistance of glass electrodes with temperature is very large, however, and the *IR* drop in the measurement may be appreciably different from that in the standardization if the two temperatures are not the same within narrow limits. The imperfect balancing of the *IR* drop in the two cases appears as an error in the measured pH.

TEMPERATURE OF THE pH ASSEMBLY

The meter with its standard cell or standard resistors is intended to be maintained near ordinary room temperature, but both electrodes should be in temperature equilibrium with the solution under measurement. Precautions to insure temperature uniformity throughout the cell are particularly important if the reference electrode and the electrode within the glass bulb are of different types, sizes, and heat capacities. Temperature gradients within the glass-calomel cell are likely to be more serious than incorrect temperature adjustment of the cell as a whole. The pH of the standard solution is, of course, a function of temperature, and the proper standard value must be selected.

Standard Cell of the pH Meter

The standard cell used in the potentiometric type of meter is often of the unsaturated type. This type of cell has a negligible temperature coefficient of electromotive force but is subject to a slow decrease of e.m.f. with time. If the drop is not too great, the potentiometer can still be balanced at the pH of the standard buffer solution, but other readings of pH or e.m.f. will be in error by amounts which increase with the scale distance from the point of standardization. The recommended procedure of standardization with two buffer solutions will reveal serious changes in the standard cell through failure of the reading for the second buffer solution to check the known value. It is also possible to remove the cell from the circuit for measurement of its e.m.f. by means of a precision potentiometer. If a calibration of the meter or slide wire, as described in the next section, shows the e.m.f. scale to be in error by more than 1 per cent, the standard cell may be at fault.

Performance Tests

From the electrical point of view, the pH meter is simply a voltmeter that has a high input resistance, high zero stability, low scale-length error, and some form of compensation for temperature effects at the electrodes. A low grid current drain, made possible by a suitable electrometer tube and freedom from surface leakage, permits accurate potential measurements to be made in systems of high resistance. In meters of the potentiometric type and many of the direct-reading type, any drift of the zero point is immediately evident and easily corrected. Scale-length error is caused by amplifier variations, by faulty calibration of the slide wire or the scale of the indicating meter, or by inadequate temperature compensation. These factors should be considered in judging the performance of a pH meter.

CLASSIFICATION OF ERRORS

Perley[36] has summarized the most important instrumental errors in percentage of the scale reading, as shown in Table 12–2. It must be remembered that a pH meter is made to read correctly at the pH of the standard and that hence these percentage errors apply strictly only to the *difference* of scale reading between standard and unknown. A direct-reading meter, standardized at pH 7, may nonetheless have a calibration error at full scale (pH 0 or 14) of 0.07 unit. However, the precision of deflection meters is being improved steadily, and measuring circuits of extraordinary stability have been de-

---

[36] Private communication.

TABLE 12-2. ERRORS OF pH METERS

(After Perley)	Maximum Limit of Error, Per Cent of Scale Reading
1. Calibration and adjustment of temperature compensator	0.1
2. Zero adjustment	0.1
Potentiometric Type	
3. Calibration of slide wire	0.3
4. Adjustment of standardizing current	0.1
5. E.m.f. of standard cell	0.05
Direct-Reading Type	
6. Calibration of deflection meter	1.0

veloped. Frequency-conversion amplifiers and expanded scales have permitted the construction of direct-reading meters capable of indicating pH with an accuracy approaching that of the best potentiometric models.

Before a new pH meter is put into service, the millivolt scale should be calibrated by comparison with a known source of e.m.f. The accuracy with which the temperature-compensating resistor adjusts the meter to the proper theoretical slope at the temperature of the measurements should also be affirmed. Inasmuch as changes in the electrical characteristics of the meter, in the resistances of the temperature compensator and the slide wire, and in the potential of the standard cell will alter the scale reading, these performance tests should be repeated at intervals of about one year. For routine work where the highest accuracy is not required, the initial examination, which may determine acceptance or rejection of the instrument, will suffice.

CALIBRATION OF THE E.M.F. SCALE

These tests are essentially comparisons of both the e.m.f. scale and the pH scale of the instrument with a known potential supplied by a potentiometer accurate to 0.1 mV. The known e.m.f. should be applied to the meter directly and again through a resistor R rated at 100 to 200 megohms. In the latter case, the IR drop in the grid circuit simulates that encountered in the measurement of the e.m.f. of a glass-calomel cell. It is a good practice to mount this resistor within a metal chield to avoid capacity pickup. The lead from the "minus e.m.f." terminal of the potentiometer should be connected through the series resistor to the glass electrode terminal of the meter, whereas that from the "plus e.m.f." terminal passes directly to the calomel electrode terminal or jack. The arrangement is shown schematically in Fig. 12–11.

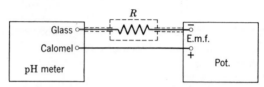

FIG. 12–11. Arrangement for calibrating the e.m.f. scale.

Line-operated instruments for testing pH meters are available commercially. They are essentially calibrated voltage sources with high input resistance.

For the calibration of the e.m.f. scale, the potentiometer is first balanced against its own standard cell and the push button locked in the closed position. The meter is brought to electrical balance in accordance with the manufacturer's instructions, and the proper e.m.f. scale is selected by means of the range switch. Commencing with a value of zero, the applied potential is increased in increments of 100 mV, and the readings of the dial of the meter at balance are noted. The cumulative error at the upper end of the scale should not exceed ±5 mV, nor should there be an error of more than 1 mV per 100 mV increment of applied voltage. A calibration curve should be constructed if the magnitude of the corrections warrants.

CALIBRATION OF THE TEMPERATURE COMPENSATOR

This check of the e.m.f. scale should be supplemented by a determination of the accuracy of the relationship between applied e.m.f. and the pH read from the dial for various settings of the manual temperature compensator. This measurement will evidently include the errors of the slide wire and standard cell revealed by calibration of the e.m.f. scale and, in addition, any inaccuracy in the temperature compensator which is introduced in the circuit when the range switch is set to measure pH.

The precision potentiometer is set again to zero e.m.f. and the push button depressed. With the temperature compensator indicating the temperature for which the scale is to be checked, the meter is brought to electrical balance and switched to the pH scale. The potentiometer is then brought into connection with the pH meter by means of the meter switch or push button and the asymmetry potential knob rotated until the meter is brought to balance at exactly the pH of zero e.m.f. for the particular instrument.

The applied potential is now increased in increments of $(RT \ln 10)/F$ V

(Appendix, Table 1) at the temperature setting of the temperature compensator, for example 59.16 mV at 25 °C, 56.18 mV at 10 °C, and 66.10 mV at 60 °C. These increments of voltage correspond, of course, to increments of exactly 1 pH unit at the respective temperatures. The reading of the pH dial is recorded for each increment of applied voltage and the error in pH computed. If the instrument is so designed that zero impressed e.m.f. corresponds to pH 7, this procedure will yield a calibration of either the low or the high pH range; the connections to the reference potentiometer are then reversed for a calibration of the other half of the scale. Table 12–3 is a suggested form for tabulating the results.

TABLE 12–3. CALIBRATION OF THE pH METER

Potentiometer Voltage	*t:* 25 °C Theoretical pH	Meter No. .... Scale Reading	Error
	Range: 0 to 7 pH		
0	7.00		
0.0592	6.00		
0.1183	5.00		
0.1775	4.00		
0.2366	3.00		
0.2958	2.00		
0.3549	1.00		
0.4141	0.00		
	Range: 7 to 14 pH		
0	7.00		
0.0592	8.00		
0.1183	9.00		
0.1775	10.00		
0.2366	11.00		
0.2958	12.00		
0.3549	13.00		
0.4141	14.00		

Instruments provided with temperature compensators of the automatic type can be calibrated in a similar manner. The resistance thermometer should be immersed during the test in a water bath or oil bath regulated to the desired temperature. If the meter is to be used to determine pH over a wide range of temperatures, it should be calibrated for at least two or three different temperature settings.

Adjustments and Repairs

A large majority of the difficulties encountered in the operation of pH assemblies can be traced to the electrodes, with worn-out batteries and defective tubes or transistors following next in order of importance. Failures of the other components of the circuits, humidity effects, and the like, are less frequent causes of trouble. According to the Beckman company, nearly all trouble stems from one or more of the following causes:[37]

1. Scratched or cracked bulb of glass electrode.
2. Cracked or broken internal assembly of the glass or calomel electrode.
3. Electrodes not properly cleaned.
4. Contamination of the potassium chloride solution of the calomel electrode or salt bridge.
5. Interruption of the flow of potassium chloride solution from the calomel electrode by clogged aperture or fiber.
6. Inaccurate buffer solution.
7. Instability of sample being tested or unusually high resistance of the solution.
8. Worn-out battery.
9. Imperfect or incorrect battery connections.
10. Defective or worn-out tubes.

LOCATING DIFFICULTIES

Unfortunately, faulty operation of a pH meter is not always evident, and the operator may not be aware that incorrect values are being obtained. For this reason it is strongly recommended that at least two reliable standard buffer solutions, differing in pH by at least two units, be available at all times and that the instrument be standardized routinely with both. A reading for the pH of the second solution, after the meter has been standardized with the first, serves as a check on the function of the entire assembly. It is well to standardize a deflection-type meter near the middle of its scale, if possible.

If this double standardization reveals operating difficulties, it must be decided whether the fault lies with the meter or with the electrodes. The most common meter troubles make it impossible to balance the amplifier, even with the electrodes entirely out of the circuit. If the amplifier of a battery-operated instrument cannot be balanced properly, the voltages of

---

[37] *Instructions for Servicing Beckman Model G pH Meter*, Beckman Bulletin 132-D; *Instructions for Servicing Beckman Model H2 pH Meters*, Beckman Bulletin 230; Beckman Instruments, Inc., Fullerton, Calif. G. Mattock, in *Advances in Analytical Chemistry and Instrumentation*, C. N. Reilley, ed., Vol. 2, pp. 35–121, John Wiley and Sons, New York, 1963, offers guidance in identifying the causes of failures in the operation of glass electrode pH meters.

the batteries should first be measured. The batteries should be tested under load, that is, with the meter turned on and all the batteries connected, and should be replaced, as a rule, when they have dropped 10 to 15 per cent below their rated voltages.

If the trouble persists, one of the tubes may be at fault. An unsteadiness or oscillation of the meter needle is a frequent indication of tube failure. Proper operation of the electrometer tube should be verified, if possible, by replacing this tube with a spare known to be in good condition. The manufacturers of several pH instruments supply circuit diagrams and instructions for locating amplifier difficulties and for making simple repairs. The literature on the particular instrument should be consulted if amplifier troubles persist.

If no difficulty is experienced in bringing the amplifier to a satisfactorily constant zero setting, the trouble may reside in the electrodes and solutions. An extra pair of electrodes, known to be in good condition, should always be kept available for test purposes. Only in this way is it easy to check on the electrode function. If this simple replacement test cannot be made, or if it is found that the electrodes are at fault, the condition of both glass and calomel electrodes should be carefully examined, in the following manner.

Inspect the glass electrode for scratches, cracks, and surface deposit. Clean the electrode by immersion in dilute hydrochloric acid followed by thorough rinsing and gentle rubbing with soft absorbent tissue. Make sure that the electrical circuit within the calomel electrode is not broken by air bubbles which appear when the level of the potassium chloride solution in the salt bridge becomes too low. A break in the calomel-mercury column itself usually does no harm. The air bubbles can sometimes be removed by gentle evacuation of the salt bridge chamber or by warming the electrode in hot water. Make sure that the potassium chloride solution can flow freely from the electrode and that a few crystals of solid potassium chloride are present in the salt bridge. A hard mass of salt is to be avoided. If contamination is suspected, replace the chloride solution.

Dry the electrodes and leads carefully, and replace the electrodes in the holder. Choose two reliable buffer solutions, preferably the standard phthalate and borax solutions. Place some of the first buffer in the sample cup, and standardize the instrument in the usual way. Insulation leakage, a defective tube, or a change in the grid bias may make it impossible to adjust the meter to the pH of this buffer solution by means of the standardization (asymmetry potential) control, even though good electrodes are used. A drift or lack of stability may have the same origin or may be caused by poor contact in the various controls. Leakage is most pronounced in humid atmospheres but may occur at normal humidities if the glass electrode terminal is not kept clean. In very dry atmospheres, grounding the case of the meter is sometimes helpful in reducing fluctuations caused by static charges.

It may happen that no difficulty is experienced in standardizing the instrument with the first standard but the reading for the second solution is in error. This is a common experience when the glass electrode is cracked or, because of unusual attack or aging, is not displaying the theoretical response. An improperly adjusted or defective temperature compensator may also be the cause. In meters of the potentiometric type, a change in the e.m.f. of the standard cell may be responsible.

REPAIRS

Remedial measures such as the replacement of batteries and tubes, the elimination of electrical leakage, and the proper care of electrodes should be the concern of each operator. Repairs of the instrument should be attempted only by those equipped with both the technical skill and experience and a full understanding of the function and requirements of an electronic voltmeter.

Determination of pH Values

The determination of pH values is guided by the following considerations:

1. The pH assembly is designed to indicate a difference between the pH values of two solutions—a standard buffer and an unknown or test solution—both of which are at the same temperature.
2. The accuracy of the instrument over the pH range of the test solutions should be demonstrated after standardization by determining the pH of a second standard solution. If possible, the two standards should bracket the pH of the unknowns.
3. Errors caused by fluctuations in temperature and by the residual liquid-junction potential are minimized by standardizing the assembly at a pH value close to that of the test solution.
4. Some glass electrodes display serious errors, increasing rapidly with rising temperature, at pH values above 10. The most suitable glass electrode should be chosen, and corrections should be applied where necessary. The errors of glass electrodes were discussed in the previous chapter.

Detailed procedures for the determination of pH will necessarily vary with the particular problem, the type of material, its stability and buffer capacity, and the accuracy required. The following recommendations, which follow closely the pH method set forth by the American Society for Testing Materials,[38] are made for the guidance of the reader.

---

[38] *Standard Method of Test for pH of Aqueous Solutions with the Glass Electrode*, ASTM Method E70-68, Philadelphia, Pa., 1968. A valuable discussion of accurate techniques for pH measurement has been given by G. Mattock, *Lab. Pract.*, **6**, 444, 521, 577 (1957).

Allow the instrument to warm up thoroughly and bring the amplifier to electrical balance in accordance with the manufacturer's instructions. Wash the electrodes and sample cup three times with distilled water, and dry gently with clean absorbent tissue. Form a fresh liquid junction, if the salt bridge is not of the continuous-flow type. Note the temperature of the unknown or test solution, and adjust the temperature dial of the meter to the proper setting.

Choose two standard buffer solutions (Chapter 4) to bracket, if possible, the pH of the unknown. Warm or cool these solutions as necessary to match within 2 °C the temperature of the unknown. Immerse the electrodes in a portion of the first standard. Engage the operating button, or turn the range switch to the proper position, and rotate the standardizing or asymmetry potential knob until the meter is balanced at the known pH of the standard, pH(S), at the appropriate temperature, as given in the table on page 73. Repeat the process with additional portions of the standard solution until the instrument remains in balance (within 0.02 pH unit) for two successive portions without a change in the position of the asymmetry potential knob. If the initial temperature of the electrodes differs appreciably from that of the solutions, several portions of solution should be used and the electrodes immersed deeply to assure that both electrodes and solution are at the desired temperature. The temperature of the electrodes, standard solutions, and wash water should be kept as close as possible to that of the unknowns.

Wash the electrodes and sample cup three times and form a fresh liquid junction. Place the second standard in the cup, adjust the instrument to balance, and read the pH of this second solution without changing the position of the asymmetry potential knob. Repeat with additional portions of the second standard until successive readings agree with 0.02 pH unit. The assembly shall be judged to be operating satisfactorily if the reading obtained for the second standard matches the assigned pH of this standard within 0.02 unit. The instrument should always be restandardized after even a short period during which the amplifier was turned off. A final check should be made at the conclusion of a series of measurements. The need for more frequent standardization should be examined from time to time.

Difficulty may be experienced in adjusting the instrument to read directly values of pH*(S) in alcohol-water solvents (see Chapter 8), as the range of zero adjustment provided may not be sufficient to compensate for profound changes in the e.m.f. of the cell. If this is found to be the case, the instrument may be adjusted, with the standard solution of known pH*(S), to a convenient arbitrary point on the scale. A measured *difference* of pH* is then determined and added to the value for the standard, pH*(S).

pH OF TEST SOLUTIONS

Wash the electrodes and sample cup thoroughly and dry with absorbent tissue. Fill the sample cup with the test solution, form a fresh liquid junction, and obtain a preliminary value for the pH. Replace the sample with other portions of the test solution until the pH values of two successive portions agree to 0.02 unit (well-buffered solutions) or 0.1 unit (water or poorly buffered solutions).

If it is desired to make use of the glass electrode to measure pH with a precision greater than 0.01 unit, special attention must be given to (1) temperature control, especially uniformity of temperature throughout the cell, (2) the electromotive efficiency of the glass electrode (which may vary with the electrode chosen; see Chapter 11), and (3) the variable liquid-junction potential. Differences attributable to the residual liquid-junction potential are found when the standard and unknown solutions contain different types of ions and have different ionic strengths. At a temperature of 38 °C, for example, an apparent inconsistency of about 0.01 pH unit between the standard 1 : 1 phosphate buffer and the standard borax buffer is observed (see page 86). This finding illustrates the difficulties that are involved in determining *useful* pH data utilizing the full measurement capabilities of the best of the modern instruments.

Furthermore, accuracy and reproducibility of a somewhat lower order than at room temperature may be expected when the temperature of measurement is appreciably different from the ambient temperature. In general, the pH of alkaline buffer solutions is much more sensitive to temperature changes than is that of acid solutions. This sensitivity to fluctuations of temperature, combined with glass electrode errors and the changing liquid-junction potential, makes measurements at high pH less accurate than those in the region near neutrality.

WATER AND SLIGHTLY BUFFERED SOLUTIONS

If the sample is in equilibrium with the carbon dioxide of the atmosphere, make the measurement as described previously with vigorous agitation. Six or more portions may have to be used to obtain a reading that drifts less than 0.1 unit in 2 minutes, particularly when the electrodes have been immersed in a buffered solution immediately prior to the measurement.

The pH of water can perhaps best be measured in a flow cell that permits a high rate of flow past the electrodes.[39] Such a cell is described in the next chapter. If the cell is in continuous use, it should be standardized daily. The

---

[39] G. A. Perley, *Anal. Chem.*, **21**, 559 (1949).

water or poorly buffered solution should flow through the cell at a rate sufficient to change the contents of the cell at least five times per minute. The pH
should not be read until the flow has continued for at least 15 minutes after
immersion of the electrodes in a buffer solution. If the pH of the flowing
solution is changing, the measured pH may lag considerably behind the true
pH. Evidently this procedure can only be used when large volumes of the
sample are available.

If the slightly buffered test solution or sample of water is not in equilibrium with the carbon dioxide of the atmosphere, the measurement should
be made in a wide-mouth flask that has been flushed with carbon dioxide-free
air or nitrogen. The contents of the flask must be protected from exposure
to air and should be agitated vigorously during the measurement.

MEASUREMENT OF THE pH OF BLOOD

Animal blood is well buffered, largely because of the bicarbonate and
carbon dioxide it contains. Hence the useful interpretation of measurements
of blood pH in terms of physiological factors requires discrimination of small
changes in pH. For this purpose a reproducibility of ± 0.003 pH unit at
38 °C is highly desirable. Moreover, the carbon dioxide tension of blood is
conveniently determined by calculation from the measured pH with the aid of
the Henderson-Hasselbalch equation.[40] This determination requires high
accuracy in the measured pH, for an error of 0.02 unit in the pH results in an
error of about 4.5 per cent in the carbon dioxide tension found.[41]

The glass electrode appears to be the best choice for pH measurements in
blood.[42] Nevertheless, for high accuracy and the best reproducibility (from
time to time as well as from place to place), particular attention must be
given to (1) careful temperature control, (2) uniform standardization procedures, and (3) choice of a suitable cell, particularly one in which pH changes
caused by escape of carbon dioxide, clotting, and time-dependent factors,
including glycolysis, are minimized.

Many cells that meet the special requirements of pH and carbon dioxide

---

[40] This equation is a combination of the mass law expressions for the solubility of carbon
dioxide and for the acidic dissociation of carbonic acid, expressed in logarithmic units.

[41] J. W. Severinghaus, M. Stupfel, and A. F. Bradley, *J. Appl. Physiol.*, **9**, 189 (1956);
O. Siggaard-Andersen, *The Acid-Base Status of the Blood*, 2nd ed., Munksgaard, Copen-
hagen, 1964; G. F. Filley, *Acid-Base and Blood Gas Regulation*, Lea and Febiger,
Philadelphia, 1971.

[42] D. B. Cater and I. A. Silver, Chapter 11 in *Reference Electrodes*, D. J. G. Ives and G. J.
Janz, eds., Academic Press, New York, 1961. Methods for determining the pH of blood
have been reviewed critically by J. Sendroy, *Symposium on pH Measurement*, p. 55, ASTM
Tech. Publ. 190, Philadelphia, Pa., 1957.

determinations have been described,[43] and several are available commercially. The syringe electrodes (see Fig. 11-10) are convenient for taking small samples of blood anaerobically. The glass electrode constitutes the plunger of the syringe. The dead space may be filled with heparin solution and the liquid junction formed at the opening of the barrel of the syringe.[44]

There is a recent trend toward the use of microcapillary glass electrodes in cells for blood pH measurement, some of which can be adapted to continuous monitoring of pH during surgery. Much work has also been done on glass microelectrodes for insertion into single cells.[45] For in vivo applications, it is of the utmost importance to avoid clotting of the blood at the glass surface.[46] It appears that deposition of proteins on the glass surface and clotting of the blood can be reduced or delayed by treating the glass with silicone preparations such as Desicote, Dri-Film, or Repelcote. Furthermore, a suitably designed liquid junction will prevent contamination and possible crenation of blood cells by leakage of the saturated solution of potassium chloride from the reference electrode. Liquid-junction potentials and abnormal suspension effects in blood pH measurements have been considered in Chapter 10. Standard reference solutions with pH values near that of blood have been described on pages 84 and 89.

Automatic Titrators

If the pH at the end point in the titration of an acid with a base is known, the titration can be conducted automatically with an ingenious instrument known as an *automatic titrator*, several forms of which are available commercially. An accuracy of 0.1 per cent can be achieved when the pH changes sharply near the equivalence point and the volume of titrant is not too small. The entire titration requires less than 3 minutes. Used in conjunction with recorders, automatic titrators will furnish a permanent record of each titration curve.

The recording automatic titrator sold by Radiometer A/S is shown in Fig. 12-12. Some commercial instruments can record either the titration curve

---

[43] See, for example, M. P. Wright in *Symposium on pH and Blood Gas Measurement*, R. F. Woolmer, ed., p. 51, Churchill, London, 1959; P. Astrup, *ibid.*, p. 81; J. W. Severinghaus, *ibid.*, p. 126; M. C. Sanz, *Chimia*, **13**, 192 (1959). Several types of cells for blood measurements have been described by G. Mattock, *pH Measurement and Titration*, Chapter 11, The Macmillan Co., New York, 1961.

[44] See Mattock, *loc. cit.*, and also D. Mendel, *Lancet*, **272**, 1125 (1957).

[45] M. Lavallée, O. F. Schanne, and N. C. Hébert, eds., *Glass Microelectrodes*, John Wiley and Sons, New York, 1969.

[46] The negative charge on the glass surface may initiate clotting. See D. Hubbard and G. L. Lucas, *J. Appl. Physiol.*, **15**, 265 (1960).

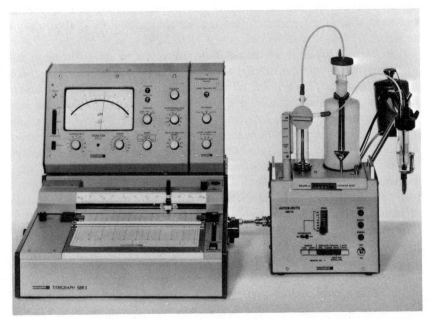

FIG. 12–12. Radiometer Titrator and Titrigraph (pH-stat). This combination functions as either an automatic titrator or pH-stat (courtesy of Radiometer A/S).

or its first derivative, and others will plot automatically the second derivative of pH with respect to volume of reagent as well.

Automatic titrators consist essentially of three parts—a pH meter, an amplifier-control unit, and an automatic buret valve. Glass and calomel electrodes are usually employed for neutralizations, but the instrument is readily adaptable to oxidation-reduction titrations and to reactions in which a precipitate or slightly ionized complex is formed.

To conduct the titration, the operator fills the buret and sets the dial to the desired end point pH. The stirrer may be automatically energized and the titration initiated by raising the vessel holder into position. At the conclusion of the reaction, the stirrer is turned off automatically and a panel light signals the completion of the titration.

Accurate automatic titrations are possible because of the so-called anticipation control, which operates after the major fraction (for example, 95 per cent) of the required volume of titrant has been added. This control governs the addition of reagent in progressively smaller increments as the equivalence point is approached. In order to afford the best compromise between speed and accuracy in each particular titration, the point at which the throttling action begins can be selected in advance by the operator. In some instruments, the rate at which the reagent is added is governed by the slope of the titration curve.

The electrode assembly of the titration cell should be standardized in the usual way with a standard buffer solution. In view of the large change of pH near the equivalence point of many neutralization reactions, however, the accuracy of standardization will often be of secondary importance.

When titrations are being conducted in nonaqueous media, the pH reading at the equivalence point cannot be ascertained in advance; hence a preliminary titration should be made to establish the pH reading at the end point inflection. When the electrodes have been used in nonaqueous media they should be washed first with the nonaqueous solvent and then with water and finally dried carefully with a clean towel or absorbent tissue. They should be immersed in water between titrations.

pH-stats

The pH-stat is a modification of the automatic titrator. It measures the amount of reagent required to maintain a constant preset pH and records this amount as a function of time. The arrangement shown in Fig. 12–12 can also function as a pH-stat. In effect, the control point is the end point of a conventional automatic titration. Rather than turning itself off at this point however, the instrument continues to operate, neutralizing acid or alkali as it is produced in the system. The position of the recorder pen is governed by the position of the piston of a syringe buret containing the reagent, so that a record of reagent consumption is provided.

Titrations at constant pH are of considerable value in kinetic studies of enzyme systems. The many uses of the pH-stat in the study of dynamic biochemical processes have been reviewed by Jacobsen, Léonis, Linderstrøm-Lang and Ottesen.[47] A convenient reaction vessel for the titration of volumes as small as 1 cm^3 has also been described.[48]

---

[47] C. F. Jacobsen, J. Léonis, K. Linderstrøm-Lang, and M. Ottesen, in *Methods of Biochemical Analysis*, D. Glick, ed., Vol. 4, p. 171, Interscience Publishers, New York, 1957.
[48] G. H. Dixon and R. D. Wade, *Science*, **127**, 338 (1958).

CHAPTER 13

industrial pH control

The dawn of the glass electrode era was attended by a great expansion in the application of pH measurements. Nevertheless, the widespread use of pH in modern industry was brought about first by the development of methods and instruments with which a continuous record of pH changes could be made automatically, and second by the adaptation of these methods to permit the pH of a process solution to be controlled at a predetermined level. Without these developments, industrial pH measurements would doubtless never have left the control laboratory.

The key elements in an industrial installation for automatic pH control are the electrode assemblies, the amplifier, the recorder, the controller, and the reagent valve or feeder. These elements are ingeniously combined not only to measure and record the pH of industrial fluids but to add the amount of reagent necessary to maintain the pH at the desired value within 0.1 or 0.2 unit. The functions of these elements will be considered further in this chapter.

Cell Assemblies

The two types of cell commonly employed in control installations are the *immersion* or *dip unit* and the *flow unit*.[1] The former is designed for direct immersion in a tank or vat, whereas the latter is mounted in a channel through which the process solution flows. Both units are necessarily of rugged construction.

The immersion cell assembly shown in Fig. 13-1 consists of a hood of metal, plastic, or hard rubber in which the glass and calomel electrodes and a resistance thermometer are mounted by means of threaded connections

---

[1] For a detailed description of the several commercial forms of these units, the reader is referred to the manufacturers' literature.

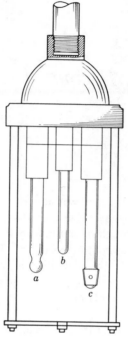

FIG. 13–1. Immersion electrode assembly. (*a*) Glass electrode, (*b*) resistance thermometer, (*c*) calomel reference electrode.

made watertight with sealing washers. The thermometer automatically adjusts the e.m.f.-pH relationship of the measuring instrument.[2] The hood is supported at the end of a metal pipe which acts as a shield for the electrode lead wires. Bar guards protect the element from damage. If the fluid contains coarse solids, a perforated metal sheath may be attached to afford added protection. Although these units are designed primarily to withstand only low external pressures, some of them will function satisfactorily at depths of 60 ft or more. Immersion and flow assemblies are shown in Fig. 13–2.

Electrode glands designed for direct installation in flow lines or processing tanks can be obtained commercially. An assembly consists of three threaded glands in which are mounted the glass electrode, the reference electrode, and the temperature compensator sensing element. A glass electrode and a

---

[2] This adjustment for the change of the slope dE/dpH with temperature must not be confused with the conversion of the pH at one temperature to that of the same solution at another. The change of pH with temperature depends on the composition of the solution.

pressurized reference electrode provided with this type of mounting are shown in Fig. 13–2*b*.

Flow cells are designed to measure the pH of a continuous sample stream piped to the electrode unit. The electrodes and resistance thermometer are fastened securely into a mounting plate which is bolted to a flow chamber of metal, glass, or plastic. The assembly may be supported from connecting piping but should be free of vibration. In order to avoid serious errors from contact potentials, all the pipe connections should be of the same metal. The electrodes, leads, and terminals are sometimes permanently sealed in plastic to make the unit waterproof. The chamber of the flow cell shown in Fig. 13–2c is supplied in stainless steel or porcelain or is made to order from other materials.

A plastic flow chamber (Fig. 13–2*d*), the top of which is threaded to accommodate the electrodes and thermometer, is also obtainable. The volume of the glass electrode compartment is small, and flow rates of from 50 to 3000 cm^3 per minute can be employed. Other flow chambers, of larger size, are made of stainless steel or Pyrex glass. A flow cell manufactured by Marconi Instruments Ltd. incorporates an internal reservoir of buffer solution for convenient standardization. An all-glass flow cell, designed specifically for boiler feed water and other lightly buffered media, is manufactured by the same company.[3]

In most applications, the rate of flow through the cell is relatively unimportant, as long as the sample is always representative of the solution in the main line or tank. Sometimes a rapid flow is desired in order to maintain a clean electrode surface, retarding the collection of finely divided solids on the pH-sensitive membrane.

Flow measurements are particularly advantageous in the determination of the pH of water or slightly buffered solutions. Simple dip measurements give incorrect results in media of these types unless the electrodes are washed repeatedly and the sample agitated.[4] The optimum rate of flow of unbuffered solutions appears to vary with the design of the flow cell. High fluctuating rates of flow disturb the liquid junction and should be avoided.

Electrodes

In general, the same forms of glass electrode are utilized in industrial cell assemblies as in pH meters. Satisfactory glass electrodes with thick walls can be made from the low-resistance glasses, and these are able easily to withstand the relatively high pressures under which flow cells sometimes operate.

---

[3] See *Hydrogen Ions*, 4th ed., Marconi Instruments Ltd., St. Albans, Herts., England, 1952.

[4] S. B. Ellis and S. J. Kiehl, *J. Amer. Chem. Soc.*, **57**, 2139 (1935); G. A. Perley, *Anal. Chem.*, **21**, 559 (1949).

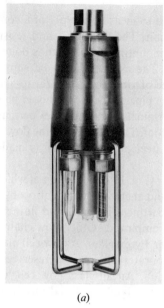

(a)

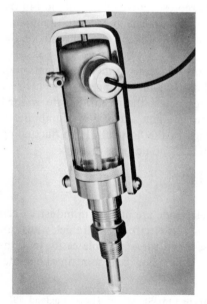
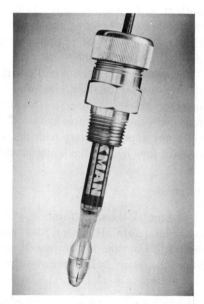

(b)

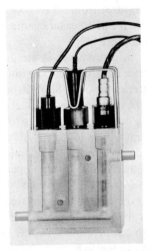

(c)   (d)

FIG. 13–2. Immersion and flow assemblies (courtesy of Leeds and Northrup Co. and Beckman Instruments, Inc.). (*a*) L & N stainless steel immersion mounting, (*b*) Beckman gland assembly with glass electrode (right) and pressurized reference electrode (left), (*c*) Beckman stainless steel flow cell, (*d*) L & N plastic flow chamber.

Calomel reference electrodes with junctions of the fiber, sleeve, palladium annulus, porous sponge, and capillary types are available for industrial cell assemblies. However, the completely satisfactory way to assure the establishment of a stable liquid junction under conditions of high pressure is still being sought. Sealed calomel electrodes that operate well at pressures below 30 psi are available. These should be kept filled with potassium chloride solution at all times.

To establish a stable liquid junction, reference electrodes operating on a bell-jar principle have been devised. Air trapped between the salt-bridge tube and its outer casing is compressed as the pressure on the external solution increases. The increased pressure of air is communicated to the potassium chloride reservoir through a breather hole near the top of the salt-bridge tube. However, the most popular and effective means of attaining free-flowing junctions at high pressures utilizes an external source of air pressure. A pressurized reference electrode is shown in Fig. 13–2*b*. For industrial installations, Polymetron AG (Zurich) provides a differential pressure regulator which automatically controls the reference electrode and salt bridge at a pressure 0.1 atm above the pressure of the process stream.

Glass electrodes of such a high corrosion resistance that they can be utilized for extended periods at 100 °C are available commercially. Some of these are well adapted to automatic control in the pH range 0 to 12.5 and at temperatures from 15 to 100 °C. Although some highly durable glasses are suitable for pH measurements above 100 °C, the useful life of these electrodes is usually brief.

For measurements at elevated temperatures the stability of the external reference electrode and the internal reference of the glass electrode is likewise a matter of concern. Slow disproportionation of calomel above 80 °C detracts from the usefulness of calomel electrodes,[5] but silver-silver chloride electrodes display excellent stability up to 100 °C. Both types of electrode behave well at temperatures as low as − 10 °C. The 40 per cent thallium amalgam-thallous chloride electrode ("Thalamid") manufactured by Jena Glaswerk Schott und Gen. (Mainz) is said to perform satisfactorily up to 135 °C but cannot be used below 0 °C. Metal-filled and amalgam-filled glass electrodes have given encouraging results at elevated temperatures.[6]

Ingruber[7] has developed a pressurized calomel reference electrode for use at high temperatures. In this electrode, shown in Fig. 13–3, decomposition of the calomel is avoided by maintaining the electrode element at a constant temperature less than 80 °C by means of a jacket of circulating water. A thermal gradient exists in the salt bridge; the Soret potential may be large,[8] but it appears to be satisfactorily constant and reproducible. In order to render the thermal junction as nearly nonconvective as possible, the stem of the electrode is packed with alternate layers of solid potassium chloride and fine glass beads or asbestos plugs which resist the movement of liquid. Ingruber was able to make reproducible pH measurements at a temperature of 200 °C and a pressure of 200 psi with this reference electrode.

Fouling of the glass electrode is a common source of trouble in industrial installations, and the solution of the difficulty may demand considerable ingenuity. In paper treatment it has been found advantageous to use a rather high rate of flow and to direct the stream against the glass electrode. The mild abrasive action of the paper pulp maintains a clean glass surface. Automatic wiping arrangements have been devised and are produced commercially, for example by W. Ingold (Zurich). Periodic automatic flushing of the flow assembly with dilute acid is also effective. The sensitivity and accuracy of the pH measurement are greatly impaired by oily materials and tar in the process

[5] J. E. Leonard, *Symposium on pH Measurement*, p. 16, ASTM Tech. Publ. 190, Philadelphia, Pa., 1957.

[6] M. LePeintre, *Bull. Soc. Franc. Elec.*, [8] **1**, 1 (1960). The glass electrode used by A. Distèche, *J. Electrochem. Soc.*, **109**, 1084 (1962), for measurements at high pressures is filled with 0.1 M hydrochloric acid, and changes in the external pressure are transmitted to the interior of the glass bulb through a layer of silicone fluid.

[7] O. V. Ingruber, *Ind. Chemist*, **32**, 513 (1956); *Pulp Paper Mag. Can.*, **58**, 131 (1957).

[8] J. N. Agar and W. G. Breck, *Nature*, **175**, 298 (1955).

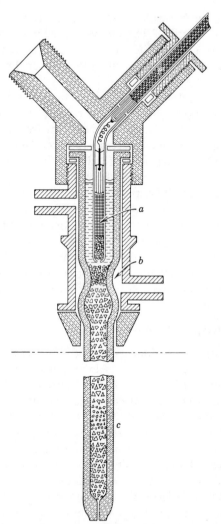

FIG. 13–3. Calomel electrode for high temperatures and high pressures (after Ingruber). (*a*) Calomel element, (*b*) water jacket, (*c*) packed stem where thermal gradient is localized.

solutions. To obtain satisfactory results it is sometimes necessary to filter the sample stream before it enters the cell. The deposition of solids on the electrodes can sometimes be reduced by adding a small amount of a detergent to the sample line or by applying a water-repellent silicone coating to the glass electrode.

An immersion assembly with self-cleaning electrodes has been developed by Polymetron AG (Zurich). The mechanical action produced by a small ultrasonic generator, mounted beneath the electrodes, removes the accumulated film without adverse effect on either the measured pH or the life of the electrodes.

It is desirable that electrode assemblies be standardized daily with a standard buffer solution. However, the conditions and requirements of control operations vary widely, and the necessary frequency of standardization will depend on the required precision and on the stability of the asymmetry potential under the particular conditions to which the glass electrode is subjected. The operator can easily determine how frequently the assembly must be standardized to achieve the necessary accuracy of control. Inasmuch as removal of the electrodes from the flow chamber is inconvenient and entails considerable danger of breakage, some operators prefer to standardize the assembly by an indirect method, as follows: The pH of a sample of the cell effluent is measured with a properly standardized pH meter, and the indicator-amplifier is then adjusted to correct for the difference between this reading and that of the recorder at the moment the sample was collected.

Amplifiers, Recorders, and Controllers

The internal resistance of immersion and flow cells, as that of all cells incorporating a glass electrode, is very high, ranging from 10 to 1000 megohms. The currents that may safely be drawn from these cells without causing an appreciable change in the cell potential due to IR drop or polarization are therefore of the order of 10^{-10} to 10^{-12} A, or too small to operate the usual voltmeter. As we have seen in the foregoing chapter, the e.m.f. of such cells can be measured satisfactorily by means of electrometers of various types. These devices amplify the output of the pH cell to such a degree that it is adequate to actuate a galvanometer or ammeter.

The conventional recorder is also not designed to operate on these very low currents. Hence in a typical arrangement for automatic recording of pH, an amplifier is used to increase the current to a level suitable for measurement by the recorder and for the operation of the various elements in the control system. The amplifier is often combined with the recorder unit. A typical arrangement of cell, amplifier, and record-controller is shown schematically in Fig. 13–4. This circuit, similar in principle to that of some direct-reading

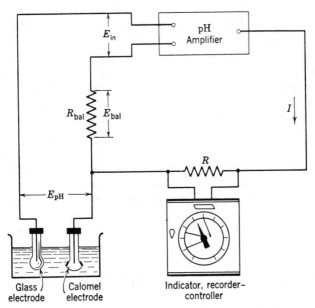

FIG. 13–4. Circuit for automatic recording of pH (courtesy of Minneapolis-Honeywell Regulator Co.).

pH meters, is of the negative feedback type. The output voltage E_{pH} is continuously opposed by E_{bal}, the IR drop resulting from the flow of a variable current I through the fixed resistor R_{bal}. This balancing action serves to keep the input signal E_{in} to the amplifier very nearly zero. Any voltage unbalance due to changes of pH in the cell is greatly amplified to produce an output current that varies linearly with the input voltage. The IR drop across another resistor R supplies the actuating voltage for the recorder. Direct-reading circuits of this type are well adapted to continuous recording of pH.

Industrial pH amplifier-meters display, in general, the diversity of designs characteristic of the laboratory instruments. However, stability, sturdiness, and dependability under extreme conditions of humidity and temperature are more important in industrial measurements than is high sensitivity. The industrial meter should not be susceptible to vibration or mechanical shock nor seriously affected by electrical interference. A design that permits rapid repairs and service may be favored. Many industrial instruments offer zero-shift compensation for temperature changes, and nearly all provide automatic adjustment of the slope factor (see Chapter 12).

The Leeds and Northrup Speedomax G and Speedomax H recorders are electronic null-balancing instruments which can be adapted to the high input impedances of pH cells. In fact, the latter is said to be capable of measuring pH with electrode systems of resistance up to 3500 megohms. Automatic temperature compensation from 0 to 100 °C is provided, and the instrument is available in round-chart and strip-chart models. Speedomax recorder-controllers are also available for industrial pH regulation.

Another industrial instrument, the Beckman Model J pH Analyzer, was designed specifically for continuous operation in rigorous industrial locations. Special provision is made for the exclusion of moisture. The circuit elements of the stabilized amplifier are contained in separate plug-in units in order to minimize lost time when repairs must be made. The Model J was designed for use with standard recorders and recorder-controllers. By changing a resistor in the meter, the range of the recorder can be shifted, compressed, or expanded when necessary to meet the changing needs of the control situation. Many other excellent industrial instruments are available commercially.[9]

With the use of a multiple switch, a single pH meter or amplifier can serve to check the pH of several separate electrode assemblies. Each station can be selected either manually or automatically. The standardization adjustment for each electrode assembly is accomplished at the automatic multiple switch.

To produce a written record of pH changes, the output of the amplifier is supplied to a self-balancing potentiometer. In the recorder, a deviation from balance caused by a pH change excites a sensitive relay which energizes a step-by-step reversible servo motor. The latter then moves a slide-wire contact until balance is restored. The contact arm is geared to a recording pen that traces a record of the measured pH on a circular or strip chart. The chart may be driven by a clock motor. Newer electronic recorders provide a continuous, stepless adjustment. The recorder is often made with a large indicating pointer visible at a distance. Multiple-record pH recorders that will furnish comparative records of pH measurements at more than one point in a process are available. Other models are designed to plot, on a single chart, two separate variables against time.

Recorders are often equipped to actuate valves, pumps, feeders, or other control elements when the pH exceeds or drops below a preset level. These recorder-controllers are commonly of two types—those in which the control impulse is transmitted electrically and those in which the impulse is pneumatic. Complex control patterns are more easily incorporated in pneumatic controllers than in those of the electric type. Pneumatic systems are accordingly simpler and less expensive and hence are used more widely than electric

---

[9] The reader is directed to the summary given by G. R. Taylor, Chapter 10 in G. Mattock, *pH Measurement and Titration*, The Macmillan Co., New York, 1961.

systems. In addition to amplifier and recorder-controller, which may be combined in a single instrument, electric control systems frequently require a separate electric control unit. The adjustments necessary to adapt both types of controller to changes in the characteristics of the process are usually simply made, as is the transfer from automatic to manual operation. The control afforded may be exceedingly sensitive and responsive.

Principles of Automatic Control

Automatic control systems depend for their operation on deviations. Before the controller can act, a deviation in the process variable must occur. Hence control functions must be evaluated in terms of dynamic rather than static conditions. For effective control, the process, the sensing element, the controller, and the final control element (valve) must coordinate properly, and all are of equal importance. A detailed discussion of the modes of automatic pH control and the associated instrumentation is manifestly beyond the scope of this book. A thorough treatment of this subject will be found elsewhere.[10]

Three characteristics are of primary importance in determining the design and effectiveness of an automatic control system. These have been termed *capacity*, *transfer lag*, and *dead time*.[11] Capacity is the ability of the system to absorb the control agent without change in the process variable. Transfer lag results from the inability of the system to supply the needed quantity of corrective agent instantly on demand; a drop in pressure across a valve, for example, contributes to transfer lag, as does the time required by a reagent, added in solid form, to dissolve in the process solution. Dead time is pure time delay in any part of the system. Measuring lag and controller lag may be considered a part of dead time.

In general, high capacity is favorable to effective control, for it levels out abrupt changes, allows time for mixing, and tends to prevent extreme changes in the position of the control valve. On the other hand, transfer lag and dead time are detrimental to good control, for they cause the response to be "too little and too late." It is highly desirable that the process solution and the corrective agent be intimately mixed as rapidly as possible, if effective control is to be achieved.

---

[10] A. L. Chaplin, *Instruments*, **22**, 579, 742, 910 (1949); *Applications of Industrial pH Controls*, Instruments Publishing Co., Pittsburgh, 1950; D. Colver Nutting, *Instr. Pract.*, **8**, 50, 123, 221, 327, 416 (1954); G. H. Farrington, *Fundamentals of Automatic Control*, Chapter 2, John Wiley and Sons, New York, 1951; M. W. Geerlings, *Plant and Process Dynamic Characteristics*, Academic Press, New York, 1957.

[11] See, for example, D. P. Eckman, *Instrumentation*, **1**, No. 3, 10 (1944).

Modern controllers are capable of four basic types or modes of compensatory response to deviations in the pH or other process variable.[12] These control patterns are termed (1) two-position control, (2) proportional-position control, (3) reset response, and (4) rate action. Reset response and rate action are usually employed in combination with proportional control.

In *two-position* regulation, of which on-off and high-low control are examples, the reagent valve is always in one or two positions. In one of these, the flow of corrective reagent is always less than needed whereas in the other it is always more than needed. Load changes can be accommodated by two-position control, provided that they do not take place at a rapid rate.

Proportional-position control utilizes a sliding-stem control valve that may assume any intermediate valve position from closed to fully open. The position of the valve stem is proportional to the value of the controlled variable at all times; hence the response is always proportional to the position of the pen of the recorder-controller. The proportionality factor is chosen by adjustment of the *proportional band* or *throttling range* of the controller. The proportional band is defined as the range of values of the controlled variable that corresponds to the full range of the final control valve. It is usually expressed as a percentage of the full-scale range of the controller. Too wide a throttling range produces stable control but sacrifices sensitivity. If the throttling range is narrowed, the magnitude of the deviations may be kept small, but the oscillations following a sudden change are increased. These oscillations ("cycling") prevent the use of too narrow a proportional band.

Proportional control initiates corrective action as soon as a deviation is apparent. Inasmuch as all intermediate valve positions are available, the flow of reagent can be adjusted to meet the process demand much more closely than in the simpler two-position mode.

In order to provide stability under a variety of load conditions, it is often necessary to superimpose *reset response*, sometimes called *droop correction*, on the proportional control pattern. The resultant is called *proportional-reset* control. Reset response advances the valve position by an additional amount determined by the magnitude and the length of time of the deviation in the process variable. Furthermore, reset action brings the variable back to the set point even though a new valve position is necessary to meet a sustained change of load.

The proportional-reset mode can handle satisfactorily most combinations of process characteristics, and it is accordingly the most widely used control

---

[12] J. C. Peters, *Ind. Eng. Chem.*, **33**, 1095 (1941); G. H. Farrington, *Fundamentals of Automatic Control*, Chapter 2, John Wiley and Sons, New York, 1951; *Fundamentals of Instrumentation for the Industries*, pp. 20–24, Minneapolis-Honeywell Regulator Co., Industrial Division, 1952.

pattern. Processes with a large dead time or large transfer lag, however, are difficult to control by this means. The addition of *rate action* or *derivative action* may then be sufficient to provide effective regulation. In pH control, rate action supplies an added valve adjustment determined by the rate at which the pH is deviating from the set point. The initial large corrective action tends to counteract the unfavorable effect of process lag. Thus rate action is able to "head off" the deviation and, acting in reverse, to prevent overshoot as the pH returns to the balance point. Both reset and rate patterns can be varied by an adjustment of the controller as needed to provide the most effective control possible,

Three types of control response are illustrated in Fig. 13–5. In two-position control (curve *a*), the pH of the process solution tends to oscillate from a value that is somewhat too high to another that is somewhat too low. Although the valve regulating the flow of corrective reagent is actuated at the positions

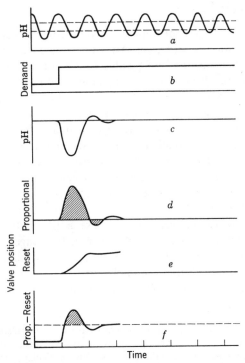

FIG. 13–5. Response patterns in automatic control (after Peters). (*a*) Two-position control, (*b*) reagent demand, (*c*) deviation pattern, (*d*) proportional-position response, (*e*) reset response, (*f*) proportional-reset response.

represented by the dotted lines, there is often sufficient lag to cause the pH to oscillate over a considerably wider range.

The other curves of Fig. 13-5 analyze the response to a sudden but sustained increase in reagent demand, shown by curve b. The changes of pH as control is once more established are illustrated in c. In proportional control (d), the valve position follows exactly the variation in the pH of the process solution. When a reset response (e) is combined with proportional position response, as shown in curve f, the new valve position (dotted line) at which the increased demand is effectively met is quickly attained.

In some applications, for example, in the neutralization of industrial waste, there are large and rapid fluctuations in the rate of flow of the process solution that is being treated. If the pH controller is unable alone to meet the demands of these variations in load, a flow rate controller may be needed. This device, operating on signals from the pH controller and a flow sensing element, apportions the amount of reagent to the flow of process solution.

Buffer Action in pH Control

The problem of pH control is quite different from the control of temperature or concentration.[13] First, the pH value is a logarithmic function of concentrations. Second, the buffer value (buffer capacity) of a given acid-base system may rise from zero to a maximum and fall back to zero within the brief span of about 2 pH units. For these reasons, pH control is highly nonlinear. Wide-band proportioning action in conjunction with reset response is frequently necessary to achieve adequate control of pH changes.

The effectiveness of automatic pH control depends to a large extent on the rate of change of pH with volume of corrective reagent near the control point. If this rate is very large, as it is near the end point of the neutralization of a strong alkali with a strong acid, proper control is very difficult, and wide cycling can easily result.[14] In cases of this sort, the control problem can sometimes be solved by using a weak acid or base as the control agent or by increasing the capacity of the mixing chamber. A two-stage treatment may be effective, the approximate pH being reached in one chamber and the close adjustment accomplished in a second chamber.

When, on the other hand, the pH is very insensitive to the addition of

---

[13] The special problems of industrial pH control have been considered by O. Petersen, *Chem.-Ingr. Tech.*, **32**, 658 (1960). Analog simulation can be a considerable aid in designing the control installation best suited to the demands of the particular process of concern; see W. B. Field, *Instr. Soc. Am. J.*, **6**, 42 (1959).

[14] S. D. Ross, Bulletin No. B51-2, Minneapolis-Honeywell Regulator Co., Philadelphia, Pa., 1950.

corrective reagent, the desired accuracy of control may also be difficult to achieve. Direct measurement of the pH of solutions of strong acids and bases, for example, can only be utilized as a means of controlling concentration near the neutral point. In this and other similar cases, control can sometimes be accomplished successfully with the aid of a duplex controlled-volume pump and a small mixing chamber in which the electrodes are mounted. The pump constantly withdraws a small sample of the process solution and delivers it, together with a metered quantity of reagent, into the mixing chamber. The strength of the reagent and the quantities of the two solutions are so regulated that the pH of the mixture lies in the range where sensitive control can be achieved.

Final Control Elements

Selection of the best means of regulating the corrective agent is one of the major problems in designing a control installation. The most common elements for final control are *sliding stem valves* and *dry chemical feeders*. These can usually be operated satisfactorily by controllers of either the electric or pneumatic type.

Because of its simplicity and precision, the pneumatic sliding stem valve is probably used more widely than any other final control element. A valve of this type is shown in Fig. 13-6. It is positioned with a high degree of accuracy by a diaphragm motor operated by changes of air pressure from a pneumatic controller. A wide range of throttling control is possible. Electric motorized valves of both the two-position and sliding stem types are also obtainable.

If the corrective agent is a slurry, there is danger that the conventional sliding stem valve may become clogged. A valve commonly utilized in installations of this kind is known as the Saunders patent valve. It throttles the flow by pressing a flexible diaphragm toward a raised seat cast in the valve shell.

Sliding stem valves are made with a variety of inner structures. These valves have widely different valve characteristics, that is, they provide markedly different curves of flow vs. stem position. Similarly, rotary stem valves and butterfly valves yield variously shaped curves of flow vs. vane position.[15] Choice of the valve best suited to the characteristics of the process and the control system often simplifies the problem of regulating the pH with the desired accuracy.

Feeders for solid reagents eliminate the need for large solution tanks and

---

[15] S. D. Ross, *Ind. Eng. Chem.*, **38**, 878 (1946); Bulletin B51-2, Minneapolis-Honeywell Regulator Co., Philadelphia, Pa., 1950.

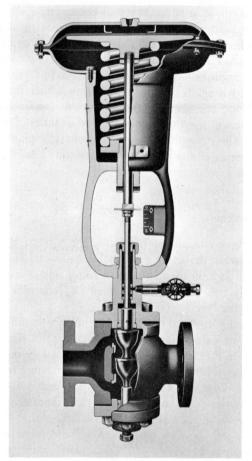

FIG. 13–6. Proportional control valve of the pneumatic type (courtesy of Minneapolis-Honeywell Regulator Co., Valve Division).

the preparation of large quantities of solutions. They also offer a means of avoiding the use of corrosion-resistant valves, fittings, and pipes. Solid agents can be added by any of several types of dry feeders. Belt feeders and vibrating feeders are common, although bin valves, conveyor belts of variable speed, and constant-weight feeders are also used. Some feeders are designed for on-off control. The rate of delivery of others can be controlled by electric or pneumatic proportional action. Some dry feeders with proportional control deliver chemicals accurately within a range from full capacity down to one-hundredth of the maximum rate.

Industrial Installations

Although flow cells must usually be installed in a bypass line, care should be taken that the linear rate of flow through the bypass is nearly the same as that through the process line, so that the composition of the sample will always match that of the solution being treated. It is desirable, further, that the cell be installed in such a way that the glass and reference electrodes are soaked continually in process solution and will have no opportunity to become dry. It should be remembered that drifts of 0.1 to 0.2 pH unit may occur if a dry glass electrode is put in service without preconditioning in an aqueous solution.

Immersion electrode assemblies are located as close as practicable to the point of addition of reagent. However, the location should be downstream and at a point where the sample is thoroughly uniform and where chemical reaction with the reagent is complete. If a dry feeder is used, vigorous stirring is usually necessary.

The extremely high resistance of the pH cell aggravates the problem of electrical losses, of which surface leakage is the most troublesome. All high-impedance leads should be shielded from electrostatic pick-up, and insulator surfaces should be kept dry. Junction boxes that can be sealed and dried with a desiccant help to eliminate losses. It is desirable to locate the amplifier

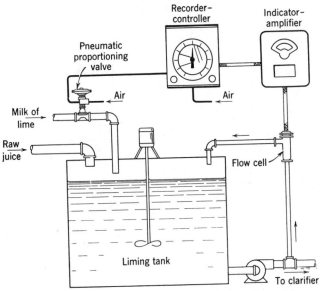

FIG. 13–7. System for control of pH in the liming of cane juice.

near the pH cell; however, the controller and amplifier may be separated by as much as 1000 ft if necessary.

An installation for automatic control of pH in sugar refining, fairly typical of industrial installations, is shown in Fig. 13–7. The electrodes are contained in a flow cell mounted in a sample line. The control signal is amplified and fed to a pneumatic recorder-controller which actuates a diaphragm motor valve of the Saunders patent type, admitting lime slurry to the tank in which raw cane juice is limed. The pH of the process solution is maintained in the range 7.4 to 8.4 and is recorded continuously. The addition of slurry to the liming tank is sometimes controlled with the use of a head box divided into two compartments by a partition. In an installation of this type a stream of slurry is diverted into the process tank or returned to the lime supply by the action of a diaphragm lever motor which moves the distributing pipe from one side of the partition to the other. Dry lime feeders are also satisfactory in this application. A control installation with immersion cell and dry feeder of the vibrator type is shown in Fig. 13–8.

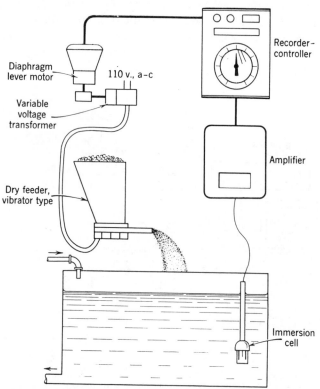

FIG. 13–8. System for pH control with a dry feeder.

appendix

TABLE 1. Values of $(RT \ln 10)/F$ from 0 to 100 °C

t °C	$(RT \ln 10)/F$ V	t °C	$(RT \ln 10)/F$ V
	(ln 10 = 2.30259)		
0	0.054197	55	0.065110
5	0.055189	60	0.066102
10	0.056181	65	0.067094
15	0.057173	70	0.068086
20	0.058165	75	0.069078
25	0.059157	80	0.070070
30	0.060149	85	0.071062
35	0.061141	90	0.072054
38	0.061737	95	0.073046
40	0.062133	100	0.074038
45	0.063126		
50	0.064118		

$R = 8.31433$ J K^{-1} mol^{-1}; $F = 96,487.0$ C mol^{-1}; T (in kelvins) $= t(°C) + 273.150$: International Union of Pure and Applied Chemistry. *Pure Appl. Chem.*, **9**, 453 (1964).

TABLE 2. Ion Product K_w of Water from 0 to 60 °C

t °C	$K_w \times 10^{14}$	$-\log K_w$
0	0.1139	14.943
5	0.1846	14.734
10	0.2920	14.535
15	0.4505	14.346
20	0.6809	14.167
25	1.008	13.996
30	1.469	13.833
35	2.089	13.680
40	2.919	13.535
45	4.018	13.396
50	5.474	13.262
55	7.297	13.137
60	9.614	13.017

Data from H. S. Harned and B. B. Owen, *The Physical Chemistry of Electrolytic Solutions*, 3rd ed., Chapter 15, Reinhold Publishing Corp., New York, 1958.

TABLE 3. Vapor Pressure, Density, and Dielectric Constant of Water from 0 to 100 °C

t °C	Vapor pressure mm Hg at 0 °C	Density g cm^{-3}	Dielectric Constant
0	4.6	0.99987	87.74
5	6.5	0.99999	85.76
10	9.2	0.99973	83.83
15	12.8	0.99913	81.95
20	17.5	0.99823	80.10
25	23.8	0.99707	78.30
30	31.8	0.99568	76.55
35	42.2	0.99406	74.83

TABLE 3. (*Continued*)

t °C	Vapor pressure mm Hg at 0 °C	Density g cm^{-3}	Dielectric Constant
40	55.3	0.99225	73.15
45	71.9	0.99024	71.51
50	92.5	0.98807	69.91
55	118.1	0.98572	68.35
60	149.4	0.98322	66.82
65	187.6	0.98058	65.32
70	233.7	0.97779	63.86
75	289.1	0.97487	62.43
80	355.2	0.97182	61.03
85	433.6	0.96864	59.66
90	525.9	—	58.32
95	634.0	—	57.01
100	760.0	—	55.72

Sources: Vapor pressure, N. S. Osborne, H. F. Stimson, and D. C. Ginnings, *J. Res. Nat. Bur. Stand.*, **23**, 261 (1939); density, values selected by H. S. Harned and B. B. Owen, *The Physical Chemistry of Electrolytic Solutions*, 3rd ed., Chapter 5, Reinhold Publishing Corp., New York, 1958; dielectric constant, C. G. Malmberg and A. A. Maryott, *J. Res. Nat. Bur. Stand.*, **56**, 1 (1956).

TABLE 4. CONSTANTS OF THE DEBYE–HÜCKEL THEORY FROM 0 TO 100 °C

	Unit Volume of Solvent		Unit Weight of Solvent	
t °C	A	B	A	B
0	0.4918	0.3248	0.4918	0.3248
5	0.4952	0.3256	0.4952	0.3256
10	0.4989	0.3264	0.4988	0.3264
15	0.5028	0.3273	0.5026	0.3272

(For values of the ion-size parameter, å, in angstrom units)

(*Continued*)

TABLE 4. (*Continued*)

t °C	Unit Volume of Solvent		Unit Weight of Solvent	
	A	B	A	B
20	0.5070	0.3282	0.5066	0.3279
25	0.5115	0.3291	0.5108	0.3286
30	0.5161	0.3301	0.5150	0.3294
35	0.5211	0.3312	0.5196	0.3302
38	0.5242	0.3318	0.5224	0.3306
40	0.5262	0.3323	0.5242	0.3310
45	0.5317	0.3334	0.5291	0.3318
50	0.5373	0.3346	0.5341	0.3326
55	0.5432	0.3358	0.5393	0.3334
60	0.5494	0.3371	0.5448	0.3343
65	0.5558	0.3384	0.5504	0.3351
70	0.5625	0.3397	0.5562	0.3359
75	0.5695	0.3411	0.5623	0.3368
80	0.5767	0.3426	0.5685	0.3377
85	0.5842	0.3440	0.5750	0.3386
90	0.5920	0.3456	0.5817	0.3396
95	0.6001	0.3471	0.5886	0.3404
100	0.6086	0.3488	0.5958	0.3415

Source: R. A. Robinson and R. H. Stokes, *Electrolyte Solutions*, 2nd ed. revised, p. 468, Butterworths, London, 1970. The values for unit weight of solvent were obtained by multiplying the corresponding values for unit volume by the square root of the density of water at the appropriate temperature.

TABLE 5. Dielectric Constants ε of Pure Liquids

Liquid	t, °C	ε
Acetic acid	25	6.13
Acetone	25	20.70
Acetonitrile	25	36.0
Ammonia	−77.7	25
Aniline	20	6.89
Benzene	25	2.274
Carbon tetrachloride	25	2.228
Chlorobenzene	25	5.621

TABLE 5. (*Continued*)

Liquid	t, °C	ε
Chloroform	20	4.806
Cyclohexane	25	2.015
Deuterium oxide	25	77.9
Dimethylformamide	25	36.7
Dimethylsulfoxide	25	46.7
1,4-Dioxane	25	2.209
Ethanol	25	24.3
Ethanolamine	25	37.7
Ethyl acetate	25	6.02
Ethyl ether	20	4.34
Ethylamine	10	6.94
Ethylenediamine	25	12.4
Ethylene glycol	25	37.7
Formamide	20	109
Formic acid	16	58.5
Glycerol	25	42.5
Hexamethylphosphoric triamide	25	30
Hydrazine	20	52.9
Hydrofluoric acid	0	83.6
Hydrogen peroxide	0	84.2
Methanol	25	32.63
2-Methoxyethanol	25	16
Methylamine	25	9.4
N-Methylacetamide	40	165
N-Methylpropionamide	25	176
Nitrobenzene	25	34.82
Nitromethane	30	35.87
Phosphorus trichloride	25	3.43
1-Propanol	25	20.1
2-Propanol	25	18.3
Propylene carbonate	25	64.4
1,3 Propylene glycol	20	35.0
Pyridine	25	12.3
Sulfuric acid	25	101
Tetrahydrofuran	25	7.6
Tetramethylenesulfone (sulfolane)	30	43.3
Tin tetrachloride	20	2.87
Water	25	78.30

Sources: Varied. See especially A. A. Maryott and E. R. Smith, *Table of Dielectric Constants of Pure Liquids*, NBS Circular 514 (August 10, 1951), and B. Trémillon, *La Chimie en Solvants Non-aqueux*, Presses Universitaires de France, Paris, 1971.

TABLE 6. Approximate pH of Some Common Reagent Solutions near Room Temperature

Substance	Concentration (mol dm^{-3})	pH
Acid benzoic	(Satd.)	2.8
Acid boric	0.1	5.3
Acid citric	0.1	2.1
Acid hydrochloric	0.1	1.1
Acid oxalic	0.1	1.3
Acid salicylic	(Satd.)	2.4
Acid succinic	0.1	2.7
Acid tartaric	0.1	2.0
Acid trichloracetic	0.1	1.2
Alum, ammonium	0.05	4.6
Alum, potassium	0.1	4.2
Ammonia water	0.1	11.3
Ammonium chloride	0.1	4.6
Ammonium oxalate	0.1	6.4
Ammonium phosphate, primary	0.1	4.0
Ammonium phosphate, secondary	0.1	7.9
Ammonium sulfate	0.1	5.5
Barbital sodium	0.1	9.4
Borax	0.1	9.2
Calcium hydroxide	(Satd.)	12.4
Potassium acetate	0.1	9.7
Potassium bicarbonate	0.1	8.2
Potassium binoxalate	0.1	2.7
Potassium carbonate	0.1	11.5
Potassium phosphate, primary	0.1	4.5
Sodium acetate	0.1	8.9
Sodium benzoate	0.1	8.0
Sodium bicarbonate	0.1	8.3
Sodium bisulfate	0.1	1.4
Sodium carbonate	0.1	11.5
Sodium hydroxide	0.1	12.9
Sodium phosphate, primary	0.1	4.5
Sodium phosphate, secondary	0.1	9.2

Data from *U.S. Pharmacopeia*, 13th ed., p. 873, Mack Printing Co., Easton, Pa., 1947, and from unpublished measurements made at the National Bureau of Standards.

TABLE 7. $p(a_H\gamma_{Cl})$ FOR SOME USEFUL SOLUTIONS FROM 0 TO 60 °C [a]

t	$m = 0.005$	0.01	0.02	0.05	0.07	0.1
			Hydrochloric Acid (m) [b]			
0	2.761	2.084	1.810	1.454	1.325	1.187
10	2.762	2.085	1.811	1.457	1.328	1.190
20	2.763	2.086	1.813	1.460	1.331	1.194
25	2.764	2.087	1.815	1.462	1.334	1.197
30	2.764	2.088	1.816	1.464	1.336	1.200
40	2.765	2.089	1.816	1.465	1.337	1.202
50	2.767	2.091	1.819	1.469	1.343	1.208
60	2.767	2.092	1.820	1.471	1.345	1.213

Potassium Tetroxalate (m)

	$m = 0.01$		0.025		0.05		0.1	
t	I	$p(a_H\gamma_{Cl})^0$	I	$p(a_H\gamma_{Cl})^0$	I	$p(a_H\gamma_{Cl})^0$	I	$p(a_H\gamma_{Cl})^0$
0	0.0181	2.206	0.0417	1.932	0.0772	1.765	—	—
5	0.0181	2.202	0.0416	1.934	0.0770	1.764	—	—
10	0.0181	2.207	0.0416	1.938	0.0767	1.765	—	—
15	0.0181	2.210	0.0415	1.940	0.0765	1.769	0.1409	1.623
20	0.0180	2.212	0.0414	1.942	0.0763	1.773	0.1404	1.627
25	0.0180	2.214	0.0413	1.947	0.0760	1.780	0.1400	1.640
30	0.0180	2.218	0.0412	1.952	0.0758	1.785	0.1396	1.643
35	0.0179	2.221	0.0410	1.957	0.0755	1.792	0.1394	1.651
40	0.0179	2.220	0.0408	1.962	0.0753	1.797	0.1391	1.660
45	0.0178	2.230	0.0407	1.968	0.0751	1.803	0.1389	1.670
50	0.0177	2.234	0.0405	1.970	0.0749	1.811	0.1387	1.681
55	0.0176	2.238	0.0403	1.981	0.0747	1.819	0.1385	1.692
60	0.0175	2.239	0.0401	1.987	0.0744	1.824	0.1383	1.702

(Continued)

[a] The data are taken in large part from the compilation of R. G. Bates and R. Gary, *J. Res. Nat. Bur. Stand.*, **65A**, 495 (1961). See this paper for references to original sources. When the solution does not contain chloride, this acidity function is properly designated $p(a_H\gamma_{Cl})°$.

[b] m = molality, I = ionic strength.

TABLE 7. *Values of* p($a_H\gamma_{Cl}$) (*Continued*)

Sodium Hydrogen Succinate (m_1) and Hydrochloric Acid (m_2); (m_2) = 0.6667m_1

t	$I = 0.01516$ $m_1 = 0.015$	0.02017 0.02	0.02517 0.025	0.03018 0.03	0.04019 0.04	0.05019 0.05	0.06020 0.06	0.07020 0.07	0.08021 0.08	0.10021 0.1
0	3.970	3.967	3.964	3.962	3.958	3.955	3.953	3.952	3.950	3.948
5	3.950	3.947	3.945	3.942	3.938	3.935	3.933	3.931	3.929	3.925
10	3.933	3.930	3.927	3.925	3.921	3.918	3.916	3.914	3.913	3.909
15	3.920	3.917	3.914	3.912	3.908	3.905	3.902	3.900	3.898	3.895
20	3.909	3.905	3.902	3.899	3.895	3.893	3.890	3.888	3.886	3.883
25	3.902	3.898	3.895	3.892	3.887	3.884	3.882	3.880	3.878	3.875
30	3.894	3.890	3.886	3.884	3.880	3.877	3.874	3.872	3.871	3.867
35	3.888	3.884	3.881	3.878	3.873	3.870	3.868	3.866	3.864	3.861
40	3.885	3.881	3.878	3.875	3.870	3.867	3.865	3.863	3.861	3.858
45	3.884	3.880	3.876	3.873	3.868	3.865	3.862	3.860	3.858	3.855
50	3.882	3.878	3.874	3.871	3.867	3.864	3.861	3.859	3.857	3.853

0.05m Potassium Hydrogen Phthalate ($I = 0.0533$)

t	p($a_H\gamma_{Cl}$)$^\circ$
0	4.090
5	4.084
10	4.082
15	4.083
20	4.087
25	4.096
30	4.104
35	4.113
40	4.125
45	4.138
50	4.155

TABLE 7. *Values of* p($a_H\gamma_{Cl}$) (*Continued*)

Acetic Acid (m_1), Sodium Acetate (m_2), and Sodium Chloride (m_3); $m_2 = 0.9624m_1$; $m_3 = 1.0243m_1$

$I = 0.01$	0.02	0.03	0.04	0.05	0.06	0.07	0.08	0.09	0.10
t $m_1 =$ 0.005034	0.010067	0.01510	0.02013	0.02517	0.03020	0.03523	0.04027	0.04530	0.05034
0	4.768	4.769	4.770	4.771	4.772	4.773	4.773	4.774	4.774	4.775
5	4.757	4.758	4.758	4.759	4.759	4.760	4.761	4.761	4.762	4.762
10	4.750	4.751	4.752	4.752	4.753	4.753	4.754	4.754	4.755	4.756
15	4.746	4.747	4.747	4.748	4.748	4.748	4.749	4.749	4.750	4.750
20	4.746	4.747	4.747	4.747	4.747	4.747	4.747	4.747	4.748	4.748
25	4.746	4.747	4.747	4.747	4.747	4.747	4.748	4.748	4.748	4.748
30	4.748	4.748	4.748	4.748	4.749	4.749	4.749	4.749	4.749	4.750
35	4.752	4.752	4.752	4.752	4.752	4.752	4.752	4.752	4.752	4.752

Acetic Acid (m), Sodium Acetate (m), and Sodium Chloride (m)

$I = 0.01$	0.02	0.03	0.04	0.05	0.06	0.07	0.08	0.09	0.10
t $m = 0.005$	0.01	0.015	0.02	0.025	0.03	0.035	0.04	0.045	0.05
35	4.768	4.768	4.769	4.770	4.770	4.770	4.770	4.770	4.770	4.770
40	4.775	4.775	4.775	4.775	4.775	4.775	4.775	4.775	4.775	4.775
45	4.781	4.782	4.782	4.782	4.782	4.782	4.782	4.783	4.783	4.783
50	4.791	4.790	4.790	4.790	4.790	4.790	4.790	4.790	4.790	4.790
55	4.801	4.801	4.800	4.800	4.800	4.800	4.800	4.800	4.800	4.800
60	4.813	4.813	4.812	4.812	4.812	4.812	4.812	4.811	4.811	4.811

Sodium Hydrogen Succinate (m) and Sodium Chloride (m)

	$I = 0.0418$	0.0681	0.108	0.158	0.217
t	$m = 0.019390$	0.03155	0.05012	0.07296	0.1
0	4.915	4.901	4.887	4.877	4.867
5	4.894	4.879	4.866	4.855	4.845
10	4.880	4.864	4.851	4.839	4.829
15	4.868	4.853	4.838	4.827	4.817
20	4.861	4.847	4.831	4.817	4.809
25	4.853	4.838	4.826	4.814	4.802
30	4.852	4.839	4.822	4.809	4.798
35	4.850	4.837	4.820	4.805	4.796
40	4.851	4.838	4.821	4.806	4.796
45	4.855	4.842	4.823	4.810	4.799
50	4.860	4.848	4.827	4.815	4.804

(*Continued*)

TABLE 7. *Values of* p($a_H\gamma_{Cl}$) (*Continued*)

Sodium Hydrogen Succinate (*m*) and Disodium Succinate (*m*)

t	$I = 0.041$ $m = 0.01$	0.101 0.025	0.202 0.05
0	5.599	5.560	5.531
5	5.582	5.542	5.513
10	5.569	5.528	5.498
15	5.561	5.519	5.488
20	5.555	5.513	5.481
25	5.553	5.511	5.477
30	5.553	5.511	5.476
35	5.556	5.514	5.477
38	5.559	5.517	5.479
40	5.562	5.520	5.481

Potassium Dihydrogen Phosphate (*m*) and
Disodium Succinate (*m*) at 25 °C[c]

m	I	p($a_H\gamma_{Cl}$)°
0.005	0.02	6.311
0.01	0.04	6.276
0.015	0.06	6.254
0.02	0.08	6.233
0.025	0.10	6.219

[c] M. Paabo, R. G. Bates, and R. A. Robinson, *J. Res. Nat. Bur. Stand.*, **67A**, 573 (1963).

TABLE 7. *Values of* p($a_H\gamma_{Cl}$) (*Continued*)

Potassium Dihydrogen Phosphate (m) and Disodium Hydrogen Phosphate (m)										
$I = 0.01$	0.02	0.03	0.04	0.05	0.06	0.08	0.10	0.12	0.15	0.20
t $m = 0.0025$	0.005	0.0075	0.01	0.0125	0.015	0.02	0.025	0.03	0.0375	0.05
0 7.226	7.196	7.174	7.157	7.143	7.130	7.109	7.091	7.076	7.056	7.029
5 7.193	7.162	7.141	7.123	7.109	7.096	7.075	7.057	7.042	7.022	6.995
10 7.165	7.134	7.112	7.095	7.081	7.068	7.047	7.029	7.014	6.994	6.969
15 7.142	7.111	7.089	7.072	7.057	7.045	7.024	7.006	6.992	6.971	6.945
20 7.124	7.093	7.072	7.054	7.039	7.027	7.005	6.988	6.973	6.953	6.927
25 7.111	7.080	7.058	7.040	7.026	7.013	6.992	6.974	6.959	6.940	6.912
30 7.102	7.070	7.048	7.031	7.016	7.003	6.982	6.964	6.949	6.929	6.902
35 7.095	7.064	7.041	7.024	7.009	6.996	6.974	6.956	6.941	6.921	6.894
40 7.090	7.059	7.036	7.019	7.004	6.991	6.969	6.951	6.936	6.917	6.890
45 7.089	7.057	7.034	7.016	7.001	6.989	6.967	6.949	6.934	6.914	6.886
50 7.089	7.057	7.034	7.016	7.001	6.988	6.996	6.948	6.933	6.913	6.885
55 7.091	7.059	7.036	7.018	7.003	6.990	6.968	6.950	6.935	6.915	6.888
60 7.096	7.064	7.041	7.023	7.008	6.995	6.973	6.954	6.939	6.919	6.892

Borax ($Na_2B_4O_7$) (m_1) and Sodium Chloride (m_2); $m_2 = 1.8548m_1$

$I = 0.010$	0.015	0.020	0.025	0.030	0.035	0.040
t $m_1 = 0.002594$	0.003891	0.005190	0.006485	0.007780	0.009080	0.01038
0 9.514	9.515	9.515	9.516	9.516	9.516	9.516
5 9.435	9.438	9.440	9.441	9.442	9.443	9.443
10 9.377	9.380	9.382	9.382	9.382	9.383	9.333
15 9.324	9.327	9.328	9.329	9.329	9.330	9.330
20 9.276	9.280	9.281	9.281	9.282	9.282	9.282
25 9.234	9.237	9.237	9.238	9.239	9.239	9.239
30 9.192	9.196	9.198	9.199	9.199	9.199	9.199
35 9.154	9.157	9.159	9.160	9.161	9.162	9.162
40 9.121	9.126	9.128	9.129	9.130	9.130	9.130
45 9.089	9.095	9.098	9.100	9.101	9.101	9.102
50 9.062	9.069	9.072	9.074	9.075	9.076	9.076
55 9.035	9.042	9.047	9.049	9.051	9.052	9.052
60 9.008	9.018	9.023	9.026	9.027	9.028	9.029

(*Continued*)

TABLE 7. *Values of* p($a_H\gamma_{Cl}$) *(Continued)*

Tris(hydroxymethyl)aminomethane (m_1) and Hydrochloric Acid (m_2);
$m_2 = 0.4961 m_1$

$I = 0.01$	0.02	0.03	0.04	0.05	0.06	0.07	0.08	0.09	0.10
t $m_1 = 0.02016$	0.04032	0.06047	0.08063	0.10079	0.12094	0.14110	0.16026	0.18142	0.2016
0	8.946	8.981	9.004	9.021	9.035	9.049	9.061	9.071	9.081	9.090
5	8.777	8.809	8.834	8.851	8.864	8.877	8.890	8.901	8.911	8.922
10	8.614	8.649	8.673	8.690	8.704	8.718	8.730	8.741	8.752	8.762
15	8.461	8.493	8.518	8.537	8.552	8.566	8.578	8.588	8.598	8.607
20	8.315	8.345	8.370	8.390	8.405	8.419	8.431	8.441	8.451	8.460
25	8.176	8.207	8.232	8.251	8.266	8.280	8.292	8.302	8.312	8.321
30	8.037	8.069	8.095	8.114	8.129	8.142	8.153	8.164	8.175	8.186
35	7.907	7.936	7.961	7.982	7.998	8.012	8.023	8.035	8.046	8.056
40	7.781	7.811	7.836	7.857	7.872	7.885	7.896	7.908	7.920	7.931
45	7.660	7.691	7.715	7.735	7.750	7.764	7.776	7.788	7.800	7.811
50	7.543	7.574	7.599	7.618	7.633	7.647	7.660	7.672	7.684	7.694

4-Aminopyridine (m_1) and Hydrochloric Acid (m_2); $m_2 = 0.4962 m_1$

$I = 0.02$	0.03	0.04	0.05	0.06	0.07	0.08	0.09	0.10
t $m_1 = 0.04031$	0.06046	0.08061	0.10077	0.12092	0.14107	0.16122	0.18138	0.2015
0	9.992	10.016	10.037	10.056	10.072	10.085	10.096	10.107	10.118
5	9.825	9.850	9.872	9.890	9.906	9.919	9.931	9.942	9.953
10	9.668	9.694	9.716	9.735	9.750	9.763	9.774	9.785	9.796
15	9.519	9.543	9.565	9.582	9.597	9.610	9.623	9.634	9.645
20	9.375	9.399	9.419	9.437	9.453	9.466	9.477	9.488	9.499
25	9.236	9.259	9.279	9.297	9.313	9.326	9.338	9.348	9.358
30	9.103	9.125	9.145	9.162	9.177	9.190	9.202	9.212	9.223
35	8.970	8.993	9.013	9.031	9.046	9.058	9.070	9.081	9.091
40	8.842	8.864	8.885	8.904	8.920	8.933	8.945	8.955	8.964
45	8.717	8.741	8.763	8.783	8.799	8.813	8.824	8.834	8.844
50	8.603	8.624	8.646	8.666	8.682	8.694	8.705	8.716	8.727

TABLE 7. *Values of* p($a_H\gamma_{Cl}$) (*Concluded*)

Monoethanolamine (m_1) and Hydrochloric Acid (m_2); $m_2 = 0.5000m_1$

$I = 0.010$	0.015	0.02	0.025	0.03	0.04	0.05	0.06	0.07	0.08
t	$m_1 = 0.02$	0.03	0.04	0.05	0.06	0.08	0.10	0.12	0.14	0.16
0	10.390	10.412	10.429	10.443	10.456	10.476	10.492	10.506	10.519	10.531
5	10.219	10.241	10.258	10.272	10.283	10.301	10.318	10.333	10.345	10.357
10	10.054	10.074	10.091	10.105	10.117	10.137	10.153	10.167	10.179	10.191
15	9.892	9.911	9.928	9.943	9.955	9.976	9.992	10.005	10.017	10.029
20	9.735	9.756	9.775	9.790	9.802	9.821	9.836	9.850	9.861	9.872
25	9.590	9.612	9.629	9.643	9.654	9.673	9.690	9.704	9.717	9.729
30	9.441	9.461	9.480	9.495	9.507	9.527	9.544	9.558	9.570	9.580
35	9.300	9.320	9.338	9.353	9.366	9.387	9.404	9.418	9.429	9.439
40	9.163	9.182	9.199	9.215	9.229	9.251	9.268	9.282	9.294	9.303
45	9.033	9.052	9.067	9.082	9.096	9.119	9.138	9.151	9.162	9.173
50	8.903	8.920	8.937	8.953	8.969	8.993	9.010	9.023	9.034	9.045

Calcium Hydroxide (m)

	$m = 0.015$		0.0175		0.02		0.0203 (Satd. at 25 °C)	
t	I	p($a_H\gamma_{Cl}$)^0	I	p($a_H\gamma_{Cl}$)^0	I	p($a_H\gamma_{Cl}$)^0	I	p($a_H\gamma_{Cl}$)^0
0	0.040	13.386	0.047	13.449	0.053	13.504	0.054	13.510
5	0.039	13.161	0.046	13.226	0.052	13.285	0.053	13.291
10	0.039	12.958	0.045	13.024	0.051	13.082	0.051	13.088
15	0.038	12.769	0.045	12.832	0.051	12.887	0.050	12.893
20	0.038	12.584	0.044	12.649	0.050	12.706	0.050	12.712
25	0.037	12.414	0.043	12.477	0.049	12.531	0.049	12.537
30	0.037	12.250	0.043	12.317	0.049	12.375	0.049	12.381
35	0.037	12.095	0.043	12.158	0.048	12.213	0.048	12.219
40	0.036	11.954	0.042	12.012	0.048	12.064	0.048	12.070
45	0.036	11.809	0.042	11.871	0.047	11.920	0.048	11.926
50	0.036	11.674	0.042	11.735	0.047	11.786	0.047	11.790
55	0.035	11.545	0.041	11.603	0.046	11.654	0.047	11.661
60	0.035	11.418	0.041	11.480	0.046	11.534	0.047	11.540

TABLE 8. pH OF THE CLARK AND LUBS BUFFER SOLUTIONS AT 25 °C[a]

25 ml 0.2M KCl, x ml 0.2M HCl, Diluted to 100 ml

pH	x	Buffer Value β
1.00	67.0	0.31
1.10	52.8	0.24
1.20	42.5	0.19
1.30	33.6	0.16
1.40	26.6	0.13
1.50	20.7	0.10
1.60	16.2	0.077
1.70	13.0	0.060
1.80	10.2	0.049
1.90	8.1	0.037
2.00	6.5	0.030
2.10	5.1	0.026
2.20	3.9	0.022

50 ml 0.1 M KH Phthalate, x ml 0.1 M HCl, Diluted to 100 ml

pH	x	Buffer Value β
2.20	49.5	—
2.30	45.8	0.036
2.40	42.2	0.035
2.50	38.8	0.034
2.60	35.4	0.033
2.70	32.1	0.032
2.80	28.9	0.032
2.90	25.7	0.033
3.00	22.3	0.034
3.10	18.8	0.033

pH	x	Value Buffer β
3.20	15.7	0.030
3.30	12.9	0.026
3.40	10.4	0.023
3.50	8.2	0.020
3.60	6.3	0.018
3.70	4.5	0.017
3.80	2.9	0.015
3.90	1.4	0.014
4.00	0.1	0.014

50 ml 0.1 M KH Phthalate, x ml 0.1 M NaOH, Diluted to 100 ml

pH	x	Buffer Value β
4.10	1.3	0.016
4.20	3.0	0.017
4.30	4.7	0.018
4.40	6.6	0.020
4.50	8.7	0.022
4.60	11.1	0.025
4.70	13.6	0.027
4.80	16.5	0.029
4.90	19.4	0.030
5.00	22.6	0.031
5.10	25.5	0.031
5.20	28.8	0.030
5.30	31.6	0.026
5.40	34.1	0.025
5.50	36.6	0.023
5.60	38.8	0.020
5.70	40.6	0.017
5.80	42.3	0.015
5.90	43.7	0.013

[a] *Source:* V. E. Bower and R. G. Bates, *J. Res. Nat. Bur. Stand.*, **55,** 197 (1955).

TABLE 8 (*Continued*)

50 ml 0.1 M KH$_2$PO$_4$, x ml 0.1 M NaOH, Diluted to 100 ml		
pH	x	Buffer Value β
5.80	3.6	—
5.90	4.6	0.010
6.00	5.6	0.011
6.10	6.8	0.012
6.20	8.1	0.015
6.30	9.7	0.017
6.40	11.6	0.021
6.50	13.9	0.024
6.60	16.4	0.027
6.70	19.3	0.030
6.80	22.4	0.033
6.90	25.9	0.033
7.00	29.1	0.031
7.10	32.1	0.028
7.20	34.7	0.025
7.30	37.0	0.022
7.40	39.1	0.020
7.50	41.1	0.018
7.60	42.8	0.015
7.70	44.2	0.012
7.80	45.3	0.010
7.90	46.1	0.007
8.00	46.7	—

50 ml of a Mixture 0.1 M with Respect to Both KCl and H$_3$BO$_3$, x ml 0.1 M NaOH, Diluted to 100 ml		
pH	x	Buffer Value β
8.00	3.9	—
8.10	4.9	0.010
8.20	6.0	0.011
8.30	7.2	0.013
8.40	8.6	0.015
8.50	10.1	0.016
8.60	11.8	0.018
8.70	13.7	0.020
8.80	15.8	0.022
8.90	18.1	0.025
9.00	20.8	0.027
9.10	23.6	0.028
9.20	26.4	0.029
9.30	29.3	0.028
9.40	32.1	0.027
9.50	34.6	0.024
9.60	36.9	0.022
9.70	38.9	0.019
9.80	40.6	0.016
9.90	42.2	0.015
10.00	43.7	0.014
10.10	45.0	0.013
10.20	46.2	—

TABLE 9. Compositions, Buffer Values β, and Dilution Values $\Delta pH_{1/2}$ of Alkaline Buffer Solutions at 25 °C[a]

50 ml 0.1 M Tris(hydroxmethyl)-aminomethane, x ml 0.1 M HCl, Diluted to 100 ml
$dpH/dt \approx -0.028$ unit K^{-1};
$I = 0.001x$

pH	x	β	$\Delta pH_{1/2}$
7.00	46.6	—	−0.02
7.10	45.7	0.010	
7.20	44.7	0.012	
7.30	43.4	0.013	
7.40	42.0	0.015	
7.50	40.3	0.017	−0.02
7.60	38.5	0.018	
7.70	36.6	0.020	
7.80	34.5	0.023	
7.90	32.0	0.027	
8.00	29.2	0.029	−0.02
8.10	26.2	0.031	
8.20	22.9	0.031	
8.30	19.9	0.029	−0.01
8.40	17.2	0.026	
8.50	14.7	0.024	
8.60	12.4	0.022	
8.70	10.3	0.020	−0.01
8.80	8.5	0.016	
8.90	7.0	0.014	
9.00	5.7	—	−0.01

50 ml 0.025 M Borax, x ml 0.1 M HCl, Diluted to 100 ml
$dpH/dt \approx -0.008$ unit K^{-1};
$I \approx 0.025$

pH	x	β	$\Delta pH_{1/2}$
8.00	20.5	—	
8.10	19.7	0.009	+0.07
8.20	18.8	0.010	
8.30	17.7	0.011	
8.40	16.6	0.012	
8.50	15.2	0.015	+0.05
8.60	13.5	0.018	
8.70	11.6	0.020	
8.80	9.4	0.023	+0.04
8.90	7.1	0.024	
9.00	4.6	0.026	+0.02
9.10	2.0	—	

50 ml 0.025 M Borax, x ml 0.1 M NaOH, Diluted to 100 ml
$dpH/dt \approx -0.008$ unit K^{-1};
$I = 0.001 (25 + x)$

pH	x	β	$\Delta pH_{1/2}$
9.20	0.9	—	
9.30	3.6	0.027	+0.01
9.40	6.2	0.026	+0.01
9.50	8.8	0.025	
9.60	11.1	0.022	+0.01
9.70	13.1	0.020	
9.80	15.0	0.018	+0.01
9.90	16.7	0.016	
10.00	18.3	0.014	
10.10	19.5	0.011	−0.01
10.20	20.5	0.009	0.00
10.30	21.3	0.008	
10.40	22.1	0.007	
10.50	22.7	0.006	
10.60	23.3	0.005	
10.70	23.80	0.004	
10.80	24.25	—	

[a] *Source:* R. G. Bates and V. E. Bower, *Anal. Chem.*, **28**, 1322 (1956).

TABLE 9 (*Continued*)

50 ml 0.05 M $NaHCO_3$, x ml 0.1 M NaOH, Diluted to 100 ml
dpH/d$t \approx -0.009$ unit K^{-1};
$I = 0.001 (25 + 2x)$

pH	x	β	$\Delta pH_{1/2}$
9.60	5.0	—	+0.02
9.70	6.2	0.013	
9.80	7.6	0.014	
9.90	9.1	0.015	+0.03
10.00	10.7	0.016	+0.04
10.10	12.2	0.016	+0.04
10.20	13.8	0.015	
10.30	15.2	0.014	+0.02
10.40	16.5	0.013	
10.50	17.8	0.013	
10.60	19.1	0.012	+0.03
10.70	20.2	0.010	0.00
10.80	21.2	0.009	
10.90	22.0	0.008	
11.00	22.7	—	

50 ml 0.05 M Na_2HPO_4, x ml 0.1 M NaOH, Diluted to 100 ml
dpH/d$t \approx -0.025$ unit K^{-1};
$I = 0.001 (77 + 2x)$

pH	x	β	$\Delta pH_{1/2}$
10.90	3.3	—	
11.00	4.1	0.009	
11.10	5.1	0.011	-0.06
11.20	6.3	0.012	
11.30	7.6	0.014	
11.40	9.1	0.017	-0.09

pH	x	β	$\Delta pH_{1/2}$
11.50	11.1	0.022	
11.60	13.5	0.026	
11.70	16.2	0.030	-0.15
11.80	19.4	0.034	-0.13
11.90	23.0	0.037	
12.00	26.9	—	

25 ml 0.2 M KCl, x ml 0.2 M NaOH, Diluted to 100 ml
dpH/d$t \approx -0.033$ unit K^{-1};
$I = 0.001 (50 + 2x)$

pH	x	β	$\Delta pH_{1/2}$
12.00	6.0	0.028	-0.28
12.10	8.0	0.042	
12.20	10.2	0.048	-0.28
12.30	12.8	0.060	
12.40	16.2	0.076	
12.50	20.4	0.094	-0.28
12.60	25.6	0.12	
12.70	32.2	0.16	
12.80	41.2	0.21	-0.28
12.90	53.0	0.25	
13.00	66.0	0.30	-0.27

Author Index

Subject Index